Crime Scene to Court
The Essentials of Forensic Science
3rd Edition

Crime Scene to Court
The Essentials of Forensic Science
3rd Edition

Edited by

P. C. White
ReForensics, PO Box 1208, Lincoln, UK

RSCPublishing

ISBN: 978-1-84755-882-4

A catalogue record for this book is available from the British Library

© Royal Society of Chemistry 2010

The RSC is not responsible for individual opinions expressed in this work.

Published by The Royal Society of Chemistry,
Thomas Graham House, Science Park, Milton Road,
Cambridge CB4 0WF, UK

Registered Charity Number 207890

For further information see our web site at www.rsc.org

Preface

Although originally conceived as a textbook for undergraduate and postgraduate students and the lay person, this book is now used by scene of crime officers, the police and the legal profession for training and as a reference book in the UK. Copies can also be found on bookshelves in many other countries, thus exceeding all our expectations, not only when the book was first published in 1998 but also when the second edition came out in 2004.

The second edition was produced to update and reflect on the changes in practice and procedure and expanded to include a further two forensic disciplines. Some high-profile trials since then have attracted public attention, not only to forensic science but to the whole judicial system, and this led to the recent government appointment of a Forensic Science Regulator. As expanded upon here in Chapter 1, the appointment was to ensure in future that the forensic science services, provided through the whole of the judicial process, was of the highest quality with trained and competent personnel using validated standard operating procedures.

Since his appointment, the Regulator has been meeting with police, crime scene investigators, forensic scientists, the Forensic Science Society, Skills for Justice and academics and some changes have been agreed and are already being implemented. With these changes and also because of the introduction of new developments and/or technology in some disciplines, which necessitated changes in practices and procedures, this prompted a third edition.

Crime Scene to Court: The Essentials of Forensic Science, 3rd Edition
Edited by P. C. White
© Royal Society of Chemistry 2010
Published by the Royal Society of Chemistry, www.rsc.org

v

Some of the contributors to the earlier editions have now retired or, sadly, died and I am grateful to the six new authors—Brian Rankin, Orlando Elmhirst, Tiernan Coyle, Cliff Todd, Mark Mastaglio, Hazel Torrance and Angela Shaw—who have accepted the challenge of updating or rewriting chapters from the previous edition. I would also like to thank the other authors who have continued to contribute.

Another reason for this new edition is that there have been some cases, including high-profile ones, where DNA evidence has not been able to assist investigations or the courts. However, convictions have been obtained from other types of evidence arising from what may be regarded as more recent forensic applications of other established professional disciplines. With the potential for further successes, and for the reader to be aware of their requirements in a forensic investigation, this edition has therefore been expanded to include three new chapters, all written by recognised specialists in their discipline. the first new chapter, on forensic ecology, is by Patricia Wiltshire. Here the reader is shown how plants and other organisms at a crime scene can provide evidence to aid both an investigation and the court. Dorothy Gennard provides the second new chapter, on forensic entomology, and this illustrates how insects on corpses or at a crime scene can assist an investigation. The third new chapter is from Tal Simmons and John Hunter, who provide the reader with a wealth of knowledge drawn from their experience in forensic archaeology and forensic anthropology respectively. These authors show how their disciplines can assist in cases where there have been burials, including war graves, major accident investigations and political incidents.

In keeping with the original philosophy, every effort has been made to keep this textbook relatively non-technical but still capable of describing forensic practices and procedures and how they are followed in the UK. Furthermore, the loss of any relevant information from previous editions has been kept to a minimum. I am grateful to all authors for achieving these goals, and for giving their support and valuable time for providing readers with their experience and expertise. Finally, I would like to record my thanks for support from the Royal Society of Chemistry, in particular Janet Freshwater who has been the commissioning editor for all editions of *Crime Scene to Court*.

Peter White

Contents

Contributors		**xxi**
Abbreviations		**xxiii**
Chapter 1	**Forensic Practice** *Brian Rankin*	**1**
	1.1 Introduction	1
	1.1.1 Forensic Science—A Definition	1
	1.1.2 Historical Background	3
	1.1.3 Forensic Science in the UK	5
	1.2 When Is Forensic Science Required?	10
	1.2.1 Has a Crime Been Committed?	10
	1.2.2 Who is Responsible?	10
	1.2.3 Is the Suspect Responsible?	11
	1.3 Duties of the Forensic Scientist	12
	1.4 Quality in Forensic Science	13
	1.4.1 Quality at the Scene—Laboratory Chain	14
	1.4.2 Laboratory Quality Procedures	15
	1.5 Accreditation of Forensic Science Facilities	15
	1.6 Personal Accountability in Forensic Science	17
	1.6.1 The Council for the Registration of Forensic Practitioners (CRFP)	18
	1.6.2 Standards of Competence	21
	1.7 Conclusion	24
	Bibliography	24

Crime Scene to Court: The Essentials of Forensic Science, 3rd Edition
Edited by P. C. White
© Royal Society of Chemistry 2010
Published by the Royal Society of Chemistry, www.rsc.org

Chapter 2   The Crime Scene   **25**
               Orlando Elmhirst

        2.1   Introduction   25
        2.2   What Is a Crime Scene?   26
               2.2.1   Crime Types Examined   27
        2.3   Why Examine a Crime Scene?   27
        2.4   Who Examines a Crime Scene?   29
               2.4.1   Background, Training and Professional
                        Bodies   29
               2.4.2   Specialist Scene Examiners   30
        2.5   How Are Crime Scenes Examined?   31
               2.5.1   Legal Framework   31
               2.5.2   Health and Safety   31
               2.5.3   Protective Clothing   32
               2.5.4   Finding Exhibits   33
               2.5.5   Selecting Exhibits   34
               2.5.6   Recovering and Packaging
                        Exhibits   34
               2.5.7   Labelling Exhibits   36
               2.5.8   Continuity   36
               2.5.9   Recording   38
        2.6   From Crime to Crime Scene   39
        2.7   Initial Actions at a Crime Scene   40
               2.7.1   First Police Officers at the Scene   40
               2.7.2   First SOCO at the Scene   40
        2.8   Preparing for a Crime Scene
               Examination   41
               2.8.1   Assessing the Scene   41
               2.8.2   Confirming the Cordons   41
               2.8.3   Scene First Aid   42
               2.8.4   Forensic Assessment   42
               2.8.5   Developing a Crime Scene Strategy   43
        2.9   Examination of a Crime Scene   44
               2.9.1   Establishing a Common Approach
                        Path   44
               2.9.2   Focal Point of the Scene   46
               2.9.3   Post-mortems   48
               2.9.4   Completing the Scene Examination   48
               2.9.5   Specialist Scene Examinations   49
        2.10   Closing the Scene   51
        2.11   Post-scene Processes   51
               2.11.1   The Elimination Process   51
               2.11.2   Submission   52
        2.12   Conclusions   52
        Bibliography   52

Chapter 3 Forensic Ecology   54
Patricia Wiltshire

   3.1 Introduction   54
   3.2 Databases   56
   3.3 Nature of the Evidence   57
       3.3.1 Algae   57
       3.3.2 Testate Amoebae   59
       3.3.3 Miscellaneous Organisms   60
       3.3.4 Fungi and Related Organisms   60
       3.3.5 Plants   61
   3.4 Palynology   65
       3.4.1 Pollen and Spores   65
       3.4.2 Identification by Light Microscopy, SEM and
             Other Techniques   65
   3.5 Taphonomy   69
       3.5.1 Production and Dispersal   69
       3.5.2 Patterns of Distribution and Residuality   70
       3.5.3 Transfer   72
   3.6 The Crime Scene   73
       3.6.1 Inside Buildings and Vehicles   74
       3.6.2 Outside Scenes   74
   3.7 Routine Procedures for Crime Scene Evaluation and
       Sampling   75
       3.7.1 Sampling   76
       3.7.2 Vegetation Survey   77
       3.7.3 Samples from a Mortuary   77
   3.8 Palynological Analysis   79
       3.8.1 Body Deposition Times and Post-Mortem
             Intervals   80
   3.9 Location of Clandestine Graves and Human Remains   81
   3.10 Differentiation of Kill Sites from Deposition Sites   82
   3.11 Cause of Death   82
   3.12 Evidence of Contact (Trace Evidence)   83
   3.13 Forensic Ecology and the Courts   84
   Bibliography   85

Chapter 4 Forensic Entomology   86
Dorothy Gennard

   4.1 Introduction   86
   4.2 Insect Succession on the Corpse   88
   4.3 Seasonal Influences   91
   4.4 Forensic Relevance of Insect Distribution   92
   4.5 Effect of Location on Corpse Decomposition   93

4.6   Effects of Chemicals on Insect Development and PMI   94
4.7   Dna Analysis in Forensic Entomology   97
        4.7.1   DNA Analysis of Insect Gut Content   98
4.8   Explosives   98
4.9   Insects and Fire   99
4.10   Investigation of Wildlife and Domesticated
        Animal Deaths   99
4.11   Modelling Time Since Death (Post-Mortem
        Interval)   100
4.12   Limitations on Interpretation of Crime Scene
        Data   102
4.13   Protocols, Procedures and Quality
        Assurance   104
Bibliography   105

Chapter 5   Trace and Contact Evidence   106
Tiernan Coyle

5.1   Introduction   106
5.2   Back to Basics   107
        5.2.1   Transfer   107
        5.2.2   Persistence   109
5.3   Casework Processes   110
        5.3.1   Case Assessment   110
        5.3.2   Contamination Prevention   111
        5.3.3   Recovery: Taping, Shaking and
                 Brushing   112
        5.3.4   Searching, Sifting and Screening   113
        5.3.5   Light Microscopy and Other Screening
                 Techniques   115
        5.3.6   Instrumental Analysis   115
5.4   Casework Scenarios   116
        5.4.1   They're Junior . . . Just Give Them The
                 Little Cases . . .   117
        5.4.2   Who Broke the Window?   118
        5.4.3   Who Climbed Through the Window?   120
        5.4.4   Who Was in the Car?   121
        5.4.5   More Complex Cases   122
        5.4.6   Intelligence   123
5.5   Trace Evidence is Like a Box of Chocolates . . .   124
        5.5.1   Polyurethane Foam   124
        5.5.2   Flock   124
        5.5.3   Condom Lubricants   125
5.6   Future Trends in Trace Evidence   125
Bibliography   126

Chapter 6 Marks and Impressions **127**
Keith Barnett

6.1 Introduction 127
6.2 Footwear Impressions 128
6.2.1 Introduction 128
6.2.2 Recovery of Impressions from a
Crime Scene 128
6.2.3 Impressions in Two Dimensions 129
6.2.4 Methods for Enhancing Two-Dimensional
Footwear Impressions 130
6.2.5 Dust Impressions 131
6.2.6 Other Deposits 132
6.2.7 Impressions in Blood 132
6.2.8 Three-Dimensional Impressions 134
6.2.9 Conclusions 134
6.3 Information Available from a Shoe 135
6.3.1 Pattern 135
6.3.2 Size 136
6.3.3 Degree of Wear 137
6.3.4 Damage Detail 137
6.4 Comparing an Impression with
a Shoe 137
6.4.1 Making a Test Impression 138
6.4.2 Comparing Impressions 139
6.4.3 Conclusion 140
6.5 Instrument Marks 141
6.5.1 Cutting Instruments 141
6.5.2 Levering Instruments 145
6.5.3 Conclusions 149
6.6 Bruising 149
6.7 Physical Evidence 150
6.7.1 An Impressed Fit 151
6.7.2 Mass-produced Items 151
6.7.3 Plastic Bags and Film 152
6.7.4 Conclusions 152
6.8 Erased Numbers 153
6.8.1 Erasure 153
6.8.2 Connecting Punches to Marks 154
6.9 Fingerprints 154
6.9.1 Why Are They Unique? 155
6.9.2 Current Developments 156
6.9.3 Enhancement of Fingerprints 157
6.10 Conclusions 159
Bibliography 160

Chapter 7   Bloodstain Pattern Analysis   **161**
              Adrian Emes and Christopher Price
              7.1   Introduction   161
              7.2   Classification of Bloodstain Patterns   163
                    7.2.1   Single Drops   163
                    7.2.2   Impact Spatter   168
                    7.2.3   Cast-Off   175
                    7.2.4   Arterial Damage Stains   177
                    7.2.5   Large Volume Stains   181
                    7.2.6   Physiologically Altered Bloodstains (PABS) 182
                    7.2.7   Contact Stains (Transfer Stains)   184
                    7.2.8   Composite Stain Patterns   185
              7.3   Evaluation of Bloodstain Pattern Evidence   186
              Bibliography   189

Chapter 8   Forensic Examination of Documents   **190**
              Audrey Giles

              8.1   Introduction   190
                    8.1.1   Qualifications and Training   191
                    8.1.2   Equipment   192
              8.2   Examinations   192
              8.3   Identification of Handwriting   193
                    8.3.1   Construction of Character Forms   194
                    8.3.2   Natural Variation   196
                    8.3.3   Comparison Material   196
                    8.3.4   Other Forms of Variation   197
                    8.3.5   Non-Roman Script   198
                    8.3.6   Expression of Handwriting Conclusions   198
                    8.3.7   Copies   200
              8.4   Examination of Signatures   200
                    8.4.1   Tracing   201
                    8.4.2   Freehand Simulation   201
                    8.4.3   Authorship of Simulation   202
                    8.4.4   Self-Forgery   202
                    8.4.5   Vulnerable Signatures   202
                    8.4.6   Guided Hand Signatures   203
                    8.4.7   Comparison Material   203
                    8.4.8   Expression of Signature Conclusions   203
              8.5   Examination of Copies   204
              8.6   Printing and Typewriting   205
                    8.6.1   Modern Office Technology   205
                    8.6.2   Word Processors   205
                    8.6.3   Laser Printers   205
                    8.6.4   Ink-Jet Printers   206
                    8.6.5   Dot-Matrix Printers   206

	8.6.6	Single-Element Typewriters	207
	8.6.7	Fixed Type-Bar Machines	208
	8.6.8	Ribbons, Rollers and Correction Facilities	208
	8.6.9	Fax Machines	209
8.7	Origin and History of Documents		209
	8.7.1	Examination of Inks	210
	8.7.2	Examination of Paper	213
	8.7.3	Development of Handwriting and Signatures Over Time	214
	8.7.4	Impressions	214
	8.7.5	Folds, Creases and Tears	216
	8.7.6	Staples and Punch Holes	216
	8.7.7	Erasures, Obliterations and Additions	217
8.8	Printed Documents		218
8.9	Procedures, Protocols and Quality Assurance		219
	Bibliography		220

Chapter 9   Computer-Based Media   221
Jonathan Henry

9.1	The Computer Crime Scene		221
9.2	Guidance on Examination of Computer-Based Evidence		222
	9.2.1	Principles	222
	9.2.2	Imaging	223
	9.2.3	Examinations	223
9.3	Storage Devices		224
	9.3.1	Ones and Zeroes, Bits and Bytes	224
	9.3.2	Magnetic Media	225
	9.3.3	Optical Media	228
	9.3.4	Magneto-Optical Media	232
9.4	Logical Structure		232
	9.4.1	Partitions and Logical Drives	232
	9.4.2	Directory Structure	233
	9.4.3	File Allocation Table and Master File Table	234
	9.4.4	Allocated and Unallocated Space	234
	9.4.5	File Structure	235
	9.4.6	Dates and Times	238
	9.4.7	Sectors and Clusters	240
9.5	Contents of Allocated Space		242
	9.5.1	Link Files	242
	9.5.2	System Swap File	244
	9.5.3	Digital Cameras	244
9.6	Contents of Unallocated Space		246
	9.6.1	Deleted Files	246
	9.6.2	Word-Processed Documents	249

9.6.3	Printed Documents	250
9.6.4	Summary	251
9.7	Internet Activity	251
9.7.1	The Internet	251
9.7.2	Internet Protocol (IP) Numbers	252
9.7.3	World Wide Web	253
9.7.4	Email	254
9.7.5	Webmail	257
9.7.6	File Transfer Protocol and Peer-to-Peer Applications	258
9.7.7	Newsgroups	259
9.7.8	Chat Rooms and Applications	260
9.8	Conclusion	262
	Bibliography	262

Chapter 10 Fire Investigation   263
David Halliday

10.1	Introduction	263
10.1.1	What is a Fire?	264
10.1.2	Ignition of Gases, Liquids and Solids	265
10.1.3	Self-Heating and Spontaneous Ignition	267
10.1.4	Flaming and Smouldering Fires	268
10.1.5	Fire Growth and Propagation	269
10.1.6	Interpreting the Physical Evidence Resulting from Fire Behaviour	271
10.2	Fire Scene Investigation	272
10.2.1	Scene Safety Assessment	272
10.2.2	Information Gathering	274
10.2.3	Visual Inspection and Strategy	275
10.2.4	Detailed Examination, Clearance and Reconstruction	276
10.2.5	Recording and Retrieval of Evidence	278
10.2.6	Interpretation and Evaluation	279
10.3	Fatal Fires	281
10.3.1	Causes of Death by Fire	281
10.3.2	Examination of Fire Victims	282
10.3.3	Lifestyle and Fire Fatalities	282
10.3.4	Post-mortem Evidence	283
10.3.5	'Spontaneous' Human Combustion	283
10.3.6	Multiple Fatalities and Major Disasters	284
10.4	Endangerment of Life	285
10.5	Laboratory Investigations	285
10.5.1	Flammable Liquid Recovery and Analysis	286
10.5.2	Examination of Clothing for Flash Burns	287
10.5.3	Ignition and Burning Tests	289

10.5.4	Examination of Incendiary Devices	290
10.5.5	Appliance Examinations	291
Bibliography		292

Chapter 11 Explosions 293
Cliff Todd, Linda Jones and Maurice Marshall

11.1	Introduction	293
11.2	Explosives Technology	294
11.2.1	What is an Explosion?	294
11.2.2	Types of Explosion	295
11.2.3	Types of Explosives	295
11.2.4	Chemistry of Explosives	296
11.2.5	Initiation and Detonation of Explosives	298
11.2.6	Essential Elements of an Improvised Explosive Device	300
11.3	Facilities Required for Forensic Explosives Examinations	301
11.3.1	Safety	301
11.3.2	Receipt	302
11.3.3	Storage	302
11.3.4	Examination	302
11.3.5	Disposal	305
11.3.6	Reference Collections and Databases	305
11.4	Forensic Questions	305
11.4.1	Was it an Explosion?	305
11.4.2	Was it an Accident or a Bomb?	306
11.4.3	Is This an Explosive?	310
11.5	Photography	317
11.6	Links with Other Forensic Disciplines	317
11.7	Case Study	317
11.7.1	The Scenario	318
11.7.2	The Prosecution Case	319
11.7.3	The Passenger's Defence	320
11.7.4	The Lorry Driver's Defence	320
11.7.5	What Really Happened?	320
Acknowledgments		321
Bibliography		321

Chapter 12 Firearms 322
Mark Mastaglio and Angela Shaw

12.1	Introduction	322
12.2	Historical Development	323
12.3	Firearms and Ammunition Terminology	324
12.3.1	Firearms Terminology	324

12.3.2	Cartridge Terminology	326
12.3.3	Ballistics	329
12.3.4	Units	330
12.4	Firearms	330
12.4.1	Long Arms	330
12.4.2	Handguns	331
12.4.3	Automatic Weapons	332
12.4.4	Air Guns	332
12.4.5	Miscellaneous Firearms	333
12.5	Firing Mechanisms	333
12.5.1	Revolver	333
12.5.2	Self-Loading	334
12.5.3	Automatic	334
12.5.4	Bolt Action	334
12.5.5	Pump Action	334
12.5.6	Lever Action	335
12.5.7	Side Lock	335
12.5.8	Box Lock	335
12.5.9	Hinged Barrel	335
12.5.10	Repeating Actions	335
12.6	Manufacturing Processes	336
12.6.1	Rifling Process	336
12.6.2	Firearms Components	337
12.7	Attendance at Crime Scene and Post-Mortem Examination	337
12.7.1	Crime Scene	337
12.7.2	Post-Mortem Examination	340
12.8	Examination at the Laboratory	340
12.8.1	Firearms	340
12.8.2	Ammunition	342
12.8.3	Clothing	344
12.9	Comparison Microscopy	344
12.10	Firearms Chemistry	346
12.10.1	Gunshot Residue (GSR)	346
12.10.2	Origin and Chemistry of GSR	347
12.10.3	Detection of GSR	348
12.11	Analysis and Identification of GSR Using SEM-EDX	349
12.11.1	Deposition and Transfer of GSR	350
12.11.2	Persistence of GSR	351
12.11.3	Recovery of GSR	352
12.11.4	GSR-Similar Particles	352
12.11.5	Contamination Avoidance	353
12.11.6	Interpretation of GSR Findings	354
Bibliography		356

Chapter 13 Drugs of Abuse   **357**
          Michael Cole

          13.1   Introduction   357
          13.2   Drug Control Legislation in the UK   358
          13.3   Drugs of Abuse and Their Sources   359
                 13.3.1   Cannabis and its Products   360
                 13.3.2   Heroin   361
                 13.3.3   Cocaine   364
                 13.3.4   Amphetamines   366
                 13.3.5   *Psilocybe* Mushrooms   367
                 13.3.6   Mescal Buttons   367
                 13.3.7   Lysergic Acid Diethylamide (LSD)   368
                 13.3.8   Barbiturates and Benzodiazepines   369
                 13.3.9   Tryptamines   369
                 13.3.10 Piperazines   370
          13.4   Identification of Drugs of Abuse   372
                 13.4.1   Physical Descriptions and Sampling   373
                 13.4.2   Presumptive Tests   375
                 13.4.3   Thin Layer Chromatography (TLC)   375
                 13.4.4   Instrumental Techniques   376
          13.5   Quantification of Drugs of Abuse   379
          13.6   Profiling of Drugs of Abuse   380
                 13.6.1   Profiling of Cannabis Products   381
                 13.6.2   Profiling of Heroin   381
                 13.6.3   Profiling of Amphetamine   384
                 13.6.4   Drug Profiling Using Metal Ion Content   384
                 13.6.5   Drug Profiling Using Stable Isotopes   385
                 13.6.6   Plant Drug Identification and Profiling
                          Using DNA   386
                 13.6.7   Numerical Methods   387
          13.7   Quality Assurance in Drug Analysis   388
          Bibliography   388

Chapter 14 Forensic Toxicology   **390**
          Robert Anderson and Hazel Torrance

          14.1   Introduction   390
                 14.1.1   What is Toxicology?   390
                 14.1.2   Forensic Toxicology in the United Kingdom 392
          14.2   Poisons and Poisoning   393
                 14.2.1   Definition   393
                 14.2.2   Factors Affecting the Toxic Dose of a
                          Substance   393
                 14.2.3   Types and Examples of Poisons   397

14.2.4   Routes of Administration and Excretion   398
14.2.5   Patterns of Poisoning   404
14.3   The Work of the Forensic Toxicologist   406
14.3.1   Role of the Forensic Toxicologist in
         Medico-Legal Investigations   406
14.3.2   The Forensic Toxicological Investigation   407
14.3.3   General Analytical Approach   408
14.3.4   Different Types of Specimen   408
14.3.5   Tools of the Trade—Methods of Analysis   410
14.3.6   Chemical Classification of Drugs   412
14.3.7   The Toxicology Report   413
14.4   Interpretation   414
14.4.1   Qualitative Results   414
14.4.2   Quantitative Results   415
14.4.3   Specific Problems of Interpretation   416
14.5   Specific Areas of Interest and Case Studies   417
14.5.1   Alcohol (Ethanol)   418
14.5.2   Road Traffic Safety   424
14.5.3   Alternative Specimens   425
14.5.4   Drug Overdose Case Studies   427
14.5.5   DUID Case   431
14.5.6   Hair Analysis Case   434
14.5.7   Fires and Explosions   434
Bibliography   437

Chapter 15   Analysis of Body Fluids   **438**
Nigel Watson

15.1   Introduction   438
15.2   Biological Evidence   439
15.2.1   Blood   440
15.2.2   Semen   442
15.2.3   Saliva   442
15.3   Tests for Blood and Body Fluids   442
15.3.1   Tests for Blood   443
15.3.2   Tests for Semen   444
15.3.3   Tests for Saliva   445
15.3.4   Determination of the Species of Origin   445
15.4   Blood-Typing   445
15.4.1   Genetics   446
15.4.2   Immunological Markers   447
15.4.3   Protein Markers   448
15.5   DNA and its Analysis   448
15.5.1   DNA Structure   448
15.5.2   DNA Analysis   451
15.5.3   DNA Amplification   452

15.6   Forensic DNA Analysis   454
         15.6.1   Short Tandem Repeats   454
15.7   Interpretation of DNA Results   459
         15.7.1   Interpretation of Mixture Profiles   460
15.8   Mitochondrial DNA   461
15.9   Developments in DNA Testing   464
         15.9.1   Low Copy Number (LCN)   464
         15.9.2   Mass Spectrometry   465
         15.9.3   Trait Identification   465
         15.9.4   DNA Microarray Technology   466
15.10   Conclusion   466
Bibliography   467

Chapter 16   Forensic Archaeology and Anthropology   468
              Tal Simmons and John Hunter

16.1   Introduction   468
16.2   Forensic Archaeology—The Theory   468
16.3   Search   471
16.4   Discovery of Human Remains   474
         16.4.1   Accidental Discovery   474
         16.4.2   Formal Discovery   479
16.5   Excavating a Burial   480
16.6   Forensic Anthropology   482
16.7   Are the Remains Human?   485
16.8   How Many Individuals?   486
16.9   Biological Profile   487
         16.9.1   Sex   487
         16.9.2   Age   491
         16.9.3   Race   495
         16.9.4   Stature   496
         16.9.5   Ante-mortem Trauma and Pathological
                  Conditions   498
16.10   Time Since Death   500
16.11   Human Identification   501
16.12   Trauma Identification and Interpretation   503
16.13   Conclusion   505
Bibliography   505

Chapter 17   Presentation of Expert Forensic Evidence   507
              Trevor Rothwell

17.1   Introduction   507
17.2   The Legal System and the Courts   508
         17.2.1   Lawyers   508
         17.2.2   Magistrates' Courts   509

17.2.3	Crown Courts	509
17.2.4	Appeals	509
17.2.5	Coroners' Courts	510
17.2.6	Scottish Courts	510
17.2.7	Civil Courts	510
17.2.8	Course of the Criminal Trial	511
17.2.9	Role of the Witness	512
17.3	Expert Witnesses	512
17.3.1	Duty of the Expert	513
17.4	Prosecution and Defence	513
17.4.1	Equality of Arms	513
17.4.2	The Forensic Scientist and the Prosecution	514
17.4.3	The Scientist Working for the Defence	514
17.4.4	Sequence of Events in a Forensic Examination	515
17.4.5	Role of the Second Examiner	516
17.4.6	The Need for Both Prosecution and Defence Experts	517
17.5	The Importance of Quality	518
17.5.1	The Individual	518
17.5.2	Setting Standards	519
17.5.3	Case Documentation	520
17.5.4	Assuring the Quality of the Work	520
17.5.5	Time Limits	521
17.6	The Forensic Scientist's Report	521
17.6.1	Statutory Duties	522
17.6.2	Format	522
17.6.3	Corroborated Evidence	525
17.6.4	Disclosure of Expert Evidence	525
17.7	Giving Evidence in Court	526
17.7.1	Preparation	527
17.7.2	Practical Details	527
17.7.3	The Witness Box	528
17.7.4	Evidence-in-Chief	529
17.7.5	Giving Expert Evidence	529
17.7.6	Cross-Examination	530
17.7.7	Re-Examination	531
17.7.8	Releasing the Witness	531
17.7.9	And Afterwards	531
17.8	Conclusions	532
Bibliography		532
Subject Index		**533**

Contributors

Robert Anderson Forensic Medicine and Science, Division of Cancer Sciences and Molecular Pathology, Faculty of Medicine, University of Glasgow, Glasgow, G12 8QQ.

Keith Barnett Forensic Science Service, Trident Court, 2920 Solihull Parkway, Birmingham Business Park, Birmingham, B37 7YN.

Michael Cole Anglia Ruskin University, Department of Life Sciences, East Road, Cambridge, CB1 1PT.

Tiernan Coyle Contact Traces, Unit 26, East Central 127, Milton Park, Abingdon, OX14 4SA.

Adrian Emes Formerly, Forensic Science Service, 109 Lambeth Road, London, SE1 7LP.

Orlando Elmhirst Forensic Science Service, Trident Court, 2920 Solihull Parkway, Birmingham Business Park, Birmingham, B37 7YN.

Dorothy Gennard University of Lincoln, School of Natural and Applied Sciences, Brayford Pool, Lincoln, LN6 7TS.

Audrey Giles The Giles Document Laboratory, Sandpipers, Hervines Road, Amersham, Buckinghamshire, HP6 5HS.

David Halliday Fire Investigation Unit, Forensic Science Service, 109 Lambeth Road, London, SE1 7LP.

Jonathan Henry Operational Support Department, Police Service of Northern Ireland, 42 Montgomery Road, Belfast, BT6 9LD.

John Hunter Institute of Archaeology and Antiquity, University of Birmingham, Birmingham, B15 2TT.

Linda Jones DSTL, Forensic Explosives Laboratory, Fort Halstead, Sevenoaks, Kent, TN14 7BP.

Maurice Marshall DSTL, Forensic Explosives Laboratory, Fort Halstead, Sevenoaks, Kent, TN14 7BP.

Mark Mastaglio Forensic Science Service, 109 Lambeth Road, London, SE1 7LP.

Christopher Price Formerly, Forensic Science Service, 109 Lambeth Road, London, SE1 7LP.

Trevor Rothwell Avonpark, Limpley Stoke, Bath, BA2 7JS.

Angela Shaw Forensic Science Service, 109 Lambeth Road, London, SEI 7LP.

Tal Simmons UCLAN, School of Forensic Investigative Sciences, Preston, PR1 2HE.

Cliff Todd DSTL, Forensic Explosives Laboratory, Fort Halstead, Sevenoaks, Kent, TN14 7BP.

Hazel Torrance Forensic Medicine and Science, Division of Cancer Sciences and Molecular Pathology, Faculty of Medicine, University of Glasgow, Glasgow, G12 8QQ.

Brian Rankin University of Teesside, Centre for Forensic Investigation, School of Science and Engineering, Middlesborough, TS1 3BA.

Nigel Watson University of Strathclyde, Forensic Science Unit, Royal College, George Street, Glasgow, G1 1XW.

Patricia Wiltshire Department of Geography and Environment, University of Aberdeen, Elphinstone Road, Aberdeen, AB24 2UF.

Abbreviations

AA	Atomic absorption spectroscopy
ABFA	American Board of Forensic Anthropologists
ABO	ABO blood groups
ABPI	Association of the British Pharmaceutical Industry
ACPO	Association of Chief Police Officers
ADD	Accumulated degree days
ADH	Acumulated degree hours
AFR	Automatic fingerprint recognition
AMG	Amelogenin
ANFO	Ammonium nitrate/fuel oil
BMK	Benzyl methyl ketone
BPA	Bloodstain pattern analysis
BZP	Benzyl piperazine
CAP	Common approach path
CE	Capillary electrophoresis
CEDAR	Centre of Excellence for Document Analysis and Recognition
CJ	Criminal Justice
CJS	Criminal Justice System
COI	Cytochrome oxidase I
CNS	Central nervous system
COSHH	Control of Substances Hazardous to Health
CRFP	Council for the Registration of Forensic Practitioners
CSC	Crime Scene Co-ordinator
CSM	Crime Scene Manager
DAB	Diaminobenzene
DNA	Deoxyribonucleic acid
DUID	Driving under influence of drugs

EAFE	European Association for Forensic Entomology
EDX	Energy dispersive X-ray analysis
EDXRF	Energy dispersive X-ray fluorescence
EMCDDA	European Monitoring Cente for Drugs and Drugs Addiction
EMIT	Enzyme multiplied immunoassay technique
EP	Electropherogram
ESDA	Electrostatic detection apparatus
ET-AAS	Electrothermal atomic absorption spectroscopy
FAAS	Flame absorption atomic spectroscopy
FAT	File allocation table
FDR	Firearms discharge residue
FEL	Forensics Explosives Laboratory
FID	Flame ionisation detector
FOA	First Officer Attending
FPF	Fibre plastic fusion
FPIA	Fluorescence polarisation immunoassay
FSS	Forensic Science Service
FSSoc	Forensic Science Society
FTIR	Fourier transform infrared spectroscopy
GC	Gas chromatography
GC/FID	Gas chromatography/flame ionisation detector
GC/TEA	Gas chromatography/thermal energy analyser
GPR	Ground penetrating radar
GSR	Gun shot residue
HMTD	Hexamethylene-triperoxide-diamine
HMX	Cyclotetramethylene tetranitramine
HOFSS	Home Office Forensic Science Service
HPLC	High-performance liquid chromatography
HVR	Hypervariable region
IABPA	International Association of Bloodstain Pattern Analysts
IC	Ion chromatography
IEF	Isoelectric focussing
ILAC	International Laboratory Accreditation Cooperation
IPM	Integrated pest management
ISO	International Organisation for Standardisation
LA-ICP-MS	Laser ablation – inductively coupled plasma spectroscopy – mass spectrometry
LC/MS	Liquid chromatography/Mass spectrometry
LCN	Low copy number

LSD	Lysergic acid diethylamine (Lysergide)
LTN	Low template number
MALDI-TOF-MS	Matrix assisted laser desorption/ionisation time-of-flight mass spectrometry
MDA	Methyledioxyamphetamine
MDMA	Methyledioxymethylamphetamine
MEOS	Microsomal enzyme oxidising system
MNI	Minimum number of individuals
MO	Modus operandi
MSER	Manufacture and storage of explosives
MSP	Microspectrophotometer
mtDNA	Mitochondrial DNA
NGO	Non-Government Organisations
NOS	National Occupational Standard
NPIA	National Policing Improvement Agency
OCF	Open case file
PABS	Physiologically altered bloodstains
PACE	Police and Criminal Evidence Act
PCM	Plastic coating mark
PCR	Polymerase chain reaction
PETN	Penataerythritol tetranitrate
PGM	Phosphoglucomutase polymorphism
PM1	Post-mortem interval
PolSA	Police Search Advisor
RAM	Random access memory
RAPD	Random amplification of polymorphic DNA
RDX	Cyclotrimethylene trinitramine
RFLP	Restriction fragment length polymorphism
RIA	Radioimmunoassay
SEM	Scanning electron microscopy
SEMTA	Skills Council for Science, Engineering and Manufacturing Technologies
SERRS	Surface enhanced resonance Raman scattering
SHC	Spontaneous human combustion
SIO	Senior Investigating Officer
SM	Standard method
SNP	Single nucleotide polymorphism
SOCO	Scene of Crime Officer
SOP	Standard operating procedure
SPSA	Scottish Police Services Authority
SSM	Scientific Support Manager
STR	Short tandem repeat

SWGSTAIN	Scientific Working Group on Bloodstain Pattern Analysis
TATP	Tri-acetone-tri-peroxide
TBS	Total Body Score
TIAFT	International Association of Forensic Toxicologists
TLC	Thin-layer chromatography
TFMPP	1-(3-trifluoromethylphenyl) piperazine
TNT	2,4,6-trinitrotoluene
TTI	Transmitting terminal identifier
UKAS	United Kingdom Accreditation Service
WDX	Wavelength dispersive X-ray analysis
Δ^8–THC	Δ^8 – Tetrahydrocannibinol
Δ^9–THC	Δ^9 – Tetrahydrocannibinol

CHAPTER 1

Forensic Practice

BRIAN RANKIN

University of Teesside, Centre for Forensic Investigation, School of Science and Engineering, Middlesborough, TS1 3BA

1.1 INTRODUCTION

Forensic scientists soon discover when talking to the general public that many people have very high expectations of forensic science. This is mainly because of their fascination with the subject, often alluded to as the 'CSI effect', which is a consequence of the proliferation of TV dramas such as *CSI* on the subject of crime scene investigation and related forensic practices. The material in this chapter sets out to address these high expectations, by providing an explanation of forensic science, a discussion of its origins and how forensic science services, more frequently now referred to as forensic practice, have been developed and operate within the UK. The duties of a forensic scientist/practitioner, and how the required high standards of analysis and behaviour are maintained, also form important aspects of this chapter.

1.1.1 Forensic Science—A Definition

If one were to ask one hundred forensic scientists to define forensic science it is possible that one would receive one hundred different definitions but it might be expected that among these the terms 'science' and 'the legal process' would have a predominance. This would, rightly, refer

Crime Scene to Court: The Essentials of Forensic Science, 3rd Edition
Edited by P. C. White
© Royal Society of Chemistry 2010
Published by the Royal Society of Chemistry, www.rsc.org

to their work in a laboratory and within the criminal justice system (CJS). A useful working definition therefore is that 'forensic science is science used for the purpose of the law'. Consequently, any branch of science used in the resolution of legal disputes is forensic science. This broad definition covers criminal prosecutions in the widest sense including consumer and environmental protection, health and safety at work, and civil proceedings such as breach of contract and negligence. The recently appointed UK Forensic Science Regulator has further expanded the definition to 'forensic science is any scientific and technical knowledge that is applied to the investigation of a crime and the evaluation of evidence to assist courts in resolving questions of fact in court'. However, in general usage the term 'forensic practice' is now being applied more widely to the use of science in the investigation of crime by the police and by the courts as evidence in resolving an issue in any subsequent trial. Hence the term 'forensic scientist' is generally accepted as meaning someone who works in a laboratory carrying out examinations on materials recovered from a crime scene, or relating to someone or something relating to an investigation. The forensic scientist can also be someone who attends actual scenes of crime in more difficult and high-profile cases, often as part of a team of forensic practitioners. The term 'forensic practitioner' is therefore more often used to include those involved within the investigative process from scene to court. Hence this would include the three core roles of crime scene examiners, fingerprint officers, and forensic scientists as well as other equally important roles, although not nearly as many actual people such as forensic medical examiners, custody retention officers, *etc.*

Confusion sometimes exists in the mind of the public between forensic scientists and those involved in forensic medicine (the latter is sometimes referred to as legal medicine). The forensic scientist, as defined above, can be involved in all types of criminal investigation but forensic medics restrict their activities to criminal and civil cases where a human body is involved. These are nearly always serious cases such as murder or rape and other sexual acts, and will require the participation of pathologists and/or police surgeons. Overall the term forensic practitioner seems to be a more acceptable term for the courts, although the more specific terms are still acceptable within the forensic community. This lends itself to a much wider definition of forensic practice to incorporate the recovery, examination, and analysis of materials as well as incorporating the various technological changes all within the court environment. This was a recent definition given by the newly appointed Forensic Science Regulator.

The narrower definition is implied in the title of this book and the following chapters will discuss the use of science in the investigation of offences such as murder, violent assault, robbery, arson, breaking and entering, fraud, motoring offences, illicit drugs and poisoning. Covering such a range is justification for restricting the definition in a work such as this. It will not attempt to resolve whether forensic science is a science, an art or neither and this chapter will not cover the probative value of contact trace materials such as fingerprints, footwear, tool marks or physical fits in the context of individualisation.

1.1.2 Historical Background

The origins of forensic science can be traced back to the 6th century, with legal medicine being practised by the Chinese. During the next ten centuries advances in both medical and scientific knowledge were to contribute to a considerable increase in the use of medical evidence in courts. Other types of scientific evidence did not start to evolve until the 18th and 19th centuries, a period during which much of our modern-day knowledge of chemistry was just starting to be developed. Toxicology, the study of poisons, emerged as one of the new forensic disciplines, and was highlighted by the work of Orfila in 1840 with his investigation into the death of a Frenchman, Monsieur Lafarge. Following examination of the internal organs from the exhumed body, Orfila testified on the basis of chemical tests that these contained arsenic, which was not a contamination from his laboratory or the cemetery earth. This evidence subsequently resulted in Madame Lafarge being charged with the murder of her husband, but more importantly raised the problem of contamination, an ongoing concern for any forensic practitioner whether they are recovering, examining, analysing various materials or interpreting findings.

During the latter part of the 19th century there was also considerable interest in trying to identify an individual. One approach, studied by Alphonse Bertillon, was to record and compare facial and limb measurements from individuals. This proved to be unsuccessful because of the difficulty of obtaining accurate measurements. However, it was the first recorded attempt in a criminal investigation to use a classification system based on scientific measurement. Interestingly, and in accord with this principle, forensic scientists today use the results from a combination of analytical measurements to discriminate between groups or to compare samples.

A more successful development in personal identification was to come from fingerprint examinations. Although Bertillon is reported to have used latent fingerprints from a crime scene to solve a case, it was Sir William Herschel, a British civil servant in India, and Henry Faulds who were credited with performing most of the early investigations.

Faulds, a Scottish physician, is also accredited with establishing the fact that fingerprints remain unchanged throughout the life of an individual. It was not until 1901, however, when Sir Edward Henry devised a fingerprint classification scheme for cataloguing and retrieving prints, that the full potential of personal identification through fingerprint evidence could be used in forensic investigations. The development in the evaluation and comparison of fingerprints has been much debated over the last 15–20 years. This will no doubt continue, as a consequence of the relatively recent removal of any specific number of points for comparison to a zero point for identification, which has placed a greater individual responsibility on fingerprint experts in exercising their interpretation.

Body fluid samples have also been found to contain information that can help to identify an individual. The progress made in this area has been dramatic, and major advances have occurred within the past decade. Until 1900 it was impossible to determine if a blood sample or stain was of human or animal origin, or to classify human blood into the four main groups: A, B, AB and O. When tests devised by Paul Uhlenhuth (blood origin) and Karl Landsteiner (blood groups) were used, discrimination between individuals was still poor. The inclusion of the Rhesus test and several different enzyme systems improved discrimination, but it has only been through recent studies of DNA in human chromosomes that there have been dramatic improvements in the confidence of identifying an individual.

To Edmund Locard (1910) is attributed an important basic principle of forensic science, this in essence being that 'every contact leaves a trace'. This phrase is both well known and universally accepted within the forensic community. The relevant passage from his 1920 publication '*L'enquête criminelle et les methodes scientifique*' translates as:

> No one can act with the force that the criminal act requires without leaving behind numerous signs of it: either the wrongdoer has left signs at the scene of the crime or has taken away with him—on his person or clothes—indications of where he has been or what he has done.

Although the examination of fingerprints or body fluid, which might be present in only trace amounts, can directly implicate a particular person in a crime, other types of trace evidence such as glass, paint, fire accelerants, gunshot residues, pollen, *etc.*, can provide links that establish contact between objects and/or people involved with a crime or present at a crime scene.

The ability nowadays to be able to analyse such a variety of materials stems from technological advances that have occurred particularly in the past 50 years. Many of the analytical techniques that have been devised offer unbelievable sensitivity and permit examination of minute quantities (traces) of material which cannot be observed directly by the human eye. To provide some indication of the amount of material being examined in these trace samples, think of a grain of sugar. This can be seen without any difficulty and weighs about 1 milligram (1 mg), (*i.e.* one thousandth of a gram or 1×10^{-3} g). Now consider one millionth of this quantity which is 1 nanogram (1 ng or 1×10^{-9} g). This amount of sugar cannot be seen, but quantities as small as this can be detected by many analytical techniques. Even lower detection limits can be obtained routinely with some instrumental methods. Although this is beneficial, extreme caution is required at every stage of any investigation and subsequent analysis to ensure a positive result is genuine and not due to contamination or any other artefact.

Since the discovery, some 25 years ago, of the application of DNA profiling to criminal investigations, the recovery, extraction and amplification of material to produce the profile has seen massive advances in sensitivity. Originally enough blood to cover a 2p coin was required, but now DNA can be recovered by swabbing or extracting an area that has no visible staining but is thought to have been touched.

Rapid developments in computer technology have also played an important role in the advancement of forensic practice. Apart from their use in controlling instruments and producing analytical data, computers permit the storage of massive amounts of information that can be searched very quickly. With increased computer capacity has come the establishment of databases for DNA recovered from body fluids and sometimes tissues and hair, fingerprints and footwear marks, *etc.*, the purpose of these being to help in the identification of an individual or items associated with an individual. These and other databases can save a tremendous amount of time and effort in a case and are beneficial both to the police in following their enquiries and to the forensic scientists in providing evidence and information for the courts. However welcome this may be to any investigation, it needs to be balanced against ethical issues surrounding individual rights. This particular point, relating to the retention of DNA profiles of individuals not convicted of a crime, was recently debated in the European court, resulting in a review of how long samples could be retained on the DNA database.

1.1.3 Forensic Science in the UK

Before specialist laboratories were established, police forces in many parts of the world relied on scientific assistance from people who, through their occupations, were able to provide the expertise required. Without a centralised system, knowing whom to approach was a problem and this resulted initially in the formation of formalised institutions, which were almost invariably established as parts of universities or hospitals. Over time, these were restricted to examining and providing expertise in a limited number of forensic science disciplines, so police forces took the step of developing their own forensic science laboratories.

Europe took the lead in this development, with the first police forensic laboratory being opened in 1910 in Lyons, France. Thereafter, police laboratories appeared in Germany (Dresden, 1915), Austria (Vienna, 1923) and other countries including Holland, Finland and Sweden, with these last three all coming into service in 1925. In the USA it was not until 1923 that the Los Angeles Police Department set up its own forensic science laboratory. The reason for this change was the failure to obtain an indictment in a case as a result of improper handling of evidence before laboratory examination. Many other police departments across America followed this lead, with the Federal Bureau of Investigation (FBI) laboratory opening in 1932.

Interestingly, the first police forensic science laboratory in the UK was not established until 1935, when the Metropolitan Police Laboratory sited at Hendon was opened. How this laboratory started is a fascinating history. It all arose from the unofficial efforts of a constable, Cyril Cuthbertson, who was interested in medicine and criminalistics (in the UK this term is usually associated with the examination of physical evidence such as footwear marks, but in the USA it has a much wider meaning and covers most of the activities undertaken in a forensic science laboratory) and became involved in applying scientific tests that helped his police colleagues in their investigations. Following his examination of a document and his attendance at court as a witness, praise for his testimony and skills soon filtered back to Scotland Yard. The Police Commissioner, Lord Trenchard, took a considerable interest in this matter as he could see the benefits of a laboratory dedicated to his police force. As a consequence, over a period of time, he persistently engaged the Home Office over this matter and this eventually paid off.

The success of this laboratory resulted in the Home Office sanctioning the development of their own forensic laboratories, under the banner of the Home Office Forensic Science Service (HOFSS), providing regional laboratories for police forces in all areas of England and Wales. The HOFSS also incorporated a Central Research Establishment at Aldermaston. These laboratories were all financed from central and local government funds until 1991 when the Forensic Science Service (FSS) became an executive agency of the Home Office. The agency comprises the five operational laboratories of the former HOFSS (located in Birmingham, Chepstow, Chorley, Huntingdon and Wetherby) and, since 1996, the Metropolitan Police Forensic Science Laboratory in London. Wherever possible, facilities are provided locally but the corporate structure allows the concentration of specialist expertise in particular laboratories so that a comprehensive service is available. Research for the FSS is carried out at one of the Birmingham sites.

With the introduction of agency status came the ability of the FSS to charge for the facilities offered on a contract, case by case, item by item or hourly rate basis depending on the circumstances. These facilities are also available to the defence in criminal cases. Where work is performed for both prosecution and defence the work from each is conducted at different laboratories and client confidentiality is maintained. Other agencies which were formerly part of government departments and offer forensic science facilities are the Laboratory of the Government Chemist (LGC), particularly in the area of drugs and documents, and DSTL, formerly the Defence Evaluation and Research Agency, operating under the Forensic Explosives Laboratory (FEL) in respect of explosives. Agency status allows the provision of services to any customer in the UK or overseas. The introduction of charging for scientific services led to the initiation and development of a commercial market. Other significant changes arose from the desire to level the playing field and not have it dominated by any one provider such as the FSS. Additional companies were formed either by amalgamation, such as LGC and Forensic Alliance to become LGC Forensics, or new companies such as Key Forensic Services Limited, and many other smaller companies and sole traders. It also resulted in an increase in the police in-house services. These major changes resulted in a competitive market for the provision of scientific services to the justice system in England and Wales and consequently gave rise to issues surrounding quality standards and consistency, not to mention pricing and the provision of research into forensic practice.

Forensic Science Northern Ireland is also a government agency and provides forensic services to the province. In Scotland forensic science facilities used to be provided by individual police forces with laboratories in Aberdeen, Dundee, Edinburgh, and Glasgow. In 2007 the Scottish Police Services Authority (SPSA) was created to provide a corporate service to the police forces for Scotland to pull together all the forensic, fingerprint and scene examiners. This model was not based on a competitive market but provided a rather more integrated scientific service to the police.

Although the laboratories referred to above can generally be regarded as the 'official' ones, there is a wide range of practitioners and practices throughout the country providing an independent forensic service to clients. These include university departments, public analysts, large and small practices and sole practitioners—collectively referred to as sole traders. Although these may undertake prosecution work, they have a particular role in working with lawyers retained by a defendant in a criminal case to explore the strengths and weaknesses of scientific evidence tendered by the prosecution. This may include the laboratory examination of original or new material in a case, but it will usually involve an evaluation of the results obtained by the original scientist and the interpretation offered. The latter may require modification in the light of further information provided by the client or discovered by the retained expert.

In summary, the introduction of charging for scientific services, *i.e.* the change to agency status, means that the 'official' and private laboratories, together with other institutions, are all competing for custom from the police or in offering a service for the defence. Unfortunately, although some of these laboratories have sought quality control and accreditation of their procedures and facilities, as described later in this chapter, there was no recognised system of accreditation or regulation of the forensic science profession in England and Wales, or in Scotland or Northern Ireland for that matter. This meant that any organisation could offer and supply forensic science services whether or not it had the technical competence and experience. This has now changed, with the introduction of a Forensic Science Regulator and a tendering process for scientific services, as will be discussed later in this chapter.

There are hidden dangers in a totally 'privatised' forensic science service. For example, commercial pressures and competition could lead to compromised standards. Constraints on budgets could also restrict both the amount of material submitted and the analytical work to be performed, not to mention how research is to be funded in a relatively immature market. The danger with this scenario is that these restrictions could prevent the forensic scientist from providing 'best evidence' in reaching a conclusion which might either provide a court with stronger evidence to support a prosecution, or show that the accused could not have perpetrated a criminal act. Therefore, it is essential that these dangers are identified and appropriate controls are put into place. The key overriding philosophy is to provide scientific services which provide assurance to the justice system and confidence to the public.

One other development in forensic science is the introduction into police forces of civilians, called scene of crimes officers (SOCOs) or crime scene examiners, to carry out the searching of crime scenes and the collection and packaging of various materials that may form potential evidence. Contrary to common belief, it is nowadays quite rare for a forensic scientist to attend a scene and one reason for introducing SOCOs into the forensic system was to reduce the amount of time forensic scientists were being called away from their laboratory work

Over the years the importance of crime scene investigation and the need for collection, packaging and transport of material of potential evidential value has been increasingly recognised. In earlier days this would have been carried out by a detective or a scientist, but it is now usually performed by specialists who have received extensive training in all aspects of crime scene examination including latent fingerprints, evidential traces and photography. This professionalism should ensure that the integrity of items received by the scientist for examination cannot be disputed. Another consequence of this was the introduction of Scientific Support Managers (SSM) to each police force. Their responsibility is to oversee the management of scientific services for the police force in terms of fingerprint identification, photography and forensic work. This has extended to include various technology advances such as digital work and CCTV. The overall role of the SSM depends on the size of the individual police force.

As mentioned earlier, with the changing market place for forensic science, the commercialisation and a number of high profile cases there were calls for regulation. Following a recommendation from the Parliamentary Science and Technology Committee, the government decided to create a new post—the Forensic Science Regulator. The first appointment to this post was in 2008. The Regulator sits within the Home Office and operates independently to ensure that quality standards apply across all forensic science services to the CJS. This involves the setting, maintaining and monitoring of quality standards. As a consequence the Regulator makes an important contribution to the continuing development of forensic science practice to the CJS. The Regulator is supported by an advisory council and a number of specialist groups, who develop and deliver the standards to the Regulator for ratification.

The forensic science work performed in the UK has for many years been regarded very highly throughout the world for its integrity. The implications of any changes, such as the change in status of the forensic laboratories, must be monitored and reviewed and actions taken where necessary to preserve the reputation of the service and, more importantly, to ensure that it is not responsible for any miscarriage of justice. The role of the Regulator is key to this vision.

1.2 WHEN IS FORENSIC SCIENCE REQUIRED?

A police officer investigating an incident will seek clarification of three issues:

1. Has a crime been committed?
2. If so, who is responsible?
3. If the responsible person has been traced, is there enough evidence to charge the person and support a prosecution?

This clarification is seldom the isolated duty of one officer and the ultimate trial will reveal the involvement of the specialist police officers and civilian staff, lawyers and scientists. Overall, forensic practice can be expected to make a contribution to the clarification of all three issues.

1.2.1 Has a Crime Been Committed?

In most cases there may be no doubt that a crime has been committed, but there are a number of occasions when only a scientific examination of items can inform the investigator that this is the case. For example, the alleged possession of an illicit drug will require identification of the seized material. Similarly, to support an offence of driving under the influence of drink or drugs a blood sample taken from a motorist will require an accurate analysis not only to establish that alcohol or a drug is present but that any alcohol present exceeds a permitted level. The presence of semen on a vaginal swab from an under-age girl is evidence of illegal sexual activity. Similarly, the demonstration of toxic levels of a poison in tissues removed at post-mortem from a body of an individual believed to have died from natural causes will be a strong indication of a crime. Doubts as to the authenticity of a document may be resolved by scientific examination and provide evidence of fraud.

1.2.2 Who is Responsible?

If a latent fingerprint is developed and recovered from a crime scene and the criminal's prints are already in a database then the person potentially responsible for that crime may soon be identified. Similarly, the existence of a database of DNA profiles may enable identification of an offender who has bled at the scene of violence or who has left other body fluids in a sexual assault. Although specific identification of an offender may not be provided by scientific examination useful leads may be produced which will enable the investigator to close down certain lines of enquiry and follow other lines to lead to the perpetrator.

1.2.3 Is the Suspect Responsible?

Irrespective of any support received from the scientist, the usual diligent police investigator often produces a suspect and the investigator will look to the scientist to provide corroborative evidence to enable a charge to be made and to assist the court in deciding guilt or innocence. The scientific examination will normally be directed towards two aspects:

1. Examination of material left on the victim or at the scene which is characteristic of the suspect.
2. Examination of the clothing and property of the suspect for the presence of material characteristic of the victim or the scene.

1.2.3.1 Materials Characteristic of the Suspect. The biological, physical or chemical characteristics of materials found on the victim or at the scene can help to confirm the identity of the suspect and/or provide evidence of their involvement or presence at the crime scene. Blood, semen, saliva, fingerprints, hair and teeth are all characteristics of an individual.

Finding fibres from clothing or the characteristic pattern of the soles of shoes worn by a suspect may provide evidence of their involvement, as can any material found that may be associated with their particular occupation. The characteristics of a vehicle used in the crime, such as oil drips or tyre marks, may indicate where it had been parked or driven over ground near to or at the scene. Paint, glass or plastic from the vehicle after a collision may help to identify the particular vehicle and hence the owner who could become a suspect. The characteristic marks that may arise from weapons, tools or other items used in committing a crime, *e.g.* a knife in a stabbing, a screwdriver used in forcing an entry or a firearm used in a robbery, especially if found on the suspect, could provide further evidence of a suspect's association with a crime. Botanical evidence such as pollen may provide links between crime scenes and suspects; finally, even insects found on decayed bodies can assist the forensic practitioner and the courts.

Clearly it is very unlikely that all these possibilities will be realised in a single case, but knowing the circumstances surrounding a case and taking into account previous experiences the forensic scientist should be in a position to exploit their skills to the benefit of the investigator and the courts.

1.2.3.2 Materials Characteristic of the Scene or Victim. The crime scene could be in a building or outdoors, but any search usually yields materials that are characteristic of the particular location, such as the following:

1. Domestic premises—external and internal painting, external and internal glass, furnishings, crockery and glassware, *etc.*
2. Commercial premises—as for domestic premises, plus process materials.
3. External scenes such as gardens, waste ground and fields—soil, vegetation and miscellaneous debris.

Where the scene involves a living or dead victim, biological and clothing characteristics discussed above for the suspect will also apply.

1.3 DUTIES OF THE FORENSIC SCIENTIST

Having established when the service of a forensic scientist might be required, we can now identify their duties as follows:

1. To examine material collected or submitted in order to provide information previously unknown or to corroborate information already available.
2. To provide the results of any examination in a report that will enable the investigator to identify an offender or corroborate other evidence in order to facilitate the preparation of a case for presentation to a court.
3. To present written and/or verbal evidence to a court to enable it to reach an appropriate decision as to guilt or innocence.

Under the adversarial system of trial used in the UK, the USA and many other parts of the world the individual forensic scientist may be regarded as, and claimed to be, an independent witness for the court; but they may not always be regarded in this way by the courts and public. It is therefore essential for forensic scientists to be able to demonstrate competence, impartiality and integrity by attention to issues such as the following:

1. Scientists should only give evidence on work carried out personally or under their direct supervision. However, an expert witness can interpret factual evidence given by another witness under oath in the light of scientific findings and knowledge.
2. Where scientific examinations are relied on for legal purposes the methods used should be based on established scientific principles, validated and, preferably, published in reputable scientific literature, so that they can be scrutinised by the scientific community at large.
3. Where the scientific findings require interpretation, the basis of any interpretation should be available to the scientific community.

It is important to recognise that the responsibilities of individual forensic scientists are personal and not corporate. Thus in giving evidence they are completely and solely responsible for their own experimental results and for the opinions expressed. However, the corporate environment will usually be a supportive structure to provide appropriate training, standardised methods and procedures, evaluation of performance and a quality management system. Attention to the latter can be a real source of assurance to the individual forensic scientist, the criminal justice system, and the public at large.

1.4 QUALITY IN FORENSIC SCIENCE

There are many definitions of quality, but for our purposes that of the International Organisation for Standardisation (ISO) is appropriate:

Quality: The totality of features and characteristics of a product or service that bear on its ability to satisfy stated or implied needs. (ISO Standard 8402: 1986)

The ISO 17025 standard was initiated specifically to give guidance to forensic laboratories on both quality management and the technical requirements for the practices and procedures in operation. This standard can be considered the technical complement to ISO 9000. As a result, any organisation that satisfies the requirements of ISO 17025 will also meet the intent of ISO 9000 requirements; the reverse is not true, however.

The ultimate 'customer' to be satisfied is the court, and it will expect there to be a total quality management system in place that will ensure the integrity of material examined by the scientist, the examination carried out and the testimony given.

1.4.1 Quality at the Scene—Laboratory Chain

The quality control system must clearly extend outside the laboratory environment and places a responsibility on everyone involved in an investigation to maintain a 'chain of custody', as often specified. (In fact, in terms of the materials used to 'bag and tag' items it is equally important that they have been produced under a quality management system to ensure the quality—particularly with items which may contain minute quantities of trace material such as DNA—hence the term 'DNA free'.) A more appropriate expression would be 'chain of integrity', since the court will need to know not only the identity of the links in the chain but also their behaviour, as illustrated in Table 1.1.

Even with this brief consideration it can be noted that apart from knowing the identity of the 'custodians' of items at each stage, the court will need to be assured of their awareness of the consequences of any deficiency in the processes in which they are involved. For this reason, in

Table 1.1 The scene–laboratory links in the chain of custody to ensure quality in this procedure.

Link	Category	Comment
1	Preservation of the scene	This can be difficult in the early stages, particularly when injury or hazard is involved, but the police must establish access control as soon as possible. Thereafter, access must be restricted to those who can make a real contribution to the investigation
2	Search for material of potential evidential value	This must be systematic, with careful records kept of the location of all material collected
3	Packaging and labelling of collected material	This must ensure that the material arrives at the laboratory as far as possible in the condition in which it is collected and that it can be related to the source
4	Storage and transmission to the laboratory	Again preservation of the condition is a priority, *e.g.* refrigeration may be appropriate

many investigations most of the process is conducted by specialist personnel with occasional assistance from laboratory-based scientists. Everyone involved will need to protect the items from the twin problems of deterioration and contamination. The latter is a vital matter when a suspect has been arrested, since all possible contact between items from two sources must be prevented with proof that the appropriate actions have been taken.

1.4.2 Laboratory Quality Procedures

Clearly, if the integrity of the articles received by the laboratory, either as a commercial forensic laboratory or an in-house police laboratory, has been maintained the responsibility is then transferred to the scientist. The methods used for the examination of various evidential materials are detailed in subsequent chapters but certain principles apply to all examinations:

1. Prevention of contamination is a prime requirement, particularly as such small amounts of material can be examined and characterised. The scientist must be able to demonstrate that the procedures used have prevented the adventitious transfer of evidential material between two sources.
2. Security of all items must be assured by recording the names of all individuals having contact with them. This is usually achieved by signing an attached label. Leaving items unattended in the laboratory must also be avoided.
3. Careful permanent records should be kept at each stage of the laboratory examination to avoid any possibility of confusion by assigning results to the wrong item.
4. All the procedures and methods used by the scientist should be fully documented. These are often referred to as Standard Operational Procedures (SOPs) and Standard Methods (SMs). Forensic science laboratories normally have a comprehensive system detailing procedures to be followed and methods to be used. However, the final source of assurance is the competence and integrity of individuals.

1.5 ACCREDITATION OF FORENSIC SCIENCE FACILITIES

There is an increasing call for third party accreditation of forensic laboratories, in common with many other scientific laboratories and industrial organisations. In the USA this call has been met by the American Society of Crime Laboratory Directors through their Laboratory Accreditation Board. Under the auspices of this board the organisation, staffing and facilities of a laboratory are subjected to evaluation and on-site inspection before accreditation. A full re-inspection is carried out every 5 years. A good proportion of the many forensic science laboratories in the United States have been accredited by this process, and some in other parts of the world, especially Asia and Australia.

Given the smaller number of laboratories in the UK it has proved more convenient to use a well established system of accreditation applicable to all laboratories offering testing services to a client. The United Kingdom Accreditation Service (UKAS) is recognised by the government as the body for accrediting all types of laboratories and in this role has established a number of standards, all of which have now been subsumed under two major standards, namely ISO/IEC 17025 and ISO 9000:2000. The former is associated with laboratory tasks and some management aspects and the latter mainly with management issues. A third standard currently under consideration is ISO/IEC 17020 for scene work. However, with a truly integrated investigative process it could be argued that one standard covering the whole process would be more appropriate. All of these processes are being internationalised through the International Laboratory Accreditation Cooperation (ILAC) orga-nisation. Clearly the wider recognition will be of great value to laboratories engaged in work supporting trade across international boundaries, but given the increasingly international nature of crime the concept of all forensic laboratories working to a common high standard has its attractions.

The accreditation process involves a series of steps. The applicant laboratory submits documentation, including its quality manual, to UKAS who then assign a Technical Officer and a Lead Assessor to be responsible for advising whether the laboratory should be accredited. This advice will be based on a pre-assessment visit for informal dis-cussion and a broad review of the quality system followed by a formal inspection of the laboratory by an assessment team. At the end of the inspection the team will discuss any non-compliance found and agree the appropriate corrective action. If the assessment team is satisfied with the corrective actions, UKAS will review all the evidence and decide whether to accredit. Following accreditation the laboratory will be subjected to regular surveillance and re-assessment visits to ensure that standards are being maintained.

A Forensic Science Working Group that took part in formulating standards for forensic science spent much time in defining what was meant by an objective test. Their definition is that:

An objective test is one which, having been documented and validated, is under control so that it can be demonstrated that all appropriately trained staff will obtain the same results within defined limits.

Objective tests are controlled by:

1. Documentation of the test.
2. Validation of the test.
3. Training and authorisation of staff.
4. Maintenance of equipment.

and where appropriate by:

1. Calibration of equipment.
2. Use of appropriate reference materials.
3. Provision of guidance for interpretation.
4. Checking of results.
5. Testing staff proficiency.
6. Recording of equipment/test performance.

Forensic scientists are required to tackle a wide variety of problems, many of which have no commercial analogue. This means that widely publicised and used methods such as those of the British Standards Institution may not be an option. The issues raised in the foregoing definition will assist forensic scientists to develop the necessary degree of objectivity in the method applied to a particular problem.

1.6 PERSONAL ACCOUNTABILITY IN FORENSIC SCIENCE

The ultimate role of the forensic scientist is the presentation of expert testimony to the court trying the issue, and in fulfilling this role the witness is completely and solely accountable for the experimental results presented and for the opinions expressed. These must be justified to the court, often in the face of fierce cross-examination, and the witness cannot shelter behind the laboratory manual or base an opinion on a consensus or majority vote. This requires the witness to be a professional in the best sense of the word: that is, to have an initial developed competence which is continuously maintained, together with a powerful sense of integrity. Clearly an employing organisation has a responsibility in this respect, but the public at large may seek the reassurance of membership of an independent professional body. The latter would seek to provide evidence of competence with a code of conduct and advice on professional behaviour.

The need for a professional body was recognised in recent times by a gathering of professional forensic scientists. The outcome of their deliberations was the establishment, with government support, of a register of competent forensic scientists. As a consequence, the Council for the Registration of Forensic Practitioners (CRFP)—an independent regulatory body to promote public confidence in forensic practice in the UK—was established in 1999. On 31 March 2009 the CRFP ceased operation, primarily due to lack of funding and unsuccessful attempts to become self-funding. However, for historical reasons it is worth devoting a section of this chapter to the CRFP.

1.6.1   The Council for the Registration of Forensic Practitioners (CRFP)

Registration of forensic scientists with the CRFP was purely voluntary but since the scheme had the support of the government and the judiciary it was anticipated that the courts would expect most forensic science practitioners to register, as a measure of their competence.

In order to become registered a forensic science practitioner had to be assessed according to a set of criteria as follows:

1. Knowing the hypothesis or question to be tested.
2. Establishing that items submitted are suitable for the requirements of the case.
3. Confirming that the correct type of examination has been selected.
4. Confirming that the examination has been carried out competently.
5. Recording, summarising and collating the results of the examination.
6. Interpreting the results in accordance with established scientific principles.
7. Considering alternative hypotheses.
8. Preparing a report on the findings.
9. Presenting oral evidence to court and at case conferences.
10. Ensuring that all documentation is fit for purpose.

The process required that candidates submit brief details of a series of approximately 60 cases that they have investigated over the previous 6 months prior to submission. An assessor would then select 6 cases from this list and request that the candidate submitted full details of these cases in an anonymised form. Collectively these cases should enable the assessor to identify compliance with the 10 criteria. Candidates who met the assessment criteria were then placed on the register in one of the following defined areas:

Anthropology, archaeology, computing, drugs, fingerprint development, fingerprint examination, firearms, fire scene examination, human contact traces, imaging, incident reconstruction, marks, medical examination, nursing, ondontology, paediatrics, particulates and other traces, podiatry, questioned documents, road transport investigation, scene examination, telecoms, toxicology, veterinary science and volume crime scene examination.

Candidates were registered for 4 years before being required to re-register. Re-registration required the submission of information on continuous professional development and maintenance of professional competence. All those registered had to comply with a code of conduct as outlined below:

- Recognise that your overriding duty is to the court and to the administration of justice: it is your duty to present your findings and evidence, whether written or oral, in a fair and impartial manner.
- Act with honesty, integrity, objectivity and impartiality: you will not discriminate on grounds of race, beliefs, gender, language, sexual orientation, social status, age, lifestyle or political persuasion.
- Comply with the code of conduct of any professional body of which you are a member.
- Provide expert advice and evidence only within the limits of your professional competence and only when fit to do so.
- Inform a suitable person or authority, in confidence where appropriate, if you have good grounds for believing there is a situation which may result in a miscarriage of justice.

In all aspects of your work as a provider of expert advice and evidence you must:

- Take all reasonable steps to maintain and develop your professional competence, taking account of material research and developments within the relevant field and practicing techniques of quality assurance.

- Declare to your client, patient or employer if you have one, any prior involvement or personal interest which gives, or may give, rise to a conflict of interest, real or perceived; and act in such a case only with their explicit written consent.
- Take all reasonable steps to ensure access to all available evidential materials which are relevant to the examinations requested; to establish, so far as reasonably practicable, whether any may have been compromised before coming into your possession; and to ensure their integrity and security are maintained whilst in your possession.
- Accept responsibility for all work done under your supervision, direct or indirect.
- Conduct all work in accordance with the established principles of your profession, using methods of proven validity and appropriate equipment and materials.
- Make and retain full, contemporaneous, clear and accurate records of the examinations you conduct, your methods and your results, in sufficient detail for another forensic practitioner competent in the same area of work to review your work independently.
- Report clearly, comprehensively and impartially, setting out or stating:
  - your terms of reference and the source of your instructions
  - the material upon which you based your investigation and conclusions
  - summaries of you and your team's work, results and conclusions
  - any ways in which your investigations or conclusions were limited by external factors; especially if your access to relevant material was restricted; or if you believe unreasonable limitations on your time, or on the human, physical or financial resources available to you, have significantly compromised the quality of your work.
  - that you have carried out your work and prepared your report in accordance with this Code.
- Reconsider and, if necessary, be prepared to change your conclusions, opinions or advice and to reinterpret your findings in the light of new information or new developments in the relevant field; and take the initiative in informing your client or employer promptly of any such change.
- Preserve confidentiality unless:
  - the client or patient explicitly authorises you to disclose something

- a court or tribunal orders disclosure
- the law obliges disclosure, or
- your overriding duty to the court and to the administration of justice demand disclosure.
- Preserve legal professional privilege: only the client may waive this. It protects communications, oral and written, between professional legal advisers and their clients; and between those advisers and expert witnesses in connection with the giving of legal advice, or in connection with, or in contemplation of, legal proceedings and for the purposes of those proceedings.

The introduction of the CRFP was a major step forward for the forensic science profession in the UK. However, the criteria used for registration are not 'standards' in the accepted meaning of the word.

With the demise of the CRFP the question was then raised about how to replace it. The Forensic Science Regulator issued a consultative paper reviewing the options for the accreditation of forensic practitioners. The outcome, although not necessarily covering all forensic practitioners within the investigative process, was to have all laboratory functions accredited to an ISO standard. This would be carried out by UKAS, which would not only cover the practices and procedures but also look at a proportion of individuals' cases, competency and training records. However there still remains a gap to be filled for the independent specialist expert not necessarily employed by a main forensic provider.

1.6.2 Standards of Competence

Another body that has been involved with drawing up standards of competence for forensic scientists in the UK is the Forensic Science Sector Committee of the Science, Technology and Mathematics Council which is responsible to one of the new government Sector Skills Councils.

Skills for Justice is the Sector Skills Council which represents employers in the Justice sector and has recently also incorporated the Fire and Rescue sector. It was formed in 2004 and has the over arching mission to help organisations and individuals within these sectors to maintain a safe and just society. It does this through workforce skills and influencing policy at all levels to ensure it takes each sector's workforce into account. So far its greatest impact on the Justice and Fire and Rescue sectors could arguably be its review and updating, in 2007, of the National Occupational Standards (NOS) for forensic science. These had previously been owned by the Sector Skills Council for

Science, Engineering and Manufacturing Technologies (SEMTA). Recently, Skills for Justice in consultation with industry practitioners have reviewed and developed new NOS in three initial areas—Crime Scene (evidence recovery from the crime, Crime Scene Investigators, SOCOs, *etc.*), Forensic Identification (fingerprints, footwear, marks, DNA, *etc.*) and Forensic Laboratories (conducting forensic laboratory investigations and analysis). The Fire and Rescue Service became a subscriber to Skills for Justice in 2009 and NOS are also available for this area. The NOS specify the standards of performance which practitioners are expected to achieve in their work and the knowledge and skills which they need to perform effectively, *i.e.* they describe competent performance in terms of the outcomes of an individual's work.

The standards are written in a generic form to enable all the different disciplines to be described. They are presented as a series of units with each unit being divided into a set of elements. The units are listed in Table 1.2 and an example of one of the elements is shown in Table 1.3. Having described standards of competence it then becomes necessary to develop a strategy for assessing scientists against such standards. Such assessment strategies are presently being developed.

Finally, in recent years the UK has seen an enormous growth in undergraduate degrees in the forensic sciences and the quality of such

Table 1.2 Professional standards of competence in forensic science.

Unit		Element	
1	Prepare to carry out examination	1.1	Determine case requirements
		1.2	Establish the integrity of items and samples
		1.3	Inspect items and samples submitted for examination
2	Examine items and samples	2.1	Monitor and maintain integrity of items and samples
		2.2	Identify and recover potential evidence
		2.3	Determine examinations to be undertaken
		2.4	Carry out examinations
		2.5	Produce laboratory notes and records
3	Undertake specialist scene examination	3.1	Establish the requirements for the investigation
		3.2	Prepare to examine the scene of the incident
		3.3	Examine the scene of the incident
		3.4	Carry out site surveys and tests
4	Interpret findings	4.1	Collate results of examinations
		4.2	Interpret examination findings
5	Report findings	5.1	Produce report
		5.2	Participate in pre-trial consultation
		5.3	Present oral evidence to courts and inquiries

Table 1.3   An example of an element associated with standards of competence.

You must ensure that you:	*You need to know and understand:*
a Make laboratory notes and records contemporaneously and that they are fit for purpose, accurate, legible, clear and unambiguous	1 Why it is important to record information contemporaneously
b Order notes and record information in a way which supports validation and interrogation	2 Why it is important to ensure that notes and records are fit for purpose, accurate, legible, clear and unambiguous
c *Uniquely* classify records and file them securely in a manner which facilitates retrieval	3 What information you need to record
d Accurately collate laboratory notes on work carried out by others into the overall records	4 Which recording systems you need to use
	5 When notes and records are complete
	6 The systems you use to order your notes and record information
	7 The importance of ordering notes and information
	8 The classification systems you use to ensure records are easily retrievable
	9 How the classification system works
	10 How to file records securely
	11 The importance of collating notes accurately
	12 The identity of others who might wish to use the notes
	13 The ways in which the notes might be used

From www.crfp.org.uk, Unit 2: Examine items and Samples, Element 2.5: Produce laboratory notes and records.

degrees has been of concern to many in the profession. For this reason the Forensic Science Society (FSSoc) began a programme of developing standards for such degrees that will be offered to the universities as part of an accreditation programme. The FSSoc has been around for over 50 years, being formed in 1959 and then more recently as a professional body in 2004. The society produces a peer reviewed journal, *Science and Justice,* as well as awarding postgraduate diplomas for individuals in a number of forensic disciplines such as documents, fire investigation, firearms and crime scene examination. The FSSoc is partnered with Skills for Justice to deliver the quality stamp Forensic Skillsmark. The award is given to a range of forensic science learning institutions and organisations such as universities and forensic science providers, and covers all levels of learning.

1.7   CONCLUSION

The discussion in this chapter has aimed to provide the reader with an understanding of the role of forensic practitioners, how they achieve professional status and how they interact with the legal process. Working as a forensic practitioner can be physically, emotionally and intellectually demanding but also intellectually rewarding. The succeeding chapters will show why this is so.

BIBLIOGRAPHY

Directory of Consulting Practices in Chemistry and Related Subjects, The Royal Society of Chemistry, 1996.

Brian H. Kaye, *Science and the Detective*, VCH, Weinheim, 1995.

Philip Paul, *Murder Under the Microscope*, Macdonald, London, 1990.

Quality standards for providers of forensic science services to the Criminal Justice System (Consultation Draft), Forensic Science Regulator, March 2009.

WEBSITES

http://police.homeoffice.gov.uk/operational-policing/forensic-science-regulator/

http://www.crfp.org.uk

http://www.european-accreditation.org

http://www.forensic-science-society.org.uk

http://www.ilac.com

http://www.skillsforjustice.com

http://www.ukas.com

CHAPTER 2

The Crime Scene

ORLANDO ELMHIRST

Forensic Science Service, Trident Court, 2920 Solihull Parkway, Birmingham Business Park, Birmingham B37 7YN

2.1 INTRODUCTION

In the year 2007–2008, the police in England and Wales recorded 759 cases of homicide, each of which generated one or more crime scenes that were physically investigated by crime scene examiners. For the general population, these homicide scenes are the epitome of crime scenes and, invariably, are the subject of media attention and television dramas. During the same period, the police also recorded 622,012 burglaries and these offences alone generated approximately a further 428,000 crime scenes that were examined. Overall, about 14% of the 5 million recorded crimes receive a crime scene examination, amounting to approximately 700,000 scenes per year.

As the title of this book implies, the crime scene is the starting point from which all other aspects of forensic science follow. Get this bit wrong, and the rest of the forensic science process is nullified—there is rarely a second chance to make good any mistakes.

Scenes are often presented to crime investigators with many unknowns that, hopefully, become clearer over time. Not only do crime scene examiners invariably work with an incomplete understanding of the case circumstances, but also they are usually confronted with a surfeit of potential exhibits; moreover, the relevance of each exhibit is

Crime Scene to Court: The Essentials of Forensic Science, 3rd Edition
Edited by P. C. White
© Royal Society of Chemistry 2010
Published by the Royal Society of Chemistry, www.rsc.org

often uncertain. Ideally, from the crime scene examiner's point of view, the whole scene should be metaphorically 'wrapped up in cotton wool' and delivered to a forensic laboratory where it can be examined under controlled conditions. For most scenes this sort of approach is rarely an option. Consequently, the crime scene examiner must select the relevant parts of the scene and get them 'wrapped up in sterile cotton wool and delivered to a forensic laboratory'.

Apart from the interpretation, physical location, recording and recovery of items from the scene, the crime scene examiner must be ever mindful of two very real threats that may affect exhibits: contamination and loss of continuity. The mere theoretical possibility that either of these two threats is present has the potential to fatally damage an exhibit's reputation within the criminal justice system. Any hint that material may have been transferred from the scene to a suspect via third parties, or *vice versa*, could invalidate any conclusions that have been made.

The wide range of scene types that are examined means that the number of people involved varies, as does the extent to which they perform their duties. It is not practical to describe how to approach *all* crime scenes in a book of this scope. Therefore, the first part of this chapter (Sections 2.2–2.5) describes the framework that is in place to enable scene examinations, and the second part (Sections 2.6–2.11) deals with the practicalities of examining scenes. In the latter sections, emphasis is placed on the most serious offences as this is when the police deploy the panoply of their resources.

2.2   WHAT IS A CRIME SCENE?

In a police investigation, a 'crime scene' is not necessarily the location of a crime; the phrase has a much wider meaning. A crime scene is any location that has a bearing on an incident under the scrutiny of an investigating body. Thus, some scenes are investigated to establish if a crime has occurred. In murder cases, there may be associated crime scenes in addition to the site where the body has been found, such as:

1. Where the murder was planned.
2. Where the murder took place.
3. Where the body had been stashed.
4. Where the weapon was hidden.
5. The vehicle in which the body was transported to the deposition site.
6. The getaway car.

7. The offender's home address.
8. The offender (usually the most important scene after the body site).

A crime scene may be as small as a bite mark or shoe impression, or a large geographical area covering several acres or even more. An extreme example of a large crime scene was that associated with the Lockerbie bombing in 1988. The debris from the explosion of PanAm Flight 103 was scattered over a distance that was almost one-third the width of Scotland and was reported by the BBC to have covered an area of over 845 square miles.

2.2.1   Crime Types Examined

The application of a scene examination could be beneficial to most crime scenes. Because of resource constraints, not all offences routinely receive the attention of a scene examiner; currently, only about 14% of all recorded crime generates a scene examination.

Two types of offence are generally recognised in policing terms: volume crime and serious crime. The volume crime types that routinely receive a crime scene examination are offences such as burglary of a dwelling, burglary of a non-dwelling, and theft of a motor vehicle. Other offences where a scene examination may be requested include criminal damage and thefts. Volume crime offence types make up the bulk of a police force's forensic examinations and are invariably conducted by a lone scene examiner.

In more serious offences, a team of examiners would be employed, the composition and size of which depends on the nature and circumstances of the offence. Murders are regarded as the most serious offences routinely dealt with by a police force. They are associated with relatively robust processes and procedures ranging from the strategic level, the Association of Chief Police Officers (ACPO) *Murder Investigation Manual*, through the middle level, ACPO's *Manual of Standard Operating Procedures for Scientific Support Personnel at Major Incident Scenes*, down to the tactical level of the investigating police force's Standard Operating Procedures (SOPs).

2.3   WHY EXAMINE A CRIME SCENE?

As stated by the United Nations Office on Drugs and Crime,

With the exception of physical evidence, all other sources of information suffer from problems of limited reliability. Physical evidence, when it is recognized and properly handled, offers the best prospect for providing objective and reliable information about the incident under investigation.

Some of the more common reasons for conducting a scene examination are as follows:

- **So that the persons involved can be identified.** Not only does the principle offender need to be identified, but also any collaborators, conspirators and potential witnesses. Moreover, the identity of the victim may need to be sought. In a homicide, it is not unusual for a body to have been stripped of all its possessions. Unless the inquiry team know the name of the body, they are severely hampered in starting their investigation into who has committed the offence.
- **To substantiate that a crime has been committed.** A suspicious death is an example of a potentially serious offence that is quite commonly encountered. In these cases, someone has died unexpectedly and there is a requirement to establish if the death was natural or not. If there is an element of doubt, the scene is treated as if it were one of murder. A fire scene is another common scene type that fits into this category. Was the fire started deliberately (*i.e.* arson, called wilful fire raising in Scotland) or accidentally?
- **So that accounts can be corroborated or refuted.** Suspects of an offence have many and varied reasons for not telling the truth, usually so as to avoid implicating themselves. Therefore, the information that arises from a scene investigation is matched against the accounts given by the suspect and witnesses. If the interviewer catches someone misrepresenting the truth, and demonstrably so, they have gained some valuable knowledge and some leverage for the next set of interviews. If, on the other hand, the account fits the scene information, then an interviewer can put a different construct on the interviewee's veracity.
- **To gather intelligence.** Intelligence gathering is one of the core functions of a police force and information about crime scenes is avidly sought. Details of criminals' modus operandi (MO), *i.e.* the way they carry out their business, can enable seemingly disparate crime scenes to be linked. The police work to a National Intelligence Model, which is consistent throughout all police forces, and there are now set ways of integrating scene information into the forces' intelligence systems.
- **So that justice can be exercised impartially.** This is the ultimate goal of all crime scene examinations. The evidence that emanates from a

crime scene should be such that it complies with 'the truth, the whole truth and nothing but the truth'.

2.4 WHO EXAMINES A CRIME SCENE?

Crime scenes are examined on behalf of an investigating body. In the UK, most crimes are dealt with by the police; so it is police personnel who conduct the examinations. Other investigational bodies that exist within the UK and who use crime scene examiners include the Serious and Organized Crime Agency, Her Majesty's Revenue and Customs and the Special Investigation Branch of the Ministry of Defence. It should always been borne in mind that whoever is examining a crime scene, they are doing so on behalf of an investigating officer.

The terminology used to describe police crime scene examiners is constantly changing to keep pace with innovative management thinking within police forces and other agencies. The term used in this chapter for police scene examiners is scenes of crime officers (SOCOs); suitable alternatives include crime scene examiners, crime scene investigators, forensic practitioners, *etc*.

External specialists may be invited to attend a crime scene and assist in the examination as experts. Their role is dealt with in detail in subsequent chapters of this book.

2.4.1 Background, Training and Professional Bodies

In the last 20 years, there has been a move away from utilising police officers and towards employing civilians. Now, the majority of SOCOs are civilians but a few forces do retain a residual number of police officers both as SOCOs and, more often, in the supervisory ranks. The most popular route for the training of SOCOs is through the National Policing Improvement Agency (NPIA) Forensic Centre at Harperley Hall in Durham. A few of the larger forces carry out their own training; chief among these is the Metropolitan Police Service which conducts its courses through its own Crime Academy. The most recent development is the introduction of university degrees offering scenes of crime training. However, there are no guaranteed jobs at the end of the courses and graduates have to apply for any vacancies advertised along with other candidates.

Police-employed crime scene examiners are not usually recognised as 'experts' by the courts. Therefore, their evidence has to be confined to facts and not opinions—unless the judge gives them express permission.

Attempts have been made over the years to standardise the work of the crime scene examiner, but none have yet achieved that aim. This may change with the inception of the Forensic Science Regulator, whom is tasked with addressing the quality standards applying to forensic science services delivered both in-house and externally to police forces. The last, almost successful, attempt at coordinating standards was through the Council for the Registration of Forensic Practitioners. But, even at its peak, only 55% of the eligible police personnel applied to register. The Forensic Regulator is aligning himself with the European Co-operation for Accreditation and the European Network of Forensic Science Institutes, who are pursuing the implementation of ISO/IEC 17020, which is the internationally recognised standard of competence of inspecting bodies, in crime scene investigation. For accreditation purposes, all scene examiners, whether internal or external to police forces, will be regarded as Type C Inspection Bodies within ISO/IEC 17020. It is the Forensic Regulator's current view that most forensic practitioners ought to be assessed through National Occupational Standards (NOS); independent practitioners should be able to demonstrate 'a level of competence commensurate with the requisite National Occupational Standards'. To this end, Skills for Justice have developed a number of NOS and the NPIA is aligning its training to fit with these standards.

2.4.2 Specialist Scene Examiners

There are circumstances where it is recognised that the skill levels of the police examiners are insufficient and that the services of specialists are required to extract more information from the scene. The decision to call upon a specialist to assist in the examination of a scene is made by the senior investigating officer (SIO) in consultation with the Crime Scene Manager (CSM). The CSM, on behalf of the SIO, has responsibility for the actions of the specialists within the scene.

Currently, there is no universal or robust accreditation system for specialists. Thus, the situation is comparable to that for police scene examiners (see Section 2.4.1). A CSM may be quietly confident, but not complacent, that a forensic scientist from one of the major forensic laboratories will have a good understanding of investigational protocols and procedures.

Although it is rare, there may be some scene circumstances where the requirement for a novel or unusual skill set is identified. Such occasional experts would, almost by definition, not be part of a forensic accreditation scheme. They may have no crime scene experience. In these cases, considerable effort must be expended in giving these individuals suitable mentoring and supervision to ensure that they do not compromise the investigation, the scene or themselves.

2.5 HOW ARE CRIME SCENES EXAMINED?

It is anticipated that a set of quality standards for examining crime scenes will be identified or defined by the Forensic Regulator, but these have yet to be drafted.

2.5.1 Legal Framework

For most scenes that are examined, the scene examiners are invited into the premises by the legal owners who, more often than not, are the victims. However, there are occasions where the owner or occupier of premises is unwilling for a scene examination to be conducted. There may be many reasons for this, but it is often because they are the suspect of an offence. It is the responsibility of the scene examiner to establish what rights they have to enter any premises. In general, there are two powers that can be exercised when permission is not forthcoming and an examination is desirable; both are defined in the Police and the Criminal Evidence Act 1984 (PACE):

1. A search warrant may be issued by a justice of the peace granting the powers of entry, search and seizure.
2. The police also have the power to search any premises without a search warrant if it is to make an arrest or after an arrest.

PACE also gives details of what powers the police have in relation to the treatment of detained persons and what samples may be taken from them.

2.5.2 Health and Safety

Crime scenes can present numerous risks to the health and safety of all persons who enter them. It is incumbent on the scene examiners and their supervisors to be acquainted with both the legislation that currently applies and the range of potential adverse effects that a scene and the examination could have upon attendees. Scene visitors can be exposed to some psychologically traumatic scenes, potentially physically injurious situations and harmful biological pathogens.

Every police force has its own detailed set of policies that are built upon relevant laws. The main provisions of these policies are under pinned by the following legislation:

1. Health and Safety at Work, *etc.* Act 1974.
2. Management of Health and Safety at Work Regulations 1999.
3. Personal Protective Equipment Regulations 1992.
4. Manual Handling Operations Regulations 1992.
5. Control of Substances Hazardous to Health Regulations 1999 (COSHH).

Crime scene examiners also have a duty in minimising hazards to others handling collected evidence further up the forensic continuity chain, by using appropriate packaging and warning labels.

2.5.3 Protective Clothing

Whenever anyone enters a major crime scene they should be wearing overshoes, gloves, scene suites, head covers and masks (Figure 2.1). There are two distinct reasons why these are a requirement: first, for the health and safety of the scene examiner and their colleagues—this clothing is part of their personal protective equipment, where it is a

Figure 2.1   A SOCO and a forensic scientist handing over exhibits at a murder scene. Note the protective clothing being worn by the two scene examiners and the inner cordon barrier tape in the background. The tent is protecting the body which is still *in situ*.

requirement of the regulations listed previously; second, as an anti-contamination measure. Making all scene attendees change into scene suits before they enter a scene, and remove them when leaving, prevents them from either bringing foreign matter into the scene or taking out some scene material and re-depositing it elsewhere.

2.5.4 Finding Exhibits

At a crime scene, a SOCO will be presented with two potential types of evidence: those that are visible and those that are latent. Visible evidence presents few problems for recovery to the crime scene examiner and should be found provided the examiner uses a systematic examination routine and the items have been recognised as being relevant to the inquiry.

Latent evidence is more problematic because it cannot be seen and has to be located and developed; *e.g.* a fingerprint examination with aluminium powder. To a degree the examiner will have to adopt a 'spray and pray' policy. This blanket approach can be refined, and the scene examiner can make some educated guesses as to where to conduct a fingerprint examination; *e.g.* on any suitable items that have been moved or disturbed by the offender and those areas that could have been touched as the intruder moved about the scene, such as the edges of windows, doors, *etc.*

For all but the most serious offences, there is invariably someone who can guide the SOCO through the scene and indicate what has been discarded, what has been disturbed and what has not. However, there should always be an element of sceptical acceptance in this approach. Extra care must be taken with certain victims who may not have attained a full grasp of what has occurred, such as the victims of distraction burglaries, who are often vulnerable individuals. Careful questioning and the spotting of confirmation signs allows the SOCO to test and confirm the victim's hypothesis about what has happened. If the hypothesis is found wanting, the SOCO can develop their own hypothesis to test. Scene examiners should continually check the relevance of potentially useful exhibits. For example, if a cigarette end is found outside a window that is the point of entry, the scene examiner needs to ask whether it belongs to the householder or might it be from the glazier who has fixed the window, and so on. The examination of a burglary scene usually takes between 30 minutes and 3 hours.

If the victim is dead or the enquiry is of a very serious nature, the element of guesswork about what items may yield evidence must be drastically reduced and, in the case of murder, all items need to be examined. For example, in a burglary, the door to a room which had been entered by the intruder would be fingerprint powdered around the door handle area and on the closing edge of the door at about the same height. In a murder, however, the entire door would be examined as well as the jambs and architraves. This work necessarily becomes very detailed, exacting and sometimes tedious as it can take days to complete.

There are numerous methods of systematically examining a crime scene and the choice of which one to use is at the discretion of the crime scene examiner. However, when a team work together there is greater room for error and it is the responsibility of the CSM to ensure there are no gaps or areas of overlap. A possible problem area might be the doorway between two rooms that two SOCOs have been asked to examine—one in each room—someone has to be given specific responsibility for the interface areas.

The development process for locating fingermarks is often messy and can be disruptive to other forensic analysis techniques. Therefore, it tends to be done as the last stage of a scene examination, hence care must be taken for the scene examiners not to leave behind their own fingermarks in the interim! The most common method employed to develop these marks is the application of aluminium powder with a fingerprint brush.

2.5.5 Selecting Exhibits

It is generally true that only a small percentage of the physical exhibits collected at a crime scene will find their way, physically, through the various processes to court. Which of the exhibits will be selected is often difficult to predict in the early stages of an enquiry. A systematic means of preserving, examining, collecting and recording all scene exhibits must be followed so that those exhibits that do complete the pathway to court have unquestionable integrity and validity. Usually, any specialists present at a scene will direct the SOCO on what exhibits to recover on their behalf.

2.5.6 Recovering and Packaging Exhibits

The array of equipment that SOCOs carry in their vehicle routinely enables them to conduct a reasonably comprehensive examination of a scene and to recover most exhibits ranging from a single fibre up to a king-size mattress.

An item must be recovered in a way that avoids damage or a change in its characteristics, or affects other evidence types which may be on that item. Thus, it is important to establish the sequence of recovery before the examination is undertaken. Fingermarks left on an article are a prime example; care should be taken when lifting an item to ensure that these marks are not being rubbed off. Swabbing an article for DNA can clear away a section of fingermarks on the same surface. Wearing gloves may prevent the scene examiner from leaving their marks, but it will not stop marks already there from being damaged. Therefore, all physical handling of items should be carefully thought through. When an item has been identified as being of potential value to an investigation it may, in a volume crime scene, or will, in a serious crime scene, be photographed *in situ* with an identifying label. The item is then recovered and packaged.

An exhibit should arrive at the place of examination and, ideally, at court in the same condition as it left the scene, without losing or gaining any additional materials; *i.e.* its integrity should be maintained. The integrity of an exhibit is paramount to any forensic interpretation that is placed upon it; consequently, it is essential that when an item is recovered from a scene it is packaged in a manner that prevents it from becoming contaminated with other materials, contaminating other materials or being misidentified with some other item. Thus, it is fundamental to use the correct method of packaging and labelling at a crime scene. Materials can now be analysed at the microscopic level. Therefore, the packaging of such samples that will be analysed using super-sensitive techniques must be able to prevent contamination at these levels. Potentially, an unsolved case may be the subject of a cold case review many years in the future so the packaging should also be robust enough to survive many years and be appropriate for not compromising any future analyses.

One of the more basic requirements for all packaging materials used is that it is free from contaminates. The 'Phantom of Heilbronn', a female serial killer who 'stalked Western Europe for more than 15 years, murdering young and old' and 'leaving her DNA at 40 crime scenes', would now appear to be a case of contaminated DNA swabs. Another potential source of prepackaging contamination can be the scene examiners themselves, who may introduce contamination from other scenes.

Ideally, items should be packaged in clear plastic or polythene containers. Such containers have the advantage of being impervious to most contaminants (excluding solvents) and of being transparent. There are a number of situations where clear plastic or polythene containers cannot be used and alternatives such as paper bags, cardboard boxes, *etc.* should be employed where necessary.

Once items are securely packaged, the packaging must be sealed. Plastic tamper-proof exhibit bags are available. The usual method of sealing a paper bag is to place a signature, using an indelible pen, over the adhesive tape which has sealed the bag. The *Scenes of Crime Handbook*, and others like it, give comprehensive information on what packaging to use and when.

2.5.7   Labelling Exhibits

Every item collected in the course of a police investigation should be given an identifying mark, which is often mistakenly called an exhibit number. An exhibit number is the unique number given to an item by a court, whereas the identifying mark is an alphanumerical identification code given to the item by whoever 'seized' it. Different police forces use different methods to generate this mark.

Every exhibit from every scene should be given a label so that it does not get confused or mixed up with other exhibits. The correct type of label is called the exhibit label or the CJ label; CJ stands for Criminal Justice (from the Criminal Justice Act 1967 which defined its use). The CJ label can either be printed on to the packaging during manufacture (Figure 2.2), or printed on a label that is attached to the packaging. The label has five sections, of which two are relevant to the scene examiner. The section marked 'FOR POLICE USE' is where a description of the item is given (usually including not only what the item is but also from where or whom it was recovered), the identifying mark and other police-related details. The other section, which is on the back of the printed label, has space for signatures.

The law requires anyone who takes possession of an item to submit a statement detailing their involvement in its continuity. In that statement they must make reference to the exhibit and sign the label, thereby confirming the identity of the exhibit referred to in the statement. It is relatively easy for the police to ensure that this process of signing the label happens as the item progresses though the chain, from scene to court; the label is signed as each new person receives the item.

2.5.8   Continuity

The continuity of an item, from when it is first seized to when it is produced in court, must be maintained at all times. If there is a break in the continuity chain the evidence will not be admissible in

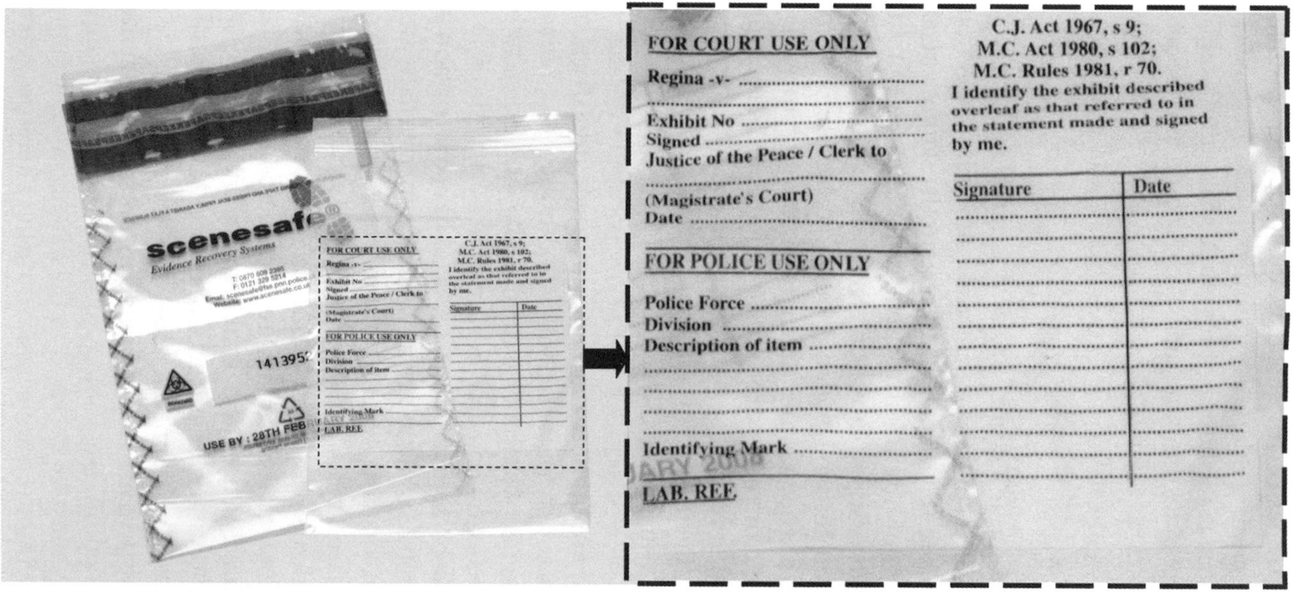

Figure 2.2 A tamper-proof evidence bag and a plastic bag with a printed exhibit label. The exhibit label has been enlarged to show the various sections on the label that are discussed in the text.

court. Because continuity is a fundamental issue in the justice system, it is closely scrutinised by the defence. The scene examiner, who is often the first person in the continuity chain, should take particular care to control the continuity of their exhibits. Therefore, every item in their care should be maintained securely; no other person should have access to the items and they should not be left in anyone else's care.

2.5.9   Recording

Accurate recording of a scene, before it is interfered with, is important for a major investigation and essential for any item recovered. The main reason is that when the scene examiner is in the witness box, sometimes years later, they must be able to fully describe the scene and the location of each exhibit recovered to within centimetres.

Still photography and video camera recording is employed to capture scenes, particularly serious crime scenes, before they are interfered with. These methods capture a huge amount of detail which can become invaluable providing both evidence and intelligence. Moreover, they are highly illustrative for anyone who was not present at the scene, such as investigating officers, forensic scientists in the laboratory and the court.

Sketches and annotation can highlight those areas that are of particular importance to a scene examiner. Although less precise, sketches have the advantage over photographs of allowing notes to be attached, comments to be made and features to be highlighted.

For every exhibit recovered there should be a description detailing the identification mark, what the item is and from where it was recovered. These will often form the basis of a statement which may be required at a later date. Consequently, the description should be full and thorough. Where practical, and this means invariably, notes and records should be taken at the time of the scene examination (contemporaneous) or, at worst, immediately after the examination. Any delay in recording information allows for errors to occur and the longer the delay the greater is the risk for erroneous recollection.

All notes and written records made at a scene and during the subsequent investigation are subject to disclosure. In broad terms, this means that they must be declared to the defence and, if appropriate, the defence has the right to take a copy of them. Therefore, everything that is written down must be defendable in court and professional in both appearance and content.

2.6 FROM CRIME TO CRIME SCENE

There is always a time delay between an incident occurring and it being designated a crime scene. This period can vary from minutes to years. Regardless of how short the interval is, there is invariably some form of disruption to the scene. Because of the nature of the disturbance, there are usually few recorded means of reconstructing what has happened. 'Evidence' can be either introduced to the scene or taken away, usually unwittingly. Typical 'evidence' that might be introduced to the scene at this stage could be footwear marks in blood, or fingermarks. Conversely, evidence can be removed from a scene, particularly small or microscopic evidence, such as parts of detonated explosive devices. It is extremely important that the history of a scene is compiled as soon as possible. Figure 2.3 shows an extreme example of scene disturbance. Once a crime scene status has been established, the scene can be managed by the police.

At a volume crime scene, the owner or person who discovers the incident can cause some disturbance of the scene. Most police forces train their call reception staff on how to advise victims on scene preservation pending the arrival of police staff.

Figure 2.3 The scene of the bombing of an army van (top left of photograph). Note the number of members of the public who have got to the scene before the police were able to impose scene control. (Courtesy of Leicestershire Constabulary)

2.7   INITIAL ACTIONS AT A CRIME SCENE

2.7.1   First Police Officers at the Scene

The first officers attending (FOA) scenes will recollect the initial key actions to take in a major investigation:

- **Preserve life.** At all times this is the primary aim and overrides anything else.
- **Preserve the scene.** Preservation of the scene is the next most important activity. One of the first actions necessary is to try and define exactly the geographical scope of the scene. Once this is established to a satisfactory level, the scene should be secured.
- **Secure evidence.** The usual method of securing a scene is by the use of crime scene tape to produce a cordon. Once the cordon is established, as few persons as is practical should be allowed into the scene until more specialist staff arrive, *e.g.* SOCOs.

At the same time as the cordon is set up, a scene log should be instigated. There should only be one log per scene and everyone who enters and departs the scene should be recorded. Ideally the log will contain:

1. Details of the logger.
2. Details of the SIO and the CSM.
3. Name and contact details of the visitor.
4. Reason for the visit and by whose authority.
5. Time in and time out of the scene.

This log will become an exhibit.

For volume crimes, the security of the scene can be left to the victim after suitable advice has been imparted by the FOA. The FOA is expected to recover vulnerable pieces of evidence that are at risk of being damaged or degraded, and place them in a secure and dry location pending the SOCO's arrival.

2.7.2   First SOCO at the Scene

Depending on the seriousness of the incident, a SOCO will be dispatched to the scene as soon as is practical. Obviously, if it is a major or critical incident, immediate attendance is expected. For a volume crime incident, the attendance request is placed on a list of other jobs for the SOCO and is dealt with in an appropriate time.

The first SOCO at the scene will consider the same key points as the FOA (*i.e.* preserve life, preserve the scene, secure evidence)—but with more specialised knowledge. Ideally, the scene should not be disturbed until the scene examination team has assembled and a CSM has been briefed. For a volume crime scene, the SOCO is the team.

2.8 PREPARING FOR A CRIME SCENE EXAMINATION

2.8.1 Assessing the Scene

Before any examination can be undertaken at a crime scene, the CSM has to assess the scene. The CSM will question as many members of the police present as possible in order to gain as complete an understanding as possible of what has occurred. The CSM should also be in contact with the police control room to establish if there are other related facts and incidents which may have an impact on their scene. The CSM then reviews what actions have been taken to date and makes any suitable adjustments.

With a volume crime, the SOCO is usually supplied with an incident log. This is a record of the incident made by the police control room, which details not only the police response but also what has occurred, names, contact numbers, *etc.* Furthermore, the victim is invariably present to describe what has occurred.

2.8.2 Confirming the Cordons

The establishment of cordons is dictated by the particular incident type, circumstances and geography. Theoretically, the chance of finding evidence is related to the area over which the evidence has the potential to be spread—the inverse square rule applies. That is, the further away from the centre of activity one goes the greater the area over which a set number of exhibits can be scattered.

In the 'real world' setting, offenders probably have taken only one or two routes into and out of the scene. If these routes can be identified, they are more likely to yield evidence than routes not taken and, consequently, should receive special attention. The same level of attention should also apply to known or inferred areas of offender activity.

The CSM must review where the cordons have initially been set up. For any scene it is much easier to bring the cordon inwards to encompass a smaller area than to enlarge it to cover a larger area so, at this stage, 'big is best'. The positioning of the cordon should be reviewed regularly as new information is received.

Usually, at a major incident, two sets of cordons are established: an inner cordon and an outer cordon. The purpose of the outer cordon is to control the movement of people in and out of the area adjacent to an incident. It is not intended to prevent the movement of all people but it does limit it to those who have a legitimate reason for being there. The inner cordon is a much more tightly controlled zone. Its purpose is to prevent anyone from entering the scene unless they have the agreement of the CSM. The scene logger will be positioned outside this cordon, but adjacent to it.

Other factors that must be considered by the CSM at this stage include:

1. Are there enough personnel to properly guard the cordon?
2. Has the rendezvous (RV) point been set up in the correct position so that it does not impact on the scene?
3. Are there any issues with animals, the press, *etc.*, entering the scene?

Volume crime scenes do not require cordons but the FOA should advise victims which areas should be left undisturbed pending the arrival of a scene examiner.

2.8.3   Scene First Aid

A large number of external factors can affect a scene and, where possible, these need to be guarded against. One of the most common and destructive factors encountered is the weather. Even if the scene is inside, there is always an external aspect that requires consideration, such as a driveway or external doors. Any developing meteorological and other impacting conditions must be monitored so that plans for mitigating action can be drawn up. One of the more common solutions is to erect tents, but this is not always possible in confined spaces or if large obstacles are present; lateral thinking then has to be applied. Occasionally, a CSM has to conduct some examinations out of sequence to lessen the impact of unavoidable situations.

2.8.4   Forensic Assessment

After any high-risk situations have been ameliorated, the examination can move on to a more proactive and standardised approach. If little information is available, a member of the SOCO team, normally the CSM, may need to conduct a flash search.

Flash searches should be avoided where possible, but when there is a lack of knowledge and time is of the essence, they can be the preferred option to the slower, more methodical approach taken subsequently. In essence, a single scene examiner will don protective clothing, enter the scene and do a relatively quick circuit of the scene ensuring that they do not knowingly disturb any visual or latent evidence; thus any disturbance of items is done by a trained examiner. The examiner should take care not to step in 'obvious' places such as the middle of stairs and hallways, to open doors via the very top edge, *etc*. On coming out of the scene, all the protective clothing worn is removed, packed and labelled up as an exhibit. Flash searches can help to determine whether there is more than one offender, if the offender is likely to be covered in blood, if it is a recent offence, if a forced entry was effected, whether the victim put up a fight, if anything is obviously missing, and so on.

The flash search can provide the enquiry team with some useful information about the scene on which to base some first, tentative, hypothesis on what has occurred. A more detailed scene and investigational strategy can be designed with this additional information.

For scenes of less serious offences, it is advisable for the scene examiner to walk through the scene before starting their examination so they have a full picture and can generate a crime scene strategy.

2.8.5 Developing a Crime Scene Strategy

The next step is to develop a broad strategy, but how detailed this is obviously depends on the known and supposed facts using information from the scene and from witnesses. If the scene is the first scene in an incident, then the strategy developed will be vague and more open to re-evaluation than if it is in the latter part of an ongoing investigation. At the start of an enquiry, information is at a premium and the examination of the scene is invariably a 'fast track action' that yields substantial data with which the investigation team can start to formulate hypotheses on which to base more developed strategies. In the later stages of an inquiry, the SIO will instigate crime scene examinations for defined reasons and anticipated outcomes. The SIO should always set some initial objectives for the examination. It is not helpful when a CSM is asked to 'forensicate' a scene!

The forensic strategy should consider whether other forensic specialists are required to assist the police resources. Invariably, if there is a body, a pathologist will need to be contacted and, where possible, they should be encouraged to visit the scene. It is always better to give specialists the opportunity to view the scene intact—photographs are a poor substitute for the real thing. All expert involvement at the scene is coordinated by the CSM who has to take the specialists' time scales into account when coordinating the scene activities. Once a strategy has been agreed, the CSM can commence the full investigation of the scene knowing what is required of them, what sort of time scales they should be working to and what resources they have available.

For less serious offences, the SOCO will formulate their own strategy but discussions with the investigating officers are desirable to establish any lines of enquiry that may be developing.

2.9   EXAMINATION OF A CRIME SCENE

2.9.1   Establishing a Common Approach Path

Once the perimeter of the scene has been secured, the next approach within the context of most strategies is, usually, to get to the focal point of the scene which, in the initial scene of a murder enquiry, means the body. A route into the scene should be devised which attempts not to coincide with the offender's. This route is called the common approach path (CAP). The CAP is a forensically cleared route that will be physically defined to ensure that people who use it do not stray off it. The two most common means of delineating a CAP are to create a forensically clean corridor between two parallel lines of scene tape or to use stepping plates.

- **Forensically clean corridor.** Once a route to the body has been agreed upon by the CSM and the SIO, a team of SOCOs will be instructed to forensically cleanse it. One SOCO will be responsible for photographing the course of the route as it progresses into the scene, from a forensically clean position. The others will be responsible for recovering any items that may lie in the path of the route. This process can be quite time consuming but is thorough.
- **Stepping plates.** This is a quicker and more often preferred method. As the name suggests, a stepping plate is a raised platform that can be stepped on. The plates are usually made of toughened plastic or aluminium and have a top surface area of less than 50 cm square which is raised about 4 cm off the ground by four corner feet, usually with rubber tips, which are designed to have a minimal contact area with the ground. To get to the focal point of a scene, a series of these plates are placed on the ground creating a series of 'stepping stones' (Figure 2.4).

By using stepping plates, the more time-consuming element of recording and recovering every item on the route can be postponed.

Figure 2.4   A forensic scientist at a murder scene using stepping plates. (Courtesy of the Forensic Science Service Ltd)

However, it is still a requirement for the photographs to be taken of the undisturbed scene as the path develops. Care is also required to ensure that the plates' small feet are not positioned such that they destroy evidence.

A diagram illustrating the various elements identified above to protect a major crime scene is given in Figure 2.5.

Whichever CAP method is employed, it is important to take a full and comprehensive set of photographs of the body and its environment. Once the body has been reached, it is often useful to clear a route round it, or to put down stepping plates to enable access to it, for the next stage of the examination.

For less serious offences, great care should always be taken when approaching a crime scene for exactly the same reason as in a major crime; *i.e.* the easiest route for the scene examiner to take is probably the one the offender has taken. For example, when approaching a broken window at a burglary, the scene examiner could stand on broken glass which may have fingermarks or shoemarks from the offender on it. Or, by walking into an office burglary where papers have been strewn across the floor, the examiner may be treading on pieces of paper that have the offender's shoemark impression on them. It is not practical to establish a

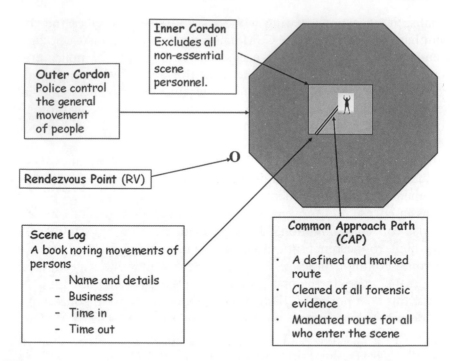

Figure 2.5   A schematic representation of the elements required to protect a major
crime scene.

CAP in a less serious offence and stepping plates would be unrealistic, so
the scene examiner will clear the route as they conduct their
examination.

2.9.2   Focal Point of the Scene

The focal point of a scene is where the most information tends to be
found and, consequently, it will receive the most attention. The first
activity that is required is to record the focal point, in as much detail as
possible. Invariably, numerous photographs will be taken. It is also
useful for one of the scene examiners to make an annotated sketch. This
should, where possible, be done by the lead examiner. Having to draw a
sketch of the incident forces the drawer to look at all parts of the scene
and it becomes a means of ensuring that a comprehensive visual check
has been made. This activity also provides valuable thinking time.

At this stage, it is helpful for the rest of the inquiry team to make a
video of the scene. A video can be an essential part of the inquiry team's
first briefing, giving all officers a feel for what they are dealing with and
avoiding the need for any unnecessary officer visits to the scene to

visualise it when interviewing witnesses and suspects. Once the focal point has been recorded, the CSM liaises with any specialists who are in attendance to determine their needs. Generally, only the police crime scene examiners have been into the scene thus far. Now that a path has been cleared, it is a suitable time to invite other essential personnel into the scene, including the SIO, forensic scientists and the pathologist. However, before they can enter the scene, they must be briefed on what they can and cannot do.

If there is a body at the focal point of the scene, its removal to a mortuary creates a significant disturbance to the immediate area and its movement can disturb evidence. Crime scenes are not an ideal place to process a corpse, but movement to a mortuary can potentially increase the risk of contamination, dislodge evidence and, possibly, allow everything to become saturated with bodily fluids. Therefore, it is becoming common practice to try and deal with some aspects of the body at the scene. In the first instance, swabbing of the body for DNA occurs; this is important if there are bite marks, *etc.* Next, zoned taping of the body is conducted, taking care not to destroy other evidence. Adhesive tape is applied to the exposed skin areas of the body; this technique should recover any loose hair and fibres that are present. The body is divided into zones for taping and each zone is taped separately. Subsequently, it may be possible to establish if any fibres present have a variable distribution over the body and then put forward a hypothesis for this distribution. The hands, feet and head of the body are placed in plastic bags to prevent the loss of material from these surfaces as the body is transported to the mortuary. Where possible, and appropriate, the outer garments are removed, photographed and packaged. Not only will this prevent loss of evidence as the body is moved but also it will prevent vital evidence, such as blood patterns, being destroyed by the victim's blood seeping from any wounds.

Once the body has been prepared, it may be placed inside a body bag which is then sealed. The body is then taken to the entrance to the scene (via the CAP) where undertakers will be ready to receive it and take it to the mortuary. The body is an exhibit and, consequently, its continuity has to be maintained; therefore, a police officer usually accompanies the body to the mortuary. When a suspect is arrested, a similar concept of 'bagging' the body can be applied, but in this instance the suspect is asked to put on a scene suit and have their hands placed in bags. The suspect is then transported to the custody cells where all their clothes can be removed in a secure and more private setting.

In volume crime scenes, the focal points of the scene are those areas where it is known the offender has been. Usually the prime areas are the point of entry, the place where the stolen item was taken from, the driver's seat in a stolen car, *etc*. In burglary scenes, the point of entry yields about 75% of all the identifications, leaving 25% for the rest of the scene.

2.9.3   Post-mortems

Ideally, the scene would now be closed down and the crime scene team would go to the mortuary to assist the pathologist in the post-mortem. Exhibits recovered by the pathologist require suitable packing and labelling; photographs are required of parts of the post-mortem for illustrative purposes; advice must be given to the pathologist about the forensic needs of the inquiry; and fingerprints may be required from the victim for identification and elimination purposes.

There are a number of advantages to the same team doing both the scene and the post-mortem:

- The team has a good knowledge of the scene and should be able to relate any significant findings from the post-mortem back to the scene.
- It is better to wait for the outcomes of the post-mortem to get a clear understanding of what should be done at the scene and to be able to relate any significant scene findings back to the body.
- If death is due to natural causes then only a minimal amount of resources will have been used.
- Only one team will have been used. So, if further scenes need examining, there will be a larger pool of unused SOCOs to respond and, thus, less opportunity for cross-contamination.

2.9.4   Completing the Scene Examination

Once the post-mortem has been completed, the investigation team should have a broad understanding of what has occurred, if not 'who-dunnit'. At this stage, there is often a full team debrief. In any serious crime enquiry there is a morning and, possibly, an evening briefing where the SIO informs the team of what has occurred and what direction they wish the inquiry to take. In the early stages of an inquiry, the scene team is an essential element of these briefings as they often have a lot of information to impart about the scene and the post-mortem, but they also have a lot to learn about suspects and the outcome of interviews. If other specialists have attended the scene, it is usual for them to be invited to the briefings so that not only can they give their opinion first-hand but also they can be questioned by the investigation team. It is also at this stage that the requirement for more forensic specialists will be discussed and agreed with the SIO as part of the ongoing forensic strategy. What specialists are required is completely dependent on the individual case circumstances.

Formal briefings do not occur with volume crimes. SOCOs make a recording of their examinations on a work sheet, which is seen by the investigating officer. If something important arises, the SOCO makes direct contact with the investigating officer and *vice versa*.

By the time of the first briefing, the investigation team should be quite advanced in its development with all the key roles appointed. For crime scenes, two important roles are the crime scene coordinator (CSC) and the exhibits officer.

The CSC is involved if there are a number of scenes to the enquiry. It is not unusual for a major enquiry to generate 10 or more scenes. Each of these scenes must be treated in such a way so as to minimise the possibility of cross-contamination. For this, the CSC develops a contamination log. This is a matrix whereby each scene examiner is matched against the scenes they have attended. It provides a ready means to highlight any risks that may arise with using the same SOCO at another scene.

The exhibits officer is another role that interacts closely with crime scene examiners. This officer, who is normally of the rank of police sergeant, is responsible for all the exhibits recovered in an inquiry. They will be trained to undertake this role but will often seek advice from SOCOs on how to package and preserve exhibits. The exhibits officer will either have possession, or know the exact whereabouts, of every recovered item pertaining to the inquiry. The number of exhibits can be in the thousands. Exhibits officers will have their own store room to which only they have the key. It is incumbent on the CSM to ensure that the exhibits officer is given the details of all exhibits arising from a scene. Volume crimes do not warrant an exhibits officer and the SOCO is expected to deal with all the scene exhibits that they collect. Either they forward the items for analysis (DNA, fingermarks and, sometimes, shoemarks), keep possession of them or place them in the police property store.

After the briefing, the CSM and scene team can return to the scene to conduct a thorough investigation with a more complete knowledge of the issues that appertain to the inquiry.

2.9.5 Specialist Scene Examinations

All aspects of scene examinations should be carried out with the knowledge and the agreement of the SIO. If a specialist is invited into a scene, the CSM must take responsibility for their actions. If the specialist is known and trusted then the burden is not too great, but if they are not known then great care must be exercised in how these people are handled and the degree of supervision they require. Difficulties can arise when experts stray beyond their specialist field and give misleading advice.

Different specialists will have different priorities, and it is the role of the CSM to arbitrate between them and devise a plan of action which will accommodate as many of the requirements as possible while ensuring none of the important elements are lost. Generally, this is not too difficult as it is common for specialists to have worked together at other scenes and all of them have a vested interest in a successful outcome. At some scenes where rapid choices have to be made due to the prevailing conditions and some of the experts are not present, the CSM has to bear in mind their potential requirements. The easiest scenes to deal with are those where planning can be done in advance. These scenes are more likely to be associated with the arrest of suspects.

There are certain scenes where the external specialist has to take the lead in advising how the examination of the scene should be conducted. Sometimes how a scene is examined can produce important information. One example would be the use of archaeologists to recover buried items. Until recently, the police would dig a hole in the ground and recover the body, and that would be it. But, by using an archaeologist, more subtle issues can be explored by looking at the individual layers of soil, their relationship to each other (stratigraphy) and what each contains. Another specialised scene type that requires guidance on how to approach it is an illicit drug factory. In this example, not only is knowledge of the production process necessary but also guidance on the health and safety aspects of a scene that may be awash with chemicals and technical equipment.

If samples are required by the specialist, it is generally better that the SOCO takes them rather than the specialist, because if the SOCO recovers the samples and records them on their work sheets, then the item's existence is captured on the exhibits inventory as a routine procedure. If the specialist takes their own exhibits and documents them, then there is a reasonable chance they will not be recorded within the system and could cause an anomaly later on in the investigation.

It is unusual for specialists to attend volume crime scenes because of costs and resource limitations.

2.10   CLOSING THE SCENE

When the examination of the scene has been completed, the scene is handed over to the SIO. Before this is done the scene examiner should remove all their equipment and clean up any mess they have created. If there are any health and safety concerns about the scene examiner's work, these should be rectified before the scene is returned to the SIO. The examiner should also advise the person to whom the scene is being handed of any potential health and safety issues that may be present at the scene; the scene examiner has a duty of care.

For volume crime scenes, the SOCO should give the owner advice on cleaning up the aluminium powder left if it is not cleaned up by the SOCO, as well as advice on such things as blood left by offenders who, quite often, have cut themselves when forcing an entry. For major scenes, it is highly advisable to arrange a walk though with the SIO so that they can see, and have explained to them, what has been done. This approach allows for a full discussion and understanding of what the SIO has asked for and what has been delivered by the CSM. At the end of this exercise, both parties can agree that the scene examination has been completed and this can be noted in the scene log.

SOCOs are scene examiners, not scene searchers. The police have trained units to undertake scene searching. Consequently, the SIO may want a PoLSA (police search advisor) team to search the scene. This is conducted with a different outlook to a scene examination; thus, fixtures and fittings may be removed, floorboards pulled up and cavities inspected with technical equipment in this specialised police activity.

2.11   POST-SCENE PROCESSES

2.11.1   The Elimination Process

Having completed the scene examination, the investigation team should now try to eliminate from the inquiry any person who may have had legitimate access to the scene. At a major scene, a list will be drawn up of persons known to have been in the scene before, during and after the incident. Police officers will be tasked to visit these people and take whatever samples are relevant for the elimination process, such as fingerprints, shoemarks, DNA, *etc.*

All SOCOs and police officers and most forensic scientists from the larger service suppliers have their DNA on an elimination database. Police personnel also have their fingerprints on record within the force's fingerprint bureau, but forensic scientists and other scene attendees may not—although this is easily rectified. For volume crime, the SOCO should take the fingerprints of persons present when they visit, and a DIY kit with a pre-paid envelope can be left for those not present. Shoemarks can be inspected at the scene. Elimination DNA is not taken.

2.11.2 Submission

All exhibits recovered at a scene should be listed with the exhibits officer. Wherever possible, the physical item should be handed over to the exhibits officer provided they have suitable means to store the items in the correct environment (e.g. refrigerators). The only general exception to this is fingermarks, which usually go straight to the fingerprint bureau to be processed immediately. What is subsequently sent off for further analysis will be the subject of a forensic strategy, which will be formulated by the forensic management team.

2.12 CONCLUSIONS

Clearly, it is essential that the examination of the crime scene must be properly managed and undertaken by following the accepted protocols, otherwise the success of the whole judicial process may be jeopardised. Although in this chapter it has been only possible to present the work required for some types of scenes, there are many more where specialist advice is required and can assist in an investigation. These will be identified in the following chapters.

BIBLIOGRAPHY

Crime Scene and Physical Evidence Awareness for Non-forensic Personnel, United Nations, New York, 2009.
Guidance for the Implementation of ISO/IEC 17020 in the Field of Crime Scene Investigation, EA-5/03, European Co-operation for Accreditation, 2008.
Manual of Standard Operating Procedures for Scientific Support Personnel at Major Incident Scenes, 1st edn, ACPO Crime Committee, 2000.
Murder Investigation Manual, National Centre for Policing Excellence, 2006.
The Phantom of Heilbronn, the Tainted DNA and an Eight-year Goose Chase, *The Times*, 27 March 2009.

Quality Standards for Providers of Forensic Science Services to the Criminal Justice System (Consultation Draft), Forensic Science Regulator, March 2009.

A Review of the Options for the Accreditation of Forensic Practitioners, Forensic Science Regulator, 2009.

The Scenes of Crime Handbook, The Forensic Science Service, 2008.

CHAPTER 3

Forensic Ecology

PATRICIA WILTSHIRE

Department of Geography and Environment, University of Aberdeen, Elphinstone Road, Aberdeen, AB24 2UF

3.1 INTRODUCTION

Ecology is the interdisciplinary scientific study of the distribution and abundance of organisms and their interactions with each other and their physico-chemical environment. It mostly involves 'whole organism' (as opposed to molecular) biology. No one person can be expert in the whole field, but the forensic ecologist must have some knowledge of animals, plants, fungi, bacteria, geology, soils and bodies of water (still or flowing). There must be an understanding of the effect of ambient environmental conditions such as temperature, humidity, oxygen tension, air flow and light intensity on communities of organisms and their habitats.

Many organisms, or parts of organisms, are capable of being retrieved from crime scenes and other relevant locations, exhibits, or cadavers. They can, therefore, provide valuable material evidence. Some organisms are highly specific in their ecological requirements, and may have restricted ecological and geographical distributions, while others are less demanding and have very wide distributions. This does not mean that they have no value in forensic investigation. In court, counsel may challenge with the question 'dandelions are found everywhere, aren't they'? Well, of course, they are not. There are many species of dandelion

Crime Scene to Court: The Essentials of Forensic Science, 3rd Edition
Edited by P. C. White
Published by the Royal Society of Chemistry, www.rsc.org

and they have particular needs and distributions. It is just that they are similar, and seem common to the non-specialist.

Ecological knowledge takes time to acquire, and the most effective practitioners are those with many years' experience. Forensic ecology has been developing over the last 16 years or so in Britain and currently there is a scarcity of appropriate and able ecologists engaged in forensic work. Hence, compared with other forensic disciplines, forensic ecology is still in its infancy. The forensic ecologist who will give the best 'value for money' will have a broad background and extensive field experience, with a high level of expertise and a working knowledge of individual organisms, biological communities and habitats—in short, knowledge of ecosystem processes. Importantly, they also need to understand the value of the discipline within the forensic context.

Some forensic analyses have become routine: DNA, fibres, footmarks, gunshot residue, fingerprints, and others. These involve techniques that are prescribed and can be learned in a relatively short time. Forensic entomology is another frequently used discipline and, again, the technique is prescribed. Botany, palynology (the study of any microscopic entity (palynomorph), which has been scattered away from its source), phycology (the study of algae, including diatoms), mycology (fungal science), pedology (soil science) and geology are, however, not applied so frequently. There are several reasons for this. First, there is a relatively small number of cases where these analyses are needed; and, secondly, these subjects are not learned quickly. These disciplines also rely less on standardised techniques and more on depth of knowledge and experience.

Over the years, much information has been accrued on the effects of earthworms, other invertebrates, foxes, badgers, rodents, birds, vegetation and fungi at crime scenes. Much is also known about decomposition processes on the surface, within various soil types and in still and flowing water. There will be some expert somewhere who can give valuable information in all these areas. However, the forensic ecologist at the crime scene must be able to highlight the potential of a wide range of organisms in any particular circumstance in order to advise on the need for other specialists.

Emphasis is given here to the importance of aspects of botany, botanical ecology and mycology. These have contributed positively to investigations involving:

1. Estimation of body deposition time.
2. Estimation of post-mortem interval.

3. Demonstration of offender pathways and events at and around an offence.
4. Linkage of people, objects and places (trace evidence).
5. Location of human remains and graves.
6. Differentiation of a kill site from a deposition site.
7. Establishment of cause of death.
8. Challenging or confirming witness testimony.

All ecologically based disciplines need to be integrated early during a criminal investigation. They should not be treated as 'stand alone' or in competition, but as parts of a whole, as they offer different kinds of information. Some will be more important in some scenarios than others, but they can be highly complementary. At the start of an enquiry, thought must be given to the potential contribution of particular specialisms and the emphasis needed for any one of them in the case.

3.2 DATABASES

DNA and fingerprint profiles are considered to be unique and databases have been constructed for comparison with unknowns. Different kinds of footwear patterns, fibres, *etc.*, are finite so, again, comprehensive databases can be constructed to which the practitioner can refer. Databases can also be constructed, and some are available, for the identification and geographical distribution of plant and microbial species, as well as for their tissues, cells, pollen grains and spores. Such information is useful for identifying and interpreting trace evidence.

Except for relatively few species, organisms occur in communities with other species of similar habitat preference; the biological community with the habitat constitutes the ecosystem. Any organism (or part of it) can act as a proxy indicator of the ecosystem from which it originates and the forensic ecologist is often called upon to reconstruct and to envisage whole environments from proxy indicators.

In nature, organisms, parts of organisms, and their physico-chemical environments, all occur in infinite combinations—ecosystems can overlap and merge one into another and there are many transitional situations. Natural, semi-natural, and artificial habitats are very varied and experience has shown that every location is unique in terms of its proxy indicators. Although the range of proxies collected from one site can be similar to others, there will never be a perfect match. Considerable effort has been devoted to constructing predictive models from proxy indicators such as soils, pollen and other markers but, for forensic purposes, no model can provide sufficient specificity to give a match.

An experienced ecologist can easily identify, for example, an oak woodland from its proxy indicators—but no two oak woodlands are identical, and identification of a specific oak woodland (or an area within that woodland), requires detailed profiles of proxy indicators relating to the specific microsite. Attempts at compiling databases and modelling will never give the required level of precision for the precise location of a site. Identification of places from proxy indicators relies on the personal experience of the operator, and there can never be an 'easy fix' for the inexperienced by reference to a database. Describing the kind of place from proxy indicators is sometimes easy, but confirmation of a particular place requires field analysis and comparator samples from that site.

3.3 NATURE OF THE EVIDENCE

There are many organisms that could, under some circumstances, be useful in the forensic context, but they have rarely been used in case-work. Table 3.1 is a summary of the main groups which can provide proxy indicators. Animals have been excluded although, on occasions, fragments of animals have provided trace evidence.

3.3.1 Algae

Algae are photosynthetic, mainly aquatic, organisms with an enormous range of morphological, molecular and reproductive variation. They are ubiquitous in fresh, brackish, and salt water, and if some moisture is periodically available, some can grow in a wide range of terrestrial habitats. Large, macroscopic algae, such as the marine kelps, can achieve great size, but most are microscopic and visible only when their populations discolour water, or form surface growths. Some can be free-floating, tangled masses around submerged plants or other objects, while others are truly planktonic. They can colonise hair and clothing of corpses in water and, in terrestrial situations, are often seen to cause 'greening' of bone once it becomes defleshed.

Diatoms are unicellular, photosynthetic, yellow-brown algae (phylum Bacillariophyta) which are distinguished by each cell having silicious outer walls (frustules) which fit together like a lidded box. Each part of the 'box' is termed a valve. The valves are very variable in morphology and surface patterning, as seen in Figure 3.1. This allows high

Table 3.1   Simplified scheme showing some organisms that may be encountered in forensic botanical and mycological investigations.

		PLANTS		
			Seed-bearing	
ALGAE[1]	*FUNGI*	*Non-seed-bearing*	*Naked seed pods (gymnosperms)*	*Seeds enclosed in ovaries (angiosperms)*
Macroscopic, e.g., kelps	Mushrooms	Mosses	Conifers	Flowering plants (Monocotyledones and Dicotyledones)
Microscopic, e.g. diatoms	Lichens	Liverworts	Seed ferns	
	Moulds, Smuts, Rusts, Mildews	Hornworts	Ginkgo (Maidenhair trees)	
	Yeasts	Ferns Club Mosses Horsetails	Cycads Gnetophytes	
Produce SPORES which germinate to form the next generation			Produce POLLEN GRAINS containing sperm cells which effect fertilisation after pollination	

[1]These categories of organisms are used in the colloquial sense as recent molecular data indicate
that many "traditional" groups of organisms comprise ones that are now classified in different
kingdoms.

taxonomic precision in identification and a huge number of species has
been recognised (approximately 100 000). They colonise surfaces quickly
and exhibit seasonal growth and, because of their sensitivity to water
quality, many species are habitat-specific. For these reasons they have
been used as proxy indicators of environmental conditions in many
areas of science, including forensic science. Other microscopic algae do
not have frustules made of silica but walls of cellulose. Although resistant to decomposition, this polymer will eventually succumb to microbial attack, whereas the diatom frustule is exceedingly durable.

   Other algae that are potentially useful to the forensic investigator are
the Chlorophyta (green algae), a morphologically diverse and species-
rich group, and the single-celled Pyrrophyta (dinoflagellates). Living
dinoflagellates show great morphological variation but it is the fossil
dinoflagellates that can be more use in forensic studies. The free-swim-
ming organism eventually produces a robust cyst whose cell walls contain sporopollenin—the same substance found in the outer walls of
pollen grains and plant spores. Such cysts became embedded into rock-
forming sediments and remain as fossils. During soil formation the

Figure 3.1 Examples of diatoms, illustrating their considerable variation in shape and surface sculpturing of silicious frustule which aid identification. ©Anna Píšková and Dreamstime.com.

particles become released from the rock and cysts find their way into the soil matrix, from which they can provide additional sources of trace evidence. Other fossil spores of unknown affinity are often referred to as acritarchs. These also show large variation in morphology and, again, provide interesting markers in soils and sediments.

3.3.2 Testate Amoebae

The cells of testate amoebae are encased in a membrane which can survive acid treatment. The outer wall (test) can be constructed of silica, calcium carbonate or foreign agglutinated particles. Some testate amoebae are common in lakes, damp sediments, or peat while others, such as the

Foraminifera, are marine. The testate amoebae have potential for forensic investigation. Again, their fossils can weather out of rocks and find their way into soils and sediments, and depending on the resistance of the test to acid treatment, they may be found in palynological preparations and provide useful trace evidence.

3.3.3 Miscellaneous Organisms

Although not covered here, animal remains that have, on occasion, been relevant to forensic investigation are Mollusca (*e.g.* snails), Ostracoda, Acari (mites), Nematoda, and cysts or eggs of other animals including birds. Mites and nematode eggs often become included in palynological samples and sometimes provide important forensic information. Large numbers of invertebrates are often observed in association with decomposing corpses and can give information on time of deposition.

3.3.4 Fungi and Related Organisms

Mycology is the study of all kinds of fungi including mushrooms, lichens, moulds, smuts, rusts, mildews and yeasts. They belong to the same major taxonomic group as animals and are not related to plants. If kept moist, fungi colonise and grow on many organic materials, from food and clothing to human remains. They can also grow on fine organic films on glass, concrete and other inorganic and inanimate objects.

Compared to the plants, which have a global estimate of 270 000 species, most of which have been described, the number for fungi is estimated to be somewhere in the region of 1.5 million species of which only about 100 000 have been described. About 14 000 species of fungi have been recorded in the British Isles, compared with only 2000 plants. Unlike most plants, many fungi are microscopic and are easily missed. Species new to science are frequently discovered, even in the UK. It follows, therefore, that fungi are more difficult to identify than plants and there are few mycologists to identify them, especially microscopic species which are the most relevant to forensic studies. Examples of some fungal spores are shown in Figure 3.2.

Due to habitat, host specificity and geographical restriction, fungi are useful as proxy environmental markers and can provide valuable trace evidence. The nature of their growth and development also makes them valuable as indicators of post-mortem interval, as well as indicators of deposition times of corpses. Identification of fungi is difficult, but molecular techniques can sometimes provide accurate information. However, only a small proportion of even the known species have any DNA sequence data available in public databases.

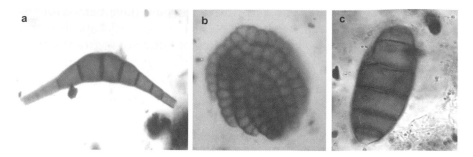

Figure 3.2   Some examples of fungal spores: (a) *Clasteroporium*, (b) *Dictyosporium*, (c) *Endophragmiella*. (Courtesy of Dr J.A. Webb)

3.3.5   Plants

Botany is especially important because communities of plants (vegetation), individual plants and parts of plants play such a huge role in everyday life. We rely on plants for much of our food and other resources. They provide essential medicines, yet can be highly toxic, and are the source of narcotics and hallucinogens. No one can avoid eating them, or contacting them, or being affected by them in some way. They have predictable life cycles, phases of which can often be linked to specific times, predictable growth responses and, to a great extent, predictable distributions. They can be found in natural or semi-natural situations but also in highly artificial ones such as gardens and parks, and in vast areas of crop monoculture.

Identification of whole plants, as well as isolated parts and fragments can eliminate or implicate certain plant species in criminal cases. Plant materials can be linked with objects and provide important trace evidence, food adulteration can be highlighted, and consumption of poisonous plants can be confirmed. It is not surprising, therefore, that identification of plant material is of particular forensic value.

3.3.5.1   Non-Seed-Bearing Plants.   These include Bryophyta (mosses), Hepatophyta (liverworts), Anthocerophyta (hornworts), Psilophyta (whisk ferns), Lycophyta (club mosses), Pteridophyta (ferns) and Sphenophyta (horsetails). Unlike the mosses, liverworts and hornworts, the ferns and horsetails have true roots and vascular sysyems. Many plants within these groups are highly specific in habitat requirement, so presence of their spores in forensic samples may help to characterise habitats. Fragments of leaves and spores of all these plants can be useful as trace evidence since many are highly habitat specific. For example, species in the genus *Riccia* (liverworts), are

characteristic of fallow fields and field edges and have helped to provenance footwear on more than one occasion.

3.3.5.2 *Seed-Bearing Plants.*   Seed-bearing plants can be plants with
naked seeds (gymnosperms) or plants where the seed is enclosed in an
ovary (angiosperms). Gymnosperms include Pteridospermophyta (seed
ferns), Gnetophyta (joint firs and others), Ginkgophyta (maidenhair
trees), Cycadophyta (cycads) and Coniferophyta (conifers). Most are
woody trees and members of the Coniferophyta are the most numerous
and widespread of the gymnosperms, they comprise about 550 species.

The angiosperms, (flowering plants), may be divided into two classes,
the dicotyledones (170 000 species) and monocotyledones (65 000 species).
They range from tiny aquatic species that consist of a leaf of 2 mm in
diameter and a single, tiny root, to huge trees over 100 m tall with trunk
girths of 20 m—and from submerged aquatics to cacti. They have specialised conducting and supporting cells in their stems, roots, and leaves
and a wide range of specialised cells and tissues can occur in all parts of the
plant. They exhibit many different morphological features.

The definitive structure in the angiosperms is the flower. The female
part has one or more carpels, each of which consists of a stigma, style

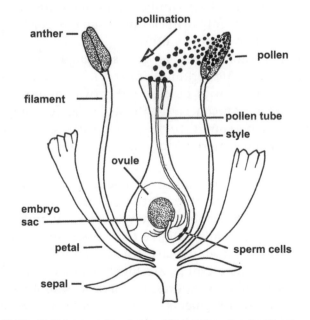

Figure 3.3   Stylised diagram of a flower illustrating the fertilisation process after
pollination.

and ovary as illustrated in Figure 3.3. Pollen grains are produced in the anther and contain cells, they effect fertilisation after successful pollination. The angiosperms produce the most diverse cells, tissues and structures in the plant kingdom. They also produce a very wide range of fruiting structures which surround the ovary and developing seeds, and these have provided valuable trace evidence in many forensic cases.

3.3.5.3 Wood, Dendrochronology and Dendrochemistry. At the beginning of a period favourable to growth, a woody plant will produce new xylem (wood) tissue just below the bark; this carries water and nutrients from the roots to the leaves. Wood cells continue to be formed throughout the growing season while conditions are favourable. During winter, or other unfavourable conditions, growth slows or ceases until favourable conditions return. The transitions are marked in the wood structure and a cross-section shows a series of rings. Generally, the wider the ring, the more favourable are conditions for growth. Narrow rings are produced when the plant is under stress or insect attack. In tropical plants, where conditions vary little, growth is more regular and rings are not produced in the same way as in more temperate species. The wood of angiosperms is more complex than that of conifers, and several kinds of cell are produced. There are many variations in arrangement and kinds of cell in wood, and samples can often be identified accurately to genus but, if sufficient growth rings are present, events may also be timed.

Since growth rings are produced annually in both stems and roots, a remarkably precise estimate of the age of a piece of wood can be obtained for some species. Ring patterns for replicate wood samples from any geographic area with similar environmental conditions can be cross-correlated, and standard time curves for different species established. Specimens of unknown age can be dated by comparison. This technique is known as dendrochronology and it is used in a wide variety of disciplines such as archaeology, palaeobotany, climatology and ecology as well as dating and matching wooden objects from crime scenes, or illegally imported artefacts. Trees will retain pollutants within the wood as it is being laid down in the tree, and chemical analysis of tree rings (dendrochemistry) can help to resolve environmental disputes by identifying the source of contamination, and putting the episode in a precise chronological framework. One technique used in dendrochemistry is energy-dispersive X-ray fluorescence (EDXRF), and it is possible to identify pollution events by targeting elements such as sulphur and chlorine for fossil fuels, and lead for leaded petrol. Other metals are useful as indicators of mining and smelting.

3.3.5.4   Identification of Plants by Molecular Techniques.   The ideal is always to identify a plant to species, but this is sometimes only possible when particular parts of the plant are present. In most forensic cases plants are identified by morphological characteristics. The techniques used—namely, ordinary light or scanning electron microscopy—require relatively simple preparation of the unknown for comparison with known reference material. Sometimes the state of the material makes this impossible, and attempts have been made to identify plants by analysis of their DNA. The results using this molecular technique have been variable and progress has been slow. If the identity of the species is suspected, then DNA studies can sometimes confirm or eliminate its presence. For example, DNA has confirmed species identity of *Cannabis sativa* (hemp) in drug-related criminal cases, and for seeds of known plants such as *Solanum lycopersicum* (tomato) in faecal matter. However, at present, DNA is less likely to be useful if the plant material is unknown.

In forensic cases, it would sometimes be useful to know if material from an *individual* plant could be identified. A diploid plant will contain DNA from both parents; if a sufficiently large database were constructed for any one species, presumably an individual plant could be identified in the same way as human individuals. However, in plants it is often difficult to determine the individual. Many readily form clones by various vegetative means: for example, it is difficult to identify an individual clover plant in a lawn. Clones are also created by horticulturalists and agriculturalists and they can be widely dispersed with identical individuals spread over wide areas, even globally. Clones can be identified by molecular techniques and the forensic potential for this has been demonstrated for cannabis. However, most plant species have not been investigated, and the extent of molecular variation within individual species is mostly unknown. An example of the potential and limitation of this approach is demonstrated by a recent limited molecular study of *Populus nigra* (black poplar) in Britain. It showed that there are least 15 clones present, and members of a single clone are dispersed throughout the country; one particular clone has genetically identical individuals in Cheshire, Yorkshire and Essex. In addition to the problem of cloning, many species regularly hybridise so the genetic profile can be very complex. Large human DNA databases are necessary because the human global population runs into billions. For most plants, each species would need a similarly large database and, even then, cloning and hybridisation could still present a problem. Consequently, in most instances, identification of unknown plants needs to be done by using conventional botanical techniques. Identification to the level of individual might never be achieved other than by physical fit (*e.g.* a splinter of wood matching a gouge), or some individual-specific phenotypic peculiarity

3.4 PALYNOLOGY

As indicated earlier, palynology is a well-established subdiscipline of botany with accepted conventions and analytical methods. It is the study of any microscopic entity (palynomorph), which has been scattered away from source. For most palynologists, palynomorphs are restricted to pollen grains and plant spores, and there are many fields where the science is used to effect: palaeoecology (the study of ancient environments), aerobiology (allergens), melissopalynology (honey), palaeobotany (ancient plants and their evolution), oil prospecting, *etc.* However, to gain as much information as possible, the forensic palynologist needs to identify anything found in a sample. The range of palynomorphs now studied also includes fungal spores and other fungal remains, trichomes (plant hairs), phytoliths (silica cell inclusions), arthropod fragments, charred fragments, fossil spores and pollen, and any other detrital material capable of identification. The forensic palynologist generally identifies as many microscopic entities as possible in a sample, and these can be of plant, animal, microbial or mineral origin. It is imperative that the palynologist has botanical, ecological and preferably, other biological training. Specialist advice should also be sought for fossil plant spores, microscopic animals, eggs, cysts and any entity outside the expertise of the palynologist.

3.4.1 Pollen and Spores

The spores and pollen grains of all plant groups have an inner wall of cellulose (intine) and an outer one (exine) in which is embedded a complex, chemically robust, but mechanically fragile polymer—sporopollenin. Pollen and spores can persist for millions of years in rocks, and thousands of years in more recent sediments. They can also persist in soil for many years, although in bioactive soils that are well aerated they can be decomposed within weeks by microorganisms.

3.4.2 Identification by Light Microscopy, SEM and other Techniques

Light microscopy is used extensively when attempting to identify pollen and plant spores. These vary in diameter from about 7 to 200 μm, with most being about 25–60 μm. Although some of the species of spores

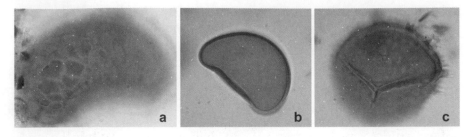

Figure 3.4   Light microscopy images of some spores from non-seed-bearing plants.
(a) and (b), ferns; (c), liverwort. Note surface sculpturing which aids
identification. (Courtesy of Dr J.A. Webb)

from moss, liverwort, hornwort, fern, club moss and horsetail have
distinctive patterning of the outer wall, discrimination is relatively poor
for the non-vascular plants. Higher resolution can be achieved for spores
of the vascular plants such as ferns, but the range of morphology is still
rather narrow. Spores may be trilete and variously sculptured, or bean-
shaped as shown in Figure 3.4 and sometimes enclosed in an outer,
variously ornamented covering—the perine. There are relatively few
diagnostic features for spore identification when compared with the
pollen of seed-bearing plants.

All seed-bearing plants produce pollen which is transported to the
female in the process termed pollination. This may be effected by vectors
such as wind, water (in the case of aquatics), insects, and floral structures
have evolved accordingly.

Standard reference keys based solely on micro-morphological features
have been produced to aid the identification of pollen grains. The pollen
grains of many gymnosperms are more-or-less spherical while others
have pollen with charateristic air sacs, which enhance buoyancy in dispersal, and others may have a hooked or curved papilla. Patterning on
the air sacs, variation in wall thickness, grain size, and variation in
sculpturing and patterning of the main body of the grain all provide
diagnostic features for identification. However, when compared to the
angiosperms, there are relatively few diagnostic characters, and discrimination is relatively poor. Images of some gymnosperm and
angiosperm pollen grains can be seen in Figure 3.5.

The flowering plants produce the most diverse ranges of pollen
morphology, but the resolution in identification is variable depending on
the taxonomic group to which the parent plant belongs. Some can be
identified to species, others to genera, while others can only be assigned
to family. Within a family, it is often possible to determine groups or
types. For example, many plants in the family Rosaceae cannot be

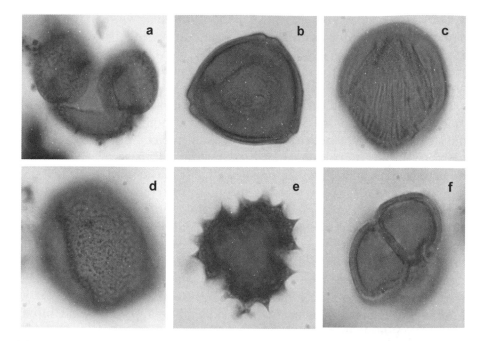

Figure 3.5   Light microscopy images of pollen grains: (a) pine, (b) hazel, (c) cherry, (d) oak, (e) daisy, (f) heather. Note surface sculpturing which aids iden-tification. (Courtesy of Dr J.A. Webb)

differentiated from one another although, for example, *Prunus*-type can be differentiated from *Crataegus*-type, and some species e.g. *Dryas octopetala* can be recognised.

Fungal spores are produced in various ways and are similarly varied depending on the group of fungi involved. Some are tiny cells with little outer differentiation, and often very abundant in the airspora. These are usually almost impossible to identify; others are multicellular and/or ornamented and lend themselves to identification, sometimes to species, based on their micromorphology.

There are several excellent texts on pollen identification, and much information available on the internet, but it is essential that the paly-nologist has a reliably identified and comprehensive reference collection, and does not rely on pictures for identification. There are no compre-hensive texts available for fungal spore identification and the palynol-ogist needs assistance from a trained mycologist with very wide experience of macro- and microfungi. Accuracy is particularly impor-tant in forensic palynology, and any identification made must be substantiated.

The forensic palynologist must also be familiar with many exotic plants. In the UK, for example, crime scenes often involve gardens that are planted with many species alien to the native British flora. Again, there is no substitute for collecting reference material and learning about unfamiliar pollen and spores.

In a single forensic case, a sample may consist of an assemblage of more than 200 kinds of palynomorph and, although a comprehensive physical database of pollen types is essential for comparison, the effective practitioner draws on a personal mental database of a very wide range of pollen and spores. In most routine forensic samples in Britain, the percentage of unknown grains should not exceed 2–5% of the total pollen recorded. Palynomorphs may be unidentifiable because of differential decay, crumpling, or excessive thinning of the exine. On occasion, an exotic pollen grain or spore may elude identification and this should then be described in as much detail as possible for future reference. If the percentage of unidentified taxa is high, the competence of the forensic practitioner should be questioned.

Scanning electron microscopy (SEM) has little application in routine forensic palynology and the experienced palynologist prefers to work with ordinary light and/or phase contrast microscopy. Correct identification often requires examination of the internal wall structure and, in the case of fungi, presence of septa. SEM only allows the outer surface of the grain or spore to be seen. Also many thousands of pollen grains and spores must be identified and counted in a forensic case and because of the sheer laboriousness of working with SEM, this technique is not suitable.

For identification by light microscopy, pollen and spores are obtained from all pollen samples by subjecting them to harsh chemical treatment to remove humic acids, cellulose, silica and any other background material likely to obscure the grains. The aim is to remove all except sporopollenin and chitin (the polymer composing insect outer skeletons, and most fungal cell walls). These treatments preclude DNA analysis, and molecular methods would be impossible for identifying palynomorphs retrieved from exhibits in the usual way. If clothing or some other object picked up pollen/spore dust of a single taxon in sufficient quantity, and it were possible to retrieve that single pollen or spore type, then DNA might be feasible. However, as discussed earlier, the dilemma of an adequate database remains and, in most cases, identification is more likely to be achieved by microscopical examination. Some success with DNA analysis has been achieved in relating pollen picked out of soil to already-identified plant material, although some results were ambivalent. Furthermore, work on mixed samples has required the manual selection of individual grains for analysis, each one taking a minimum of 20 minutes. Some forensic cases have involved the identification of 50 000 or more pollen grains, so such techniques would be infeasible in reality. Raman spectroscopy has been used in an attempt to characterise phylogenetic relationships through pollen grains but, again, the prospect of using this technique for routine forensic palynology is unlikely.

3.5 TAPHONOMY

Palynological taphonomy may be defined as being "all the factors that influence whether a palynomorph will be found at a certain place at a certain time". This is a complex matter and only salient factors are presented here. The mixture of pollen and spores in the air is known as the 'airspora' and it is envisaged to be falling as 'pollen rain'. Pollen and spore production by plants, the concentration of palynomorphs in the airspora, and patterns of distribution, are varied. Palynologists continue to devote much effort into measuring these variables, and the literature contains many models of temporal and spatial variation which have been constructed to facilitate interpretation of complex palynological data. Such models, and those derived from experimental trials testing pollen movement, transfer and retrieval, are of interest. However, since they relate only to specific sets of experimental parameters, the models are geographically limited, and are based on relatively small data sets they are of limited use to the forensic palynologist. Whatever the nature of the airspora and its palynomorph load at any place and at any time, the only data of concern to a forensic palynologist are (1) those obtained from comparator samples from the crime scene or other pertinent place, and (2) what is retrieved from exhibits in a criminal investigation. Every forensic case is unique and interpretation must be made on the basis of sound sampling, analysis, and interpretation of specific data relating to each investigation.

3.5.1 Production and Dispersal

Pollination is achieved mainly by wind and animal vectors (mostly insects). Wind-pollinated flowers produce vast amounts of pollen, to enhance the probability of locating a female. For example, a single inflorescence of *Alnus* (alder) can produce up to 6 million grains in one season. Flowers are often grouped into dense inflorescences, such as catkins, that can shake in the wind and maximise the movement caused by air currents. Plants that are pollinated by various animals produce less pollen because the female is targeted. Pollination by vectors leads to exchange of genetic material between individuals but, despite the advantages of out-crossing, more than half of all flowering plants in the temperate region regularly undergo self-pollination—even before the flower bud opens. It is a common phenomenon in more extreme environments where animal vectors are rare or unreliable, or where a given population is well-suited to a particular habitat and genetic diversity would not be particularly advantageous.

Wind-pollinated plants tend to be well represented in the airspora, whereas insect-pollinated and self-pollinated taxa tend to be under-represented or not at all. Some plants such as *Urtica dioica* (nettle) and *Ilex aquifolium* (holly) are dioecious (sexes on separate plants). In some instances there might be a large population of these species at a site but, if the population is largely female, it might not register at all in the pollen record. In most plants, nearly all of the pollen produced will fall close to the parent and, in many (particularly insect- and self-pollinated species), it is only when the plant dies and falls to the ground that pollen is released from the parent. Such plants are poorly represented in palyno-logical samples, even when they grow close to the sampling site, and they can provide valuable forensic evidence of contact with specific places.

3.5.2 Patterns of Distribution and Residuality

Analysis of thousands of surface samples over a number of years has shown that palynomorph assemblages at ground level are highly het-erogeneous and unpredictable, especially where people have manipu-lated the environment, as in towns and gardens. The patchiness of pollen and spore concentration in any sample taken from the ground is, among other things, a function of:

1. Heterogeneity in the seasonal and annual pollen rain.
2. Vulnerability or recalcitrance of palynomorphs to decomposition.
3. Effects of physical barriers.
4. Action of people and animals.

Any surface open to the pollen rain will have a unique palynological signature although it might be similar to others in close proximity. Similarity will decline with distance.

In most situations, wherever decomposer microorganisms are present, pollen and spores will eventually decay, and an experienced palynologist can sometimes differentiate between current year and older residual pollen from the state of the exine. Some studies have also demonstrated the variable resistance of certain palynomorphs to decomposition. Taxa such as *Alnus* (alder), *Tilia* (lime), *Polypodium* (polypody fern), and *Lycopodium* (clubmoss) are particularly robust, and can remain on the ground for years, while others such as Juncaceae (rushes) decompose very quickly indeed and are rarely found. However, analysis of modern soils does not always support experimental findings, and it is imprudent to adhere strictly to the few published results. Palynomorphs will accumulate in the environment, and on surfaces, and remain for varying amounts of time—often for years. Thus, resistant palynomorphs can be over-represented because of accumulation, and the palynological profile may not reflect the local vegetation as closely as might be expected.

In locations where pollen and fungal spore release is seasonally variable, prediction of periodicity in production and dispersal for various plant and fungal taxa can be useful, especially for allergy studies. Pollen calendars have been constructed for some areas, and these can be used for the places where the monitoring has been carried out, and for providing general guidelines by extrapolation. However, they are imprecise, do not have wide applicability, vary annually and from place to place and hence they do not have the precision required for the interpretation of forensic data. On occasion, comparator samples from crime scenes are taken from surfaces where palynomorphs are periodically removed. Where lawn clippings, or other vegetation, constitute the comparator sample it will represent a palynological 'snapshot'. The profile from that sample will depend on the frequency of removal of the receiving surface (*e.g.* grass cuttings) and the nature of the subsequent pollen rain. Seasonality can be of great importance here, and the sample might not be representative of conditions at the time an offence was committed. Wherever possible, in such cases, it is prudent to obtain comparator samples on the anniversary of an offence. Analysis must be carried out on the crime scene samples and pollen calendars should not be used for interpretation.

In most cases, the comparator sample will be a surface soil or plant litter, and these may have accumulated pollen and spores for at least 1 year, or more. Such an accumulated mixture will mask the seasonal highs and lows of species abundance in the airspora and the timing of sampling is less critical. Residuality can be important in forensic contexts, especially where attempts are made to time events. Although the pollen rain varies throughout the year, the presence of one or more years' previous input makes temporal estimates problematical.

Where crime scenes involve bodies of water, it must be remembered that much of the pollen and spore load in the surface sediments and muddy foreshore can be waterborne rather than airborne. These sites are capable of receiving pollen and spores from the whole catchment area of a river or stream and their profiles are often highly diverse, containing many taxa exotic to the local flora or even the country's native flora. Their profiles are often so mixed that they appear to be highly artificial and can produce highly specific trace evidence, especially where a crime scene is in an urbanised area.

Another factor which is often ignored is the vulnerability of pollen and spores to palynological processing in the laboratory. Although sporopollenin and chitin are resistant to the highly corrosive reagents used to dissolve away background detritus in palynological samples, prolonged exposure can result in disappearance of the palynomorph. This is particularly true of very thin-walled palynomorphs; Juncaceae (rushes) and many of the less robust fungal spores probably fail to survive the process.

3.5.3 Transfer

Pollen and spores are present virtually everywhere. They are easily picked up from some surfaces but can stick tenaciously to others, or become deeply embedded in them. They settle on to soils, plant litter, foliage and inanimate objects; they are breathed into nasal passages, and get trapped in clothing, hair, fur, feathers, *etc.*; they are an important component of household dust. Soils and other palyniferous materials can be picked up by items such as footwear and transferred to vehicle pedals and any other surfaces contacted by the wearer. There can also be secondary transfer from one item to another, although the pollen assemblage will be diluted in the process. Palynomorphs can adhere for considerable periods even when the mineral component of the soil has been shed during wear or usage.

It is often difficult to retrieve pollen and spores from surfaces in which the grains can become embedded, *e.g.* rough wooden boards, coarse paper and coarse fabrics. They can also be held by electrostatic forces, and the tenacity with which they are held is variable. For example, pollen picked up by shiny leather shoes can be rubbed into carpet, but it is not so easily transferred from the carpet to other objects; textiles often act as pollen and spore traps and it requires considerable effort to release them for analysis. They offer interstices as well as electrostatic forces, and 'dressings' on fabrics reduce both effects. For example, dressed or starched cotton is much less tenacious than synthetic fleece, nylon or woollen fabric, and the clay coating on high-quality paper reduces its attraction for palynomorphs. Remarkably, even articles machine-washed with strong detergents, or subjected to dry cleaning, have yielded profiles of palynomorphs which could be related back to crime scenes.

Washing and dry-cleaning machines appear to be relatively ineffective in removing pollen and spores from the weave of fabric and the interstices of washable footwear. Soft and hard plastics also attract pollen and spores, and apparently 'clean' plastic objects have yielded important trace evidence after treatment.

3.6   THE CRIME SCENE

The role of the botanical ecologist and mycologist can be pivotal at a crime scene, and critical information can be lost if someone with appropriate skills is not consulted early in an investigation. Terrestrial plants and fungi are not mobile, and they are less affected by changes in the weather than, for example, flies. In most cases, plants continue growing even in suboptimal conditions and are not so affected by ambient temperature, humidity, wind speed and light intensity. An event such as trampling, or deposition of a corpse, can affect their growth and appearance, but such changes can be beneficial for crime scene inter-pretation. Instead of failing to function when conditions are unfavour-able, as is the case with flies and other invertebrates, the changed performance or appearance of plants and vegetation provide valuable clues as to previous events.

A competent botanist will immediately recognise species growing anomalously, and characterise plant communities from proxy indica-tors. For example, native British oak woodland can be envisaged from twigs, leaves, or seeds of oak, wood anemone, bluebell, honeysuckle and bramble. If litter from that woodland floor contained fragments of hay-meadow plants, the botanist would realise the anomaly and look for reasons—such a find could indicate dried horse dung. But there might also be fragments of plants that are alien not only to the oak woodland, but to the native flora. These would have been brought in by some outside agency and might be pertinent to a criminal enquiry. The identification of premeditated digging of a grave for a murder victim has also been possible by careful examination of severed roots, recovery of plant parts and anomalous distribution of leaf litter in grave-fill.

Crime scenes are varied, and it is impossible to construct ecological protocols applicable to all. However, there are several kinds of infor-mation that can be obtained by ecological techniques and crime scene practice should facilitate them:

1. Direction of offender approach and departure pathways.
2. Timings and order of various events.

3. Traces of evidence transferred from offenders and victims, and traces transferred to offenders and victims.

The first two can be evaluated only by inspection of the crime scene, while trace evidence is obtained from samples collected at the crime scene and used later.

3.6.1   Inside Buildings and Vehicles

Contrary to common belief, ecology can be helpful when crimes have been committed inside buildings as well as outside. The inside of a room or vehicle can act as an incubator for insect and microbial growth. The need for forensic entomology is accepted when deaths occur indoors, but few realise that fungi may give equal, or even better, information under some circumstances. Indeed, fungal growth has given accurate post-mortem interval estimates in several recent cases. Soil and dusty foot-prints found at indoor crime scenes can also yield both botanical and mineralogical trace evidence that might link offenders with other specific and pertinent places.

3.6.2   Outside Scenes

Forensic ecology is mostly applied to crime scenes outside buildings. It can provide intelligence to aid the police in interviewing suspects. It can also yield various kinds of evidence wherever a body has been deposited or buried, or where an offence against the person has occurred, parti-cularly if the offender has contacted the ground or vertical objects such as fences, hedges and trees during an attack.

The ecologist should be called as soon as possible after the offence has been reported, and before other forensic scientists or search teams are allowed to sample the site. The order in which the various specialists are invited to carry out their inspection and sampling will depend on the nature of the case, but information can be maximised if access to the crime scene is prioritised and scientists work in a hierarchy. For various reasons, there is usually some police pressure to remove the body to the mortuary as quickly as possible. This should be resisted until an ade-quate ecological survey can be conducted, and the kind of survey undertaken will depend on the deposition site. There are no fixed pro-tocols, and each case must be treated individually.

One of the first tasks may be to identify an offender approach path. Often, there will have been few options for access because of natural barriers (banks of bramble or nettle, deep streams, barbed wire fences, *etc.*) and, once these have been recognised, anomalies in surface litter, foliage, stems and soil should be noted. Impressions in fresh foliage, broken stems, impressions in wet leaf litter and soil, compensatory growth of specific plants, and much other botanical evidence should be noted. Vegetation at any crime scene, but particularly beneath and surrounding human remains, should not be touched until inspected by the ecologist. Trampling, breaking and cutting of vegetation; setting up common approach paths in inappropriate places; removal of corpses before the ecological examination; inappropriate placing of equipment such as generators and lights; and construction of tents can all result in the loss of critical information.

On occasion, it is possible to estimate the amount of time since a site was visited by an offender and, sometimes, the number of visits made. Plants and fungi are pivotally important in establishing timings, since both exhibit compensatory growth and reorientation if disturbed or displaced. Revival times for broken plants, amount of etiolation (growth in partial or complete absence of light) in previously covered shoots and leaves, discolouration of specific lichen species, soil disturbance and many other factors can help to time events and show direction of access. The use of tents and other coverings can result in ambient environmental conditions being greatly altered in a short time. Any kind of covering creates an incubation effect which can enhance the performance of some organisms, such as insects and microorganisms, but adversely affect plants. Tent poles and the inevitable trampling during the erection of a tent can damage critical environmental evidence. If a tent is unavoidable, one side should be left completely open to the elements to minimise the incubation effect.

3.7 ROUTINE PROCEDURES FOR CRIME SCENE EVALUATION AND SAMPLING

After being briefed by the police, the ecologist should put on full protective clothing, including hair and foot covers and mask, and access the crime scene via the common approach path. The methods used to determine the offender route have been outlined above, but will involve assessment of standing vegetation, ground vegetation, ground litter and soils. A comprehensive photographic record should be obtained of vegetation at the crime scene, both in a 360° panorama and close up where necessary. Photographs should also be provided from any other place pertinent to the case, such as the garden of a suspect's home.

3.7.1 Sampling

Often, the aim in a conventional ecological evaluation is to determine the abundance, distribution and performance of various plant and animal taxa within the environment. Here, sampling methods are designed to maximise the kind of information required, with statistical analysis invariably forming an intrinsic part. The aim in a forensic ecological investigation is to evaluate the environment in terms of what can be compared with items retrieved from an offender, or what can be useful for timing events. The kind of sampling employed does not lend itself easily to statistical techniques, partly because, due to constraints on time and resources, the sampling has to be targeted and closely focused. Soil and/or vegetation will be taken to provide samples for comparison with any future trace evidence obtained from suspects, or to estimate temporal aspects of corpse deposition. These 'comparator' samples are not controls *sensu stricto* since the aim is to target the most likely places contacted by the offender; sampling is systematic along paths, and in any place likely to be pertinent.

Any sampling protocol will be a compromise in that the material collected will represent a fragment of the whole environment. It is imperative, therefore, to collect as many fragments as possible to obtain the most comprehensive palynological profile feasible within the time and resources available. The larger the sample area, the more accurate will be the assessment of the ecological status of the location. In every case, as large a scrape area as possible should be evaluated and, for an individual sample, a minimum area is typically 30×30 cm. The sample is then homogenised in the laboratory and aliquots taken for analysis. A subsample can thus represent an area of at least $900 \, \text{cm}^2$. Since there is often only one opportunity for collection, as many such samples as possible should be obtained from the crime scene.

Only material that is likely to have actually been contacted should be collected. Thus, only the very surface layers should be taken; soils should be lightly scraped and certainly no deeper than 0.5 cm. In the case of buried remains, an offender will probably have stood in the grave-fill. Organic content and biological activity vary greatly with depth in soils and, in such cases, it is important to obtain bulk samples of the grave-fill for homogenisation in the laboratory, as well as samples of the whole profile. Whole vertical grave profiles can be obtained with stainless steel monolith holders. The soil samples obtained can be stored in plastic bags and frozen. Litter or plant material must be placed in paper containers and kept as dry as possible to prevent fungal growth. Should plant material be needed for identification or as evidence at a later date, it should be laid flat and pressed between absorbent paper. As much information as possible should be recorded for each sample, *e.g.* a GPS reading, the type of material, date, time, collector, *etc*. Samples are then given to the exhibits officer for storage and transport to the laboratory as appropriate.

If a body of water is involved at the crime scene, samples should be obtained of the water's edge, the water itself at various depths, and the basal sediment at whichever locations are deemed appropriate by investigating officers. Water and mud samples can be frozen for storage.

3.7.2 Vegetation Survey

It is essential to list as many plant species as possible at a scene of crime, or any site pertinent to the investigation. Since comparator samples will represent a fragment of the place, it is possible that plants growing there might not be represented in the comparator profile, and yet be found on a critical exhibit. It is important to know the potential range of palynomorph taxa that could be found in a palynological preparation and, although sampling would be targeted in an attempt to include actual locations contacted by an offender, there is no guarantee that they would be sampled. The actual profile obtained from a range of comparator samples, combined with a list of all potential taxa that might have been transferred to an offender, is the best that can be achieved. Since some pollen is airborne, it is often necessary to extend the botanical survey to as wide an area as possible around a crime scene.

This approach has the benefit of economy since, in many instances, places can be eliminated from, or implicated in, an enquiry by comparison of their vegetation and the palynological profiles obtained from exhibits. Rapid scanning of preparations from exhibits will allow the palynologist to assess the feasibility of the item having contacted specific places. However, this can only be done effectively if the field survey has been carried out by the ecologist/botanist/palynologist personally responsible for the analysis and interpretation of the samples. Because of complexities of taphonomy, it is not possible to interpret forensic palynological or other botanical information adequately unless the relevant sites are visited by the analyst.

3.7.3 Samples from a Mortuary

When a corpse is involved, samples from the mortuary should be obtained. The cadaver can be a source of important botanical and mycological evidence. It sometimes provides the best comparator sample of all. The hair, skin, nails and nasal passages can be sampled for palynomorphs and other evidence, and sampling techniques have been devised to obtain the maximum evidence. Fungal growth on skin, bone and items associated with a corpse can give important information on conditions in which the corpse had been stored before deposition, as well as post-mortem interval, and time since deposition of the remains. Any botanical or mycological remains retrieved by the pathologist should be made available to the botanist/palynologist, and samples of gut contents should be considered for analysis.

During a post-mortem examination, the pathologist will routinely obtain a sample of stomach contents. Vomit may also be found at a crime scene. Such material may be analysed chemically for substances such as narcotics and poisons, but it is also often necessary to determine, or to confirm, the nature of the last meal of the deceased, or what a survivor had eaten. Invariably, plant food is included in a meal. Knowledge of plant anatomy is crucial in gut content analysis since the material is often comminuted into tissues, and even into individual cells. Plant cells have walls made of cellulose, which cannot be digested, so they are usually well preserved. It is often relatively easy to identify seeds, fruit tissues, leaf and stem fragments and even root-crop food.

When food is comminuted down to aggregations of cells, identification becomes more difficult, although some cell types are distinctive. Starch grains (amyloplasts) are nearly always present in plant foods, and they may be seen intact inside cells, or floating freely in stomach fluids. Starch grain analysis can aid the identification of the plant with considerable precision, but digestive enzymes (amylases) in saliva and the stomach start to degrade them as soon as they are eaten. Partially degraded starch grains are frequently found in stomach fluids and this may give information on the length of time they have been resident. When starch-rich foods are cooked at high temperature, they are converted to amorphous, gelatinised starch. It is often the case that most, if not all, the starch found in stomach contents has been derived from highly processed food and little other information is then obtainable from its presence.

After working on a large number of cases involving analysis of stomach contents, a botanist learns to recognise patterns and timing of degradation of plant and animal tissue. There is no absolute benchmark for giving time estimates, but it is often possible to give an opinion that the food has been in the stomach for more than, or less than, a specific period. The estimates will always be relatively crude but they can be useful to the investigator.

Although they are not always sampled by the pathologist during the post-mortem examination, the contents of the lower gut and faeces can be informative. Data are available on the speed of transit of food through various parts of the gut for healthy individuals, so the presence in the lower gut of food items known to have been eaten by a victim can help provide a post-mortem interval. Identification of plant material can also help determine whether there has been poisoning.

3.8 PALYNOLOGICAL ANALYSIS

There is no simple method or standard protocol for retrieval of palynomorphs from comparator samples or the huge array of items that have been analysed over the years. Each palynologist will have their preferred methods. However, once material has been obtained, the aim is to remove all background detritus by the use of a large array of corrosive reagents and to have an homogenised, concentrated assemblage of palynomorphs stained and mounted on several microscope slides for each sample. The palynologist must have access to an extensive and properly provenanced reference collection, and there should be no reliance on pictures or images from the internet for identification.

Initially, each slide is rapidly scanned. This gives a good idea of pollen concentration and state of preservation as well as the main features of the community from which the sample was obtained. On occasion, the scanning procedure can provide rapid intelligence for police, and detailed analysis necessary for evidential purposes can follow if required.

Analysis is typically carried out using high-power phase-contrast microscopy. Conventionally, a minimum of 300 palynomorphs are counted and this is considered by some to constitute a 'statistically sound sample'. This is based on the fact that palaeoecologists working on ancient peat deposits found that the law of diminishing returns operated at counts of over 300 grains. Every effort is made to count many more than 300 palynomorphs in order to obtain as much information as possible from the sample. When counting is finished, the slide is then scanned further to pick up any 'rare' taxa, although such grains cannot be included in the pollen sum. Any unusual or unknown grains are photographed for further reference, and their position is recorded on the slide. This is invaluable when material is provided to palynologists working for the defence.

The pollen and spore counts are expressed as percentages of the total sum of palynomorphs. On occasion, where certain palynomorphs are overabundant, others can appear to have disproportionately low representation. In this case, it is standard practice to present the data in two ways: with overabundant taxa either included or excluded from the pollen sum. Because of the complexity of the dataset, and the large numbers of variables, no suitable statistical technique has been universally accepted for forensic palynology. Where the data involve very many comparators and large numbers of exhibits, the dataset can be unwieldy. In such cases, multivariate statistics can be used to draw attention to certain relationships and patterning, but interpretation must be made using botanical criteria. Another approach is to use likelihood ratio analysis, but palynomorphs are not randomly distributed in any environment and such tests are based on this assumption. It is difficult to provide graphic representation of forensic palynological data and results are usually provided to the court in tables of percentage data. Some attempt has been made to construct pollen diagrams similar to those used in palaeoecology, but these can rarely be presented in sufficient detail and their interpretation can be troublesome in a court.

3.8.1 Body Deposition Times and Post-Mortem Intervals

Accurate estimates of post-mortem interval, or the length of time a corpse has lain at a deposition site, are necessary since they aid identification of unknown victims by narrowing the search in missing persons registers. There are occasions where entomology may fail to give the required level of accuracy in post-mortem intervals or body deposition times. These are usually when a victim has been dead for a long period and cycles of the first waves of insects have been completed, or where insects have failed to gain any access to the body. In several recent cases, analysis of fungal growth rates on cadaver skin and bone, and on associated materials such as carpets and fabrics, have supplied accurate information.

If a victim has been in position for a few years, plant roots or stems growing over or through the remains are capable of being dated by dendrochronology, and they can provide a minimum time for deposition. In the shorter term, the relationship of the remains to successive falls of leaf litter, growth of grasses, and other herbs can also provide time estimates. In one recent case where skeletonised remains were found in January, deposition time was calculated fairly accurately from the continued growth of *Calystegia sepium* (bindweed) underneath the corpse, and the inhibition by body fluids of fruit-ripening in female *Urtica dioica* (nettle) plants. The experienced field ecologist will simultaneously absorb a very large number of indicators; in some cases, combined observations on earthworm and molluscan activity; fungal growth; presence of certain microarthropods beneath the corpse and scattered bones; root, shoot, and leaf growth; fern life cycle; recovery of cut and trampled stems; decomposition of various fabrics; growth of algae and mosses; differential orientation, growth, and discolouration of lichens; and activity of voles, have all contributed to the accuracy of estimates. Sequential palynological analysis of various bones, including those of the skull, has also provided useful information. In a case where a skeleton was found in February, it was possible to narrow the time of deposition to a window of about 6 weeks in the previous autumn from shoot growth patterns of bramble stems, and the presence of autumnal spores of rust fungi on bramble leaves in the grave-fill.

3.9 LOCATION OF CLANDESTINE GRAVES AND HUMAN REMAINS

Elimination of unlikely places for body deposition or graves is of great value to investigators, since time and resources are saved for search teams. As discussed in more detail in Chapter 16, large areas of landscape can be eliminated by virtue of their geology, shallow soils, unfavourable topography and relative access for vehicles and people. If the grave site is suspected to be in a prescribed area, the plant ecologist can give information on the recent history of the vegetation and soils. Areas that would have been inaccessible around a critical period, or ground that is now supporting dense vegetation but was accessible around a critical period, can be identified. Variation in hydrology can also be determined from surface vegetation, so that waterlogged and unsuitable burial places are highlighted. Estimates of the ages of trees, fallen wood, thickets of brambles and nettles, ant hills and other features can help.

It is relatively easy to eliminate irrelevant places, but more difficult to locate the burial or the corpse. Responses of surface vegetation and fungi can sometimes highlight disturbed ground; *e.g.* if grasses colonise the surface of a grave, they can be enhanced by the release of ammonia during deamination of body proteins. The nature of the herbaceous sward will depend on very many factors and can be highly unpredictable and subtle. Its development does not necessarily follow a neat and predictable pattern.

If a suspect has been identified, palynological analysis of belongings can yield detailed information regarding the kinds of place contacted. From the assemblages and profiles retrieved from footwear, digging implements and/or vehicles, the skilled palynologist is able to 'see' communities of vegetation, sometimes in remarkable detail. Assemblages of pollen grains, plant spores and fungal spores can give information on the vegetation, geology, soil type, soil fertility, altitude, hydrology, amount of leaf and woody litter on the ground, and degree of artificiality of the ecosystem. If the botanist/palynologist is unfamiliar with details of the vegetation of a part of the country that is relevant, a detailed description given to a local ecologist can result in the location of the grave.

3.10   DIFFERENTIATION OF KILL SITES FROM DEPOSITION SITES

Frequently, murder victims have been found in places other than where they were killed. If there are sufficient palynological differences between the two sites, and hair, clothing or footwear of the victim contacted the ground at the kill site, it is usually an easy task to differentiate between them. Furthermore, the kill site may be so distinctive that it can be located by reconstruction from palynological evidence. The same methods as for finding clandestine graves can be employed. Irrelevant places can be eliminated by visual inspection since if the plant communities do not resemble those retrieved from the victim, they are unlikely to be important.

3.11   CAUSE OF DEATH

The irrigation of the turbinate bones inside the skull releases palyno-morphs that have been inhaled in life. Even after death, when mucosal membranes have decomposed, palynomorphs can remain stuck to the bony complex. The mucosae of the nose must be ignored since they could represent contamination before and after death. Cases where the turbinates have been analysed have provided evidence to show that:

1. A victim was likely to have drowned in bathwater rather than a natural body of water.
2. Dust retrieved from a neonate indicated that it had drawn breath in the bag of peat in which it was found.
3. A young child had possibly been suffocated by a pillow with a cotton pillowcase.
4. A young man inhaled soil from the surface of his grave before being buried.
5. A young girl had inhaled pollen from a bank of nettles as she was being carried to her grave.

Causes of death can also be established from other palynological samples. For example, the analyses of swabs from wounds have shown the presence of pollen from the garden from which the 'weapon' was obtained whilst in another case an analysis of gut contents confirmed the presence of cannabis pollen and *Psilocybe* (magic mushroom) spores.

3.12   EVIDENCE OF CONTACT (TRACE EVIDENCE)

Fragments of glass, paint, fibres, *etc.* are regarded as typical trace evidence samples. However, pollen and spores are also further examples as they can also provide links between people, objects and places. Offenders walk on soil, dust, vegetation and leaf/woody litter; lean against buildings; sit on seats; climb across lichen and moss-covered walls and roofs; brush against vegetation and other palyniferous surfaces. Invariably, these can be directly transferred to anything with which they come in contact.

The whole assemblage of both plant and palynomorph taxa must be considered when comparing and interpreting the palynological profiles of places and objects. Some taxa will generally be considered to be more meaningful than others. A high pollen-producing, easily dispersed, wind-pollinated plant such as *Alnus glutinosa* (alder) is more likely to be picked up than an insect-pollinated plant such as *Salix cinerea* (grey willow), or self-pollinated one such as *Triticum aestivum* (wheat). In most places in the UK, a value of 5% for alder would certainly be less significant than the same value for willow or wheat. However, a high abundance of any palynomorph in a comparator sample may not be reflected on an exhibit because of subsequent wear and abrasion, or even attempts to remove trace evidence by an offender. Furthermore, the actual place contacted by an exhibit can never be known exactly and comparator samples are taken from 'most likely places of contact'.

The distribution of palynomorphs at any location is heterogeneous, and a single comparator sample will contain only a 'fragment' of the true palynological profile of that site. To gain a meaningful 'picture of place', as many spatially separated samples as possible should be obtained and the results combined. Exhibits such as footwear and vehicles will also have fragments of the true palynological profile. Because, if an offender has walked about, there may be a combination of multiple fragments from one critical location on the footwear, and these can be compared with combined comparators. But the taphonomy can be more complex than that.

In some circumstances, the palyniferous material, or accumulated palynomorphs themselves, can be secondarily transferred to other objects, possibly over several vehicles, or several places. Such subsequent secondary transfer will be diluted, and only a third-order fragment may be retrieved, but there can still be important trace evidence in partial profiles if the original profile was distinctive enough.

There can never be a perfect 'match' between any two comparator samples, although the closer sampling sites are to one another the more similar they will be. Likewise, because of fragmentary profiles and multiple depositions, an absolute match between comparator samples and exhibits can never be expected. The degree of confidence in determining whether there had been contact depends on the richness of the respective palynological profiles and is enhanced by the presence of rare taxa.

For links between exhibits and crime scenes to be acceptable to the court, there needs to be either a complex assemblage, with many points of similarity between place and object, or some exceptionally unusual or rare component. If both are present, palynological trace evidence can be very powerful. This makes the forensic palynologist a useful member of the investigative team, and in spite of all the caveats that apply, palynological profiles distinctive enough to demonstrate convincing links between items and places have been repeatedly demonstrated.

3.13   FORENSIC ECOLOGY AND THE COURTS

The forensic ecologist/botanist/palynologist must observe forensic science protocols for keeping adequate records, continuity of possession, submission of reports, making statements, disclosure, making the required declaration of understanding and providing indices of unused material.

Evidence from forensic ecological investigations has been tried and tested many times in the British court system. It has been included successfully in cases which have resulted in conviction of offenders and, indeed, on occasion provided the only forensic evidence. However, it is unlike many areas of forensic science in that procedures and protocols are not standardised and, at the time of writing, practitioners do not come within the remit of any formal accreditation body.

Some aspects of forensic ecology are easily tested. For example, estimations of post-mortem interval, or deposition times, can be confirmed by testimony of witnesses, or confession of an offender. If clandestine graves are located, then their very discovery verifies the data and, if the data were palynological, then by default, these would implicate the owner of the exhibit(s) from which the palynological information was obtained. Presentation of palynological data as trace evidence is more difficult. A jury composed of non-scientists has to be introduced to the concept of taphonomic variability and a plethora of unfamiliar terms and ideas. These data do not lend themselves easily to graphic representation, and the palynologist has to explain, in great detail, the significance of the various values for palynomorph taxa. The palynological report must be written as clearly as possible to make it accessible to the non-specialist, but palynological data are rarely simple and over-simplification can lead to spurious arguments.

The best approach for dealing with ecological and palynological data would be for both prosecution and defence specialists to work together for a consensus to be presented to the court, but this does not appear to be in prospect.

BIBLIOGRAPHY

H. M. Coyle (ed.), *Forensic Botany: Principles and Applications to Criminal Casework*, CRC Press, Boca Raton, 2005.

J. Jansonius and D. C. McGregor (eds), *Palynology: Principles and Applications*, Vols 1–3, American Association of Stratigraphic Palynologists Foundation, 1996.

D. M. John, B. A. Whitton and A. J. Brook, (eds), *The Freshwater Algal Flora of the British Isles: An Identification Guide to Freshwater and Terrestrial Algae*. Cambridge University Press, Cambridge, 2002.

G. D. Jones and V. M. Bryant, A comparison of pollen counts: light versus scanning electron microscopy, *Grana*, 2007, **46**, 20–33.

J. D. Mauseth, *Botany: An Introduction to Plant Biology*, 2nd edition, Saunders College Publishing, Philadelphia, 1995.

D. C. Mildenhall, P. E. J. Wiltshire and V. M. Bryant, Forensic palynology: why do it and how it works, *Forens. Sci. Int.*, 2006, **163**, 163–172.

P. D. Moore, J. A. Webb and M. E. Collinson, *Pollen Analysis*, Blackwell Scientific Publications, Oxford, 1991.

E. F. Stoermer and J. P. Smol, (eds), *The Diatoms: Applications for the Environmental and Earth Sciences*, Cambridge University Press, Cambridge, 1999.

P. E. J. Wiltshire, Consideration of some taphonomic variables of relevance to forensic palynological investigation in the United Kingdom, *Forens. Sci. Int.*, 2006, **163**, 163–172.

P. E. J. Wiltshire, Forensic ecology botany and palynology some aspects of their role in criminal investigation, In: *Criminal and Environmental Soil Forensics*, ed. K. Ritz, L. Dawson, D. Miller, Springer Science, 2009, pp. 129–149.

CHAPTER 4

Forensic Entomology

DOROTHY GENNARD

University of Lincoln, School of Natural and Applied Sciences, Brayford Pool, Lincoln LN6 7TS

4.1 INTRODUCTION

Insects and other arthropods, such as mites or spiders, are often the silent indicators of criminal or civil misdemeanour. They have a role comparable to that of other biological evidence. Insects respond to opportunities to gain food, a reproductive site, a place to lay eggs and a place to develop through the stages of their life cycle. Their presence at crime scenes must be interpreted in the light of knowledge about life cycles and patterns of insect succession, combined with how environmental conditions influence growth, in order to provide an estimate of the time since colonisation of a dead body.

Forensic entomologists work on cases of suspicious death, alleged abuse and neglect, or infestations of urban contexts such as food preparation areas, building sites or hospitals. The materials examined by them can range from food sold for human consumption, to animals and humans living and dead. In each instance, forensic entomology can contribute to the evidence and assist courts in reaching their conclusions. This is particularly valuable for the 48–80 hour period after death when the body's physiological changes are no longer readily interpretable by forensic pathologists.

Crime Scene to Court: The Essentials of Forensic Science, 3rd Edition
Edited by P. C. White
© Royal Society of Chemistry 2010
Published by the Royal Society of Chemistry, www.rsc.org

Civil cases rely more on knowledge of a specific order of insects and their ecology than on insect succession, and could be the result of an individual seeking the help of a forensic entomologist to bring a claim for reimbursement for personal loss or inconvenience. For example, infestation of buildings by insects may not only be an issue for environmental health officers but can also, on occasion, involve forensic entomologists where the matter forms part of a court case.

Such a civil case might be in the context of pest control in rented holiday flats. For example, a person who has become ill because of eating tainted food may file a case in the civil court alleging that there is a cockroach infestation and that a pest control company should have been employed by the property owner to remove the pests. A forensic entomologist may be employed by the person to investigate the nature and size of the cockroach infestation at the flat and whether control measures allegedly put in place by a company employed by the landlord had been adequate.

Fundamentally, any insect, or indeed arthropod such as a mite or spider, can provide evidential information of importance in forensic investigations. However, greatest attention has been paid to those insects recorded in association with a corpse (Figure 4.1) and this has become the subject of ecological investigation by forensic entomologists.

The insect species that are valuable in forensic terms in a particular case may vary geographically. In general, in Europe and the USA, they include species known to be necrophagous (feeding on dead bodies) such as *Calliphora vicina* Robineau-Desvoidy (Figure 4.2a), *Lucilia sericata* (Meigen) (green bottle, *Phaenicia sericata*) (Figure 4.2b) and *Calliphora vomitoria* Linnaeus (Figure 4.2c) or *Sarcophaga carnaria* (Linnaeus)

Figure 4.1 Greenbottles attracted to the wounds on a decomposing body.

(a)   (b)

(c)   (d)

Figure 4.2   (a) *Calliphora vicina* Robineau-Desvoidy illustrating the yellow basicosta
characteristic of this member of the Calliphoridae. The marker shows
the feature used to identify this species. (b) *Lucilia sericata* (Meigen).
(c) The head of *Calliphora vomitoria* (Linneaus) and the marker illustrates
the characteristic golden hairs on the jowl. (d) Member of the Sarcophagidae (fleshflies).

(Figure 4.2d). Detritivores (feeding on organic matter), usually associated with compost and manure, such as the black soldier fly *Hermetia
illucens* (Linnaeus) have also been used to determine post-mortem
intervals on some occasions. All that is required to use a species is
experimental data relating to the length of time to develop through the
individual life stages from egg to adult emergence at constant temperatures. Determination of size in terms of length of larvae in each
larval stage may also provide important information that can be
translated into information on time since death.

4.2   INSECT SUCCESSION ON THE CORPSE

As it decomposes, a predictable succession of insect species colonise the
unburied corpse. Although the stages of decomposition are not dis-
cretely discernable, the speed of decomposition of a particular body is

(a)   (b)

Figure 4.3   (a) An example of a sepsid fly (*Orygma luctuosum* Meigen). (b) A member of the Staphylinidae (*Creophilus maxillosus* (Linnaeus)).

environmentally dependent and strongly influenced by temperature and the rapidity of insect colonisation.

In arid and semi-arid environments, such as Colorado (USA), the decomposing body may be a source of nutrient and water for insects. A number of necrophagous species will be found on bodies left in such environments. They can include members of the blowflies (Calliphoridae): *Phormia regina* (Meigen), the black blowfly; *Calliphora livida* Hall; *Calliphora vicina* Robineau-Desvoidy; *Cochliomyia macellaria* (Fabricius); *Lucilia sericata* (Meigen); *Calliphora coloradensis* Hough; and *Sarcophaga bullata* Parker, a species of flesh fly (Sarcophagidae). Sepsid flies, and dung (Scathophagidae) flies would also be expected to be present (Figure 4.3a).

The beetle colonisers could include clerid beetles such as *Necrobia rufipes* De Geer; silphid beetles such as *Nicrophorus marginatus* Fabricius; dermestid beetle species such as *Dermestes frischii* Kügelann, *Dermestes marmoratus* Say, and *Dermestes caninus* Germar; histerid beetles such as *Saprinus* sp.; and members of the Nitidulidae *Nitidula ziczac* Say and *Omosita colon* (Linnaeus). Predatory rove beetles (Staphylinidae) such as *Creophilus maxillosus* (Linnaeus) could also be present because eggs and larvae, rather than the body itself, provide food for this secondary coloniser (Figure 4.3b).

In more temperate conditions the families of insects would be similar, although the species may differ slightly depending upon where the investigations are taking place. Cosmopolitan species such as *Calliphora vicina* and *Lucilia sericata* are found in many countries. *Phormia regina* would be present in the USA, but its presence in Europe (including the UK) would be unexpected.

Disturbance of a body at a crime scene might potentially cause disruption of this sequence of insect succession. However, research suggests that this is not an important influence on the different species of insects colonising the body.

Succession on buried corpses is slower, not least because the soil restricts the numbers of insect species that can gain access to the corpse and also because the temperature fluctuates much less and tends to be lower than the ambient air temperature. Decomposition is restricted as a result. The rate of decomposition will vary dependent on conditions in the grave. An example of the speed and decomposition state is presented in Table 4.1. However, even after 10 years, a buried body can be adequately preserved and the skin and organs may still be recognisable. Preservation will depend upon the burial conditions, including soil type and moisture.

Much is known about the insects colonising bodies (including those arthropods found on bodies in graves) because of the work of a Frenchman called Jean Pierre Mégnin (1828–1905). He recorded four species of flies on buried bodies: *Calliphora vomitoria*, *Muscina stabulans* (Fallén), *Anthomyia* sp., and a scuttle fly (a member of the Phoridae) *Phora atra* (Meigen). He also noted a species of beetle *Rhizophagus parallelocollis* Gyllenhal, as well as springtails (Collembola) such as *Achorutes armatus* Lameere and *Heteromurus nitidus* (Templeton).

On an exhumed body scuttle flies are most likely to be the dominant family, or indeed may even be the sole family to be recovered, represented by a single species: species such as *Conicera tibialis* Schmitz *Triphleba hyalinata* Meigen, *Megaselia rufipes* (Meigen) or *Megaselia scalaris* (Loew) are candidates likely to be present. Adult female *Conicera tibialis* have been shown to move down below the soil surface to colonise a corpse buried at a depth of 2 m in friable soil. It is thought that this species has at least two generations per year and that the second generation can recolonise the buried corpse apparently without coming to the soil surface.

The level of insect infestation on the corpse is dependent upon the depth of the soil covering the body. Scuttle flies can appear on a body

Table 4.1 An example of potential duration of decomposition stages of the buried body.

Decomposition stage	Time after burial
Early decay	5–48 days
Moderate decay	8 days–8.7 months
Advanced decay and partial skeletalisation	5.7 months–10 years

around 1 year after interment, whereas beetles such as *Rhizophagus parallelocollis* Gyllenhal (Rhizophagidae) and the rove beetle (Staphylinidae) *Philonthius ebeninus* (Gravenhorst) are most likely to colonise the interred body in the second year after it was buried.

Today, unless the death is suspicious, most bodies are rapidly removed from the place in which death occurred and are maintained in refrigerated conditions, or embalmed. These actions reduce the likelihood that insects will invade a body. This is in contrast to earlier centuries when the body was buried after being retained for 3 days in the home of the deceased, or in a religious building. Under such circumstances, the opportunities for insects to gain access to the body were much greater then than they are in the 21st century. In consequence, less new information is available about the succession of insects on buried bodies.

4.3 SEASONAL INFLUENCES

Insects are cold-blooded. They require a level of solar radiation to provide a minimum external temperature to which the body is warmed before the insects are able to utilise energy for activities such as feeding, flying, mating and egg laying. The majority of forensically relevant insects are active in three of the four seasons in the UK and Europe. With rare exceptions such as *Calliphora vicina* in more southern counties of the UK, the majority of necrophagous insects apart from scuttle flies are absent in the winter months. In the future the effects of climate change may influence this situation. *Calliphora vicina*, for example, has been recorded from the Kerguelen islands in the Antarctic; so opportunity, and not temperature alone, plays a role in determining the availability of a species to colonise a body.

In countries other than the UK, seasonal and geographic variation result in the same species being considered inactive in a particular season in one specific geographic location but not in others. For example, in the southern state of Texas the black blowfly *Phormia regina* is inactive in the summer months but further north this species is present throughout the summer. Similarly, in those states where spring and autumn tend to be cool seasons *Calliphora vomitoria* is more commonly recorded then than in the other seasons.

Weather conditions play a large role in determining insect activity. Rain prevents insects such as flies from flying, although flesh flies are known to be able to fly in the rain. Wind speed also has a strong influence on whether flies are able to fly. Blowflies such as *Calliphora vicina* are thought to have an optimum wind speed for flight of 0.7 m/s.

Higher wind speeds tend to limit any necrophagous insects' ability to colonise a body, unless they are able to gain access by walking. The season and weather conditions, including the presence of rain and high winds during the period prior to finding the body, have to be taken into consideration when calculating the post-mortem interval (PMI).

Within a particular temperature range, insect growth is directly related to temperature. This fact enables the forensic entomologist to relate the insect's larval growth and development to the temperatures in the environment concerned. In general, at higher temperatures, above this optimum range, the speed of growth and development reduces until the point of death. In temperate countries these upper limits are not important; the lower temperatures are significant, however, and must be taken into consideration and subtracted from the ambient temperature in any determination of time since death. Such base temperatures are the minimum temperatures at which the insect can survive but will not grow. Base temperatures are known to vary, and insects are genetically acclimated to particular locations. For example, in the southern counties of England a base temperature of 1 °C is considered appropriate for *Calliphora vicina*. In contrast, in the northern counties and on the Pennine hills the base temperature for the same species is considered to be 3.5 °C. Base temperature is therefore a significant component of the equation used to determine time since death, using accumulated degree hours (ADH) or accumulated degree days (ADD).

4.4   FORENSIC RELEVANCE OF INSECT DISTRIBUTION

Forensic entomology has a role to play in interpreting whether an insect (and thus a body on which it is feeding or ovipositing) has originated from a place other than the location in which it was found. Knowledge of entomology can also place a suspect or victim at a crime scene. The presence of species characteristic of a particular type of habitat, which differs distinctly from the one in which the body was found, may indicate that the body has been moved. For example, *Lucilia sericata* is known to prefer open, sunny habitats and to lay eggs on a body only under such conditions. In contrast, *Lucila caesar* (Linnaeus) (Figure 4.4) prefers more shaded conditions such as those found in hedgerows, woodlands and copses. To find *Lucilia sericata* on a body in woodland provides cause for questioning whether the body had been moved.

The altitude at which the person dies is also of significance in terms of the potential for insect colonisation and calculating time since death. At high altitudes such as on mountains, there is little food available to sustain large populations of insects that could potentially colonise a

Figure 4.4   *Lucilia caesar* (Linnaeus).

corpse. The greater the height above sea level, the less the insect diversity. However, there have been cases where a species of blowfly often found at lower levels, such as *Calliphora vicina*, has been recorded at high altitudes from the dead bodies of climbers and walkers.

The presence on a corpse of human ectoparasites such as fleas and lice is also of significance. Both families ideally require the body to be alive or to be maintained at a temperature little different from that of a live body, for them to remain on the corpse. Fleas leave a body very soon after its host (person or animal) has died. Fleas are of particular importance in cases of drowning, as they are likely to survive no more than a day submerged in water. Therefore, they provide an indicator of time since submergence of a body that was likely to be still at or near normal living body temperature when the person went into the water. On the other hand, lice such as the human ectoparasite *Pediculus humanus* var. *corporis* DeGeer are thought to survive from 3 to 6 days on a submerged corpse. Consequently, if all the lice recovered from a body are dead it is reasonable to conclude that the person could have died at least 3 days earlier.

4.5   EFFECT OF LOCATION ON CORPSE DECOMPOSITION

Insect succession is also influenced by whether the corpse is kept in a closed, insect-proof environment such as a cupboard; sealed beneath floorboards; or tightly wrapped in a bag, blankets or clothing. All of these are well-known places for hiding corpses. Bodies can also have been stored in a freezer or large refrigerator, or in a car from which insects have been excluded. In each of these conditions the initial colonisation will be delayed. When the body is found, therefore, it will not necessarily have decayed as rapidly as it would if it had been left exposed on the soil surface. The history of the body since death is an important aspect to take into consideration when calculating the time since death.

In addition to environmental data and the condition of the corpse, the possible causes of death are important pieces of information that should be provided by the forensic pathologist. In some instances, they may be obvious from the condition or location of the body, *i.e.* when a body is found in a car in which a tube leading from the exhaust pipe has been pushed in through the car window.

Comparison of the differences in insect succession and corpse decomposition rate in cases of carbon monoxide poisoning by car exhaust fumes in relation to bodies retained on the soil surface has shown that speed of decomposition will be faster in a corpse retained in a car. The sequence of insect colonisation differs, often by a day after death, in contrast to bodies laid on the soil surface where insects would be expected to appear within the first hour or so and lay eggs shortly afterwards. Similar delays in oviposition of up to 3 days might be expected where a body was stored in the boot of a car.

4.6   EFFECTS OF CHEMICALS ON INSECT DEVELOPMENT AND PMI

The possibility that insects are able to survive on tissue from a corpse where death was due to poisoning has been known for at least 110 years. Larvae of flies from the Piophilidae, Psychodidae and Fanniidae families feeding on bodies poisoned with arsenic have been recovered from some crime scenes (Figure 4.5). The larvae are able to bioaccumulate the element. This provides forensic entomologists with a source of further evidence that can be passed to the forensic toxicologist for further study.

Arsenic is not the only element that has been recovered from insect larvae. Mercury can be extracted from larvae, puparia and adult blowflies that had been feeding on fish that contained methylated mercury. Mercury passes along the food chain. Rove beetles (Staphylinids), which feed on the fly larvae feeding on such contaminated fish, will also bioaccumulate any mercury retained by the maggots. These secondary predators can also provide a means of confirmation of the nature of any drug consumption by the corpse if collected from the crime scene.

Insects can indicate the presence of drugs in individuals who have taken such substances for medical or recreational purposes, or in cases

Figure 4.5 A member of the Piophilidae, which have been known to lay eggs on bodies poisoned with arsenic.

of suspected sexual assault such as date rape. Such drugs can potentially be extracted from the entire larva, the maggots consuming the bodies of individuals, from either the larval crops, or from the puparial stage. For example, opiates (codeine), tropane alkaloids (cocaine, its metabolite benzoylecognine), benzodiazepines (triazolam and oxazepam), barbiturate (phenobarbital) and tricyclic antidepressants (alimemazine and clomipramine) have been recovered from blowfly larvae previously fed on tissue intoxicated with these drugs. Using the larvae as a means of investigating the presence of any drugs can be a more sensitive method of determining the identity of the drug than using tissue from the corpse.

Attempts to relate the amount of drug consumed by the victim to the amount of drug recovered from the insect have however, met with varying degrees of success. The particular drug being investigated appears to be an important factor in whether there is a successful outcome or not. Even using gas chromatography (GC) and gas chromatography–mass spectroscopy (GC-MS) to investigate the presence of cocaine and its dominant metabolite benzoylecognine, in both the larvae and decomposing skeletal muscle, may not allow the accurate determination of drug concentration in the corpse. More success has been achieved in correlating levels of the opiate codeine in larvae of the green bottle *Lucilia sericata* with levels in the corpse.

Drugs may have an effect on larval life cycles. This has to be taken into account when estimating time since death, a consideration that has been shown to be particularly true of flesh flies. Larvae of *Sarcophaga pere-grina* (Robineau–Desvoidy) fed on tissue drug-intoxicated with cocaine and its metabolite at levels of the median lethal dose and double the median lethal dose have accelerated development. This drug reduces the time in the larval stages, and causes early pupation and adult eclosion (emergence). Similarly, consumption of heroin increases their speed of larval growth by up to 29 hours. In contrast, a different species of flesh fly, *Sarcophaga tibialis* Macquart, when fed on heroin-laced food spends considerably longer in each life stage than do larvae fed on food without heroin. Although some drugs speed up the life cycle, others will lengthen it. Morphine, in contrast to the previous examples, slows the rate of development to the adult stage in the blowfly *Lucilia sericata*. If these effects are not taken into consideration then the time since death could be underestimated by up to 24 hours. This is because *Lucilia sericata* are capable of excreting the drug during the post-feeding stage. The effect of a particular drug or its metabolites (or both), on the speed of development therefore appears to be species specific, at least in the flesh flies, because of variation in levels of metabolism of the particular drug.

The stage at which the insects start to feed on the body, and where on the body they feed, is important. Flies such as the scuttle flies, blowflies and flesh flies, the earlier colonisers of a body, are likely to provide a rich source of information on the drugs present as they feed on the soft tissue that has not undergone decomposition. Species such as the dermestids, which feed on drier tissue later in the succession, may be a less useful source of information where drug concentrations are required as well as information about their identity. Larval shed skins and faecal waste of the insects colonising in these later stages may provide a measure of drug concentration that is less than was actually present in the body.

Nonetheless, puparial cases and the pupae themselves are an alternative source of information to analysing corpse tissues for drugs if the latter is not present. Puparia, in particular, are able to remain in a suitable state of preservation for a long period. Techniques such as fluorescence polarisation immunoassay are very valuable in determining drug concentrations from puparial tissue, provided that the quantity is greater than $10\,\mu g/g$.

Poisoning by agricultural chemicals is a common occurrence in rural communities, and consumption of pesticides is a frequent means of suicide. It can even be used as a means of attempting to hide the presence of the corpse, or delay colonisation by the insects. The effect of insecticides on larval growth and development has been investigated to determine their value in investigations of such poisonings. Identification of the organophosphate malathion in corpses has been successfully achieved using larval stages of two species of *Chrysomya*, *C. mega-cephala* (Fabricius) and *C. rufifacies* (Macquart). This indicates that insect larvae have a role to play in this aspect of forensic investigation.

Customs officers impound goods that have an insect infestation both for reasons of environmental health and in cases of illegal contraband. The presence of insects can indicate the geographical origin of such impounded material. For example, entomologists from the National Museum of New Zealand examined insects present on drugs that had been seized at the point of entry. Three families of beetles, the Carabidae, Bruchidae and Tenebrionidae were identified. These particular beetles were not native to the place where the ship originated, but from an area between the Andaman Sea and Thailand. The beetles led the customs officers to the source of the drugs because of their geographic distribution. As a result of this forensic evidence, the officers were able to successfully prosecute the case.

4.7 DNA ANALYSIS IN FORENSIC ENTOMOLOGY

DNA analysis is a valuable tool in forensic entomology in a number of ways. First, it is important as a means of determining the species of larvae on a corpse. Flesh fly larvae are often difficult to identify from their morphological features without considerable expertise and the advantage of access to a museum collection of identified specimens. Determination of their DNA from larval specimens can often reveal their identity easily. This is equally true of trying to identify the species from their eggs.

The approach requires species-specific primers in order to extract DNA specific to the Calliphoridae. Once extracted, a number of techniques, from polymerase chain reaction—restriction fragment length polymorphism (PCR-RFLP), random amplification of polymorphic DNA (RAPD), mitochondrial DNA (mtDNA) analysis and microsatellites have all been used to examine insect DNA from crime scenes. Insect mtDNA subunit cytochrome oxidase I (COI) is used for analysis because it is the biggest of the cytochrome units and the protein sequence combines variable and highly conserved regions. Since mtDNA is resistant to degradation and can be analysed rapidly, it is also an efficient means of distinguishing between insect species.

Knowledge of the within-species variation of the genetic profile is important for the interpretation of the identity of insect specimens recovered from crime scenes. For example within the COI subunit in both the clerid beetles *Necrobia rufipes* and *Necrobia ruficollis* (Fabricius) (Figure 4.6) there is a great degree of heteroplasmy (difference in mitochondrial genomes), which is not found in the species *Necrobia*

Figure 4.6   *Necrobia ruficollis* (Fabricius); an example of a clerid beetle.

violacea (Linnaeus). Heteroplasmy is therefore valuable for intraspecies discrimination.

4.7.1   DNA Analysis of Insect Gut Content

Investigation of dried remains of insects can provide an important source of information in a number of areas of scientific investigation. For example, insect remains found in the gut of a predator have been used to determine their diet. This was exemplified by the Reverend Frederick W. Hope, an entomologist who in 1842 identified insect remains in a mummified ibis and contributed to ornithological knowledge of a topic then much discussed—the diet of the ibis. This area of entomology has also played a role in forensic interpretation based on serological analysis and the use of DNA.

An interesting application of gut analysis is where intelligence suggests that the body of a missing person is held at a particular location. If the body is absent but a maggot mass or even an individual larva remains, the entomological evidence may provide a surrogate. The tissue retained in the crop provides a source of DNA from the person upon whom the larvae had been feeding. Comparison of DNA from the crop with DNA known to come from the missing person will provide conclusive evidence of their previous presence.

4.8   EXPLOSIVES

Where someone dies in an explosion it may be possible to determine the nature of the explosive by analysing the gut content of any insects found feeding on the remains of the body. Third instar larvae (stage before pupation) of the green bottle *L. sericata* previously fed on TNT (trinitrotoluene) contaminated liver can be used to reveal by-products and by-components of TNT if analysed by a scanning electron microscope and energy dispersive X-ray spectroscopy (SEM/EDX). Consuming tissue contaminated with TNT can increase the speed of the normal life cycle in this species by up to 41%.

4.9 INSECTS AND FIRE

There has been much discussion about the ability of colonising larvae to withstand burning. A rapid response to a fire, if noted soon after it has been started, can provide a fixed point for determining time since death accurately, even if all of the insects died in the fire. The larval instar at the point when the fire was set can be determined under circumstances where the fire is rapidly controlled. This is not true where the fire is not extinguished rapidly, *e.g.* when cars or buildings burn themselves out. In this instance a fixed end point for the fire may be unknown.

The degree of heat generated, the level of protection the insects receive from the body and the presence of blankets or other forms of insulation, will dictate whether or not insect remains are present. Therefore, it is worth considering examining fire debris and any other burnt remains in case entomological evidence is present.

The attractiveness of burnt tissue to flies as an egg-laying site has also been questioned and the issue of a possible delay in colonisation raised. Again, the condition of the body and level of burning will dictate whether it is attractive to flies. The odours released during and after burning are attractive to flies. The availability of cracks and edible flesh suitable for egg laying and subsequent larval feeding and development will dictate whether the burnt flesh is colonised. Evidence and unpublished observations suggest that burnt flesh is more attractive to both *Lucilia sericata* and *Calliphora vicina* than unburnt flesh and so colonisation of burnt flesh is more likely to take place first. The condition of the flesh may also allow more rapid development. This should be taken into consideration when determining the post-mortem interval.

4.10 INVESTIGATION OF WILDLIFE AND DOMESTICATED ANIMAL DEATHS

Forensic entomology has a role to play not only in cases of human death but also in cases of poaching and wildlife crime. The UK Animal Welfare Act 2006 requires that animal welfare is paramount and cases of fly strike, *e.g.* in domestic and farm animals such as pigs, sheep, dogs and rabbits which have been neglected leading to infestation by maggots, is punishable by law. In such cases the forensic entomologist may be asked to offer advice or an interpretation of time since infestation, or act as a defence expert reviewing the data provided by the prosecution. The animal infested with the maggots will have been alive and living with the effect of maggots consuming its flesh, although it may subsequently have had to be humanely destroyed. The predominant species initiating this form of infestation or strike are from the genera *Lucilia*, *Calliphora*, *Chrysomyia* and *Phormia*. Such species of fly will most likely initially infest the perineum. Flies such as the house fly may then secondarily colonise any areas of lesion or dermatitis that the original flies engendered.

4.11   MODELLING TIME SINCE DEATH (POST-MORTEM INTERVAL)

The larval stage is the predominant stage used to calculate the postmortem interval (PMI). Ideally, locally generated development rates provide the most accurate way of finding out the length of the life stages at a fixed experimental temperature. This information is combined with an accurate estimation of the temperatures at the crime scene for the time when the body lay undiscovered. Correction factors are calculated from temperatures at the crime scene in relation to records from recognised meteorological stations nearest to the crime scene. The temperatures at the crime scene when the body laid undiscovered are then back-calculated using the corrected temperatures from the meteorological station(s).

The most frequent models used for determining time since death are the linear models based, between upper and lower temperature threshold limits, upon the direct relationship between temperature and time for the insect to develop. These models are called the accumulated degree hour (ADH) and accumulated degree day (ADD) models and are also commonly used in integrated pest management (IPM) strategies for controlling insect pest infestation on crops. The ADD and ADH models require that the corrected temperature values between the two thresholds are multiplied by a series of set time periods and the totals accumulated until they give a measure of the physiological energy budget needed to reach the life stage of the oldest specimens discovered on the corpse.

The egg stage is often the shortest life stage. The presence of eggs is frequently considered to indicate death within a day to only a few days. It is important to consider whereabouts on the body the eggs were laid. If it is inside some of the body orifices, then the temperature could be greater than ambient and this must be corrected for in any calculations. Larvae will tend, by the second instar, to gather together in groups. These maggot masses raise the group temperature above ambient and so can increase the rate of larval growth. Maggot mass temperatures should be considered when calculating ADH or ADD.

In addition to using growth of the larvae, larval weight and larval length has been used to calculate PMI. Isomegalen and Isomorphen graphs have been developed. Isomorphen graphs relate growth to a particular instar at a particular stage. Such graphs have been produced for a small number of species, notably the blowflies *Lucilia sericata* and *Calliphora vicina, Chrysomya albiceps* (Wiedemann), *Protophormia terranovae* (Robineau-Desvoidy) (Figure 4.7) and *Liopygia argyrostoma* (Robineau-Desvoidy).

Isomegalen graphs are graphs of length of maggots for a range of different temperatures in relation to their age in days. These graphs allow the age of the maggot to be read directly off the graph if the temperature is constant. Where the temperature is not constant it is impossible to determine the age range accurately.

The use of larval width has been proposed as an alternative to length for the determination of the PMI. The measurement of body width between the ventral and dorsal surfaces at the junction of the fifth and sixth larval abdominal segments, as shown in Figure 4.8, is considered comparable with body length for predicting the age of the larva. These measurements can be translated into length with 95% accuracy. Such an approach can be advantageous where larvae from a crime scene have been damaged, or poorly preserved.

Figure 4.7 *Protophormia terraenovae* (Robineau-Desvoidy).

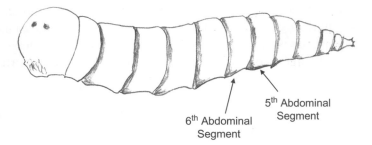

6th Abdominal
Segment

5th Abdominal
Segment

Figure 4.8   Diagram of a third instar larva showing the fifth and sixth abdominal
segments where the larval width is measured.

4.12   LIMITATIONS ON INTERPRETATION OF CRIME
SCENE DATA

A number of features should be considered when interpreting entomological data to calculate time since death. These include the conditions
under which the samples were kept when transported from the crime
scene to the forensic entomologist's laboratory. Such aspects include the
storage fluid and means of 'fixing' the samples as well as the accuracy of
the meteorological data, the availability of information about the
duration of life stages at particular fixed temperatures for the species
concerned and the base temperatures for that geographic region. It is
also important to be sure that the oldest/biggest specimens available
from the body are used for the calculations. The estimation of time since
death is determined as the minimum duration of the time since death.

In order to fix the maggots satisfactorily it is important to expose
them for at least 30 seconds to water at a temperature of more than
80 °C; this prevents later decomposition of the samples. (If this is not
done the larvae tend to blacken due to internal bacterial decomposition
and may not be in an acceptable condition for review by a defence
lawyer's expert at a later date). Fixing the larvae in this manner allows
length comparison with published data from samples that have also
been treated in the same way. This is particularly important where
samples are being compared to growth tables (isomegalen and isomorphen graphs) which relate period of growth to a particular instar.
Alternatives to the use of length for the determination of the PMI have
been proposed because head-curling (bending of the head towards the
ventral surface during killing and fixing) may affect the accuracy of the
measurement of larval length.

Where the temperature is relatively constant, *e.g.* in a flat or house
with functioning central heating, it is possible to read off the time since

oviposition based on the length of the maggot. It is important to ensure that the method used to obtain the maximum length of the specimen is consistent with that used to produce the isomegalen graphs. Such aspects as shrinkage due to killing, shortening due to prepupation, *etc.*, must be taken into consideration.

Where temperatures fluctuate or it is necessary to use data from the local meteorological station it is also important that the relationship between internal and external temperatures be determined. To do this temperature data loggers are located in the position where the body was found and the temperatures are recorded for 5–10 days from the time of when the body was found.

A second aspect to be considered is the influence of the conditions in which the corpse was kept prior to the examination. Mortuary refrigeration units do not necessarily prevent the growth of the larvae already feeding on the body and this can cause some misinterpretation of the time since death, which may hence be longer than is immediately evident.

Equally, the length of storage or transport time under refrigerated conditions (4–5 °C) will influence the length of development of the insect and the effects differ with individual life stages. For example, the duration of storage at a temperature of 4 °C can influence development through the length of life cycle of the cultures should these be returned to the higher temperatures of the crime scene. In *Protophormia terraenovae* (Robineau-Desvoidy) the length of the first instar and prepupae will decrease the longer the specimens are kept under refrigeration, up to a period of 10 days. Keeping the second instar and puparial stage refrigerated for intervals up to a similar length of time causes the time of emergence to be delayed. Overall refrigeration can result in an error of more than 6 hours in the length of the time since death where larvae are stored at 4 °C for a period of 10 days.

In cases where animal cruelty is alleged, knowledge of the conditions under which the animal was kept, prior to analysis by a forensic entomologist, is crucial. This information is particularly important where an animal is killed and the corpse is retained, often under unspecified circumstances, at a veterinary establishment before the body can be sampled. Surgeries tend to be warm places, open for 24 hours a day, and the corpse may be covered by a cloth and left in a room without any regard for the need to take temperature measurements during this period. Hence, the speed of larval development may be influenced or the changes in temperature remain unacknowledged and so the determination of the time since death becomes potentially inaccurate.

In cases involving sexual assault, insect activity on corpses may cause changes that are misleading. The degree of disarray of the clothing is

Figure 4.9   Maggot masses can disrupt clothing, simulating the effects of sexual assault, as the larvae move within the mass to regulate their temperature and feed.

primary evidence in cases where the bodies of victims of sexual assault are found, particularly when the body is in an advanced state of decay. Maggots in a mass can relocate clothing, particularly on the underside of the body, in a very short time (Figure 4.9). The position of clothing, particularly on large corpses, can alter daily between the decomposition stages of bloat and advanced decay. This can cause a misleading interpretation of the existence of sexual assault in a case if the presence and potential effect of maggot masses and the environmental conditions are not considered sufficiently.

4.13   PROTOCOLS, PROCEDURES AND QUALITY ASSURANCE

Forensic entomology is a recognised and dynamic scientific discipline. Its practices and procedures are the result of worldwide discussions and the work of a number of researchers. In 1986, Kenneth Smith produced the *Manual of Forensic Entomology*, outlining procedures at the crime scene and urging the recovery and laboratory culturing of live specimens to confirm species identification. In 1990, Paul Catts and Neil Haskell published a procedural guide and 'death scene case study form' intended to help crime scene investigators to recover insects effectively. More recently, Wyss and Cherix described procedures at crime scenes in their book *Traité D'Entomologie Forensique*, and Magni and her co-workers in *Entomologia Forense* provide guidance on procedures and protocols for use at Italian crime scenes.

To standardise good practice for all the forensic disciplines in which scientists collect entomological samples, the European Association for Forensic Entomology (EAFE) developed a protocol that, if followed, helps maintain at least minimum standard operating procedures in forensic entomology. The intention of the professional body was to ensure that risk of error is minimised and that courts would be better able to interpret and compare entomologists' reports.

BIBLIOGRAPHY

J. Amendt, C. P. Campobasso. E. Gaudry, C. Reiter, N. LeBlanc, and M. J. R. Hall, Best practice in forensic entomology—standards and guidelines, *Int. J. Legal Med.*, 2007, **121**, 90–104.

G. S. Anderson, Effects of arson on forensic entomology evidence, *Can. Soc. Forens. Sci.*, 2005, **38**(2), 46–67.

T. K. Crosby, J. C. Watt, A. C. Kistemaker, and P. E. Nelson, Entomological identification of the origin of imported cannabis, *Journal of the Forensic Science Society*, 1986, **26**, 35–44.

E. M. El-Kady, Problems facing application of forensic entomology, *Pakistan J. Biol. Sci.*, 1999, **2**(2), 280–289.

D. E. Gennard, *Forensic Entomology: An Introduction*, John Wiley & Sons, Chichester, 2007.

F. Introna, C. P. Campobasso, and M. L. Goff, Entomotoxicology, *Forens. Sci. Int.*, 2001, **120**, 42–47.

P. Magni, M. Massimelli, R. Messina, P. Mazzucco and E. Di Luise, *Entomologia Forense*, Edizioni Minerva Medica, Torino, 2008.

C. Wyss and D. Cherix, *Traité D'Entomologie Forensique*, Presses Polytechniques et Universitaires Romandes, Lausanne, 2006.

M. Wood, M. Laloup, K. Pien, N. Samyn, M. Morris, R. A. A. Maes, E. A. de Bruijn, V. Maes and G. De Boeck, Development of a rapid and sensitive method for the quantitation of benzodiazepines in *Calliphora vicina* larvae and puparia by LC-MS-MS, *J. Anal. Toxicol.*, 2003, **27**, 505–511.

CHAPTER 5

Trace and Contact Evidence

TIERNAN COYLE

Contact Traces, Unit 26, East Central 127, Milton Park, Abingdon OX14 4SA

5.1 INTRODUCTION

'Every contact leaves a trace' is probably one of the most effective soundbites ever composed and it has become a mantra for modern forensic science. However, while conveying a general forensic principle, perhaps it does more harm than good to trace evidence as a discipline. Every contact may indeed leave a trace—but what does that trace mean? The human condition, the search for meaning, is at the heart of forensic science and in no other field of forensic science is the human condition better expressed than in the field of trace evidence.

Trace evidence is ubiquitous because it comes from the environment in which we interact as part of our daily lives. Although the environment in which we interact is populated by manufactured materials supplied at the transient whims of the global manufacturing economy, our behaviour as human beings remains highly individual. The strength of trace evidence, its value to the court, lies in being able to put the finding of these materials into the context of human behaviour. It is largely for this reason that a DNA-style statistic is not the appropriate means to express the strength of trace evidence in the courtroom.

This chapter deals with the principles shared by all types of particulate trace evidence, focusing on the types most commonly examined in forensic science laboratories, typically fibres and glass. Laboratory

Crime Scene to Court: The Essentials of Forensic Science, 3rd Edition
Edited by P. C. White
© Royal Society of Chemistry 2010
Published by the Royal Society of Chemistry, www.rsc.org

processes are described and some case studies are presented to illustrate particular points.

5.2 BACK TO BASICS

To understand trace evidence, we must appreciate that it comes from an interaction with a human being's own unique, individual environment. We gather trace evidence on to ourselves by having contact with our environment in a manner that enables microscopic fragments of material to be transferred. That material is carried around until it has either been transferred to something else, lost, or destroyed. The first important issue to consider is the transfer of materials; the second, closely related to the first, is the persistence of those traces in the context of human behaviour.

5.2.1 Transfer

In chemistry, in order for two molecules to react, they must collide while possessing sufficient energy to facilitate the reaction. Transfer of trace evidence can be described in similar terms. As an 'evidence type', fibres retain a unique position within the trace evidence portfolio in that human beings are habitually and frequently in contact with textiles. The transfer of fibres from one garment to another, in the classic two-way transfer scenario, requires each garment to have surface characteristics which allow loose fibres to be removed from it while also gathering loose fibres from the surface of the other garment. The nature of the surface of a garment plays a significant role in ascertaining the degree to which fibres will transfer, if indeed at all. Forensic scientists have, over the years, invented the term 'sheddability' to describe this characteristic. It is a characteristic that can be difficult to quantify, but it is crucial in assessing whether or not a case involving fibre evidence should proceed and it is very important to get it right. It is not possible to assess the sheddability of a garment without a proper scientific examination. A waterproof nylon fabric may not appear to shed fibres, but if the fabric has been damaged fibres will shed from the damaged areas as illustrated in Figure 5.1. Attempts to pre-screen clothing without a scientific examination of the fabric are likely to result in a large number of false negatives, particularly where the assumptions made in the screening process lack any kind of scientific rigour.

At any time, all surfaces of garments contain loose fibres that are not part of the garment's construction, but have been picked up by that garment as a result of previous contacts with other textiles. If contact

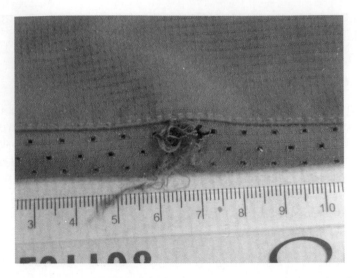

Figure 5.1   An example of damaged clothing showing fibres that can be shed.

between two garments occurs in an area where these fibres are present, they, together with fibres from the donor garment, may transfer onto the other garment's surface. It must be noted that these types of 'foreign' fibres are a fact of life and they exist on all garments at all times. The numbers and types of these fibres are highly individual, as they are a record of the recent activity of each individual garment.

Most clothing sheds fibres to some degree or another. Even if a garment does not shed its own fibres, it is likely to have 'foreign' fibres present on its surfaces. It can be said with confidence that in the majority of situations where contact occurs between one fabric and another, fibres will transfer and that with the exception of airborne materials, all transfer is the result of direct contact. Identifying whether or not the source of the transferred fibre is likely to be the garment constructed of the particular fibre or something else that happened to have those fibres on its surface is another matter entirely.

Transfer of other particulates such as glass is less complicated than that of fibres in the sense that glass is a smooth material and does not have a population of loose or broken fragments on its surfaces available for transfer. It is when glass is broken that fragments small enough to physically interact with a human being can transfer. Breaking a window causes airborne fragments, similar in size to those shown in Figure 5.2, to transfer to those in close proximity, whereas walking on fragments of glass could cause fragments to transfer to the sole of footwear.

Figure 5.2   A pin surrounded by glass fragments of a size range typical of those examined in forensic science.

5.2.2   Persistence

Once trace evidence has been transferred on to a surface, it is important to understand the factors involved in its remaining there. Unless the material is so chemically unstable that it degrades or evaporates, trace evidence will remain in place until it is physically removed from the surface. A general principle covering all particulate-based trace evidence is that the majority of individual particulates transferred are lost relatively quickly from fabric while it is being worn 'normally'. It can be said for most particulates that approximately 80% of the traces are lost during this period (2–4 hours) while the surface of the garment is active. The general trend of persistence is one of exponential decay, where the initial loss is rapid, but those particulate traces more strongly bound to the surface are retained for longer periods.

In a similar way to transfer, persistence is often viewed solely as a tool in assessing whether or not a case requires a trace evidence examination. Often insufficient information is available to assess scientifically whether or not a trace examination is likely to assist in this decision, particularly with reference to the activity of the garment in the intervening period between the alleged offence and the seizure of the garments. Merely calculating a time between the offence and the seizure of clothing as the sole means of assessing likelihood of finding trace evidence is entirely unscientific.

The most important aspect about persistence is that it is not merely a tool to be considered in pre-case assessment; it describes the very nature of the microscopic world in which we exist. The persistence curves show us that surfaces are in a constant state of change. Particulates are being transferred, deposited and lost every minute of every day. It is this key component of persistence which delivers the evidential strength of contact traces, that the particulate population on our clothing and bodies is a function of the recent history of an individual.

Taking the elements of transfer and persistence into account leads inevitably to the conclusion that when interpreting the presence of trace evidence, it is not merely the fact that sources other than the questioned item exist that is pertinent, but whether or not another source of the trace is likely to have made contact and transferred the traces within the circumstances of the case itself. M. C. Grieve and K. G. Wiggins have summarised the situation as follows

  . . . the textile population in a person's wardrobe and sur-roundings is based on personal taste and is therefore, highly indi-vidual. It would be extremely unusual to pass a person in a street wearing exactly the same items as yourself from head to toe. In addition to this, fibres carried on clothing reflect the owner's per-sonal environment, and if transferred to other surfaces can provide a link back to their origin.

5.3   CASEWORK PROCESSES

Cases involving trace evidence share a similar process with other evi-dence types. Resources involved in modern forensic laboratory pro-cesses are very precious and resources need to be prioritised in order to best meet the demands of a modern criminal justice system.

5.3.1   Case Assessment

Like all other evidence types the processes involved in examining trace evidence begin with a gathering of information. Depending on the complexity of the case, this may involve pooling information from a variety of sources such as the investigating officers, the crime scene investigators and/or other investigative authorities. Attention to the detail of the seizure of the items is important, and ascertaining the potential for inadvertent contamination of the exhibits during the arrest of suspects and seizure of clothing, *etc.*, is paramount.

In recent times attempts have been made to provide a Bayesian approach to case assessment with a view to predicting the likely success of a case prior to examination. Those advocating the approach describe the consistency that a framework approach brings to a process which otherwise is entirely subjective and relies on information that is at best incomplete and at worst unreliable. Others believe that a framework does not compensate for the quality of the information provided to make the assessment and all that this framework succeeds in doing is condensing judgement to a speculative statistic that has little real meaning.

5.3.2 Contamination Prevention

A strategy for examination must be devised to minimise any contamination from the laboratory process. Modern forensic laboratories have adopted very sophisticated contamination prevention systems and processes to minimise biological and trace contamination. In some laboratories in the UK, the wearing of medical-style scrubs with disposable laboratory coats, hairnets, masks, gloves and overshoes has become standard practice. In other organisations, knowledge and experience gained from DNA processing facilities and the examination of explosives has influenced their design of search laboratories, by introducing a positive pressure environment.

Whatever the precautions a laboratory undertakes to minimise contamination, it must be appreciated that forensic science laboratories cope with hundreds of items of clothing a year from a large number of different sources. The items examined vary in condition from clean to heavily bloodstained and items of clothing taken from badly decomposed bodies. The very fact that these items need to be exposed to a laboratory environment, physically examined, perhaps taped or shaken, means that the laboratory will certainly become contaminated with material from each exhibit examined there. It is therefore very important to adopt a strict cleaning regime in laboratories so that their environment can be continually monitored. Furthermore, it is important that records are kept of the activity on each laboratory bench, particularly if a laboratory contains multiple benches where items from different cases may be examined by different people at the same time. These precautions minimise the potential for cross-contamination and provide a basis for a scientific investigation of cross-contamination should it become a focus of pre-trial debate. For cases with particulates, the recovery of the population of traces on the bench and on the outside of the exhibit bag is performed. Additionally, the use of a disposable surface (such as paper), to act as a barrier between the exhibit and the bench, is important since it both reduces direct contact between the item and the laboratory and facilitates the recovery of any debris that falls from the item during laboratory processes.

In developing the laboratory strategy it is important to separate the items in a case depending on their origin. Items in a case relating to a victim should not be examined in the same laboratory, or on the same day, as those from a suspect. This is also true for items seized from different locations, and for items of which the wearer is unknown or in dispute. Each forensic science laboratory has a finite number of search rooms and on that basis there are going to be cases that will stretch the laboratory's resources—such as those with multiple suspects and victims and large numbers of items. Developing a strategy for examinations of items in these cases requires a delicate balance to be struck as the impact on laboratory resources can be great.

5.3.3   Recovery: Taping, Shaking and Brushing

The type of surface of the item and the potential presence of other evidence types such as body fluids, fingerprints, gunshot residues (GSR), *etc.*, will have an impact on how trace evidence should be recovered. Recovery of trace evidence from clothing in the laboratory is probably the most straightforward. The method of choice in Europe for the recovery of fibres and hairs involves taping the surface with strips of low adhesive tape. The recovery of glass, paint flakes and other particulates involves shaking or brushing the fabric over a disposable paper and recovering the debris.

The use of small squares of tape (mini-tapes) has been advocated as the optimum means of DNA recovery. This method is used extensively in the UK; however, the impact of this technique on the recovery of other types of trace evidence, with the exception of GSR, has been largely ignored. This can be detrimental in respect of fibre evidence, because the areas targeted for mini-taping tend to be precisely the same as those of interest for fibre evidence. There have been cases where mini-taping has been performed and yet the results of the DNA analysis have not provided any further assistance. Subsequent fibre examinations in these cases have often provided a link, with the caveat that fibres recovered from the areas that were mini-taped were lost in the DNA process. Potentially, these fibres could have provided further evidence strengthening the textile fibre link between the parties, particularly if the number of fibres found from the areas of the garment not mini-taped were low. A key test of any validated process must be its impact on other processes. Since the impact on the recovery of trace evidence by using mini-taping can be severe, it is imperative that a method for recovery of trace evidence from mini-tapes be developed so that it does not constrain the ability of forensic scientists to explore solutions other than DNA. At present forensic science is failing to assist the court in the proportion of cases where a 'DNA only' approach has been taken. Hopefully, this issue will receive attention in current reviews being carried out by the Regulator.

The recovery of particulates from items such as footwear and weapons can be difficult because of the complexity of damage on the surfaces, blood pattern analysis and fingerprints. A recent method of recovering fibres from non-textile items, involving the use of a charged plastic rod, avoids the need to use tape. Adequate photography of such items, usually with the aid of a microscope, is vital for the interpretation of trace evidence. In particular, with weapons such as knives, the location of the trace on the weapon (*e.g.* on the cutting surface, as shown in Figure 5.3, or on the handle) is very important contextual evidence. Fibres present in blood or in damage on the soles or toecaps of footwear are likely to be more significant than those recovered from a general recovery process such as taping.

Vacuuming as a means of recovery of trace evidence is still in use in some laboratories, but it causes operational difficulties later in the process by causing multiple types of trace evidence to become entangled, making it difficult to isolate and analyse.

5.3.4 Searching, Sifting and Screening

Trace evidence from the scene or from items examined at the laboratory tends to be retained on tapings, in debris pots or in other appropriate receptacles such as paper pockets. It is likely that the 'trace harvest' will encompass over a dozen tapings from the outside of a garment, together with several debris pots from outside surfaces and pockets. The focus of the examination moves from the item submitted to the laboratory, to the population of fibres and debris in tapes and pots.

The search for trace evidence remains entirely a human task, primarily a manual, visual search using a stereomicroscope. The ability of the human eye to discriminate particulates using a stereomicroscope remains unsurpassed by technological attempts to automate the process. The technology exists to do the job; however, the risks to commercial manufacturers of instrumentation currently outweigh the benefits. The

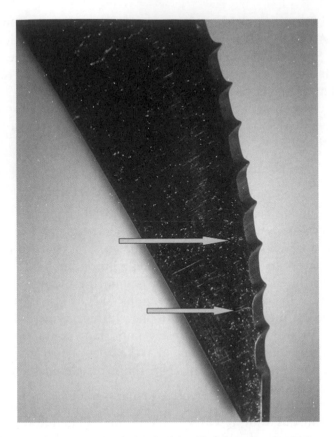

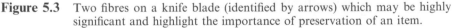

Figure 5.3 Two fibres on a knife blade (identified by arrows) which may be highly significant and highlight the importance of preservation of an item.

demand for trace evidence worldwide has decreased dramatically since the advent of DNA analysis. The appetite for investing in trace evidence in both the public and private sectors has evaporated to such a degree that only a few suppliers exist for some key specialist equipment. This, together with the more rigorous demands that a regulated forensic environment brings, creates a significant barrier to innovation in this area.

   The process of searching tapings and debris pots is largely dependant on the purpose of the examination. It may be that the target of the search is known, such as for glass particles, a particular type of fibre or colour of paint. Alternatively, it may be that the target of the search is entirely unknown and that the purpose of the search is to characterise the trace material to provide intelligence to an ongoing investigation. Either way, the particulates of interest are isolated from the matrix of debris for the next stage of analysis. Fibres are recovered and mounted permanently on microscope slides, whereas paint may be mounted on wax or plasticine for microscopy.

5.3.5 Light Microscopy and Other Screening Techniques

Microscopy of trace evidence is a crucial part of the process of identifying and comparing traces. The physical and morphological information obtained by a microscopic examination of particulates is of such high quality that it imparts a very high level of discrimination. Polarised light microscopy enables specific types of fibres to be identified and also discriminates between glass, quartz and plastic particles. Comparison microscopy enables a side by side comparison of individual traces with a reference material, or perhaps other traces recovered from areas of significance. The importance of comparing traces visually in this manner cannot be understated. If traces are indistinguishable it provides a powerful indication of a positive association, although it remains possible that instrumental analysis may provide further discrimination between microscopically indistinguishable traces. Although microscopic analysis is undoubtedly powerful, its weakness lies in its reliance on the human ability to visually identify and compare traces. As a technique it is somewhat constrained by its subjective nature, but this is overcome in most laboratories by having the microscopy checked by another suitably qualified, trained and competent scientist.

Glass particles tend to be screened first by measuring each individual particle's refractive index. Although this provides a high level of discrimination, it has been noted that improvements in glass manufacturing in recent years are making it less effective as a screening tool.

5.3.6 Instrumental Analysis

Instrumental analysis is vital in the examination of trace evidence because it provides an objective means for identification and comparison of traces. However, it does not provide the same information that a microscopic examination provides and on that basis using an instrument as a substitute for microscopic observations is discouraged.

Trace evidence can be characterised instrumentally in an almost infinite number of ways; however, forensic laboratories are constrained by their laboratory equipment, the skill of the staff and an ever more regulated environment. Forensic scientists are also constrained by the destructive nature of some sampling processes required by instrumental analysis. The ideal instrumental analysis procedure for trace evidence is one that fully characterises the material without destroying it so that it is available for other scientists to test if needed.

Instrumental analysis is conducted globally in a wide variety of ways depending on the resources of each laboratory. The type of instrumental analysis most frequently used for trace evidence tends to be spectroscopic. The instruments are now more commonly attached to microscopes or are built as an integral microscope system. Microspectrophotometers (MSPs) for measuring spectra in the UV and visible ranges and Fourier-transform infrared (FTIR) systems are the most traditional of techniques employed for fibre analysis. SEM with either EDX (energy dispersive X-ray) or WDX (wavelength dispersive X-ray) analysers and microprobe analysis are used for glass and paint analyses and additionally laser ablation–inductively coupled plasma mass spectrometry (LA–ICP–MS) is used for the analysis of glass. Other techniques for the elemental analysis of glass are currently being researched, although they are yet to make the transition into operational laboratories.

Raman instruments coupled to microscopes have emerged as an important tool for the characterisation of trace evidence in recent years. The cost of this instrumentation has fallen considerably in recent years with costs more in line with FTIR instrumentation. The scope of spectral information provided by a combined FTIR and Raman approach provides better levels of discrimination than a single spectroscopic approach. The combined approach also provides the trace evidence examiner with considerable flexibility when facing difficult samples— such as the analysis of condom lubricants in the aqueous supernatant generated from swab extraction, or a paint smear on a tool such as that identified in Figure 5.4. Increasingly, the non-destructive nature of Raman spectroscopy has made it attractive for laboratories seeking to improve the quality of their services without impacting on their ability to deliver services in a timely fashion.

5.4   CASEWORK SCENARIOS

Modern trace evidence casework usually begins with a request by the investigating authorities. Specific examinations are dictated by the case circumstances, such as the search for glass if a window has been broken, the search for fibres in cases where contact between two textiles has been alleged or the search for paint in a hit and run vehicle accident case. In some cases, however, the investigating authorities are unable to specify their requirements precisely. It is these cases that provide an environment where trace evidence examiners have the freedom to express themselves and where the best work in trace evidence can be found. It is

Figure 5.4 Crushed paint on the business end of a jemmy. The paint is the light area of the right-hand side of the bevel and ideally should be analysed *in situ*.

regrettable that the impact of DNA in forensic science on investigating authorities, government and even on forensic science institutions has pushed trace evidence to the forensic equivalent of a 'weapon of last resort', when there tends to be little time left to trial and the opportunities for trace evidence to be explored have been reduced.

5.4.1 They're Junior . . . Just Give Them The Little Cases . . .

These days a trace evidence examiner's life begins with a job that is so interesting that it provides the trainee with great enthusiasm for a long, rewarding and interesting career ahead. Laboratory organisations generally view new forensic scientists with cautious optimism and the best organisations invest considerable resources into their training. The 'little cases' provide the new examiner with challenges that enable the examiners to develop and hone their scientific skills. The bigger, high profile cases tend to provide challenges related more to logistics and the management of expectations, rather than challenges of a scientific nature. It is somewhat disappointing that laboratory organizations, while recognising the 'little cases' as being the ones suitable for junior staff, tend to consider them beneath the level of their senior staff, no doubt influenced by economic factors. A laboratory where the senior scientists are tied up solely in long, high profile cases and the junior staff are busy with a high volume of cases that contractually demand a very tight turnaround is not one where the long-term aims of the individual forensic scientist or the laboratory will be served. Happy trace evidence examiners are those who have a caseload varied in size and scope within their chosen specialism, so that their hard-earned skills are maintained and improved, while a healthy trace evidence department is one where the junior and the senior scientists are able to work together on the same types of cases. Working in a properly regulated quality environment should, in theory at least, allow for the same level of quality of service irrespective of whether the individual scientist is junior or senior, while of course recognising the need for a certain level of experience in particular cases.

On to the 'little cases', which like all trace evidence cases usually start with a simple question such as, 'Who broke the window?' These are precisely the same questions asked in the serious cases, and the challenges and the level and quality of the work required to meet them are generally the same.

5.4.2   Who Broke the Window?

For the investigator the answer is simple; the person who broke the window has particles of glass from that window on their clothing. Find the glass, find the offender.

Scientifically speaking, it is of course not that simple. If a person breaks a window, glass will transfer to that person and their clothing by means of the deposit of airborne particles. Glass will also transfer to clothing if it is in direct contact with fragments of broken glass, or if the person is in the vicinity of the window when it is broken by something or someone else. The breaking of glass is something that most people will have experienced as part of their normal lives; some, such as those working in the demolition or recycling industries, may experience it on a daily, occupational basis.

All of these factors and more need to be considered by the forensic scientist when reaching a conclusion. The conclusion reached in cases examined in the UK tends to be phrased in terms associated with the Bayesian approach to the interpretation of evidence. A typical interpretation section in statements begins like this:

'In assessing the evidence, I have considered two propositions:

1. The person broke the window
2. The person did not break the window'

The scientist will then list the factors that they have taken into consideration, which will vary in each case depending on the information provided to the scientist. In cases involving glass this includes (among others):

1. The number of particles of glass found on the clothing.
2. The number of the particles that are indistinguishable from the window glass.
3. The number of groups of glass particles (if any) present on the clothing.
4. Whether or not the scientist would expect to find the number of particles matching the window by chance.

The conclusions from the interpretation of the findings are weighed up and placed on an evidential scale, *e.g.*:

Taking all of this into consideration, the findings are more likely if the first proposition (*that of the person breaking the window*) is true, rather than the second (*that of the person not breaking the window*). On that basis they provide very strong support for the proposition that the person broke the window. The scale of the support is: none, limited, moderate, moderately strong, strong and very strong.

The Bayesian interpretation sits well with glass evidence, perhaps more than any other type of trace evidence, because of the ease with which glass appears to fit into a robust statistical treatment. Frustratingly perhaps for the non-scientist, the number of different groups of glass on clothing has a pronounced effect on the likelihood ratio, and if the findings show evidence of many different types of glass in the debris from the clothing, this can reduce the strength of the evidence. There are significant assumptions in these arguments, but the most common criticism levelled at the Bayesian approach is that in order to take issues into account, the likelihood ratio requires to be modified, but in a way that perhaps betrays the purely statistical foundations of the expression. Depending on the scientific organisation, some use the likelihood ratio in these instances as a guide, others apply it quite slavishly. Recent trends amongst providers in the UK appear to be favouring a far more slavish approach where possible. Where there is no empirical data to rely on, a 'fudge factor' is applied to the likelihood ratio to modify it. The 'fudge factor' is largely at the discretion of the individual scientist, although it is applied within a quality framework based on peer review.

This debate has raged for a long time and will continue to do so. Globally, each jurisdiction has its view on the role of the forensic

scientist in court. Some apply strict constraints on the level of interpretation a scientist can provide, others grant considerable freedom of
expression to the expert. Some believe that the Bayesian approach
delivers precisely what the court demands of the forensic scientist—their
opinion. Others believe that the Bayesian approach moves the forensic
scientist from the witness box to the bench or the jury box and that the
impartiality of the forensic scientist is undermined by it. Whatever the
criticisms of the Bayesian approach, perhaps its greatest achievement
has been the formation of the verbal scale, a common language for
forensic scientists in which they can feel comfortable expressing their
own individual opinion, 'fudge factor' or not.

5.4.3   Who Climbed Through the Window?

Breaking the window is one thing, entering the premises is another. If
there are a number of suspects involved in a burglary, they may all have
glass on their clothing; perhaps one of them broke the window and the
others got glass on them by being in close proximity or by climbing
through. In this case, the presence of fragments of glass on their clothing
may not assist in assessing who climbed through.

The presence of fibres on broken glass is highly significant, not just
because of the fibres *per se*, but because of where they were found. In
Bayesian terms the proposition and the evidential strength will be
affected by many factors, including the types of fibres found, whether
they were transferred as a clump or as single fibres, and if there were any
signs of recent damage on the garment that could have been caused by
contact with a sharp surface such as broken glass.

If no glass is found on the clothing or fibres on the glass, it may be
considered that this in some way supports the proposition that the
clothing was not in contact with the window. Of course, it is possible for
an offender to climb through a window without being in contact with the
broken glass; or perhaps their clothing was seized so long after the
offence that any glass particles were lost. In theory the framework exists
that enables these factors to be taken into account by the Bayesian
approach, but the difficulties in so doing are obvious. In jurisdictions
where the burden rests entirely on the prosecution to prove their case, it
could be argued that the Bayesian approach can never provide support
for the defence in the same quantifiable manner that it does for the
prosecution, because the defence is under no obligation to assist in
the calculation of the likelihood ratio in the first place. Recently some
organisations have provided guidelines for how the likelihood ratio
should be framed in situations where the defendant provides 'no

comment' type defences, including adopting scenarios on the behalf of the defendant. It will be interesting in the future to see how well this will be received by the courts, particularly with reference to a defendant's human rights.

5.4.4 Who Was in the Car?

Motor vehicles are one of the most common items used by criminals to commit their offences. They provide means of transport to and from crime scenes, together with being the focus of the crime itself. Modern vehicles are almost impossible to drive without the keys and incidences where premises have been burgled with the sole purpose of obtaining car keys have been on the rise in recent years.

Of course DNA profiles can be obtained from stationary areas within vehicles such as steering wheels and gear sticks; however, arguments over how the DNA could have been transferred to the surfaces can reduce the value of the finding of the profile in the case. Although some prosecuting authorities are content to proceed merely on the basis of a DNA profile in these cases, others are reluctant to go forward without forensic evidence directly relating to contact of the individual within the vehicle.

One of the myths about linking clothing to vehicles is that if the clothing of the suspect does not shed or is of low evidential value then there is no point in proceeding further. There are few more 'fibre rich' environments than the interior of a modern motor vehicle. Car seats provide a harvest of fibres from the textile environment of the most frequent or habitual users of the vehicle. Together with the use of specifically engineered automotive fibres, such as car seat fibres and flock, this means that it is virtually impossible to have contact with the interior of a car without automotive fibres transferring from the inside to the clothing.

In some cases, placing the suspect inside the vehicle is sufficient for the investigators, in others the investigators require to know the identity of the most recent driver of the vehicle. In some instances, several suspects may have driven the vehicle since it was used in the offence, some perhaps taking turns driving it during a journey. In these cases the distribution of fibres within the vehicle is significant, and requires the surface from each seat to be examined. The surface of the seat belts can provide assistance in this matter. The interpretation of a distribution of fibres can prove to be difficult, but it has proved very effective in providing the answer to the query in numerous cases.

In road traffic accidents, where collisions have occurred between occupants of the vehicle and its interior, the use of a type of trace evidence called 'fibre plastic fusions'(FPF) has proved successful in assessing the likely position of the driver and passengers. Friction as a result of high-speed contacts between clothing and surfaces can be sufficient to cause the fabric to melt and fuse to the plastic surface causing an FPF. Likewise, the plastic surface can melt and transfer to the clothing causing a plastic coating mark (PCM). Examination of the interior of the vehicle for FPFs and clothing for PCMs, can aid reconstruction of the events within the vehicle at the time of the crash.

5.4.5   More Complex Cases

Contact is one thing, the nature of the contact is another. In some cases, the presence of suspects at a crime scene is not in dispute, but their activity within the scene is the focus of the investigation. Either the suspect has not provided any comment, or they have provided investigators with an account of their activity.

In cases such as this, novel crime scene techniques such as fibre mapping have provided a means for exploring the nature of the contact, in a way that no other forensic evidence type can. Fibre mapping involves the use of single strips of adhesive tape in a one-to-one fashion covering the entire surface of interest (the body at the scene and other areas of interest). Photography at the scene enables the precise location of each single fibre to be plotted on the photographs, showing in photographic terms where the fibres were deposited; these maps can be presented to the jury as visual aids. While the photographic presentation is important, the ability to correlate fibre distributions with other evidence types, in particular the findings from the post-mortem, is particularly useful. One prosecuting barrister who experienced a case involving fibre mapping said that:

> One of the most powerful weapons in the prosecution's armoury is the ability through expert evidence to reconstruct the factual scenario which existed at the time of the commission of the crime. The fibre evidence in this case did exactly that. It enabled the prosecution to provide the jury with an insight into the nature and extent of the physical interaction between the deceased and his assailant. At the conclusion of the trial the presiding judge went out of his way to comment favourably on the fibre evidence which he described as both fascinating and compelling.

Some have advocated the use of zonal approaches at the crime scene. The zonal approach describes taping a particular area of interest with a single tape, rather than a one-to-one approach. Although a zonal approach can give indications of a distribution of fibres, it cannot precisely locate the position of each individual fibre and neither can it associate fibres with other evidence types *per se*. However, in cases where the use of one-to-one taping is not appropriate, such as where paramedics have been in contact with the body, a zonal approach is a far superior approach to fibre recovery at a scene than bagging up clothing at the post-mortem and submitting it to the laboratory.

5.4.6 Intelligence

In the early stages of an investigation the most important first step is to locate a suspect, so it is not surprising that the emphasis of an investigation at this stage is on traditional policing methods, fingerprints, DNA and the newest technological sources of information from CCTV, mobile phones and computers.

Trace evidence rarely finds itself included in the initial strategy beyond a protocol for preservation during the processing of items for body fluids, but it can play a role and should not be discounted merely based on the perception that it is time consuming, expensive and likely to provide little value.

Where fibre tapings have been performed at the crime scene, either as one-to-one tapings or via the zonal approach, areas of the body clearly associated with the activity of the offender can be prioritised and assessed for fibre populations. Intelligence from these searches may inform the investigators as to the likely colour and type of clothing worn by a suspect, assisting CCTV operators in screening potential suspects, or highlighting accounts from eyewitnesses perhaps otherwise treated as insignificant.

Debris obtained from the victim's body may assist also. In the case of Gary Ridgway (the 'Green River killings'), hundreds of distinctive microscopic spheres of paint were found relating to the bodies of six of his victims. Gary Ridgway worked spray-painting trucks. It had taken 20 years for a trace evidence link to be properly explored. The scientist who discovered the particles, Skip Palenik, stated that

Had I looked at the evidence in the 1980s, I would have found the same paint particles. When they interrogated Ridgway in 1987 and he told them he painted trucks, they would have been able to say: 'We have some more questions for you.'

Trace evidence can also assist the court in issues beyond that of mere contact. In the case of the manslaughter of Damilola Taylor, for example, the presence of a single polyester fibre in a bloodstain which bore distinctive damage characteristics allowed the scientist to assess the proposition that the fibre was part of the blood droplet while airborne and that the likeliest source of the fibre (and the blood droplet) was the weapon used to kill Damilola Taylor.

5.5   TRACE EVIDENCE IS LIKE A BOX OF CHOCOLATES ...

Each taping or debris pot has its own story to tell. Discriminating between the ordinary and the evidentially significant particulate requires skill, experience, talent and an investigative mind. The presence of a type of trace on one surface may be evidentially significant in one case but not in another.

It is simply not possible to anticipate the types of particulates that might be of assistance in a given case; for that reason trace evidence examiners often find themselves in a state of catch-up in cases where the particulate is unusual, there is no body of work, no standard operating procedure, no background information, and perhaps no obvious means of analysis for a new particulate. Recent cases highlight the importance not only of casework-based research and development, but of a quality framework that allows for some experimentation. Case studies involving fragments of polyurethane foam, cosmetic glitter and even breakfast cereal have been published in recent years. What follows are some types of trace evidence that have been researched by caseworkers in recent times on the back of specific cases.

5.5.1   Polyurethane Foam

Fragments of polyurethane foam on the surfaces of clothing come from a variety of sources in our environment and their presence may be evidentially significant. As particulate matter, foam appears either as clumps of closed cells or as single or L-shaped rods and can be analysed in a similar way to a traditional fibre examination. Recent studies have shown a high level of discrimination and a background study of the incidence of polyurethane foam in clothing has been undertaken.

5.5.2   Flock

The use of flock fibres has grown over recent years; they are widely used in the automotive industry as an textile for engineering and luxury finishes. As illustrated in Figure 5.5, flock fibres are an excellent target fibre because of their high degree of sheddability. A recent study has shown

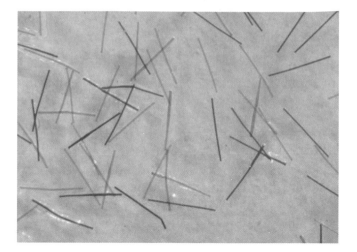

Figure 5.5   Flock fibres which are used as an automotive textile and are an excellent target fibre because of their high degree of sheddability.

that the discrimination power of certain types of flock fibre is very high using traditional methods.

5.5.3   Condom Lubricants

The analysis of condom lubricants has been the subject of considerable attention in recent years. However, the issue has been largely one of competition for the swab between the demand for extraction of biological material and the demand for chemical extraction of the lubricant. A compromise position has been adopted whereby half the swab is applied to one process and half to the other, but it is not ideal by any means. A recent study has shown that by using Raman spectroscopy to analyse the swab *in situ* that condom lubricants can be detected without the need for extraction. Further work has shown that condom lubricants can be detected by the analysis of aqueous supernatants taken from swabs extracted for biological purposes by Raman spectroscopy and FTIR mapping of the meniscus.

5.6   FUTURE TRENDS IN TRACE EVIDENCE

Economic demands, together with increasing regulation of modern forensic science worldwide, are putting trace evidence examination under severe pressure. The justice system demands quicker, cheaper processes, while expectations for the levels of intelligence and evidential value are insatiable. If trace evidence examination is going to survive the emerging environment of the 21st century, it will need to change at a fundamental level.

Current technology exists which, if brought to bear in the right way and under the right conditions, could allow forensic scientists perhaps to meet at least some of the high expectations of the criminal justice system. However, if trace evidence is to survive, the role of the trace evidence examiner will also require reform. The demands of the regulatory environment will put increasing pressure on the generalists to demonstrate competency across their wide portfolio of evidence types, and perhaps it is time now to move to a new model of trace evidence provision.

BIBLIOGRAPHY

R. D. Blackledge (ed.), *Forensic Analysis on the Cutting Edge*, John Wiley & Sons, Chichester, 2007.

B. Caddy (ed.), *Forensic Examination of Glass and Paint*, Taylor & Francis, London, 2001.

M. C. Grieve and K. G. Wiggins, Fibers under fire: suggestions for improving their use to provide forensic evidence, *J. Forens. Sci.*, 2001, **46**, 835–843.

M. M. Houck, *Mute Witnesses Trace Evidence Analysis*, Academic Press, London, 2001.

M. M. Houck, *Trace Evidence Analysis More Cases in Mute Witnesses*, Academic Press, London, 2004.

M. M. Houck (ed.), *Identification of Textile Fibers*, Woodhead Publishing, Cambridge, 2009.

J. Robertson and M. Grieve (eds), *Forensic Examination of Fibres*, Taylor & Francis, London, 1999.

CHAPTER 6

Marks and Impressions

KEITH BARNETT

Forensic Science Service, Trident Court; 2920 Solihull Parkway, Birmingham Business Park, Birmingham B37 7YN

6.1 INTRODUCTION

Everything we do leaves a mark on the world. It could happen while walking, driving, working or sitting in the chair. When carrying out these tasks our shoes, clothes, hands and even the tools we use can leave unique telltale signs to the expert eye. A car will leave tyre tracks, the shoes on our feet will leave footwear impressions and even the screwdriver used to take the lid off the paint tin will leave its mark. The simple principle proposed by Locard was that 'every contact leaves a trace', and in the field of marks and impressions if you put two items together they are likely to leave a mark on one another.

Sherlock Holmes said in *A Study in Scarlet* that 'there is no branch of detective science which is so important and so much neglected as the art of tracing footsteps'. The same is also true for the examination of tyre tracks and the comparison of plastic bags or instrument and glove marks. In all these cases there are features present that can be used to form a unique connection between suspect and scene or stolen property. The following chapter gives an insight into the methods used to recover and present evidence at court in some of these areas.

This evidence can be divided into two distinct groups: damage-based and non-damage-based. In the first group, which includes evidence from

Crime Scene to Court: The Essentials of Forensic Science, 3rd Edition
Edited by P. C. White
© Royal Society of Chemistry 2010
Published by the Royal Society of Chemistry, www.rsc.org

footwear and instruments, an item has to acquire damage in order to leave behind a unique impression. In the second group, as with fingerprints, a combination of inherent features provides the unique link. Any damage to the features of the fingerprint, such as a scar, is not used in its classification as it may fade with time. Just as it is true that no two fingerprints have been found to be the same, even in identical twins, so in the examination of damage features no two items, no matter how similar they are initially, are likely to acquire the same random damage features during use.

In the first part of this chapter (sections 6.2–6.8) the evidence provided by damage and wear will be addressed, specifically evidence that can be obtained from footwear, instruments and mass-produced items. In section 6.9, fingerprints and the evidence they provide will be discussed.

6.2 FOOTWEAR IMPRESSIONS

6.2.1 Introduction

Every time a person takes a step, whatever the surface they are walking on, they will leave behind a footwear impression. An impression could be defined as the retention of the characteristics of an item by another object. Hence, on soft ground shoes impress themselves into the ground and leave behind their characteristic impression. The impression left behind is not always obvious and can be difficult to find without the aid of a specialist technique. It is also vital that as many impressions as possible are recovered from the scene of a crime as this will increase the chance of finding an area corresponding to the area where the damage features are located on the undersole of a shoe. It is possible that the undersole has only one or two of these damage features present and they may all need to be found to provide a significant link between a suspect's shoe and the scene of a crime.

6.2.2 Recovery of Impressions from a Crime Scene

There are several mechanisms by which a shoe can leave an impression. On a two-dimensional (flat) surface, such as a tiled floor or a piece of paper, material from the undersole can be deposited and remain for a considerable length of time. This may be due to a static electrical charge produced on the undersole transferring particles to a surface, or the wet deposits on an undersole being left behind on the surface. A three-dimensional impression is formed if the surface over which the shoe passed was soft and the undersole sank into it before moving on. The different types of impression require a variety of techniques to recover them, as no single method can cover all eventualities. Recovery can be achieved simply, by casting with dental stone or by applying a gelatin lifter, but often the impressions are fragile or transient in nature and these methods will not be successful. Recently, a number of new and novel techniques have been applied to the problem, with startling results. Some of these techniques were developed to aid fingerprint enhancement and recovery, others specifically for the enhancement and recovery of footwear impressions.

6.2.3 Impressions in Two Dimensions

In all cases where an impression of this type has been left behind on a flat surface, it will be visible only when there is contrast between the background and the material of the impression itself. To use a simple example, a footwear impression made on plain white paper with talcum powder would be almost invisible. Treat this with a black powder, say finely powdered charcoal, that adheres preferentially to the white talc, and the impression will become visible because of the black and white contrast created. To enhance most impressions so that they are fully visible requires the production of this contrast. The impression found can be photographed successfully, with a scale alongside it to provide a permanent record for court purposes. Often it is possible to seize the item on which the impression has been made and submit it to the laboratory for examination. On other occasions the enhancement has to take place *in situ* and a suitable method has to be used at the crime scene to improve the visibility of the impression. The detail produced has to be of such a quality that it can be photographed.

Porous surfaces such as paper, cardboard and carpet cannot be treated in the same manner as non-porous surfaces such as glass, laminate flooring, linoleum and tile. On most of the latter surfaces dyes or chemicals can be applied, but on the former these would be absorbed into the surface and obliterate the impression.

Impressions in dust need to be treated with special care because any contact with a brush or spray would destroy them. The recent development of methods involving electrostatic treatment has allowed the full recovery of even the most delicate impressions of this type on the most unlikely surfaces.

Impressions made in blood are likely to be the most important of all those found at a crime scene because they can be relevant to a particular offence. Often a person has legitimate access to the premises where the crime was committed and therefore any dust or dirt impressions are of little relevance as they may have been there for some time.

If, however, someone has bled during a crime, the suspect's presence at the relevant time can be proven if they have stepped in that blood and left a subsequent footwear impression. Heavy impressions left by an undersole that has made contact with a pool of blood contain little for the forensic scientist other than the possibility of determining the size and pattern of the shoe concerned. They contain none of the fine damage detail that is required to make a significant comparison. It is now possible to enhance the detail of even the faintest of impressions left in blood by using some of the new techniques available.

6.2.4   Methods for Enhancing Two-Dimensional Footwear Impressions

The search at the crime scene for footwear impressions is a systematic business. Apply only one technique and the chances are that just a few impressions made by one particular mechanism will be found. The more techniques used the greater the chance of recovering a large range of different types of impression that may be present at the scene. It is usual to start the search with techniques which will cause little or no damage to the impressions present and then progress to those which are likely to cause irreversible change to them. However, the techniques all have one thing in common, they provide a contrast between the impression and the surface on which it has been made.

The simplest and in some cases the most effective starting point is to shine a light obliquely across the surface of interest. Oblique means a light source positioned close to the surface giving a low angle of incident light. This will cause shadows to be produced by the deposits adhering to the surface and in turn this shadow provides the vital contrast. The method will show an array of impressions formed by a number of different mechanisms without any disruption occurring to them. Crime scene examiners now have a number of portable white and coloured light sources that they can use in this manner to quickly find footwear impressions. Portable equipment that throws a flat light pattern over a surface is available and highly effective.

It is possible to illuminate the surface with a number of different light sources other than normal white light. Some of these are certainly more practical for use in the laboratory than at the crime scene. Now, however, it is possible to take a portable low-powered laser to a scene and use it, with great care, to illuminate surfaces of interest. Light from a laser is intense, monochromatic, coherent and polarised. The intensity of the radiation and the fact that it is monochromatic give rise to fluorescence and phosphorescence of some materials either present in an impression or in the background on which the impression has been made. In addition to this, impressions can be stained with a suitable dye to make them fluoresce under the effects of laser light. The monochromatic properties of the radiation also enable considerable enhancement to be made of faint impressions. In effect a laser acts like an extremely powerful light source that can be tuned to give out light at a few specific wavelengths. The main drawback is the amount of energy it produces: place a piece of paper in front of the light beam and you could soon have a fire on your hands. A method is required to dissipate this energy, and this can be achieved by using a fibre or a liquid optical light guide. This has the added advantage that the direction and spread of the light can be controlled.

If a difference in fluorescence is created between the deposit and the surface then the impression becomes visible and a photographic record can be produced. Ultraviolet and infrared light sources can be used in a similar way and again the results can be photographed.

6.2.5 Dust Impressions

The most fragile impressions present at a scene are likely to be those in dust, therefore it is important that the scene is examined for these next. In 1981 research was instigated to determine if it was possible to recover impressions electrostatically. This proved successful and a portable device, an electrostatic lifter, was produced that could be used at the scene of a crime.

The modern devices use only an easily available 9 V battery, but they are capable of producing a variable high voltage. This voltage, which could be anything up to about 15 000 V, is then applied to a thin conductive film made up of a sandwich that includes an upper layer of thin aluminium foil and a lower layer of black insulating plastic. When the high voltage source is turned on, an electrostatic charge is produced in the aluminium layer of the lifting film. This charge causes the dust or other particles of the footwear impression to jump on to the black underside of the lifting film. When carefully removed and turned over, the film can hold a complete dust impression as it was on the originally examined surface. Enhancement is achieved because most of the dust impressions contain large amounts of dead skin that drops from our bodies, so the impressions are light in colour and on the black background of the lifting film a colour contrast is obtained. The technique has been used successfully on a variety of surfaces including linoleum, carpet and even car body panels.

6.2.6 Other Deposits

Most of the impressions that would remain after this treatment are resilient and require more vigorous methods to recover them. These methods of enhancement commonly include lifting using slightly sticky gelatin sheets. These sheets are able to pull more firmly held material from the surface on which the impression has been made. The sheets are normally black, and as the deposits lifted are lighter a contrast is now obtained between the two making the impression visible. Figure 6.1a shows a piece of paper with a very faint dirt impression present. Figure 6.1b shows the impression gelatin lifted from the top left-hand corner of the paper.

It is still possible to treat footwear impressions that remain after lighting and electrostatic lifting with fingerprinting powders. They will adhere to the fine material still stuck to the surface and may make just enough contrast to show them for photography. Powders can even be used before an attempt is made to gelatin lift an impression.

6.2.7 Impressions in Blood

One special material left at the crime scene is worth considering in a little more detail. Blood shed at a scene can fall on to two types of surface mentioned earlier, porous and non-porous. When an undersole comes into contact with blood it will pick up some of it and then deposit it at various places until the blood becomes dried or is all used up. It is best thought of as a rubber stamp being inked and then being used to produce a number of similar stampings and then occasionally being re-inked.

6.2.7.1 Blood on Non-porous Surfaces. Blood marks on a non-porous surface can be enhanced in two ways, either with a protein dye such as acid black 1 or with a haemoglobin stain such as diaminobenzene (DAB). The latter stain is difficult to use at the crime scene on the large areas that require treatment. Blood contains a large amount of protein, so enhancement is achieved by choosing a protein dye that will stain the blood impression a totally different colour in contrast to the surface on which it was deposited. Protein dyes are the preferred enhancer at a crime scene as they are easy to work with and cause fewer health risks than treatments like DAB. The impressions found can diffuse when they are treated in this manner but this can be avoided by including a blood fixative, 5-sulphosalicylic acid, with the dye when it is applied. Obviously substances other than blood such as milk, eggs and meat contain protein, so it is necessary to show that

RE: Cuddly Toy Raffle

On behalf of "The Lennox", I write with grateful thanks for your very kind donation of £44 which was raised by raffling/naming one of our soft toys.

It is wonderful fund raising efforts like this that help us continue in our work to provide grants for families with a child suffering from cancer. Parents often need to give up work to provide the care necessary for their child. These grants are an invaluable resource when finances become stretched at this time due to reduced income and additional expenses their child's illness incurs.

The raffles have been a great success and we hope that you will be in a position to help us again in the future.

Once again our heartfelt thanks goes to you for your efforts and of course to each and every individual that entered into the raffle.

Yours faithfully,

Martin Horne
Project Manager.

(a)

(b)

Figure 6.1 (a) A letter recovered from the floor at a crime scene with no discernible visible print found on surface of paper. (b) Following a gelatin lift, a dust impression of a shoemark was revealed near the top left-hand side of the letter.

the impression which has been enhanced contains other blood constituents such as haemoglobin.

The presence of haemoglobin can be demonstrated by treating the small areas of impression with DAB. Research has recently shown that some of the latest DNA profiling techniques, like low copy number DNA (LCN) can produce results from even the minutest amounts of DNA that are still present after fixation and staining have been carried out.

6.2.7.2 Blood on Porous Surfaces.   Recently much has been made of a number of staining agents. Some, such as luminol, need viewing in the dark but others, such as ninhydrin, can be viewed in normal light. Ninhydrin reacts with the amino acids present in blood and the mark produced does not leach out from the porous material in which it was made. It can be therefore used to good effect on sections of carpet or clothing. Stains such as luminol have become the method of choice for blood impressions on porous surfaces at crime scenes, even though they can be difficult to photograph properly. Large areas can be treated relatively safely and if necessary areas with significant detail can be recovered for further laboratory examination.

6.2.8   Three-Dimensional Impressions

The only recourse for the satisfactory recovery of impressions of this type is by casting. This was traditionally carried out using plaster of Paris, but the problem with this material is that it remains fairly soft even when set and therefore it was always difficult to clean the cast afterwards. Now, however, plasters are available that become extremely hard when set so that it is possible to clean them off under running water.

If a mark at the scene is impressed into soft soil it can easily be cast with one of these new plasters. The plaster is weighed out into a heavy-duty plastic bag and mixed with the correct amount of tap water. The corner of the bag is then cut off and the plaster squeezed gently out into the mark in the same way that you would ice a cake. Once set, the cast can be pulled up along with any adhering soil and returned to the laboratory for examination.

6.2.9   Conclusions

The methods given here are a a small selection of the many that are available to the forensic scientist in their search for footwear impressions both at the crime scene and on items submitted to the laboratory.

Once a footwear impression has been treated to obtain the maximum amount of fine detail, the forensic scientist has to consider the other significant exhibit, the shoes from the suspect!

6.3 INFORMATION AVAILABLE FROM A SHOE

What information can be obtained from a suspect's shoe that will be of significance when it comes to comparing it to a footwear impression recovered from the scene of a crime?

There are four main aspects of the undersole of a shoe that the forensic scientist will consider when making their final opinion:

1. The pattern.
2. The size.
3. The degree of wear present.
4. The random damage present.

6.3.1 Pattern

A pattern match between the undersole of the shoe and the impression at the scene is imperative because without it the comparison needs to go no further. Today, there are a myriad of different undersole patterns on shoes because they are not just something to keep your feet warm and dry but an important fashion accessory. Only a small proportion of shoes today have plain leather undersoles. By far the largest section of the footwear market is dominated by training shoes that have increasingly complex undersole patterns. Each of the major manufacturers tries to make the pattern of the undersole specific to them and each new upper style manufactured may have a new undersole to go with it.

Collections of the different undersole patterns encountered are kept as a database by laboratories. These have been established in an effort to answer three vital questions:

1. What make of shoe could have made the impression at the scene of a crime?
2. How common are these undersoles among arrested people?
3. How do their pattern components and wear vary with time?

The collection allows the forensic scientist to give vital early evidence to the police about the style of shoe a suspect may be wearing. It also provides assistance to the scientist in gauging the strength of their evidence by showing how common a particular undersole pattern is and how this pattern may vary with size. With the large number of impressions in a collection the scientist may even be able to determine small differences in the design of the items used to produce different undersoles within a particular size. This will be investigated later in this section.

The methods employed to manufacture an undersole can also impart some degree of individuality to the pattern present. There are two methods commonly used to make undersoles: one is to cut a number of units from a large piece of pre-moulded rubber, the other is to inject molten material into a pre-formed metal mould.

With the 'cut outs' variation can occur as each unit is individually cut from a large piece of rubber using an undersole-shaped cutter. The operator has some latitude in the positioning of the cutter on the sheet and therefore variations can occur in the pattern present, particularly around the edge of the undersole.

With the injection moulding process every undersole unit that is produced by a single mould should be identical. However, because of the large quantities of shoes that will be produced in a particular size, the manufacturer can often have a number of moulds to make any one given size of shoe. This allows the manufacturer to produce several undersoles at the same time in an automated process. There can be considerable variation in the fine pattern detail from one mould to another, even within the same size, as the moulds are often handmade.

6.3.2 Size

Size is felt to be an important feature of a shoe. However, to the scientist determining this, it is a two-edged sword. The investigating officer always wants to know 'What size shoe am I looking for?' And the scientist finds this a very difficult question to answer. Why? Well, it is not the size of the undersole that determines the size of the shoe printed on the label but the space inside the upper that accommodates the foot. There can be a large variation in undersole size among the shoes worn even by one person. Look at the variation in your own shoes. The other problem is knowing how the shoe undersole varies with size if only a partial impression of the complete undersole pattern has been recovered.

The length of an undersole is, however, one of the factors to be taken into account when judging if a significant connection is present. Some sizes are much more common than others; in adult males in the UK the average shoe size is 8 or 9 and only a small percentage of that population wear size 6 or size 12 shoes.

6.3.3 Degree of Wear

The pattern on an undersole changes as a shoe wears. These changes are only small, but if an unworn undersole is compared with one that has been worn for a few months a distinct difference will be noted in the pattern features. Some bars or blocks may have become wider and some detail may even have disappeared altogether. This is often seen with the fine surface stippling present on the undersoles of some new shoes such as Dr Martens boots.

People walk in different ways, placing their weight on different parts of the undersole. Some have deformities of the foot, or maybe a limp, and all these peculiarities will show up in the wear pattern on the undersoles of their shoes. More rapid wear usually occurs in the areas where the weight upon the undersole or scuffing has been greatest, and this causes a more rapid change in the pattern. Some of these wear features can be specific to a person's shoe undersole, but more often they can provide a further significant connection between a shoe and a crime scene. Figure 6.2 shows how a training shoe undersole pattern can change as it is worn from new until it is a few months old.

6.3.4 Damage Detail

Continual contact between a shoe and the ground causes the undersole to acquire cuts, scratches and other damage features that happen completely at random. As a consequence two shoes that start with identical undersole patterns will, over a period of time, start to gain damage features in different places and so obtain a degree of individuality. Each impression they leave can now be unique to that particular shoe. As time passes this damage record will continue to change, with some marks being removed and others added. By the examination of this unique set of fine damage features the forensic scientist can, even from some small fragment of an impression, tell if a particular shoe did or did not make a given footwear impression.

Using the knowledge gained it is now possible look at how a forensic scientist carries out their comparison between a shoe's undersole and the impression recovered from the scene of a crime.

6.4 COMPARING AN IMPRESSION WITH A SHOE

To compare the undersole of a shoe with an impression made at the scene of a crime the scientist has to produce a new impression from that shoe's undersole in the laboratory. Without these test impressions the

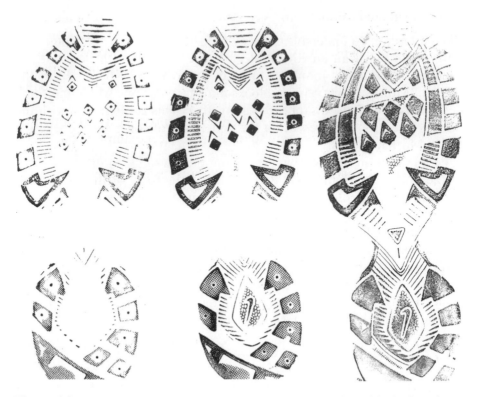

Figure 6.2   Increasing wear to an undersole with time, starting with the brand new
undersole on the left.

contact points of the shoe may not be clear and the same can be true of
any damage or wear features.

6.4.1   Making a Test Impression

There are several methods available for the scientist to make these
impressions. One of the simplest, quickest and most commonly used
ways is to apply a film of light oil uniformly to the undersole by pressing
it on a sheet of oil-impregnated foam rubber. The undersole is then
pressed on to plain white paper and the oily mark produced dusted with
a powder mix of fine black powder and magnetic iron filings, using a
magnetic brush. Because the brush is magnetic it does not have to come
into contact with the paper. The impression produced will show all the
wear and damage features present on the undersole and, if required,
other test impressions can easily be made to try to accurately reproduce
the manner in which the impression was made at the crime scene.

If a three-dimensional impression has to be compared with a suspect shoe then it may be necessary to reproduce the test impression in a similar way. Perhaps, for example, the impression at the scene was made in soil. This effect can easily be achieved using a tray filled with fine damp sand. The resulting impression of the shoe in the sand can then be cast and used to carry out the comparison.

As a general rule the scientist should try to reproduce as accurately as possible the mechanism used to make the 'scene impression' when making their own test impression. This can sometimes be difficult as there are numerous variables that may cause slight variations in the impressions that are left at the scene.

6.4.2 Comparing Impressions

Evidential strength is easily determined where damage marks in the undersole of a shoe appear in an impression left at a crime scene. The combination of even a small number of these damage features can be considered to provide a unique link between a shoe and a crime scene. The likelihood of two or three similar shaped marks occurring in an identical place on an undersole is extremely unlikely. Figure 6.3 shows a

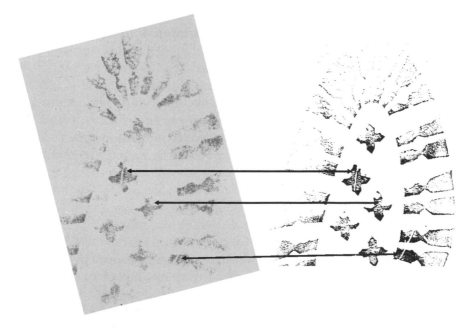

Figure 6.3 Damage features used for footwear comparison: scene mark on the left and test mark from the undersole on the right. The arrows indicate the corresponding damage features.

crime scene mark with a number of damage features, arrows indicate the corresponding damage features on the undersole of the shoe which made the crime scene mark.

However, life is not always that simple. As has been shown, shoes can wear and change with time to such an extent that these damage features can be removed or new ones added. Therefore only a limited percentage of comparisons give this level of evidential certainty. Sometimes a court will need to investigate the level of evidential significance in the scientific comparison obtained. To do this scientists often refer to Bayes' theorem, a method of describing the probability of something occurring. This is normally used to test two contrasting propositions, one that a shoe made a particular impression and the second that some other shoe made it. It is possible to consider how likely one or other of these is to occur by determining its likelihood ratio. The higher the likelihood ratio, the more likely the first proposition is to be true; the lower, the more likely the second is to be true. Take a simple example where we know from our pattern statistics that a particular shoe is very common, say 1 in 10 of the shoes seen. The likelihood of encountering these shoes is now known. If a size marking is visible, say an 8, this applies to about 1 in 6 of the shoes of the adult male population. Finally, by looking again at the pattern collection it may be possible to see that about 1 in 20 shoes with this type of pattern have worn to the same extent. Now, because the three things that have been measured are independent of each other it is possible to multiply the numbers together and come up with a likelihood ratio of 1 in 1200. If, however, one of these numbers changes—say the pattern that has now been found at our crime scene is extremely uncommon, say 1 in 200—the likelihood ratio changes to 1 in 24 000. This is much greater than that on the first occasion and the proposition that the shoe made the crime scene impression is more likely to be true than with our very common shoe. The higher the likelihood ratio, the more damning the evidence.

6.4.3 Conclusion

These pages have only touched the surface of footwear evidence both at the crime scene and in the examination of a person's footwear. To make footwear work as an investigative tool five things have to be in place: effective quality scene impression recovery, the generation of good intelligence from these, the prompt identification of suspects and recovery of their footwear, evidential strength assessment, and finally the quality comparison of footwear and impressions. If one fails, the whole system fails. Footwear commoness and a Bayesian view of evidence form the basis for intelligence generation and evidential assessment, although this chapter does not allow space to cover this in detail.

6.5 INSTRUMENT MARKS

As with footwear, the presence of damage on any instrument used in the commission of a crime can be utilised to form a connection between that instrument and the mark it has left behind. The instruments normally encountered fall into two main categories: those that cut and those that act as levers. The first category includes bolt croppers, drills and knives, the latter, jemmies, screwdrivers, and chisels. One or other of these groups are often used to gain access to a crime scene, to carry out a burglary or a murder. The following sections deal with each group of instruments and the special requirements for their examination and comparison to marks left behind at the scene.

6.5.1 Cutting Instruments

The commonest cutting instruments now examined by the forensic scientist are bolt croppers. In the past drills were used to drill doors or window frames, but the introduction of window locks has largely negated their use. No matter what the cutting instrument encountered, the method of comparison is normally the same. A test cutting is produced with the item in question and any damage features it contains are compared with those on the mark recovered from the scene. The final examination is usually carried out with a comparison microscope.

6.5.1.1 Construction. It is important to know how the particular instrument has been constructed, as this knowledge can be invaluable when carrying out a comparison. It is not always necessary to know how the whole item is constructed but it is important to know how the final cutting edges are formed during manufacture.

A pair of bolt croppers has two long metal handles that are joined at a fulcrum, and these then connect to the cutting blades. This form of construction allows the blades to be adjusted to give the best action and for the blades to be replaced as they become damaged or worn (see Figure 6.4). Bolt croppers can vary in length. The smallest that can easily be hidden in an inside pocket are about 12 inches (30 cm) long, the largest up to 4 feet (1.2 m) long. The length of the handles determines the amount of force that can be generated at the blades, but the size of the item that can be severed is limited by the hardness of the blades.

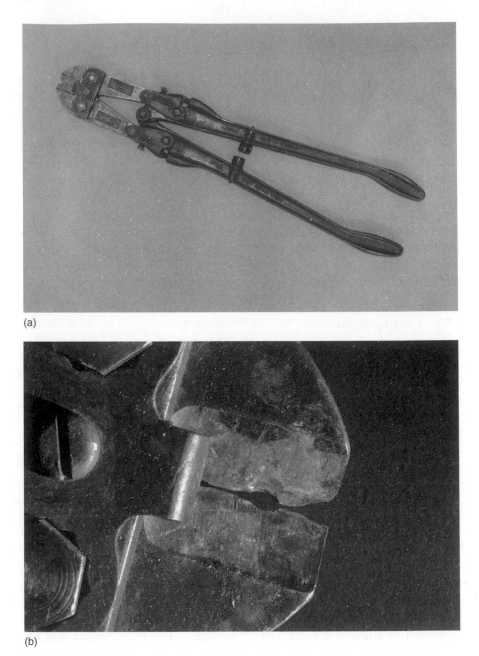

(a)

(b)

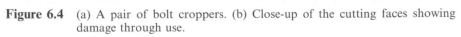

Figure 6.4   (a) A pair of bolt croppers. (b) Close-up of the cutting faces showing damage through use.

As criminals are not bothered about the size of the item they are cutting, the blades of the bolt croppers often suffer severe damage when they are used.

Even though most bolt cropper blades start with a forged and ground cutting edge, within a short period of time they acquire damage features that are unique to them. Even the grinding of the final cutting edge when the item is manufactured may impart some areas of individuality.

Drill auger bits also have this final finish and the same 'manufactured' areas of individuality may be present. Surface damage to the bit during use may also produce some areas of individuality.

6.5.1.2 Marks at the Scene. The items severed by cutting instruments at crime scenes are most likely to be shackles of padlocks, chain link fencing and telephone wires. Drill bits were normally used to cut holes in window frames to allow the release of internal catches or to drill holes around door locks to weaken the door so that it became easier to force.

In all these instances the damaged items should be seized if possible and submitted to the laboratory for examination. As the blades of the bolt croppers shear through the smooth metal of a shackle or the auger bit cuts through the painted wood of a window frame, the damage features on their edges will cause a series of parallel grooves which produce scratches on the item. Some of these scratches may only be a fraction of a millimetre in width and they are known collectively as 'striation marks'. This pattern will be produced again and again by the cutting item until it acquires further damage and the pattern changes. The damage features can be used to connect the severed edge of a padlock shackle, as seen in Figure 6.5. Here the right-hand side shows the damaged padlock, with the test mark produced by the suspect's bolt croppers on the left-hand side.

If the damaged item cannot be recovered to the laboratory to make a direct comparison with the instrument in question, then taking a cast of the damaged portion can produce a record of the mark. It must be stressed that casting should only be used as a last resort because it may not produce a clear record of all damage features present. The object should be cast rather than photographed as a photograph cannot be printed with enough accuracy to reproduce the relationship of all the fine striation detail normally contained within the damaged surface. Illumination of the mark is also an important factor when carrying out a comparison and it is often difficult to produce good lighting conditions

Figure 6.5   Images of views through a comparison microscope. The image on the left
of the vertical line is from a test cut mark made with the tool in the
laboratory which is being compared with the image on the right, a cast of
the mark taken from the crime scene.

even in the laboratory. Therefore, the production of a properly illuminated image at the crime scene would be almost impossible.

Casts are normally made with a type of silicon rubber similar to
that used by dentists to take an impression of teeth. The product consists
of a hardener and a base that combines in a similar way to the fillers
used to repair damaged motor vehicle body panels. For the casts to be
effective the casting material has to be sufficiently runny to flow into the
mark but robust enough to be handled when set. A point of note here is
that the marks, however they are produced, are a negative (inversion)
of the item that made them. Therefore when a comparison is carried
out the scientist must make test marks of their own so that they are
comparing like with like. It is not possible to compare a cast directly
with a mark.

6.5.1.3 Comparison. Probably the most difficult part of the comparison examination for the scientist is the production of control marks from the submitted instrument. Often they will not have seen the mark *in situ* at the scene and will therefore have to produce a number of test marks to give the full range of the detail that could be produced by the damaged cutting edge of the instrument. The production of test marks with bolt croppers and auger drill bits does not present as much of a problem as those to be made with a levering instrument. The difficulty will be discussed later in the section on levering marks.

It is also important when carrying out a comparison that the forensic scientist illuminates both the questioned and test marks in the same manner. Oblique lighting can be used to make the various damage features of an instrument mark cast shadows in a similar way to the oblique illumination of a surface for footwear impressions. The light striking both questioned and test mark must originate from the same direction, or the shadows produced will appear different. It would not be possible to obtain any match if the marks were not illuminated in the same manner.

Initial comparison of the test and scene marks is often carried out using a standard bench search microscope with the items obliquely illuminated. In the final examination of the test and scene marks a comparison microscope is normally used. A comparison microscope consists of two identical sets of microscope lenses linked together by an optical bridge. The questioned sample is placed on an adjustable stage below one set of optics and the test mark is placed on a similar arrangement below the other. With the same magnification in use on each side the fine striation detail present in the marks can be set side by side in the viewing field produced by the optical bridge.

Figure 6.5 shows two photographs taken using a comparison microscope. The split between the different sides of the microscope can be seen running from the top to the bottom of the picture. The object on the left is a test cut mark made in the laboratory. The object on the right is a cast of the mark taken from the scene of the crime. On the scene mark the striation detail is readily visible and it will be seen that some of this also occurs on the test mark. Given the number of fine striation details that agree, there is no doubt that the bolt croppers used to make the test mark also made the mark taken from the scene.

6.5.2 Levering Instruments

The levering instruments usually encountered by forensic scientists in their work are jemmies and screwdrivers. Whatever the type of levering instrument, the method of comparison used is normally the same as with a cutting instrument. A test mark is produced using the item in question and the detail present compared with the detail in the mark recovered from the scene. The initial comparison is carried out using a standard bench microscope and the final one using a comparison microscope. The use of a comparison microscope is not always necessary as photographs can often be produced that show the unique characteristics required to prove a match. An overlay or side-by-side montage could perform the job equally as well.

6.5.2.1 Construction. Once more it is important to know how the particular instrument has been constructed as this knowledge can be invaluable when carrying out a comparison. It is not always necessary to know how the whole item is constructed, but it is certainly important to know how the blade and edge of the lever are made.

Jemmies and screwdrivers are normally forged from a length of metal bar to produce a flat blade and the final finish to form the edge is usually produced by grinding; some close-up photographs are shown in Figure 6.6. Some modern screwdrivers are cast and then ground to

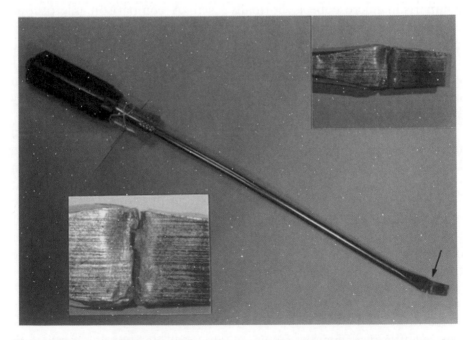

Figure 6.6   A screwdriver broken at the scene of the crime; inset enlargements show the grinding detail as well as the physical fit between the two broken items.

provide the finished edge. This difference is normally unimportant except where a mark has been caused by impression of an instrument into a soft surface, because this may show the face of the blade as well as its edge. If this is the case the detail shown on the blade may not be random but, as with a moulded shoe, may be the same for all screwdrivers produced from a particular mould. If a comparison has to be carried out on these features then this limitation in the strength of evidence should be borne in mind and marks examined for damage rather than mould features.

Screwdrivers and jemmies are made in a wide variety of shapes and sizes, but once again the criminal is not concerned with this. Often the item will be used for a purpose for which it was not designed and become damaged; pieces may be broken from the tip or it may be dented. When this damage occurs the item has obtained its individuality and the mark it produces, as with a cutting instrument, will be unique and will not change until it acquires further damage.

6.5.2.2 Marks at the Scene. Items commonly attacked with a levering instrument at a crime scene include window and door frames. These surfaces are relatively soft, so the levering instrument tends to impress its detail into the surface rather than scratch it. However in some cases, where the levering item has slipped in use, a combination of both scratch and impressed marks can be produced. If possible the items that have been levered should be recovered from the scene and submitted to the laboratory for examination. The following section will concentrate on the comparison of impressed detail as has already been touched on with scratch mark comparison earlier in the section on cutting instruments.

The mark produced with a levering instrument takes the form of the edge or, in some cases, the full blade of the object used to carry out the levering, therefore any damage present on this part of the item can be impressed into the surface. The unusual shape of a damaged tip or blade can be also recovered by casting, as with cutting instruments. With an impressed mark the cast produced is likely to show good detail for comparison, whereas this is not always true of casts of cutting marks.

6.5.2.3 Comparison. The comparison can be carried out in a similar way to that employed for cutting instruments. Test marks are produced in a way that tries to mimic the mechanism by which the mark was left at the scene. The tip of a blade can be quite thick and any alteration to the angle in which it is presented to the surface can cause a significant variation in the detail produced, so a number of test marks will be required. If striation detail is also produced then test marks have to be produced with the blade at various angles to the surface. This is necessary as the combination of striation detail may only have been made with one small area of the tip.

Again the test and questioned mark can be placed on the stages of a comparison microscope and the details compared side by side. Some microscopes allow the two images produced to be superimposed one on the other and with impressed detail this is often the better method for carrying out the comparison.

Since the detail in the marks may be quite large, and in some cases the marks may be almost flat, it should be possible to produce good, accurate photographs of the damage detail present. A photographic transparent overlay can then be produced of one of the marks and this can be placed over the photograph of the other mark to accurately show the corresponding damage features. Sometimes it may be sufficient to put the two photographs side by side to be able to highlight the similarities in the test and questioned marks.

Casts of test marks for a jemmy taken from a suspect and from the window frame at the scene of a crime are shown in Figure 6.7. If the edge of both marks is followed then it will be seen that they both agree with each other, with large areas of damage being present. There is no doubt that the jemmy from the suspect made the impression found at the scene.

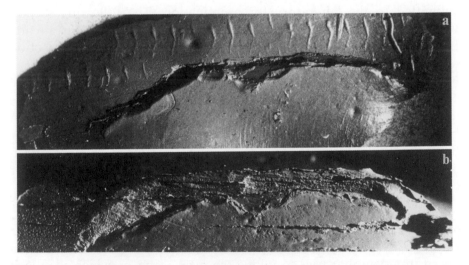

Figure 6.7   Marks made by a levering instrument: photographs of silicon rubber casts: (a) Test mark made in the laboratory. (b) Marks from a damaged window frame.

6.5.3 Conclusions

The value and strength of the evidence gained from the examination of instruments and instrument marks can be enormous. Given the presence of matching damage detail it can, like in the examination of footwear and footwear impressions, provide a unique connection between an instrument and the scene of the crime.

6.6 BRUISING

One special type of mark not already mentioned is the bruise. It is quite common to find bruising on the body of someone who has been assaulted. Sometimes these bruises are caused by the victim's clothing being forced into the skin by a blow, but often they are caused by the impact from the undersole of a shoe or an instrument used by their attacker. Car and lorry tyres can also leave behind their telltale traces.

An impact from by the undersole of a shoe or an instrument causes a temporary deformation of the skin to occur. However, a record of that brief contact may still be found in the form of a bruise that will exist for some time. This bruise can show characteristics from the pattern of a shoe's undersole if the person has been stamped on. A bruise may also show the shape of the instrument used to strike the victim. These marks can be photographed with a scale alongside them and used for comparison with the shoe or instrument thought to be responsible. Colour photographs are the best choice because it may not be possible to distinguish a bruise from blood smears in black and white photographs.

The degree of detail present in the bruise will depend on the force with which the blow was struck and the shape or pattern of the object used. A light impact causes some of the blood vessels at the point of impact to be broken and show a reddening of the area. Usually the mark on the skin appears as the negative of the object that struck it. This happens because the blood released from the damaged blood vessels is forced away from the contact point into the non-contact areas. To illustrate this, Figure 6.8a shows a bruising pattern on the forehead of a deceased person. Figure 6.8b shows an overlay of the impression over the bruise, the red of the bruise lining up with the non-contact area of the shoe impression.

A similar comparison could be carried out when an object is used to strike the body and produce a bruise. The visibility and information present in the mark can improve with time and a second visit to the mortuary to look at the bruising on a body, a few days after the initial post-mortem, can prove very revealing.

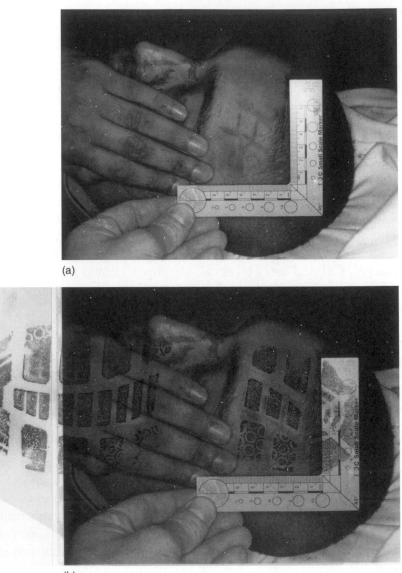

(a)

(b)

Figure 6.8   In this assault up to four persons punched and kicked the victim. (a) This mark was recognised as having been caused by stamping and photographed. (b) Control footwear mark from a boot of one of the suspects overlaid on the victim's head.

6.7   PHYSICAL EVIDENCE

Another aspect of forensic science that can provide significant evidence is a physical fit. This occurs when it is possible to fit two pieces of a broken item back together like a jigsaw. For example, if a screwdriver breaks during a burglary and one piece is left at the scene and the other part, usually the handle, is taken away, it can be recovered from the accused (see Figure 6.6). The break that occurs has an uneven surface and by showing that the detail of each piece can be matched together there will be no doubt that they were originally one complete item. Any attempt to fit items directly together should not be carried out until all other records and comparisons have been completed, as this may damage the delicate surface at the break.

6.7.1 An Impressed Fit

Sometimes it is not the fact that the item has been broken that is important but that the two items have pressed against one another for a period of time. This may produce dirt or discolouring patterns that can be used for comparison purposes.

For example, a pickaxe handle was used to strike a man during an aggravated burglary, and was left by the offender as he fled from the scene. A short while later the police, acting on information they had received, searched the house and garage of a suspect. In the garage they found the head of a pickaxe and the question was asked, 'Did this at one time fit on to the handle left at the scene?'. Examination of the inside of the hole into which the handle fitted revealed forging and hammering marks caused as the item was hand-produced. The inside of the head was cast and seen to have detail present which was a negative image of that on the pickaxe handle. There was no doubt that the two items had at one time been fitted together.

6.7.2 Mass-produced Items

The products from almost all mechanical processes, even those in which items are mass-produced, can be examined to show a degree of evidential similarity between them. Cases as diverse as those involving the comparison of stolen metal waste with the stamping machine that produced it, to the comparison of stolen silver ingots and the corresponding mould used to cast them, have been undertaken in the laboratory. Even small items such as a pierced ear sleeper manufactured by a jeweller can provide telling evidence. One such item was found in the boot of a motor vehicle of a suspect smash-and-grab raider. The jeweller had made a shaped die to his own individual design and the shape and pattern of a sleeper it produced was compared with the one found in the vehicle. The fine detail agreed and therefore gave a conclusive link between the suspect's vehicle and the scene of the crime.

6.7.3 Plastic Bags and Film

The presence of manufacturing extrusion detail on plastic bags and metal pipe can also be used to link items together. With the increase in drug trafficking the need to be able to connect items that have been used to package drugs has become increasingly important. The packages are often made up from lengths of cling film or plastic bags and sealed with adhesive tape.

Recently, methods have been developed to compare some of the physical properties of cling film and plastic bags such as the die lines present on them when they are made. Cling film is produced by a method known as blown film extrusion. The process involves molten plastic, usually polyethylene or polyvinylchloride (PVC), and this is extruded through a circular die in the form of a continuous tube. The tube is then slit to form a single sheet. In the manufacture of plastic bags the extruded tube is heat sealed and perforated at regular intervals. The cling film or bags are then wound on to a cardboard former and cut into individual rolls.

During the course of a production run, solidified fragments of plastic build up around the edges of the die and cause faint lines or striations to be present in the finished product. These striations vary in position, number and intensity during a production run and will differ between production runs.

Comparison of the striation patterns may then be used to assess the likelihood that products have arisen from the same production batch when evidential links between items are being sought. These striation lines can be best seen by producing shadow graphs of them to give a photographic record of the results.

Heat seals that form the ends of plastic bags can be treated in a very similar way to other impressed detail. Any damage or faults in the seals of the seized packages can be compared with the bags taken from the suspected source in order to prove a connection.

6.7.4 Conclusions

The examples here are just a sample of the many different types of unusual comparisons that can be undertaken. It is worth emphasising again that any items that come into contact with one another can leave a mark behind. This almost insignificant occurrence can be used to form a unique connection between the items and this may be all that is required to solve the most serious of cases.

6.8 ERASED NUMBERS

The identification of stolen property and its recovery is an important aspect of the police investigation of crime. When personal identification by the owner is not possible it may be necessary to resort to the restoration of an object's obliterated serial number to prove its identity. This happens most frequently with stolen cars, when engine numbers are erased by the thieves and substituted with new ones to give the vehicle a new identity. On most items the manufacturer's serial number is stamped into the metal, but now serial numbers can also be stamped into the plastic of such items as video recorders and mobile phones.

The aims of the investigating scientist and current areas of interest will be looked at as well as how punches and stamps can be linked to the restored numbers.

6.8.1 Erasure

When marks punched into metal have been erased they can usually be restored by etching the surface with acid. The strength of acid used depends on the chemical reactivity of the metal involved.

All metals are crystalline, and when an indentation is punched into the metal's surface a large localised stress is produced that alters the crystal structure immediately around and below the mark. There is a plastically deformed area around the mark and an elastically deformed area surrounding that. Therefore, even if the surface has been removed by filing or grinding in order to obliterate the mark, it is likely that the plastically deformed area will still exist. With the application of a suitable acid etching reagent, this area, because it is more electrochemically active, will be more readily attacked than the surrounding one.

With cast-iron engine blocks from motor vehicles, heat can be used to restore the erased numbers. Again the heat acts preferentially on the area where these numbers had been stamped. An oxyacetylene torch that is moved over the surface normally provides the heat. After the heating is completed and the surface has cooled, it can be rubbed down to reveal the erased numbers.

Other surfaces can also be treated in an effort to restore stamped serial numbers. The identification numbers in polymers are usually produced with a heat stamp that causes the polymer to shrink when it is applied. Experiments have revealed that the heated regions possess a higher swelling capacity under the action of solvents than the unheated areas. As more and more electrical items, such as mobile phones, are being stolen this method of number restoration will continue to be of great significance.

A similar method has allowed the restoration of numbers on wooden items where steam is used as the swelling agent.

6.8.2   Connecting Punches to Marks

Methods have already been discussed to show how to connect two items together by using their damage features. During use a punch will often become damaged, or the chromium plating may deteriorate and flake off, producing areas that are unique. These unique damage features will be stamped into the object and may be used to connect the punch and object together. Comparison can be carried out in a similar way to that used to compare cutting and levering marks left by bolt cutters and jemmies.

6.9   FINGERPRINTS

The unique nature of fingerprints was noted around the turn of the 20th century and first used in a trial in the UK in 1902. However, long before that the Chinese were aware of the qualities of fingerprints. Sir Edward Henry is credited with producing the modern fingerprint classification system and setting up the first fingerprint bureau at Scotland Yard. Since the first fingerprints were used as a unique method of identification several million have been taken and catalogued, and in that time no two prints taken from different fingers, thumbs and palms have been found to be the same.

The ridges on the hands, fingers and thumb and the patterns that they form are unique to each individual; even identical twins have different fingerprints. The patterns are present at birth and are formed in the early stages of pregnancy. They remain unchanged for life unless some deep damage occurs to the skin such as a burn or severe scaring. They are one of the last features to be lost from the skin as it decomposes and therefore can be used to identify dead bodies months or even years after death.

Fingerprints can be left at a crime scene when surfaces are touched or objects handled. They are produced from a mixture of natural secretions that come from the various glands in the skin and these are then set down by the ridge detail on the surface of the skin. Other areas of the body, such as the soles of the feet and the palms of the hands, also have ridge detail and therefore they can also leave behind a characteristic pattern. The next time you kiss yourself in the bathroom mirror, just look to see the lip ridge detail present! These patterns are also believed to be unique to an individual but occur much less often at a scene or on a recovered item.

Fingerprints will also be caused when the finger has been contaminated with another material such as blood and others can be produced when the finger has been pushed into a soft material like putty.

Methods for enhancing latent fingerprints will be discussed in more detail later.

6.9.1 Why Are They Unique?

The individuality of fingerprints is still the subject of some controversy, especially in some countries that do not require a minimum number of 'points' to show proof of a connection. A 'point' is a classifiable characteristic and will be discussed later. Some countries require only a few of these 'points', while the British system is based on the '16 points of comparison' rule which was adopted in 1953. It is said that if such a number is found then there is no likelihood of another finger having made that print. Recently this '16 point rule' has been relaxed so that an experienced fingerprint officer can, like a footwear examiner, give an opinion as to the likelihood of a finger having left a particular fingerprint at a crime scene without 16 points being present.

Fingerprints can be split into three groups by their general pattern. These patterns are arches, loops and whorls. The words quite accurately explain the shapes you would be looking at. These large groups can be further subdivided by smaller differences in this basic pattern. For arches, these are plain or tented, and for whorls, small elliptical, twin, composites, lateral pockets or accidental. Again the terminology explains the different patterns fairly clearly.

In all patterns, other than arches, there are points where the ridges break or bifurcate known as deltas: loops have only one delta each and whorls have two or even more. Deltas give rise to further characteristics based on the number of ridges between a delta and the core of the pattern. By taking these properties of the fingerprint into account they can be classified, catalogued and held on a fingerprint card or database (see Figure 6.9).

To be able to take a comparison further, a series of classifiable characteristics of 'points' are required. These are described as identifiable peculiarities that occur in certain positions in the print. There are four basic characteristic bifurcations: where a ridge divides, a ridge

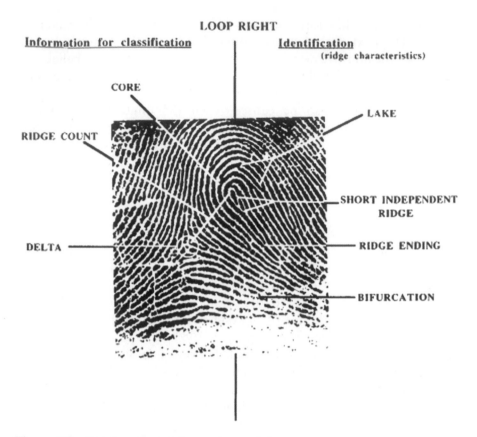

Figure 6.9   Details extracted from a fingerprint.

ending, a lake and an independent ridge. These are the points referred to earlier which, when taken in combination with each other, can provide the unique fingerprint every police officer wants to have recovered from their crime scene.

6.9.2   Current Developments

A new fingerprint system, 'Ident 1' has now been taken up by most of the police forces in England and Wales. The computer system allows details of fingerprints to be checked quickly and simply by and between the various fingerprint bureaus in the country. After the fingerprint officer has marked a number of significant points the computer uses a special algorithm to determine the closest matches between a suspect's fingerprint, held in the database, and a scene print. These are given in the form of a list and the fingerprint expert then has to decide which of these possible matches is correct. Until systems like this and AFR were developed, all of a bureau's fingerprint records were kept in a massive series of card indices. Some 7 million sets of fingerprints are held in the National Fingerprint Collection. They may have been divided into smaller categories by age of offender or offence type, but they had to be manually examined for each fingerprint comparison.

A fingerprint expert is likely to carry out a comparison using a photograph of the crime scene mark and a photograph of the suspect's fingerprints viewed side by side on a piece of equipment that is a variation of the comparison microscope. A report of the result will be issued and if necessary the expert will be called to give their evidence in court.

6.9.3 Enhancement of Fingerprints

Like footwear impressions, latent fingerprints can also be enhanced to make them visible. In the early years they were often directly photographed or powdered with a variety of materials to make them visible and then photographed and presented as evidence. This meant that only a limited number of prints were available for comparison purposes. Now it is possible to use a wide range of chemical treatments or different types of light source in order to visualise a greater number of prints. With fingerprints, unlike footwear impressions, it is possible to adopt a sequential approach to the enhancement of any print present. When a fingerprint is laid down it normally contains a number of different compounds that are excreted through pores in the skin; therefore, a search for one set of compounds such as lipids will not necessarily be to the detriment of other materials present such as amino acids. The proportion of the chemicals produced may vary from one person to the next.

6.9.3.1 Enhancement Techniques. Latent fingerprints left at a crime scene or on an item that has been handled, such as a cheque, contain a mixture of natural secretions that come from the various glands in the skin. Prints may also have been caused by contamination of the hand by another material or because the hand has been pressed into a soft material.

There are three main types of sweat gland in the human body; these are the eccrine, sebaceous and apocrine glands. Each of these excretes different compounds along with a large amount of water. Most natural fingerprints contain a mixture of substances from the eccrine and sebaceous glands. Some of the prints made from these materials will persist on the touched item for a considerable time; others may be transient or even decompose. Water is the first component to be lost from a fingerprint, so there is little to be gained in trying to look for this in fingerprints that are a few days old, but fatty material is likely to be found for several days or possibly longer.

A number of circumstances can alter the quality of a print left behind during a crime. These can include the lapse of time and the nature of the surface on which the print was made, along with factors such as storage of the item, its exposure to light and humidity after it was recovered.

Some of the techniques employed by the scientist are specific for individual compounds while others detect any oily or fatty products. Since the chemical and physical nature of the print is not known when it is examined, a systematic approach must be adopted for the full recovery of any prints present. A number of different complementary routes have now been determined that should facilitate the maximum recovery of the latent prints.

The following are just a few examples of the techniques that are currently available. Apart from the application of aluminium powder, which is frequently seen in use by the TV detective, the other commonly used technique is to treat the object with ninhydrin.

Ninhydrin is a compound that will react primarily with amino acids. It produces a purple coloured print, the development of which may be speeded up by the use of heat and humidity. This heat treatment can also cause further fingerprints that are present to become visible. Some prints, however, can require a longer period of time to appear. The fingerprints found are then individually photographed and recorded for comparison. The ninhydrin technique is very effective on paper and other porous surfaces but it is not successful on wetted items or silk-finish painted surfaces. Like most enhancements, ninhydrin can interfere with any further forensic examinations that are intended including those for indented impressions, body fluids, hair and other particulate material. The person carrying out the treatments must be aware of the implications for the detection of other scientific evidence that may be present and decide on the order of examination of the particular object in each case.

Fluorescence techniques, such as exposure to laser or ultraviolet light, can also prove very beneficial. The resulting print can be enhanced further by the application of a dye such as gentian or crystal violet. The dye will stain the fatty constituents of sebaceous sweat to produce an intense purple coloured print that can then be illuminated with laser light which will cause an increase in contrast.

Gentian violet is a very useful dye even when used without special illumination. It can be effective on adhesive and pressure-sensitive tapes as well as crockery and light-coloured metals such as aluminium, particularly if there is any oily or greasy contamination of the metal surface.

The vapour given off by superglue, either ethyl or methyl cyanoacrylate, will produce a white deposit with some latent prints. The process is humidity sensitive and items to be treated in this way are often humidified as part of the treatment. Fluorescent dyes can be applied to an object after exposure to superglue has taken place to make visualisation easier as any print present is now effectively fixed in place.

Vacuum metal deposition can be used on smooth, non-porous surfaces like polythene, leather, photographic negatives and prints. Thin films of gold and zinc are deposited on a fingerprint when only monolayers of fats are present. The use of this technique is limited as it requires extremely expensive equipment and there is also a limit to the size of the items that can be placed in the vacuum chamber. The treatment is effective because of the disturbance in the physical and chemical nature of the surface by the fingerprint. This is shown by different growth rates of the gold film on the surface of the object. The contrast is improved by exposing this gold film to vaporising zinc in the same chamber.

One surface from which most investigating police officers would love to be able to recover fingerprints is human skin. A number of methods have been reported in the literature as having potential but none appears to have fully cracked the problem, so the search is still on to solve the issues encountered with this vital surface.

These sections on fingerprints and their enhancement can only touch on the value of prints to the forensic process. Fingerprints can now be recovered for examination from the most unlikely surfaces, increasing their value in linking a suspect to a crime. The development of these new techniques has led to a resurgence of work with fingerprints, and police forces have now formed enhancement laboratories increasing the number of fingerprints that can be recovered. This applies not only to items submitted for examination but also by attending the crime scene where fingerprints can be treated *in situ*.

6.10 CONCLUSIONS

The comparison of a mark or impression, whether from an instrument or from the undersole of a shoe, is probably one of the most difficult and challenging areas of forensic science in which to work. From the brief outline of the techniques and the examples given here, this may not seem to be the case. However, the results that are obtained from these examinations rely heavily on the knowledge and experience of the forensic scientist who carries out the work. In view of this, the forensic science provision, in England and Wales, is now covered by the Forensic Science Regulator's office. Standards for the delivery of this vital science are being drawn up at the present time but it is envisaged that any testing establishment will reach BS EN ISO/IEC 17025:2005 for testing laboratories and the ILAC G-19 guidelines for forensic science laboratories (based on ISO/IEC17025:1999). As a significant part of this, even now quality trials are undertaken by individual scientists to cover the range of marks and impressions they are likely to encounter. The scientist who produces the report used in court will have undergone suitable training and have several years of experience in these types of comparisons before they are allowed to report their findings to a court.

This chapter has given at least an insight into the type of work a marks and impression examiner can be asked to perform, both at the scene of a crime and also on items submitted to their laboratory. So, the next time you take the lid off that paint tin, just remember what you could be starting!

BIBLIOGRAPHY

R. O. Andahl, The examination of sawmarks, *J. Forens. Sci. Soc.*, 1978, **18**, 31–46.

W. J. Bodziac, *Footwear Impression Evidence*, 2nd edition, Elsevier Science, New York, 2000.

J. F. Gowger, *Friction Ridge Skin: Comparison and Identification of Fingerprints*, Elsevier Science, New York, 1983.

P. S. Hamer, B. Gibbons and D. A. Castle, Physical methods for examining and comparing transparent plastic bags and cling films, *J. Forens. Sci. Soc.*, 1994, **34**, 61–68.

S. H. James and J. J. Nordby, *Forensic Science—An Introduction to Scientific and Investigative Techniques*, CRC Press, Boca Raton, 2003.

H. C. Lee and R. E. Gaensslen, *Advances in Fingerprint Technology*, Elsevier Science, New York, 1991.

P. McDonald, *Tire Imprint Evidence*, Elsevier Science, New York, 1989.

P. D. Pugh and S. J. Butcher, A study of the marks made by bolt cutters, *J. Forens. Sci. Soc.*, 1975, **15**, 115–126.

L. W. Russell, C. J. Curry and D. A. Castle, A survey of case openers, *Forens. Sci. Int.*, 1984, **24**, 285–294.

CHAPTER 7

Bloodstain Pattern Analysis

ADRIAN EMES AND CHRISTOPHER PRICE

Formerly, Forensic Science Service, 109 Lambeth Road, London SE1 7LP

7.1 INTRODUCTION

Bloodstain pattern analysis (BPA) is the term most commonly used to describe the examination, identification and interpretation of patterns of bloodstaining in relation to the actions that caused them. For many years BPA has played an important role at scenes of violent crime, where it forms an integral component of the reconstruction of events. DNA profiling now has extremely high discriminating power and this has had the effect of focusing the attention of police, scientists and the courts more frequently on questions concerning how bloodstaining was caused. A consequence of this is that BPA is now applied as often in the laboratory to the distribution of blood on items such as clothing and weapons as it is at scenes of crime.

The identification of bloodstain patterns is, by its nature, subjective, but it is underpinned by sound scientific principles. Each and every bloodstain pattern is unique, but by applying these principles, together with the application of experience based on experimental work, the identification and interpretation of patterns can be made sufficiently objective to be used with confidence in legal situations.

Crime Scene to Court: The Essentials of Forensic Science, 3rd Edition
Edited by P. C. White
© Royal Society of Chemistry 2010
Published by the Royal Society of Chemistry, www.rsc.org

The following are examples of how BPA may assist the investigation at scenes where assaults are known to have occurred:

1. It may prove possible to establish the relative positions of assailants and victims and the possible sequence of events.
2. Disturbance to the scene subsequent to bloodshed is often apparent in the staining.
3. The presence of spattered clotted blood gives a good indication of an interrupted or prolonged assault having occurred.
4. Recognising patterns of staining greatly increases the chances of locating blood from different sources and it is this understanding that frequently determines the samples that are taken for DNA profiling.
5. It may be possible to assess the degree to which the assailant would have become bloodstained during the assault.

There are limitations, however. For example, it often proves impossible to determine the direction of a trail of dripped blood between two sites of attack from an examination of the resultant stains. Determining the number and sequence of blows to a victim is usually much better achieved by a medical examiner or pathologist than by use of BPA.

Looking at the overall usefulness of BPA to an investigation, at one end of the spectrum police may, for example, be confronted with just a bloodstained room and little or no information as to how the blood got there. Analysis of the bloodstain patterns will usually be limited to addressing such issues as whether the staining is the result of an accident or a crime and, if the volume of blood can be estimated, whether the person who shed the blood is likely to be dead or alive. On the other hand, in situations where a defendant or suspect offers a version of the events that caused the bloodshed, there is the opportunity to test defined hypotheses that relate directly to the core concerns of the prosecution and of the defence.

In 1983 in the USA a group made up of scientists, police investigators and lawyers, all with a common interest in BPA, set up an organisation that continues to grow to this day. The International Association of Bloodstain Pattern Analysts (IABPA) has been the main forum for BPA practitioners around the world, and it established a terminology that is widely used. This terminology helps to clarify communication within the subject, and it has necessarily been modified as research and experience have led to a greater understanding. In the USA in 2002 the FBI established a Scientific Working Group on Bloodstain Pattern Analysis

(SWGSTAIN). They have refined a BPA terminology, published in 2009, which is now being taught in the UK. Some of the essential terms are described in this chapter, but a complete list is not appropriate here.

This chapter gives a brief outline of the basics of BPA and illustrates how it may be used in the investigation of violent crime. For the reader who wishes to explore the subject in more detail, a number of textbooks have been published in recent years and an excellent bibliography of over 500 titles was compiled in 2009 by SWGSTAIN.

7.2 CLASSIFICATION OF BLOODSTAIN PATTERNS

Only by understanding the dynamics of the actions that cause bloodstain patterns can the scientist give reasoned answers to the questions asked by investigators and the courts. Over the years bloodstain pattern analysts have developed a number of systems for classifying the various patterns of staining encountered and the following is a broad classification that is widely used among those involved in teaching BPA:

1. Single drops.
2. Impact spatter.
3. Cast-off.
4. Arterial damage stains.
5. Large volume stains.
6. Physiologically altered bloodstains.
7. Contact stains (also referred to as transfer stains).
8. Composite stain patterns.

Groups 1–4 can be considered as the spatter groups, in that they involve the distribution of drops of blood. Groups 5–7 are non-spatter groups. Group 8, the composite stains, involve a combination or overlapping of stains from either or both spatter and non-spatter groups. We now consider these eight categories of bloodstain patterns in turn.

7.2.1 Single Drops

This category concerns primarily the formation and properties of blood drops falling vertically under the effect of gravity. This includes the dripping of blood from wounds and weapons, trails of blood, and the phenomenon of satellite spatter, also referred to as secondary spatter. Consideration is also given to the characteristics of stains formed by blood drops landing at an angle on to a surface. Many of the characteristics of falling drops of blood are applicable to understanding the behaviour of droplets of blood projected by forces other than gravity, such as by impact.

7.2.1.1 Drops of Blood.   Blood can drip from any surface where the force of gravity can overcome the surface tension retaining the blood at the dropping site. In free fall, a drop of blood is spherical because this is its form of least energy. A single drop of falling blood will not break up in flight unless it is acted upon by another force.

The volume of a drop is determined predominantly by the shape and size of the dropping site. If there is a large surface area, such as the base of a saucepan, the surface tension holding the blood tends to be greater than if the surface area is small, such as the tip of a knife. Consequently the drops formed will also tend to be larger. If the available surface is extremely small (*e.g.* a single hair), then it may be impossible for gravity alone to overcome the surface tension, so that no drops can form.

Single drops of blood formed under the influence of gravity alone may vary in volume, from maybe as small as 15 microlitres (μl) up to 100 μl or more, depending on the nature of the dropping site and the amount of blood available.

7.2.1.2 Stains from Single Drops.   When a drop of blood falls on to a smooth surface the resulting stain is circular, with no distortion at its edges. A rougher surface will result in an approximately circular stain with irregular edges. If the surface is sufficiently rough, the surface tension of the impacting drop will be overcome sufficiently to allow the production of spines radiating from the stain, or satellite stains in the form of discrete secondary stains (sometimes referred to as secondary spatter). Wet blood itself provides an example of a surface that promotes very pronounced satellite spatter when other blood drops impact on it, and the effect of the underlying substrate is often obscured.

The size and shape of the stain produced by a drop of blood falling under the influence of gravity perpendicular on to a target surface will depend upon three variables.

- **The volume of the individual drop.** The greater the volume the larger the stain for the same dropping height and the same target surface.
- **The dropping height.** Increasing height produces an increase in stain size until the terminal velocity is reached, which is approximately 7 m for a 50 μl drop.
- **The nature of the target surface.** Non-absorbent surfaces will lead to the formation of larger stains as the entire volume of the drop spreads over the surface, whereas absorbent surfaces will tend to

produce smaller stains as a proportion of the blood is absorbed within the substrate.

In general terms, therefore, the size and shape of a stain can give no indication of the dropping height unless the other variable parameters of the drop's volume and the effect of the target surface are known.

7.2.1.3 Trails of Blood. A trail of bloodstains on the ground is the result of movement of a source of dripping blood, such as may occur when a bleeding individual moves about. Of particular interest to police investigators are those trails leading away from the scene of crime, indicating that the perpetrator may have bled. The stains within a trail may be regularly spaced when from an unchecked blood flow, but are more often irregular or complex if dropped from an object which has variable movement such as a swinging arm or hand. Individual stains in trails may indicate the direction of movement by exhibiting irregularities at their leading edge, though in practice, the directions of many trails are impossible to interpret because of an absence of such detail.

7.2.1.4 Satellite Spatter. Satellite spatter consists of small droplets of relatively uniform size, usually 1–2 mm in diameter. The drops are projected at low velocity, and on non-absorbent surfaces their stains often retain a dense, globular appearance when dry. The distance travelled by the droplets depends on the volumes of blood involved, the dropping height and the nature of the target surface. Experiments have shown the limits of travel for satellite spatter caused by the extreme case of blood into blood to be approximately 1 m horizontally and 50 cm vertically, with the density of spatter decreasing away from the origin. Thus it is a common occurrence when an injured person drips blood on to the ground for satellite spatter to land on the footwear and lower clothing of a person nearby in the form of small blood stains (Figure 7.1).

If DNA tests show this to match the injured person, then the presence of the satellite spatter is strong evidence to support the view that the two people were close together (within 1 m) when bleeding occurred. On the other hand, the satellite spatter itself says nothing significant about the events which caused the bleeding.

7.2.1.5 Case Example. This case illustrates the importance of understanding the behaviour and significance of dripping blood. It

Figure 7.1   Blood that has dripped on to the ground, and satellite spatter staining the footwear of a nearby individual. The large stain on the ground is the result of several drops that have pooled together. (The Forensic Science Service® © Crown copyright 2009)

concerned a murdered young boy, found lying on the floor of a garage after being repeatedly stabbed. The suspect had a deep cut to the palm of one hand which he claimed was received when he disarmed the boy who was attacking him with a knife. He also stated that he had not stabbed the victim and that the boy was still standing when he left the garage. Many bloodstains on and around the body matched the victim, while some of the staining by the body and in a trail leading from the garage matched the suspect. Of particular interest were stains on the upper back of the boy's shirt that were clearly the result of blood drops falling vertically under the effect of gravity. These matched the suspect and showed clearly that he had stood over the victim dripping his own blood.

In addition to the strong evidential value of the stain pattern, this case illustrates two important aspects of scene work and the preservation of evidence. First, the importance of regarding the body as an integral part of the scene; and secondly, the importance of removing the clothing before moving the body. If the shirt had not been removed before transporting the body to the mortuary, the vital pattern of staining would almost certainly have been obscured by additional blood leaking from the wounds.

7.2.1.6 Stains from Drops at Angles Other than Perpendicular. The shape of a stain produced by a vertically falling drop of blood striking a target surface at an angle other than perpendicular will depend upon the factors outlined above, together with the following (Figure 7.2):

- Landing at an angle, an impacting blood drop produces an elliptical, elongated stain with the tapered end of the stain pointing in the direction of travel. The tapered leading edge of the stain may show considerable distortion, whereas the back edge is more smoothly elliptical.
- The angle of impact that the blood drop makes with the target surface will affect the shape of the stain. This angle is the internal angle between the plane of the target surface and the path of the impacting blood drop. The more acute the angle, the greater the elongation of the stain formed.
- Drops landing at angles of less than 40° often produce satellite stains by a process termed wave cast-off. These satellite stains tend to be very elongated and can be at some considerable distance from the parent stain, though still in line with it.

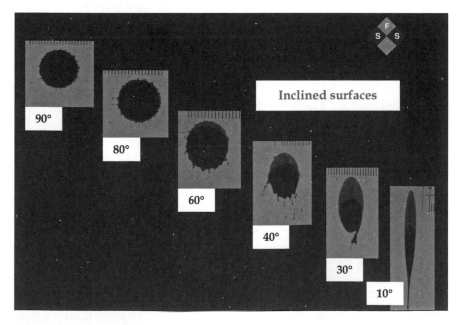

Figure 7.2 Stains made by single drops of blood landing at different angles to the surface. (The Forensic Science Service® © Crown copyright 2009)

Study of the shape and measurement of the width-to-length ratio of a stain enables the angle of impact to be calculated. These relationships apply equally well to spattered droplets of blood striking a surface at an angle, and are thus of great practical use in determining the area of origin of impact spatter (see next section).

7.2.2 Impact Spatter

This is the most common type of pattern encountered in casework and can be caused by a wide range of actions, such as kicking, stamping, beating, punching, and shooting. The cause of impact spatter is a force impacting directly into wet blood that breaks the liquid blood into small droplets of varying volume (Figure 7.3). The droplets are dispersed radially from the impact site along the paths of least resistance. They have various trajectories and velocities. Generally speaking, the greater the force applied the smaller the average size of the droplets formed. It is important to remember that the force referred to is that applied at the

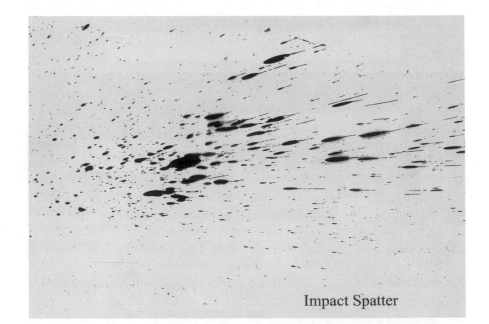

Impact Spatter

Figure 7.3   An example of impact spatter on a horizontal surface, such as this page when laid flat. The spatter has originated from an area just above the left edge of the photograph. The individual droplets have radiated from the origin, and formed stains of various sizes randomly positioned within the pattern. (The Forensic Science Service® © Crown copyright 2009)

site of impact and this does not necessarily imply the application of violence or extreme energy in the overall causative action.

7.2.2.1 Factors Influencing Impact Spatter. A number of variable factors will affect the distance, direction and quantity of blood spattered by an impact. These include the amount of wet blood at the impact site, the position of the impact site relative to the attacker, the shape and size of the weapon, and the speed and angle of the weapon at the moment of impact.

Assuming the droplets have the same initial velocity and trajectory, the smaller droplets will travel less distance than larger ones because of the effects of air resistance. There will be considerable variation in the sizes of the stains in the pattern formed on a target surface; most will be quite small, usually 3 mm or less, although occasionally stains as large as 10 mm may be found. As the distance from the impact site increases, the density of the stains within the pattern decreases. The stains from a single impact will show directionality from a common area of origin, although the number and distribution of the stains within the pattern will be random. Estimates can be made of this area of origin, and the methods used to do this are described later.

In the 1970s, Professor Herbert MacDonell, working in the USA, classified impact spatter on the basis of the velocity of the impacting object. However, experience shows that in some circumstances this can be restrictive and possibly misleading. It can be demonstrated that the effects of the varied geometry that occur at the area of contact between a weapon and the bloodstained surface often have a greater effect on the range of drop size than does the velocity of the impacting object. Consequently many workers have adopted the following classification proposed by Terry Laber which is based on the observed stain sizes within a pattern (not just impact) and does not link these to specific velocities or actions:

- **Large.** A bloodstain pattern consisting of individual stains that are predominantly 6 mm or larger. Blood dripping from objects typically show stains in this size range.
- **Medium.** A bloodstain pattern consisting of individual stains that are predominantly 2–6 mm in diameter. Cast-off bloodstaining is characteristic of this size range.
- **Fine.** A bloodstain pattern consisting of individual stains that are predominantly 2 mm or less in diameter. Impact spatter resulting from a medium-velocity impact such as a beating and spatter from a high-velocity impact such as a gunshot are both capable of producing stains of this size.

- **Mist.** A bloodstain pattern consisting of individual stains that are predominantly less than 0.1 mm in diameter. Because of the pronounced action of air resistance on these minute droplets, they will travel only a very short distance from the area of origin. Spatter arising from a high-velocity impact such as from gunshot is characteristic of this size range.

7.2.2.2   Information from Impact Spatter.   Detailed examination of the patterns can provide significant information to the investigation. Recognising impact spatter allows the identification of sites of attack and determination of the relative position of objects or people at the scene. The patterns may also offer information on the nature of the impact that caused the spatter. Most importantly, they can give an indication of the likelihood of bloodstaining being present on the assailant or others present at the scene, thereby helping to determine the significance of bloodstains found on a suspect's clothing.

Figure 7.4 shows the shirt of a man who bludgeoned his victim with a club hammer. This is an extreme example involving many blows with a heavy weapon and considerable spatter has been directed back towards the assailant. The nature of the weapon has a significant effect on the

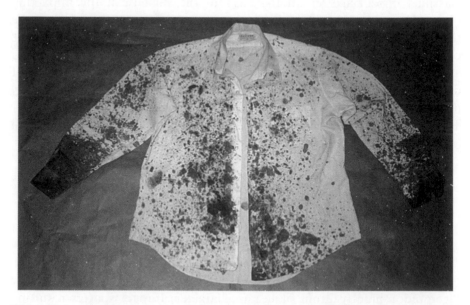

Figure 7.4   The heavily spattered shirt of an assailant who bludgeoned his victim with a club hammer. Note the dense spatter on the cuffs, the parts of the shirt that would have been nearest the origin of the spatter. (The Forensic Science Service® © Crown copyright 2009)

amount of blood that is projected back to the assailant. In contrast to the hammer, a beating with a large frying pan will probably result in little or no blood being projected back as the weapon only allows blood to be spattered to the sides. In situations where there have been a great many blows it is very likely that the assailant will be bloodstained to some extent, but with a single blow or just a few it may be that no blood at all is directed on to the assailant.

With kicking and stamping a wide range of patterns may be encountered, typically on footwear and trousers. Figure 7.5 shows a 'classic' kicking distribution on a shoe. There is evidence of forceful contact staining where the blood has been forced into the crevices around the seams and other recessed areas. This is associated with some degree of spattering. Stamping with the heel may well produce spatter that travels up inside the lower part of the trouser leg. This is a most significant finding, as is the presence of contact stains caused by hair.

With punching, the amount of blood on the assailant will depend on the nature of the injury and the amount of blood available to be spattered. Punching into wet blood will cause impact spatter that may be seen around the cuff, and is particularly significant if it is on the inside. It may also commonly be found on the upper sleeve, across the chest area and other parts of the front. Note that in the example shown in

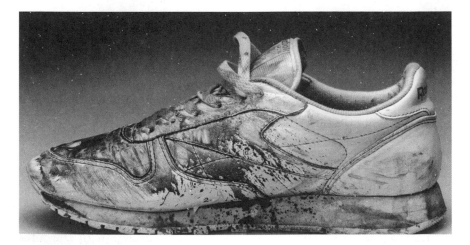

Figure 7.5   Impact spatter on a shoe caused by kicking the head of a person who was already bleeding. The large area of contact staining is associated with a small amount of impact spatter on the side of the shoe. Most of the spatter produced at the moment of impact would have been directed away from the shoe. (The Forensic Science Service® © Crown copyright 2009)

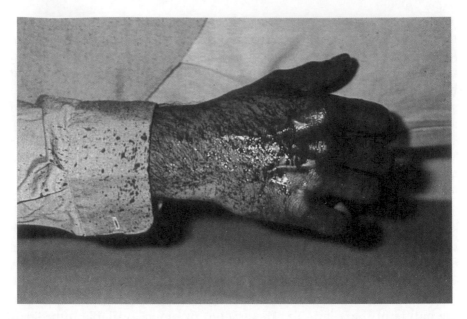

Figure 7.6   Bloodstained fist and shirt cuff caused by punching into a heavily
                 bloodstained surface. (The Forensic Science Service® © Crown copyright
                 2009)

Figure 7.6 the most significant evidence will be lost as soon as the hand is
washed, leaving only a partial pattern on the cuff. It should also be
remembered that a bleeding nose may produce a significant amount of
blood but soon afterwards there may be no visible injury.

With an assault by stabbing, a single stab to the body rarely causes
blood to be transferred to the assailant, as most of the bleeding will be
internal or absorbed by the victim's clothing. However, if there is subsequent contact with the victim or multiple stabbing occurs, there is
likely to be a transfer of blood, often from impact spatter produced from
the build up of blood on the surface of the skin.

When a person is shot from close range the bloodstain patterns that
may be produced are affected by a number of factors. These include the
site of the injury, the type and calibre of the weapon, and the distance
between the muzzle and the skin.

A contact or near-contact injury from a shotgun will probably include
extensive spattering of blood, tissue and bone fragments. Most of this
will travel away from the victim in the general direction of the initial
shot, away from the person firing the weapon. This is referred to as
forward spatter and is usually dramatic in appearance. However,
because it is directed away from the shooter, that person may receive
little or no blood on their clothing.

A contact or near-contact injury to the head from a firearm may also produce an aerosol of blood droplets that travel back towards the person firing the weapon, and this is called backspatter. The mist-like blood droplets comprising backspatter are unlikely to travel more than 1 m from the injury site, and will not easily be seen on clothing.

7.2.2.3 Determining the Origin of Impact Spatter. Locating the position from which an impact spatter originated may be an important element in reconstructing a scene of crime. In many instances an experienced BPA examiner can estimate this position by eye. In some cases a more accurate measurement of the origin is required, *e.g.* to locate the position of a victim at the moment he was shot. For many years the technique known as 'stringing' has been used, with varying success. The principles of this technique are soundly based on trigonometry, but the method has practical difficulties, and cannot easily be checked by others.

The following illustrates the principles of 'stringing' for an impact spatter onto a horizontal surface. A straight line is drawn through the long axes of a number of selected stains, and where these lines intersect is termed the area of convergence, as shown in Figure 7.7.

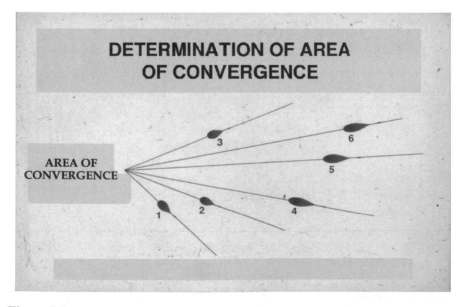

Figure 7.7 Determining the area of convergence of an impact spatter—top view showing the convergence of lines through the long axes of selected stains. (The Forensic Science Service® © Crown copyright 2009)

The actual impact site will lie somewhere on a perpendicular above the area of convergence. To determine this, the width and length of the selected stains are measured. This allows the angle at which they landed on the surface to be calculated—the ratio of the width to the length closely approximates the sine of the angle of impact. Using a protractor and lengths of string, these angles are projected from the stains to the perpendicular to define the area of origin as shown in Figure 7.8.

This method assumes that the trajectories of the blood droplets are straight lines. In reality all the droplets are affected by gravity and their trajectories are curved downwards. Therefore the strings in this method will project back to an area higher than the actual impact site. It is a cumbersome technique and is no longer in routine use in the UK. However, the principles of stringing form the basis for the method of determining the area of origin that is now recommended.

A Canadian physicist, Professor Fred Carter, became interested in BPA in the 1980s and in collaboration with the Royal Canadian Mounted Police, developed computer programmes called Backtrack™ and Images™ to determine the area of origin. These programmes have been modified and improved over the years into systems that are quick

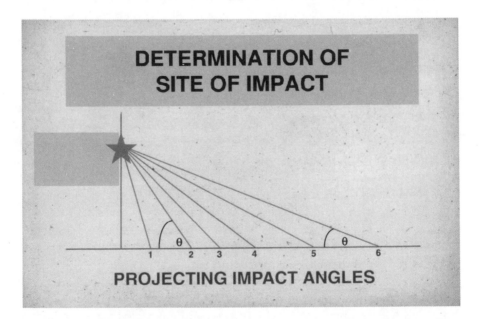

Figure 7.8   Determining the area of origin of an impact spatter—side view showing strings from the selected stains set at calculated angles to a perpendicular marker above the area of convergence. (The Forensic Science Service® © Crown copyright 2009)

and easy to use at scenes of crime. A number of selected stains are photographed using a digital camera, and the Images™ programme calculates the angle at which the stains landed on the surface. This data, together with data on the position of the stains within the pattern, is processed by Backtrack™ to produce a graphical representation of the area of origin.

The ease of use of these programmes means that the analysis of the data can be carried out quickly and, if necessary, remotely from the scene. Importantly, the results can be checked by other scientists. Although these methods use a more sophisticated understanding of the physics involved than in the old 'stringing' method, it must be remembered that at the present time it is impossible for any method to calculate the exact trajectory of a blood droplet in the vertical plane. However, the computer methods work at a level of accuracy sufficient for practical purposes and allow the confident estimation of areas from which blood could and could not have originated.

7.2.3 Cast-Off

The term cast-off is applied to blood thrown from the surface of a moving object, either by the action of centrifugal force (swing cast-off) or by the object being brought to an abrupt halt (cessation cast-off). The patterns created can be varied and complex.

7.2.3.1 Swing Cast-Off. This is most often associated with the swinging of a weapon, although wet blood on the hands is also a common source of cast-off. Centrifugal force will cause wet blood on the surface of the weapon to run towards its far end, allowing it to pool at one or more sites, dependent on the shape of the object under consideration. When the momentum applied to the mass of blood is sufficient to overcome its surface tension, the blood will be cast from the surface, rapidly forming a series of spherical droplets. When these land on a surface they will form a pattern which is generally linear in shape and comprised of stains which are similar in size—an 'in-line staining' pattern as shown in Figure 7.9.

The size of the droplets cast off from a swinging weapon will depend on the shape of the object, the nature of its surface, the velocity with which it is swung, and the amount of blood that is present. Observations from many experiments have identified certain aspects that are relatively constant.

Firstly, it is true to say that pronounced cast-off staining at any distance from the site of attack, *e.g.* on ceilings, is usually only seen with

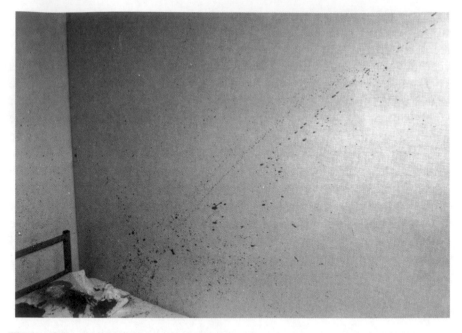

Figure 7.9   Several lines of swing cast-off can be seen on the wall, cast-off from the
swinging pickaxe handle used to attack the victim on the bed. Most of
the staining on the lower part of the wall near the bed is impact spatter.
(The Forensic Science Service® © Crown copyright 2009)

long, relatively light, weapons. Short, heavy weapons tend to be swung
more slowly and in shorter arcs, using the weight to cause the damage,
and are less likely to produce cast-off.

Secondly, as more force can usually be applied on the forward swing
of a weapon than on its back swing, it is generally easier to generate cast-
off when a weapon is swung in the forward direction. Also, as more force
is being applied, the average drop size from the forward swing tends to
be smaller than from the backswing. In practice one sees more cast-off
due to back swing than forward and this is simply because there is more
blood available. After blood is cast-off on the backswing little or none is
left to be cast off on the subsequent forward swing. Particularly large
forces are generated when a weapon rapidly decelerates and accelerates,
as at the end of its backswing, and this often leads to pronounced cast-
off, sometimes of particularly large drops that can travel quite considerable distances from the site of attack.

The physics of the cast-off process is complex, and the trajectory of
the drops leaving a swinging object is determined by competing vector
forces. Thus, depending on the velocity of the blow and the amount of

blood present, drops can leave the surface at any angle between radial and tangential to the arc of the swing. Generally speaking, all of the cast-off blood is projected away from the arc of the weapon's travel, so it is not surprising that in most instances no blood lands on the wielder, who is inside the arc. A person kneeling to attack may be an exception, as the lower leg is then often positioned outside of the arc and thus can become stained. It is also possible for cast-off blood to be found on the back of the attacker.

Considering that there are so many variables affecting cast-off, it is not surprising that much variation is seen in the resulting patterns. 'In-line staining' is the only truly distinctive pattern resulting from swing cast-off and, in practice, a great deal of cast-off blood at scenes and on clothing does not form patterns that allow the mechanism of origin to be recognised. Even when dealing with in-line staining patterns, the presence of so many variables should make for caution when interpreting the causative events.

7.2.3.2 Cessation Cast-Off. The term cessation cast-off is used for those commonly encountered situations where the swing of a weapon or other object is abruptly arrested, causing rapid deceleration and a large amount of cast-off. In this situation the arresting force does not act directly on the blood pool, but is transmitted to it, *e.g.* through the length of a weapon's handle. Although, by definition, this is a form of cast-off, the force acting on the blood acts in a way similar to a direct impacting force, and the resulting stain patterns often show more characteristics of impact spatter than of cast-off.

The rapid deceleration of the weapon will often produce a characteristic pattern in the blood remaining on its surface. During the swing phase blood will tend to form into runs towards the distal end as centrifugal force acts upon it. The rapid cessation of the swing then tends to produce 'feathering' along the leading edge of these runs, and fine spines are projected forwards. This characteristic pattern is termed percussive staining and an example is illustrated in Figure 7.10.

7.2.4 Arterial Damage Stains

This category is unique in that the causative factor is internal rather than external, *i.e.* the pressure within the circulatory system forcing blood to exit from a damaged artery. A wide range of patterns can result, depending on the extent and site of the injury, the direction of spurting, and whether the victim was stationary or moving.

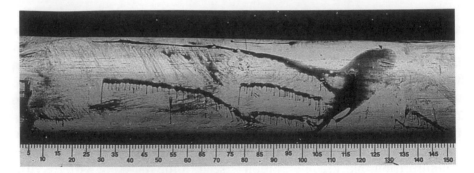

Figure 7.10   Percussive staining on part of a white pole that had been used as a weapon. A number of larger stains were deposited on the pole, one of which is shown. The pole has then been swung, causing the stains to run towards the distal end (from right to left). The downward feathering was produced when the pole hit its target and stopped suddenly and the blood continued to move in the direction of the swing. (The Forensic Science Service® © Crown copyright 2009)

7.2.4.1 General Features.   Most arteries in the body are generally afforded good protection against damage during normal day-to-day activities. However, in violent assaults and other incidents they may become damaged by sharp-edged instruments or, where the vessels are near the surface, by blunt instrument trauma. Arteries commonly damaged in assaults include the following:

- **Temporal artery.** Both sides of the forehead.
- **Carotid artery.** Both sides of the neck, and commonly used for checking the pulse.
- **Radial artery.** Near the surface of the inner aspect of the wrist, near the base of the thumb.
- **Aorta.** The main vessel leading from the heart, deep in the chest cavity.
- **Femoral artery.** In the thigh running from the groin to the knee.
- **Brachial artery.** On the inner aspect of the upper arm.

When an artery is damaged the blood is emitted in a column which soon breaks up into individual droplets of approximately equal size. Consequently, when the droplets land on a surface they form stains which are also equally sized and parallel to each other. The droplets are readily affected by gravity and when they land on vertical surfaces they give patterns showing a markedly downward direction. Individual droplets frequently coalesce on the surface, causing pronounced runs of blood. The blood leaves the artery in a series of spurts that correspond to the beating of the heart, and if the injured person is moving this can result in a V- or W-shaped pattern. An arterial gush results from a stream of blood hitting a surface, producing a large stain, maybe several centimetres in diameter. This may show satellite spattering, and is usually associated with a lack of movement by the victim. 'Arterial rain' is the term used to describe blood that has spurted into the air and fallen to the ground under gravity.

Other factors that affect the range of patterns formed include the extent of the damage—small wounds produce small droplets, larger wounds produce larger ones. Small wounds allow blood to be projected further than from larger wounds. The appearance of the staining will be affected by the site of the injury, and particularly by whether or not it is covered by clothing. Variation of the angle at which the arterial blood strikes the target surface will also significantly alter the appearance of the staining.

7.2.4.2 Simulated Arterial Bleeding. In order to carry out realistic experiments and demonstrations of arterial bleeding, an arterial pump that mimics the process was first constructed by Anita Wonder in Sacramento, California. This was further developed by us in London and has been used successfully in many training courses for forensic scientists. The operation of this pump is shown in Figure 7.11.

Blood is drawn from a reservoir and circulated through silicone tubing by a peristaltic pump. The rhythmic action of the heart is simulated using a solenoid-driven arm that squeezes and releases the tubing, giving two pressures approximating a pulsing heart. The blood is pumped through a manifold and directed along one of several lengths of tubing, each of which has a preconstructed hole or slit to mimic a wound of a particular size. As the blood spurts, the tubing can be directed at surfaces at varying distances and angles to demonstrate the characteristics and range of patterns produced by arterial bleeding. A typical pattern is shown in Figure 7.12.

7.2.4.3 Case Examples. Damage to a carotid artery will result in blood spurting from the wound, and although there will be dramatic and probably fatal blood loss, this is highly directional so that it does not always follow that the perpetrator will become heavily bloodstained. In a case in London some years ago, a victim urinating in a public toilet was attacked from behind and slashed across the throat with a craft knife, severing the carotid artery. There was extensive bloodstaining on the walls and floor at the scene and the victim died as

Figure 7.11   The arterial pump being demonstrated by the authors during a training
course. (The Forensic Science Service® © Crown copyright 2009)

a result of the injury. When arrested shortly afterwards the suspect was
found to have numerous small spots of blood on his shoes that were
thought to have resulted from satellite spatter from the blood spurting
on to the floor. The remainder of his clothing was virtually unstained.

Assaults by beating that cause massive head injuries may well result in
bleeding from a temporal artery. Attempted suicide by slashing the ulnar
or radial artery in the wrist is another cause of arterial bleeding. Some
incidents involving arterial bleeding are particularly well documented,
such as the death in London in 2000 of Damilola Taylor, who was
stabbed in the thigh just above the knee causing bleeding from the
femoral artery.

An unusual situation occurred on a busy main road in London when
two vehicles collided head on. The driver of one survived but the other
was dead at the wheel and appeared to have sustained a major wound
causing extensive blood loss. During the subsequent post-mortem
examination he was found to have a penetrating wound to the arm, but
there was no obvious cause of this wound from anything within the
vehicle. The dead man's vehicle was black, and blood was not obviously
visible; but a detailed examination revealed that bloodstains caused by
arterial spurting were present on the outside. This staining had arisen
before the accident, and it was concluded that the driver had been the

Figure 7.12 A pattern caused by simulated arterial bleeding. The blood was spurting perpendicular to the surface and the 'artery' was moving from left to right, resulting in five or six spurts indicated by the upper areas where the stains have run together. (The Forensic Science Service® © Crown copyright 2009)

victim of a stabbing before escaping in his car. While driving down the main road he became unconscious and lost control of his vehicle.

7.2.5 Large Volume Stains

Arterial damage can cause large volume stains, but also included here are those stains caused by prolonged bleeding or by the sudden release of a large quantity of blood, typically from the mouth or directly from a wound. Stains in this category can also arise after death when the accumulation of gases within a putrefying body eventually causes the emission of blood and other body fluids from wounds or orifices.

Sometimes, scenes with large volumes of blood are found with no victim present and little or no accompanying information. It may then be necessary to determine the volume of blood present in order to give an indication of whether the missing victim is alive or dead. Often there will be both dry and still wet blood present and several different surfaces of varying porosity will be involved. Consequently, in practice, calculation of the original blood volume by area and depth measurement, or by weighing samples dried to constant weight, will at best produce very rough estimated values.

7.2.6   Physiologically Altered Bloodstains (PABS)

This term is used to cover distinctive patterns that may arise when blood undergoes a physiological or biochemical change before or shortly after it is shed. Examples of this include clotting and admixture with other body fluids such as saliva or with vomit or excreta.

7.2.6.1   Clotting.   The process of clotting involves a complex series of chemical reactions that occur in shed blood. This process occurs more quickly if the injury is accompanied by damage to the surrounding tissues. The actual process begins about 15 seconds after the injury is caused, but at this stage there will be no visible change to the blood. At some time, which may be several minutes later, the blood thickens and the products of clotting are deposited around the injury. In extreme situations clot retraction will occur, and a clear separation of clot and serum becomes apparent. Clotted blood has a distinctive appearance when it is spattered, and thus indicates that some time has elapsed since the original injury was caused. This time interval usually cannot be determined with certainty in any particular case as so many unknown factors will have affected the timing of clot formation.

When an impacting force acts upon clotting blood, the clot and serum components may be apportioned differently in the resultant droplets, producing some irregularly shaped stains with clumps of clot and other pale coloured stains of a watery appearance. These latter stains are easily confused with stains of diluted blood.

The detection of stains arising from clotted blood can be extremely important when reconstructing events. Care must be taken to distinguish clotting phenomena from those resulting from the chromatography of blood seen on some porous surfaces and those resulting from the presence in the stains of pieces of body tissue.

Over the years, situations have been encountered where blood has become admixed with all manner of other body fluids and subsequently been spattered to form patterns. In practice, however, only a few of these are regularly encountered.

7.2.6.2   Expired Blood.   Blood will become mixed with saliva when there is a wound to the mouth or throat. The resulting stains have a diluted appearance (and the presence of saliva may be shown by other tests). There can also be limited frothing, leaving traces visible in the resulting stains. Penetrating chest injuries allow blood to become mixed with air that produces marked frothing. With injury to the lungs, surfactant from within the lungs may also become mixed with the blood and air mixture, and this can cause pronounced frothing and bubbling visible in the stains. There may also be mucus present, producing a distinctive bead-like appearance to the stains. The act of coughing, sneezing or snorting when blood is present in the mouth or the airways will often produce a discrete pattern that may well resemble impact spatter. An example of expired blood caused by coughing is illustrated in Figure 7.13.

7.2.6.3 Flyspeck. Another type of biological action that can produce stains of distinctive appearance is that caused by flies, which are

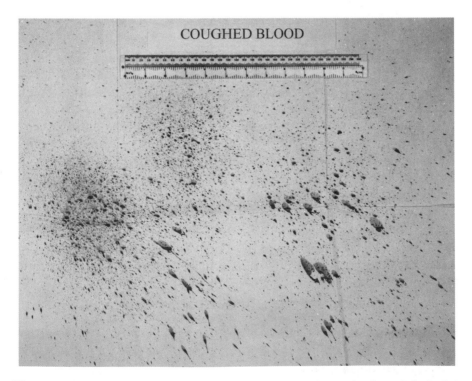

Figure 7.13 An example of coughed-up blood. Detailed examination of the individual stains shows the presence of bubble rings, some degree of dilution with saliva, and some irregular stains. (The Forensic Science Service®
© Crown copyright 2009)

frequently found associated with a corpse a few days after death. The fly feeds on the blood and tissues of the corpse, regurgitating some of this from its proboscis, leaving small round stains. These stains are called flyspeck and may have a small tail like a comma, or they may have a diluted appearance. Occasionally they may bear the impression of the end of the fly's proboscis. These stains may be mistaken for blood spatter, although they are frequently found in warm places that do not appear to fit with the rest of the scene, such as around a light fitting or on a window frame.

7.2.6.4   Vomited Blood and Products of Decay.   Injury to the abdomen or ingestion of blood may well lead to vomit being mixed with blood. The acid environment in the stomach causes the red blood cells to form small clumps so that blood in vomit is characterised by its brown granular appearance resembling coffee grounds. The biochemical products of decay may also lead to stains with a distinctive appearance. Even though they contain blood, on close inspection they can usually be distinguished from normal blood by their dirty brown colouration.

7.2.6.5   Nonphysiological Actions.   It is not uncommon for someone to mix or apply a substance to bloodstains with the intention of disguising the presence of blood. Cleaning fluids of all kinds have been used in attempts to remove bloodstains from surfaces, and occasionally, in desperation, a covering coat of paint may also be applied. Washing or wiping blood may disguise its presence, but in nearly all instances will leave traces detectable by the use of specialised light sources and chemical reagents.

When blood is spattered on to a wet surface the blood will diffuse and form stains with an indistinct perimeter. This is in contrast to where the bloodstain has been allowed to dry, even momentarily, and then partially removed by wiping. A distinct halo or ghost stain will result, consisting only of the blood at the periphery.

7.2.7   Contact Stains (Transfer Stains)

This category includes stains whose appearance is due to direct contact between an object wet with blood and another surface. During the examination of a bloody scene or of bloodstained clothing it is probable that any number of stains will be found that have no clearly defined shape, and nothing of significance can be said about how these stains

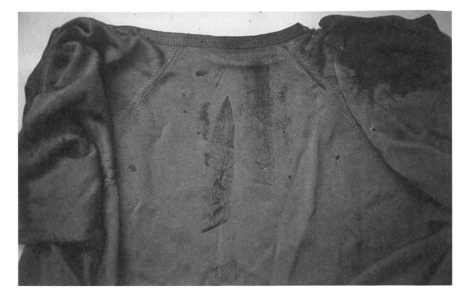

Figure 7.14 A contact stain made by the murder weapon being wiped on the back of the victim's jumper. A number of knives were recovered during the murder enquiry and test wipes were made, but none was found that matched. (The Forensic Science Service® © Crown copyright 2009)

were caused. Contact stains are of particular interest if there is information about the bloody object that left the pattern, as, for example, with a fingerprint, shoemark, fabric mark, or weapon impression. An example of such a stain is shown in Figure 7.14.

Finger and footwear marks are types of evidence that tend to be included as part of their own forensic disciplines rather than a BPA matter. Suffice it to say that a great deal of technological expertise is nowadays available to detect, enhance and retrieve such evidence (see Chapter 6). In serious crime where the effort is justified, this is carefully integrated within complex strategies that aim to maximise the marks, DNA, and other forms of evidence retrieved.

7.2.8 Composite Stain Patterns

This category includes both superimposed stain patterns on a target surface caused by different actions, together with patterns caused by a complex action. An example of the latter is a person with arterial damage to their wrist who waves their arm around—arterial bleeding and cast-off would then be combined in a single simultaneous action and the resultant stain patterns may have features of both.

Composite patterns can be difficult to interpret, even for the most experienced analysts. Anita Wonder has published the results of her systematic study of many hundreds of spatter patterns caused by impact, cast-off, and arterial damage, and has produced a list of objective criteria for the identification of the main pattern types. These criteria take account of the shape of the overall pattern and the size, alignment, and density of stains within it. This provides a useful key to refer to when faced with any pattern whose mechanism of formation is not readily apparent.

The objective criteria for identifying the main spatter groups are summarised in Table 7.1.

7.3 EVALUATION OF BLOODSTAIN PATTERN EVIDENCE

Police, forensic scientists, and the courts have always been interested in the evidence provided by bloodstains. The ready transfer of blood from one person to another or from a person to a scene has, over the years, been a major stimulus for the development of techniques to individualise bloodstains. During the 1950s and 1960s this was limited to antigenic blood groups such as the well-known ABO system. The late 1960s and the 1970s saw the development of many biochemical blood grouping methods, and in the late 1980s DNA came to the fore. Developments in DNA technology have since been rapid and spectacular, and details of these techniques can be found in Chapter 15. These methods have given us the ability to discriminate between individuals to the extent that if a DNA profile from a bloodstain is found to match a particular

Table 7.1 Objective criteria for the main spatter groups.

	Impact	*Cast-Off*	*Arterial*
Shape of pattern	Wedge, star, trapezoid, fan	Linear, columnar, curved	Broad, columnar or linear. S, V or W shape
Alignment of stains to whole pattern	Radiate from area of origin	Subtle changes in angle	Markedly parallel, affected by gravity
Alignment of stains to each other	All at angles to each other	In-line, parallel, or slightly changing angles	Markedly parallel
Density of spatters	Decreasing density from centre	Uniform or slight decrease	Uniform
Distribution of stain sizes	Decreasing range of different sizes from centre	Constant or slight changes	Uniform

individual, the chance of a match if the bloodstain had come from someone else (unrelated to that individual) is quoted as being of the order of one in a billion (a thousand million).

This strength of evidence for the transfer of blood from, say, a victim to a suspect has meant that increasingly suspects and defendants may feel obliged to explain how the blood came to be on their clothing. For example, where an individual is suspected of assault by kicking it is not uncommon to find the victim's blood on their shoes. Faced with this evidence the suspect may offer an explanation, such as admitting to being present, but providing first aid to the victim. How do the courts assess the validity of this explanation when the prosecution alleges that the blood got on to the defendant's shoes because it was he who did the kicking?

In this hypothetical case it is no longer necessary just to show that the blood on the defendant's shoes matched the victim. The issue for the court is not now about the 'source' of the blood but about which version of an alleged 'activity' was correct—was it the prosecution's allegation of kicking or the defendant's version of rendering first aid? In order to help the court in this situation, the forensic scientist needs to evaluate the appearance and distribution of this staining. This is an example of the forensic scientist offering more help to the court by operating at what has been called 'activity' level rather than at 'source' level.

Before carrying out this evaluation the relevant items need to be examined to locate any blood present, and stains selected for DNA analysis. It is extremely unwise for forensic scientists to discuss the significance of bloodstains without first establishing that they are indeed blood and from whom that blood may have originated.

In recent years the question of how forensic science evidence should be evaluated has been considered at length, and in the Forensic Science Service an Interpretation Group has been at the forefront of this work. The three fundamental principles of the evaluation process, which can be applied to all areas of forensic science, are as follows:

1. Evidence is evaluated in the light of other relevant information in the case, often referred to as the 'framework of circumstances'.
2. To evaluate evidence it is necessary to consider at least two propositions which are mutually exclusive—these are usually the prosecution and defence propositions.
3. It is necessary to consider questions such as, 'What is the probability of the evidence if the prosecution proposition is true, compared with the probability of the evidence if the defence proposition is true?' The ratio of these probabilities is known as the likelihood ratio.

To apply this approach to the evaluation of bloodstain patterns, it is necessary to have an expectation of what might be found in given circumstances. Returning to the case of alleged kicking, let us assume that blood matching the victim has been found on the defendant's shoes. The prosecution proposition is that the defendant kicked the victim, and the defence proposition is that the defendant did not kick the victim but offered first aid. What would we expect to find if the prosecution proposition is true, and similarly, if the defence proposition is true?

During many training courses we have carried out experiments whereby people wearing different types of footwear have simulated violent attacks by kicking a bloody surface. Various characteristic features of the bloodstaining are regularly seen, indicative of the forceful contact. These include the presence of contact staining in the crevices of the shoe and directional, spattered blood associated with the contact stains. Therefore, if the suspect has this type of staining on his shoes, the evidence will support the prosecution.

But what might be expected if, as the defendant says, he was just giving first aid? Other experiments show that such incidental contact will produce contact bloodstains, but usually not of the type indicating that any force was used, and with no associated spatter. So, contact staining in the crevices together with directional, spattered blood associated with contact stains would be much more likely to be seen in kicking-type activities rather than in first aid-type activities. Therefore, if the suspect has this type of staining on his shoes, the evidence would provide support for the prosecution allegation. On the other hand, the absence of this pattern of staining on the shoes would provide support for the defence proposition.

This is a simplified example of what is known as a Bayesian approach to evaluating evidence, and for BPA it is still in its early days. For this to become more effective, more information, collected in a systematic fashion, is required on the bloodstain patterns that are seen in different sets of circumstances. This information would provide more reliable estimates for the frequencies with which features are likely to be present in these different scenarios. There are practical difficulties in establishing such a database and the work done to date has, perhaps, only established the experimental protocols that will need to be followed to make the data collected fully usable.

The ability of forensic scientists to assess and interpret evidence has become central to the development of many areas of forensic science, not just that of DNA where frequency data is readily available. BPA is proving to be a discipline where these Bayesian principles can be applied effectively, and we expect this approach to be central to the development of BPA in future years.

BIBLIOGRAPHY

T. Bevel and R. M. Gardner, *Bloodstain Pattern Analysis With an Introduction to Crime Scene Reconstruction*, CRC Press, Boca Raton, 2002.

A. L. Carter, The directional analysis of bloodstain patterns theory and experimental validation, *Can. Soc. Forens. Sci. J.*, 2001, **34**(4), 173–189.

A. Emes, Expirated blood—a review, *Can. Soc. Forens. Sci. J.*, 2001, **34**(4), 197–203.

I. W. Evett, G. Jackson, J. A. Lambert and S. McCrossan, The impact of the principles of evidence interpretation on the structure and content of statements, *Science and Justice*, 2000, **40**, 233–239.

G. Jackson, S. Jones, G. Booth, C. Champod and I. W. Evett, The nature of forensic science opinion, *Science and Justice*, 2006, **46**, 33–44.

S. H. James (ed), *Scientific and Legal Applications of Bloodstain Pattern Interpretation*, CRC Press, Boca Raton, 1999.

T. L. Laber, Bloodspatter classification, *IABPA News*, 1985, **2**(4).

H. L. MacDonell, *Bloodstain Patterns*, Laboratory of Forensic Science, Corning, NY, 1993.

A. Y. Wonder, *Blood Dynamics*, Academic Press, San Diego, California, 2001.

A. Y. Wonder, *Bloodstain Pattern Evidence: Objective Approaches and Case Applications*, Academic Press, San Diego, California, 2007.

CHAPTER 8

Forensic Examination of Documents

AUDREY GILES

The Giles Document Laboratory, Sandpipers, Hervines Road, Amersham, Buckinghamshire, HP6 5HS

8.1 INTRODUCTION

The identification of forgeries is one of the oldest of the forensic sciences with references made to it in Roman law in the 3rd century AD, and forgery being a statutory offence in England in the 13th century. Notwithstanding this, the examination of documents in modern forensic science laboratories is an up-to-date science exploiting modern technology to cope with the ever increasingly sophisticated documentation of modern living.

The forensic scientist specialising in the examination of documents will have a number of areas of expertise. These areas include the identification of handwriting and signatures; a knowledge of modern office printers; the composition of inks, papers and the materials from which documents are produced; and techniques such as electrostatic detection of impressions (ESDA) and infrared imaging techniques which allow the origin and history of documents to be studied.

Forensic document examiners are frequently misnamed 'handwriting experts', even though this is only part of their expertise. In recent years this term has been hijacked by graphologists. However, graphology has

Crime Scene to Court: The Essentials of Forensic Science, 3rd Edition
Edited by P. C. White
© Royal Society of Chemistry 2010
Published by the Royal Society of Chemistry, www.rsc.org

no relevance in forensic document examination. Graphology is a pseudo-science purporting to determine personality from handwriting. Its methods of interpreting handwriting are very different from the objective approach taken by the forensic scientist and are demonstrably unreliable.

8.1.1 Qualifications and Training

There is no doubt that it is extremely difficult to become a fully trained forensic document examiner. The majority of forensic scientists specialising in document examination in the UK were traditionally trained and employed in the government laboratories. Although opportunities in such laboratories have been reduced in recent years, graduates entering this field still need to train for up to 2 years alongside highly experienced document examiners before handling their own casework. This is a broad and complex specialty and there is no substitute for on-the-job training on real casework. Self-training or learning through correspondence courses is inadequate.

There few formal qualifications upon which the legal profession or the public can rely when wishing to employ a forensic document examiner. The Forensic Science Society offers a diploma in the field, but problems in setting standards have not gained it recognition amongst the majority of practitioners. A number of laboratories in both the private and public sectors have looked to external accreditation with standards such as ASCLAB, ISO 9001 and ISO 17025 which have successfully provided benchmarks for assessing the quality of work in these laboratories. New regulations set in place by the Forensic Science Regulator will follow this form of quality assurance.

It is still very much up to the individual lawyer or member of the public to take the trouble to ascertain that the expert is competent and credible. This is a very difficult job for the non-scientist and it is not assisted by the various so-called 'Registers of Experts'. Inclusion in such registers generally requires payment of a fee and a reference from a single satisfied customer.

The properly trained and qualified forensic document examiner will, however, be able to demonstrate a solid scientific background at appropriate degree level, a period of training in an established forensic science laboratory, possession of a fully equipped laboratory, a record of continued active participation in research and development in the field and appropriate registration with an external quality assurance assessor.

8.1.2   Equipment

Most of the techniques employed in forensic document examination have been developed to extract as much information as possible from the document without damaging or altering it in any way. This is usually a prerequisite of any proposed examination since a disputed document is of considerable value—indeed, its very existence may itself be evidence.

The forensic document examination laboratory will have the following basic equipment:

1. Good lighting sources, including daylight which is essential.
2. Low-power stereo microscopy allowing magnification of ×5 to ×40.
3. Infrared, ultraviolet, high intensity and transmitted light sources for studying inks, alterations and latent marks.
4. Accurate measuring grids and graticules.
5. Electrostatic detection apparatus (ESDA) or equivalent for the study of impressions.
6. An oblique light source.
7. Methods of recording visual results by either photographic, thermal imaging or computer image capture techniques, as well as facilities for preparing demonstration materials for presentation in court.

More sophisticated laboratories will have available advanced techniques for ink analysis and electronic imaging, desktop publishing facilities for producing illustrated reports, as well as databases of background literature, handwriting, typewriting, transmitting terminal identifiers of facsimile machines and ink data. Forensic document examiners working within a general forensic science laboratory will also have access to yet further sophisticated equipment such as β-radiography, high performance liquid chromatography (HPLC) systems, Raman spectroscopy and lasers, *etc.*, which are occasionally used in the examination of documents.

8.2   EXAMINATIONS

Forensic document examiners are asked to carry out examinations to provide information in a number of areas:

1. The identification of individuals from their handwriting.
2. Identification of signatures as genuine or forgeries.

3. Determination of the origin and history of documents—where and how they were produced and what has happened to them in the course of their existence.
4. Dating of documents.
5. Identification and interpretation of alterations, deletions and additions to documents.
6. Identification of counterfeit documents.

It may well be that the forensic document examiner will be asked to study a particular aspect of a document such as a disputed signature. This will not deter the experienced examiner from checking that other aspects of the document under examination are consistent. A disputed signature on the final page of a multipage document may be identified as genuine. However, if there are unobtrusive differences between the paper and type style of a signature page and that of the previous pages, this may give a considerably different view of the origin of the signature.

The full and proper examination of documents therefore takes place in a properly equipped forensic document laboratory. There is still a tendency for lawyers to misunderstand the level of sophistication available today in such laboratories. There is a commonly held belief that the 'handwriting expert' can merely look at the document in a lawyer's office or in the public area outside court and, possibly with the aid of a magnifying glass, give a full and positive opinion regarding the authenticity of the document. Examinations carried out without the benefit of proper laboratory facilities will always be to some extent inconclusive and may be entirely misleading. There is no substitute for proper examination in a laboratory.

The second most popularly held erroneous belief is that adequate examinations can be made using copies. Copies do not show all the details of the original documents and the conclusions which can be drawn from the examination of such material will nearly always be restricted.

8.3 IDENTIFICATION OF HANDWRITING

With the exception of fingerprinting and DNA profiling, the majority of forensic science sets out to establish links between individuals and places or objects. The identification of handwriting is one of the few forensic sciences that actually identifies the individual.

It is an established fact that handwriting can be recognised. Most adults can recognise the handwriting of their immediate family and close friends. Every character of the alphabet, both block capital and cursive, and the numerals can be constructed in a number of different ways. Each person's handwriting therefore displays a particular combination of character forms which gives that handwriting much of its individuality. Support for the implied individuality of handwriting has come from studies in the USA where software developed by State University of New York (SUNY) Centre of Excellence for Document Analysis and Recognition (CEDAR) has been refined to extract and measure handwriting features. In a study of 1500 handwritings the software was able to distinguish between handwritings of different individuals with 95% accuracy.

The basic shapes and construction of handwritings are taught in school. In some countries handwriting systems are adhered to rigidly, but in the UK, children tend to learn their teacher's version of handwriting. It is difficult to determine the nationality of handwriting with certainty. However, certain character constructions are more likely to be found in some nationalities than others.

Individual handwriting features begin to be introduced during adolescence when the young person begins to experiment with the appearance of handwriting and mimics attractive features from sources other than the standard taught systems. The handwriting of any individual tends to attain maturity in early adulthood and remains consistent in shape, structure and proportion over the years until changes are introduced as a result of old age.

8.3.1 Construction of Character Forms

Handwritten characters can differ in their construction, shape and relative proportions; of these, construction is probably the most important feature used to distinguish between handwritings. Some of the block capital forms, such as 'E,' 'G', 'H' and 'K', can be written in several distinctly different forms as shown in Figure 8.1. It is unusual to find one person using more than one of these forms.

Different constructions of other characters are more subtle. For instance, the point of entry and exit of the character 'O' or the numeral '8' can be of crucial importance in distinguishing two different handwritings. The shape of the individual character form is also an important consideration. Characters may be generally angular or rounded in their formation.

Both the internal proportions of the individual character forms and their relative proportions will be taken into consideration. An individual may consistently introduce a large form of a particular character into handwriting. Similarly, only part of a character may be relatively large,

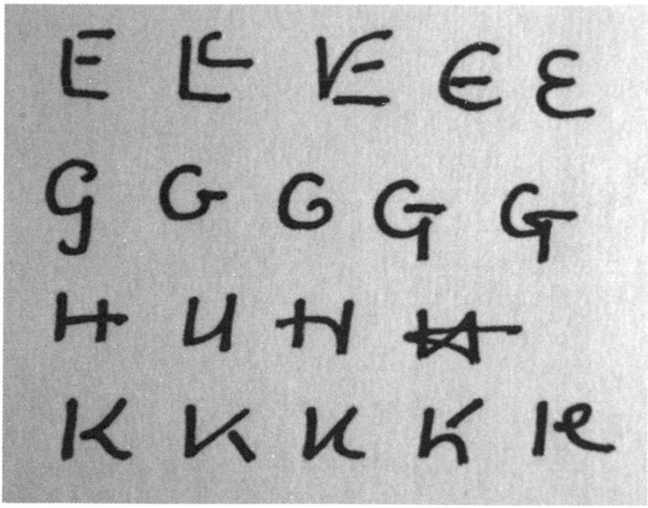

Figure 8.1 Different constructions of block capital character forms E, G, H and K, as found in UK writings.

such as the bowl of the 'P' or 'R'. Equally important are such features as the point at which a crossbar is made across a vertical in such characters as 'H' and 'T'.

Frequently, the detailed construction of character forms cannot be determined accurately without magnification of the pen lines. This is particularly the case where it is necessary to determine the order and direction of movement of the pen lines. In some circumstances, the direction of movement of the pen can be used to determine if the writer is left- or right-handed. There are a number of different features of pen lines which can be used to determine the direction in which the pen is travelling.

Ballpoint pen lines show features which are particularly useful in determining direction. They frequently show striations which follow the direction of curves and in addition the pen often deposits ink after a change in the direction of the pen line. Lightening of pressure at the endings of strokes indicates the pen leaving the surface of the paper fluently, whereas a definite spot of ink on the paper can indicate the position of the beginning of the pen stroke. Some writings made with a ballpoint pen showing striation marks are illustrated in Figure 8.2. Other writing instruments produce pen lines that are more difficult to interpret.

The use of a good stereoscopic microscope and proper lighting is essential to study the character forms in detail. A hand lens may provide some assistance in determining structure but is an inadequate tool for

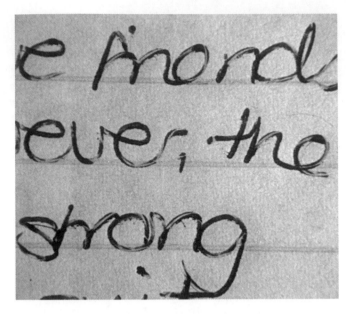

Figure 8.2   Ballpoint pen lines showing strong striation marks.

detailed analysis since, unlike the stereo microscope, the ordinary lens permits no perception of depth in the field of view.

It is essential that handwritings are compared 'like with like'; a character 'a' with a character 'a', the character 'B' with a character 'B'. Similarly, cursive writing must be compared with cursive writing, block capitals with block capitals.

8.3.2   Natural Variation

Very few people write every character in the same manner on every single occasion. All handwritings will exhibit some natural variation, most often in the shape and proportions of characters but also occasionally in structure. It is essential that sufficient handwritings are available for examination to enable this range of natural variation to be determined. If insufficient handwritings are available for examination there will always be the problem of not knowing if any differences have arisen because they were written by different people or merely the result of natural variation in one person's handwriting.

8.3.3   Comparison Material

The choice of comparison handwritings may well determine the scope of the final analysis. The best possible handwritings for an examination are those made in the course of day-to-day business as close in time as possible to the writings in question. This is particularly important when the writings of young or elderly people are under consideration. Specimen writings produced specifically for the purpose of comparison are of limited assistance in forensic examinations. It is difficult to produce entirely natural handwriting under such circumstances. Furthermore, writings made all on one occasion will not show the normal range of variation that will occur from day to day. Sufficient comparison material must be examined in order to establish the full range of natural variation in the person's handwriting under examination.

If only specimen handwritings are available, great care must be taken in their production to limit the possibility of copying the questioned document or including unnatural features. This is most easily achieved by dictating passages at intervals with breaks in between and removing the specimens from sight as they are produced.

Software programs such as Write-On and CEDAR have been developed to assist in cases involving large volumes of handwritten material. These programs assist in identifying the location and frequency of character forms but require all of the documents to be scanned in.

8.3.4   Other Forms of Variation

Handwritings can vary involuntarily for a number of reasons: illness, increasing age, difficult writing positions, use of alcohol or drugs, stress or tiredness. In general these types of problems are evinced by lack of pen control and do not result in fundamental differences in structure which are seen in the writings of two different individuals.

Handwriting can also be disguised. Such handwritings are often detected by their inconsistency: the slope of a piece of handwriting may vary, or unusual construction of character forms may appear. Often some characters are formed in eccentric ways, or features such as loops and curls are added.

Conversely, an attempt can be made to simulate another person's handwriting. Any attempt at simulation is a compromise between accuracy and fluency. To simulate another person's handwriting it is necessary to suppress one's own natural handwriting characteristics and adopt those of the person whose handwriting is the target of simulation. This is a difficult process and one that cannot be maintained over long periods. Inevitably the simulated handwriting will show characteristics of both the simulator and the target handwriting, with the balance changing as time proceeds. The simulation will also be identifiable by differences in detail. The simulator may achieve a superficial similarity to the target handwriting, but it is unlikely that the detailed structure of the individual character forms will be correctly perceived or reproduced.

Handwritings undergo the greatest degree of change in adolescence (when the individual style is forming) and old age (when pen control becomes increasingly difficult). However, mature handwriting can also undergo developmental changes over relatively long periods. By using this variation it is occasionally possible to place a piece of handwriting within a specific time frame.

8.3.5 Non-Roman Script

The principles of consistency in construction and natural variation are the same for all handwritings whether they are written in Roman, Arabic, Chinese or any other script. A forensic document examiner trained in the study of UK handwritings will be able to detect significant differences between handwritings in different languages and scripts but will be handicapped by not being familiar with the expected range of variation for any particular problem. In such circumstances the forensic document examiner will be cautious in attributing significance to similarities and differences.

8.3.6 Expression of Handwriting Conclusions

Having completed the comparison of two handwritings, the forensic document examiner will assemble the observations of differences and similarities in the detailed structure, shape and proportions of the component character forms of the handwritings, and from these draw the appropriate conclusions. Unlike fingerprint analysis, which is based on a specific number of points of similarity, the forensic document examiner's conclusions are based not only on the number of similarities but also on their quality. The variations in construction of character forms appear in different frequencies in the population. For instance, in the UK population the radial form of the block capital 'K' is far more common than the propped 'V' form.

In all handwriting comparisons the presence of differences is of profound significance. There will always be similarities between handwritings because there are a finite number of character forms that can be used. However, the presence of even a single, consistent difference between handwritings must be explained since it is a strong indicator of different authorship. Where there is a sufficient quantity of both questioned and genuine comparison handwritings, and they match in all respects without any significant differences being detected, then a safe, firm conclusion of common authorship can be given.

There are, however, many occasions when the quantity of handwriting, either questioned or genuine, is restricted. There may also be clear indications that one or other of the handwritings under examination is not natural or has been altered by an outside influence. In these circumstances a qualified conclusion can be given.

The practice of forensic science laboratories, in the UK and around the world, is to express conclusions on a qualitative scale describing the strength of the evidence. Typically this is:

Positive		*Negative*
Conclusive evidence		Weak evidence
Very strong evidence	Inconclusive evidence	Strong evidence
Strong evidence		Very strong evidence
Weak evidence		Conclusive evidence

The term 'no evidence' should not be interpreted as 'did not write'. There are, however, some circumstances, such as when considering if a very young or infirm person could have been responsible for a particularly fluent, well-developed piece of handwriting, that exclusion of an individual as a possible writer is acceptable. Nevertheless, because handwriting is, to an extent, a voluntary act and can be consciously changed, it is unusual to be able to say with total confidence that a person was not the author of a particular piece of handwriting.

The term 'balance of probabilities', often used by lawyers, is not used by forensic document examiners. The expression of probability requires a mathematical basis which cannot be satisfied by qualitative examinations such as handwriting comparisons.

The reliability of handwriting opinions was challenged in the US courts in 1995 and temporarily likened to a technical skill rather than a science. This challenge galvanized the forensic document examiner community in the US into an overhaul of background research, validation of procedures in handwriting examinations and proficiency testing of examiners. The reliability of evidence in this field has been put on a firmer footing through the work of Dr Moshe Kam who, with the backing of the FBI, established error rates for the process of handwriting comparisons and demonstrated that forensic document examiners have identifiable skill in this field. By 2002 handwriting evidence was re-established in the US courts as a firm, and fitter, member of the forensic sciences.

8.3.7   Copies

Forensic document examiners are often asked to examine copies of documents. The skilled examination of handwriting and signatures requires an analysis of the fine detail of the handwriting including stroke direction and order, crossings between strokes, and pressure. These details are lost during the copying process and hence the conclusions that can be drawn from the examination of copies will be restricted. However, where the amount of handwriting under examination is large and its image clear, it is possible to determine sufficient characteristics of the handwriting for strong, and in some cases, very positive conclusions to be reached.

Carbon copies, having been made as a result of pressure from the writing instrument, frequently show more detail than photocopies and again there are circumstances where positive conclusions regarding authorship can be drawn.

The process of facsimile transmission greatly distorts the handwriting image and in general only qualified conclusions can be drawn from their examination.

8.4   EXAMINATION OF SIGNATURES

Signatures are very specialised pieces of handwriting and specific problems are involved in their identification. Our signature is the piece of handwriting which we all use most frequently. Even individuals who write infrequently can produce a consistent signature. Because we use our signature so frequently it becomes more or less unconsciously produced each time it is written.

The greatest problem in identifying signatures as genuine or forged lies in the small amount of comparable material that is contained within any signature. This factor, combined with the natural variation which is inherent with any signature, makes signature identification one of the more testing areas of forensic science.

Because of these difficulties, it is essential that the original signatures are examined. Photocopies are not adequate substitutes. The amount of available information contained within a signature is so restricted that every available feature must be used. Fluency cannot be assessed accurately from a photocopy, nor can pen lifts or guide lines be detected. Only restricted conclusions can be drawn from the examination of copies.

The basis of signature identification is very much the same as that for handwriting. However, since signatures are used for personal identification they are frequently the targets of forgery. Considerable effort will be expended in attempts to simulate another person's signature. This can be achieved in a number of different ways.

8.4.1 Tracing

If a document bearing a genuine signature is placed on a window or a sheet of glass over a light source and overlaid by another document, it is possible to trace the outline of the genuine signature on to the forged document. The tracing, often made in pencil, can then be inked over. The resulting forgery, although superficially similar to the genuine signature, will be detectable by its lack of fluency as the simulator laboriously follows the line of the original signature, and by mistakes in the detailed construction of the character forms. These will be superficially similar to those in the genuine signature but the simulator is generally only interested in obtaining a pictorial, passable simulation and will not necessarily use the correct number or order of strokes to achieve this.

Guide lines may be left on the document and even if only fragments remain these can be detected by viewing the signature under specialised lighting conditions. Pencil lines remain opaque when viewed in the infrared region of the spectrum, whereas many inks can be transparent at these wavelengths.

Guide lines may also be in the form of indented impressions on the target document. These are produced by placing a genuine signature on top of the target document. The genuine signature is written over heavily so that impressions are transferred to the document beneath. These are then inked in to produce the simulation. The impressions can be readily detected in light shone at a shallow oblique angle to the document, as seen in Figure 8.3. In addition the signature is likely to lack fluency and contain mistakes in a similar manner to the direct tracing.

Tracing may be more subtle, with sections of several signatures being incorporated into a single simulation. Several instances have been seen of simulations produced from templates prepared from genuine signatures.

8.4.2 Freehand Simulation

The freehand simulation is likely to be the most fluent, but the simulator still needs to achieve a compromise between accuracy and fluency. Unusual hesitations and pen lifts in the pen line will occur whilst the simulator pauses before embarking on the next section. Often these types of simulation show the greatest degree of deviation from the genuine signatures at their ending, when the simulator relaxes concentration and becomes more confident. This usually results in a

Figure 8.3   Traced simulation of a signature as seen in obliquely directed light.

reversion to the simulator's natural handwriting rather than that of the target signature. Again the simulation is detected by its superficial similarity and its differences in the detailed shape, structure and proportions of its component character forms.

8.4.3   Authorship of Simulation

Simulated signatures are not natural writings. To produce a simulated signature the forger's own natural characteristics of writing must be suppressed and those of the person whose signature is the target of forgery adopted. Hence it is not normally possible to identify the author of a simulated signature by comparing it with natural handwriting.

8.4.4   Self-Forgery

Certain signatures are written by individuals with intent to deny them at a later date. These signatures are generally different from normal natural signatures in some very obvious feature, often the form of initials or slope, but the detailed structures of the less obtrusive features are unchanged.

8.4.5   Vulnerable Signatures

Signatures which are short and simple and which contain a number of natural pen lifts are vulnerable since they reduce the requirement of maintaining fluency while executing unfamiliar and possibly complex character forms. Signatures that naturally demonstrate a wide range of variation are also vulnerable since it is difficult to assess if differences have arisen merely as a result of this variation or because of simulation by another person.

Signatures of elderly people often naturally contain many of the features which are the hallmark of simulation, such as lack of pen control, undue variation and poorly formed character forms. These signatures are, therefore, particularly vulnerable.

8.4.6   Guided Hand Signatures

An infirm, elderly or partially sighted individual may be assisted by another person in signing documents. Guided hand or assisted hand signatures are a particularly difficult area for the forensic document examiner. The characteristics of such a signature may be a mixture of those of the signator and the assistant.

The forensic document examiner is often asked to comment on the degree of influence exerted by the assistant. Clearly, if the signature deviates strongly from the genuine signature and contains a predominance of the assistant's signature characteristics a degree of influence has been exerted. However, the forensic document examiner must be cautious in commenting on the intent of the assistant and the person signing. It may be that the signator, while unable to exert sufficient pen control to write, may have been completely clear in their desire to sign the document in question.

8.4.7   Comparison Material

The requirement for adequate comparison material in signature examinations is similar to that required in all handwriting comparisons. However, it is worth stating again that like can only be compared with like—in other words, signatures are only comparable with other signatures in the same name. 'Smith' cannot be compared with 'Jones', nor can signatures generally be compared with day-to-day handwritings.

The ideal comparison material would consist of genuine signatures made in the course of day-to-day business close in time to the date of the signature in question.

8.4.8   Expression of Signature Conclusions

Given a reasonably stable and mature signature and an adequate range of genuine signatures for comparison, the forensic document examiner can expect to identify a genuine signature or a simulation. However, there will be occasions, particularly in this very testing area, where only a qualified conclusion can be given. These qualified conclusions are generally expressed in a similar manner to those for handwriting examinations.

8.5   EXAMINATION OF COPIES

Copies of documents can be produced in a number of ways, *e.g.* simple photocopying or electronic scanning. The limitations of examining copies have already been explained. However, it is often the case that a copy is the only evidence of a pre-existing original. The forensic document examiner is then charged with the task of determining from the copy if it was indeed prepared from an authentic contemporaneous original document or if the copy has been recently manufactured or manipulated to misrepresent the facts.

The copy may be a montage produced, either physically or electronically, partly from genuine documents. The simplest demonstration of this is the image of a genuine signature transferred by photocopying on to a fraudulent document. The image may be transferred intact and without apparent disturbance, in which case often the only way of demonstrating the montage is to locate the original signature from which the image was taken. However, often it is necessary to manipulate or retouch the image to make it fit the new document. These processes are detectable as inconsistencies in the image of the pen lines. Modern computer technology facilitates the manipulation of documents and makes the forensic document examiner's task yet more challenging.

It may be important to demonstrate the origin of a particular photocopy document. There are two types of mark which appear on photocopies that can be used to do this: trash marks and drum or mechanism marks.

Trash marks are transient and produced by dust particles or debris on the glass surface of the photocopier. These will appear as a series of dots or marks on the resultant copy. The same configuration of trash marks on a number of documents would indicate that they have a common origin. However, it is important to remember that if a copy bearing trash marks is copied, the daughter copies will also bear the same configuration of marks.

Marks made by the drum or the mechanism of the photocopier will persist for a longer period and can be used to identify a specific machine. Drum marks do not necessarily appear in the same position on every page but can be detected by the constant intervals at which they appear. The interval is directly related to the circumference of the drum and not the dimensions of the document being copied.

The development of colour photocopiers raised concerns as to their possible use in counterfeiting. Many photocopier manufactures (and also colour laser printer manufacturers) incorporate a code of dots into every copy produced. These codes may assist in identifying the brand of copier, model, serial number and, in some cases, the date of a particular copy.

8.6 PRINTING AND TYPEWRITING

8.6.1 Modern Office Technology

The replacement of the old-style manual typewriter with fixed type bars, first by electric typewriters containing golf-balls and print wheels, then by dot-matrix printers and today by the ubiquitous ink-jet or laser printer, has completely revolutionised the forensic examination of typescript in the last 30 years.

8.6.2 Word Processors

The term 'word processor' actually refers to the computer program used by the keyboard operator to prepare a computer file of the document which is to be printed. The printing of the document from the computer can be made using a number of different types of machine, most commonly laser printers, ink-jet printers or dot-matrix printers. The appearance of the printed characters that finally appear on the paper is to a large extent governed by the computer program. More primitive word processor systems were connected to electronic typewriters fitted with daisy wheels. The number and nature of the type styles which could be prepared from these machines was much more limited.

Changes in software operating systems can be very important in dating computer-generated documents. The introduction of the Times New Roman typeface in the early 1990s by Microsoft, and its subsequent replacement as a default typeface by Calibri in 2007, are particular markers which have been important in the authentication of documents. It is, however, very difficult to distinguish between the myriad of type styles available, some of which may differ in only very subtle features. The use of computer programs such as Adobe Photoshop can be helpful in comparing the detail of type styles, line spacing and baseline alignment of characters.

8.6.3 Laser Printers

The laser printer works very much on the same principle as a photocopier. It contains a photosensitive drum which becomes charged as the laser light hits it, producing an 'electrostatic negative' of whatever image is to be printed. The drum is sprinkled with negatively charged toner which clings to the charged areas originally scanned by the laser beam. A sheet of paper is passed under the rotating drum and given a charge greater than on the drum, allowing transfer of the toner from the drum to the paper where it is then fused in place by heat or pressure. Images produced by laser printers are generally of very high quality. They are also very reproducible and it is extremely difficult to distinguish between the work of two different laser printers. However, it has been established that, on occasions, drum faults may occur on laser printers in the same way as they do on photocopiers.

As difficult as it is to distinguish the work of one photocopier from another, it is similarly difficult to identify any document as having been produced by a particular computer/printer system. The number of fonts available and their versatility is immense.

8.6.4   Ink-Jet Printers

In an ink-jet printer the only moving part in the print head is the ink itself. The printer uses a grid of tiny nozzles to which specially formulated ink flows from a reservoir. There are two main types of ink-jet print technology: continuous drop, and drop on demand, with the latter now being more common. These technologies rely on the ink being heated to transfer it to the paper. However, there are many variations on the basic technology. Epson has developed piezoelectric ink-jet systems which rely on the deformation of crystals to force the ink from the ink chamber on to the paper.

The main difference between the laser and the ink-jet printer is the state of the ink. When the ink from the ink-jet printer hits the paper it spreads into the fibres and gives the print a slightly ragged appearance. Some systems require the use of specially adsorbent paper. Figure 8.4 shows a high-magnification image of a section of an ink-jet printed signature which has been inserted into a laser-printed document—the fused appearance of the toner of the laser printing is clearly different from the ink.

Although the work of ink-jet printers can readily be identified it is as difficult to identify a particular machine or differentiate between the work of several machines as it is with laser printer copy.

8.6.5   Dot-Matrix Printers

Dot-matrix printers work by striking the paper through a ribbon with a number of pins. Each time a pin hits the ribbon it transfers a dot of ink

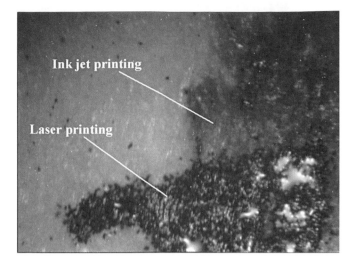

Figure 8.4 High-magnification image of a section of an ink-jet printed signature which has been inserted into a laser printed document.

from the ribbon to the paper. The characters are formed from a matrix of pins arranged in a head. These printers are relatively slow and noisy when compared with ink-jet and laser printers. However, they are often retained for processes which involve producing several copies of a document all at the same time such as invoices, packing notes, letters of credit or shipping manifests.

These types of printers have a relatively large number of moving parts and there is greater scope for examiners to identify the print. Examples have been seen where individual pins have been damaged or have not functioned. However, these tend to be relatively rare since the eccentric character forms produced are very noticeable and rapidly repaired.

8.6.6 Single-Element Typewriters

There are still a substantial number of documents which are produced on electric typewriters equipped with either a daisy wheel or, more rarely these days, a golf-ball. These single elements can be easily removed from the machine and replaced using another of a different style. The elements can also be transferred between machines. One of the most important features of these single elements is that they deteriorate with increasing use. Faults develop on the element and these are transferred to the print on the paper. Figure 8.5 shows typing produced with a print wheel on which the underlining bar is beginning to deteriorate.

Figure 8.5   Typing produced using a damaged print wheel.

These faults can be used to distinguish between work from two typewriters, or a particular document can be identified with a particular print element. However, these faults are relatively transient. Once the fault begins to develop on a daisy wheel in particular, the wheel will deteriorate rapidly, revealing a number of different faults until it needs to be replaced. Once it is replaced, these faults no longer exist. This particular feature can be used to identify the date on which documents were produced.

8.6.7   Fixed Type-Bar Machines

In this traditional type of typewriter each individual character is fixed to the end of a type bar. The number of moving parts within these machines is relatively large and in consequence, usage results in the variation of the relative alignment of individual characters. Furthermore, the individual characters can become worn or damaged so that the printed character is imperfect. These faults and imperfections tend to become progressively worse over the period of usage. The presence or absence of such faults can be used to identify the work of an individual typewriter.

8.6.8   Ribbons, Rollers and Correction Facilities

Where older-style typewriters are involved, study of the typewriter itself and any carbon paper or correction papers can reveal useful information. The text of documents typed using carbon film ribbons can be detected fairly easily. These machines function by stamping out the carbon image of the character from the ribbon on to the paper. The resulting negative remains on the ribbon and the text of the document can be followed using a moderately powerful magnification system. The process is, however, very tedious especially when the ribbon mechanism involves complex vertical and horizontal movements. A computerised optical system of transcribing ribbons has been developed and is available in a small number of laboratories.

The roller or platen of a typewriter may also contain information. The image of the text may have been inadvertently transferred to the roller. The roller itself may contain physical defects which affect the text of the document.

Careful examination of corrections, correcting ribbons, correcting papers and transfer or correction fluid onto fabric ribbons can all provide additional information to link a questioned document with a particular typewriter.

8.6.9 Fax Machines

Based on casework experience, fax machines are still ubiquitous in both homes and offices. Many manufacturers market multifunction machines which incorporate scanning, printing, copying and faxing facilities and may use a variety of printing technologies including laser, ink-jet or thermal transfer. The printing of received facsimile transmissions is usually low resolution, precluding useful handwriting and signature comparisons. However, faxes can be manipulated in the same way as photocopies. The transmitting terminal identifier (TTI), which appears as a line of data at the top or bottom of a received facsimile transmission, is generated by the sending terminal and contains data to identify the sender. The American Society of Questioned Document Examiners has developed a data base of TTIs which can be used to identify the make of the sending terminal. However, the TTI can either be deliberately deleted from transmissions or information such as telephone numbers, date and time can be changed at will. The TTI on a received facsimile transmission may, therefore, contain deliberately modified information.

8.7 ORIGIN AND HISTORY OF DOCUMENTS

There are occasions when documents used in criminal activities or in the course of civil litigation are either created specifically for the purposes of deception or altered for similar ends. For example, entries in diaries, or occasionally whole diaries, are prepared to show that a particular sequence of events occurred in the past. Similarly, files can be reconstructed and correspondence backdated.

The forensic document examiner is asked to assist in determining the authenticity of such documents and entries. The dating of documents and writings is a difficult forensic science problem. Since there are no reliable techniques available which allow the absolute dating of inks, it is necessary to use other features of documents to determine where and when they were created and their history.

8.7.1   Examination of Inks

The ink used to write particular entries on a document may be of interest for several reasons. Principally the forensic document examiner will be interested in establishing if inks are similar or different. In this way it is possible to determine if entries have been added to or altered.

The examination of inks on questioned documents is generally confined to non-destructive comparative analysis. Inks that appear similar to the unaided eye can be very different when viewed in the infrared region of the spectrum. Similarly, illuminating an ink with high-energy light at one wavelength can promote it to release light energy at different wavelengths. The degree to which this occurs depends on the chemical nature of the ink. Figures 8.6a–c show a set of black ballpoint pen inks as viewed under different conditions. Several instruments have been developed which allow inks to be viewed under a wide range of different lighting conditions to enhance some of the less visible effects. The fact that many inks can be rendered transparent in the higher wavelengths of the infrared region of the spectrum is an essential tool in the examination of questioned signatures. Pencil markings remain opaque at these wavelengths and can easily be detected under the transparent ink.

Raman spectroscopy of inks is a largely non-destructive process which has proved useful in distinguishing between inks which cannot be distinguished using more traditional infrared and fluorescence techniques. The Raman response can be considerably enhanced by a technique called surface enhanced resonance Raman scattering (SERRS) spectroscopy. This involves the application of a very small quantity of a silver or gold colloid on to the ink line under examination. The extent of any damage is minimal and can be confined to 1–2 mm within the width of the pen line. SERRS spectra obtained from two black ballpoint inks are shown in Figure 8.7.

On some occasions further destructive techniques can be employed to compare inks more rigorously. In fact this destructive analysis can be carried out using relatively tiny amounts of ink which are carefully removed from within the pen line of a single character form. Thin-layer chromatography (TLC) or HPLC of these ink samples is used to

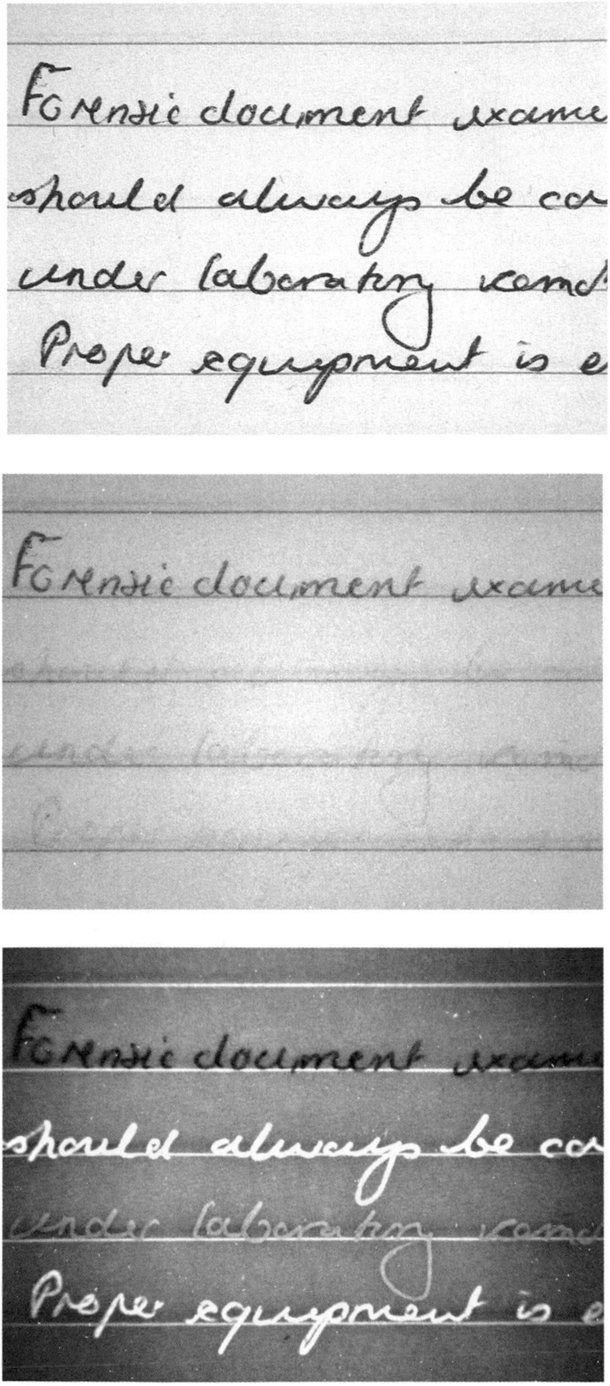

(a)

(b)

(c)

Figure 8.6 Black ballpoint pen inks as seen (a) in normal light, (b) in infrared light, (c) under conditions that cause fluorescence.

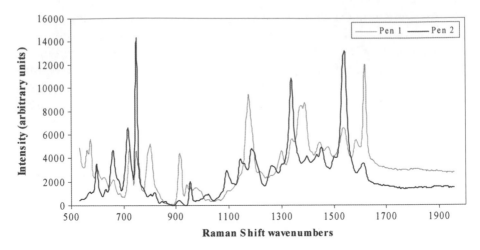

Figure 8.7   Raman spectra of two different black ballpoint inks.

separate and compare the dye components. In two studies reported in 2000, combining various techniques (thin-layer chromatography, infrared and fluorescence techniques and SERRS spectroscopy) allowed over 90% of the ballpoint pen inks under review to be distinguished.

Most of the ink analysis carried out in the UK is comparative. If parts of a document can be demonstrated to have been written using two different inks, then clearly two pens have been used. However, the opposite is not equally true. If the inks used in two writings are indistinguishable they could have been written either using the same pen or pens containing similar inks. Since most ink dye manufacturers do not change basic ink compositions very often, there are likely to be several million similar pens available for any one type of ink.

The Secret Service Laboratory in Washington, DC has maintained an extensive library of inks made in the USA. However, it is rare that absolute identification of an ink is required. This certainly would be necessary if an attempt is made to date the ink. Unfortunately, as indicated previously, there are still very few methods available for reliably dating inks.

For a number of years in the 1990s certain manufacturers of inks in the USA cooperated in an ink-tagging programme where rare-earth trace components of fluorescent dyes were added and changed regularly to allow inks to be dated. This practice was revived in 2002 by a major US ink manufacturer. Some confusion has arisen from the fact that the markers used in the earlier program may have been reused in later years.

Other laboratories claim success in dating inks by comparing the relative extractability of ink components from paper. Again the method is comparative, but the inks must be identified to make sure that they are directly comparable. More alarming has been the attempt to date inks using artificial accelerated techniques. There is substantial disquiet in the forensic document examination community concerning these techniques. Current thinking is that they are fundamentally flawed. The promising work in ink dating has focused on the measurement of volatile components of ink and their duration. However, the method is limited by the fact that the volatile components are detectable for a relatively short period of about 2 years. Where ink dating evidence has been offered in courts of law there has, very frequently, been disagreement between the experts. Although ink dating services are offered by a very small number of laboratories in the USA, they are rarely used in other countries.

8.7.2   Examination of Paper

The size, thickness, density, colour and finish of papers can be compared to determine if they are from a similar or different origin. Optical brighteners are incorporated into some papers and give differing reactions when viewed in ultraviolet light. The biological fibres used in a particular paper's manufacture can be identified by microscopy and can be used to give a broad geographical origin of the paper. However, since wood pulp is transported on a global scale, this is often inaccurate.

The origin of a paper is most accurately identified by its watermark. Some paper manufacturers periodically introduce changes into watermarks. Repairs to the dandy roll used in the manufacturing process to produce the watermark may be identifiable in subsequent production runs. However, not all manufacturers keep detailed records of changes made over the years and this has been exacerbated by the takeover and absorption of some of the older brands.

In a similar manner the chemical content and colour of papers can be used to establish when they were manufactured, if the manufacturer can be identified and holds the necessary records.

Pages torn from a pad of paper often show a specific pattern along the top edge where the adhesive has been applied unevenly to the backing sheet. The pattern may vary through the thickness of a pad so that the pages can be identified as being from the same pad, different parts of the same pad or different pads altogether. Similarly, sheets from the same batch of paper may show similar guillotine marks along their edges, although these are more difficult to identify.

8.7.3 Development of Handwriting and Signatures Over Time

As noted previously the handwriting and signatures of individuals vary over the years. If sufficient examples of genuine handwritings and signatures are available to demonstrate the different stages of development, questioned writing and signatures can be matched to a particular period. This type of examination is particularly applicable to the writings of young and elderly people.

8.7.4 Impressions

Impressions of handwriting may be left on a document as a result of writing on another document which at the time of writing was overlying the first one. These impressions are rarely visible to the naked eye and need to be detected using specialised equipment and techniques.

Deep impressions can be detected using light directed at a low oblique angle to the document. In this way the surface of the document is thrown into relief and the impressions can often be read. Where the impressions are small and fragmentary the best results are often obtained using a low diffuse light or by viewing the document in natural daylight in the early morning or evening.

Fainter impressions can be detected using the technique of electrostatic detection (ESDA). This technique was developed jointly by the London College of Printing and the Metropolitan Police Forensic Science Laboratory in London in 1979 and revolutionised forensic document examination. The long and tedious hours piecing together fragments of indented impressions in dimly lit rooms were replaced with a relatively fast and highly sensitive technique which could develop impressions made through several layers of paper. Such impressions are too faint to be detected by any other methods. It is thought that the ESDA process detects areas where the paper fibres have been deformed by the movement of the writing implement or by two documents moving against each other. The technique is not always successful. Impressions cannot be detected on documents which have been treated with solvents. It is therefore imperative that any document which will eventually be examined for fingerprints should first be submitted to the forensic document laboratory before treatment with fingerprint reagents.

Recently a number of reports have been made of 'secondary impressions'. These impressions are caused not by the action of writing but by storing two documents together such that the fairly deep embossing of actual handwriting on one document causes detectable impression marks on the second document. Such 'secondary impressions' can be very confusing, particularly if they appear on the same document as those impressions caused by the act of writing.

In the last few years the utility of the ESDA technique has been extended and it can sometimes now be used in the determination of the order of writings. The ESDA 'lift' bears not only black traces caused by the development of the impressions but also traces caused by the ink lines present on the document, as illustrated in Figure 8.8. The lower set of impressions shows the black ESDA trace continuous over the white lines of the writing of the document being examined. This indicates that the writing that caused the impressions was made after the writing of the document itself. In the upper set of impressions the black ESDA trace is broken by the white lines of the writing of the document being examined, and the sequence cannot be confirmed.

It has been shown that where impressions and ink lines on a document intersect, the appearance of the intersection can be used to determine if

Figure 8.8　ESDA 'lift' showing black traces of impressions and white traces caused by writings on the document.

the writings or the impressions were made first. Situations where the black impression line can be seen to be continuous over the white ink line indicate that the impressions were made after the writings. The technique can only be used when the ESDA lift is of good quality and there are a substantial number of intersections which are clear, since where impressions have been made after writings on a document only a proportion of the intersections will show the black line continuous over the white ink trace. Experiments where the ink lines of writing were made after impressions on a document resulted in the black impression trace being broken on every occasion. Similar reactions occur between impressions and writing on opposite sides of a piece of paper. The mechanisms of the response are not fully understood and great care must be exercised in interpreting the results of these analyses.

Although ESDA has now been in routine use in forensic document laboratories for over 30 years the underlying mechanisms of the process are still not fully understood and the ESDA response can be variable. However, it is one of the most powerful weapons in the armoury of the forensic document examiner.

8.7.5 Folds, Creases and Tears

The physical condition of a document may betray much of its history. The arrangement of folds and creases in a document may indicate if it was folded into an envelope or folded in conjunction with other documents.

Torn edges of a document may be matched to those of another document, particularly if they were torn together from the same pad. Portions of torn documents can be matched by studying the mechanical fit of the torn edges.

Envelopes which have been opened and re-sealed may contain fragments of torn paper fibres within the seal. These can be detected by transmitted light through the document or X-ray techniques. The act of re-gluing an envelope flap may require the use of additional adhesive which can be detected by similar techniques.

8.7.6 Staples and Punch Holes

The act of stapling one document to another causes physical damage to the document as the staple ends are punched through. If the staples are removed the staple holes remain obvious. Further damage can be caused to the back of a document where the ends of the staples are forced in, often distorting the paper between the staple holes. The relative size and the positioning of staple holes can be used to identify documents which have been stored together.

The study of staples is also important in considering if sections of documents have been removed, such as in notebooks. Prising open staples often causes distortion to the staple holes which become widened. Microscopic examination of the staple legs will show marks made by pliers or other implements used to prise the staple out. Besides examining the condition of the legs of the staple in the central fold of a book it is important not to neglect the exterior of the spine, where the act of pulling out the staple or replacing it may have left detectable marks on the document.

Punch holes also cause irreparable changes to a document. Punch holes become worn and distorted over a period of time as a result of movement of documents against the holding prongs of a file or clip. Often documents stored together for a period of time will show a similar pattern of damage to the punch hole.

8.7.7 Erasures, Obliterations and Additions

The act of erasing an entry from a document, be it typewritten, handwritten or printed, will damage the document. The extent and nature of the damage will be different depending on whether the erasure was carried out by mechanical or chemical means.

In general, mechanical removal of an entry will physically damage the document. Using an eraser to remove entries results in the paper fibres of the document being disturbed. This disturbance, if large enough, may be detected using obliquely directed light which throws the surface of the document into relief. Less obvious erasures made with a traditional rubber eraser can be detected by the use of a very fine powder of dyed lycopodium spores. The powder will adhere to fragments of rubber left on the surface of a document in the process of erasure. The powder is spread over a document held at an angle and the excess shaken off gently. The powder can then be brushed off the surface of the document leaving it undamaged. However, the action of some plastic erasers cannot be detected in this way. Consequently, the lack of reaction to lycopodium powder cannot be taken as proof that no erasure has taken place.

Chemical erasure involves the use of a solvent to remove an ink entry. This may be comprehensive and leave no trace visible to the unaided eye. The effect of the solvent on the document, however, can be detected by viewing it in ultraviolet light or under conditions which promote infrared fluorescence.

Traces of entries removed by erasure can be enhanced by viewing the document under different optical conditions. Pencil entries are opaque in the infrared region of the spectrum and can therefore be enhanced using an instrument which combines this lighting facility with powerful magnifying capabilities. Similarly, traces of an ink which has fluorescent properties can be enhanced by promoting the fluorescence and magnifying the image.

Obliterated entries are detected by separating the original entry from the obliterating material. This can be achieved non-destructively by exploiting the optical characteristics of inks and correcting fluids. A black ink entry obliterated with another black ink can be recovered if the inks react differently, for instance, in infrared light. Entries obliterated with white correcting fluids can be recovered using a combination of lighting conditions including viewing from the back of the document. Today's modern range of instrumentation in forensic document laboratories allows images of such features to be captured electronically and manipulated to give a clear view of the original obliterated or erased entry.

The detection of added ink entries again involves the non-visible properties of the inks. Anyone making a fraudulent addition to a document will make an effort to choose an ink of similar colour. However, these visually similar inks may well have very different properties in infrared light and can easily be demonstrated as having been made with different pens.

There are differences in the manufacture of pencils which can be detected using techniques such as scanning electron microscopy (SEM). However, such investigations have the disadvantage of being destructive and expensive.

Unlike a pen tip, the point of a pencil alters as writing proceeds. It is only necessary to change the position of a pencil in the hand by 90° to fundamentally change the appearance of the drawn line it produces. In consequence, it is extremely difficult to detect tampering of pencil entries.

8.8 PRINTED DOCUMENTS

Most practising forensic document examiners will have a working knowledge of printing techniques sufficient to identify the process by which a document or parts of a document have been produced. However, some fields of printing required a highly specialised knowledge.

High-value documents, *e.g.* bank drafts and identity documents such as passports, are specially printed using methods to deter alterations or counterfeiting. Documents can be printed on papers which are sensitive to the application of chemicals or mechanical erasures and which change colour. Coloured or fluorescent fibres can be added to the paper. Security backgrounds can be produced with fugitive inks which run on application of a liquid, and complex patterns of printing can be produced which are difficult to imitate. Sections of documents can be printed using inks with special spectral properties or which change colour under different lighting conditions. The techniques used for studying alterations to documents can also be used to detect breaches of a document's security.

A knowledge of printing processes is frequently needed in the identification of counterfeits. The counterfeit must be compared directly with the genuine article, be it a passport, cheque or perfume packaging. The forensic document examiner needs to be able to identify the differences in printing between the products to show differences in their mode of production.

8.9 PROCEDURES, PROTOCOLS AND QUALITY ASSURANCE

In forensic document examination, as in any forensic science, certain criteria need to be applied to ensure a correct interpretation of the evidence.

Any document, either questioned or of known origin, submitted to a forensic document laboratory needs to be handled properly and securely to ensure it retains its integrity throughout the examination. Documents need to be properly identified, labelled and recorded. The appropriate testing must take place and this should always be the most rigorous available. The results of examinations and tests should be recorded clearly and unambiguously in the laboratory notes along with details of standards and controls.

The forensic document examiner's report should be comprehensive, setting out the examinations employed, the results and the conclusions drawn. Forensic document examination is a science but this is no bar to clear, concise and understandable reporting. These processes will at least ensure that a full examination is carried out in a properly equipped laboratory. It is, however, extremely difficult to ensure that the results of an examination are interpreted correctly.

Properly trained forensic document examiners working in fully equipped laboratories can generally be expected to reach broadly similar conclusions from the examination of the same material. Problems arise when two experts are given different material, often in the form of control handwritings or signatures. More serious disagreements arise when inappropriately trained individuals, or those lacking proper examination facilities, are employed as experts. Provided the forensic document examiner has the opportunity to do so, it is generally not difficult to refute the evidence of such individuals. Visual presentation of results is the most effective method of presenting the evidence, particularly where there are conflicting views. Indeed it is always wise to work on the premise, 'If you can't show it, don't use it'.

BIBLIOGRAPHY

V. Aginsky, Measuring ink extractability as a function of age: why the relative ageing approach is unreliable and why it is more correct to measure ink volatile components than dyes, *Int. J. Forens. Doc. Examiners*, 1998, **4**, 214–230.

D. M. Ellen, *The Scientific Examination of Documents*, 3rd edition, CRC Press, Boca Raton, 2006.

W. R. Harrison, *Suspect Documents*, Sweet & Maxwell, London, 1958.

R. A. Huber and A. M. Headrick, *Handwriting Identification Facts and Fundamentals*, CRC Press, Boca Raton, 1999.

J. S. Kelly and B. S. Lindblom, *Scientific Examination of Questioned Documents*, 2nd edition, CRC Press, Boca Raton, 2006.

A. S. Osborne, *Questioned Documents*, Boyd, Albany, NY, 1929.

Proceedings of 2nd European Academy of Forensic Science Meeting (European Document Experts Working Group), Cracow, 12–16 September 2000 (papers by C. Neumann and W. D. Mazella, and by E. Wagner).

CHAPTER 9

Computer-Based Media

JONATHAN HENRY

Operational Support Department, Police Service of Northern Ireland, 42 Montgomery Road, Belfast, BT26 9LD

9.1 THE COMPUTER CRIME SCENE

The use of computers to commit crimes and computers as the subjects of crime has redefined where and what a crime scene can be. The computer crime scene may well exist as a traditional scene with the evidence clearly visible, but more often the evidence is not in an obvious form and requires the actions of a suitably qualified examiner to preserve it and to interpret its meaning.

Once upon a time the documentary evidence of a business fraud might have been found in the locked filing cabinet of the accountant's office: that locked filing cabinet is now more likely to be a folder on their computer hard drive or in the remnants of deleted files thought long gone. Overseas 'hackers' and their computers, penetrating the security of a UK network, may be well beyond our physical grasp but the records of the penetration may be available to us, and properly preserved and interpreted can lead us to the source. In its most extreme form the crime scene may exist only fleetingly as a conversation in a chat room, with unknown traces of the conversation remaining on the participants' computers.

The computer crime scene exists in terms of the traces of actions performed on a computer that remain on the hard disk drive or

Crime Scene to Court: The Essentials of Forensic Science, 3rd Edition
Edited by P. C. White
© Royal Society of Chemistry 2010
Published by the Royal Society of Chemistry, www.rsc.org

removable media, a single line in an email header, or even as entries in a server log file showing access to files or an email account.

The rule of 'every contact leaves a trace' is just as applicable to computer-based evidence as any other. This applies to the actions of the examiner also. Such is the nature of computer-based evidence, the contact is usually date and time stamped and can prove compelling evidence for the prosecution or the defence. Similarly it is just as fragile as other forensic evidence and without proper treatment can be lost.

In order to understand the forensic evidence available to an examiner it is necessary to have a basic understanding of the methods of data storage on computer media, both in physical terms, *i.e.* the devices used for storage, and logically, *i.e.* the structure and format of the data on those devices. The following information is intended as an introduction to the principles of how data is stored physically on computer media and logically in Windows-based computer systems, and how the actions of the user produce these data. The principles behind the physical media are independent of operating systems. Some of the principles relating to the actions of users may be equally applicable to operating systems other than Windows, *i.e.* Macintosh; Linux *etc.* This chapter is not intended as a guide to performing computer-based evidence examinations, nor as a technical reference on the format and location of data commonly forming the basis of such examinations.

9.2 GUIDANCE ON EXAMINATION OF COMPUTER-BASED EVIDENCE

9.2.1 Principles

There are four overriding principles that form the basis of any examination of computer-based evidence. These have been encapsulated within the Association of Chief Police Officers Good Practice Guide for Computer Based Evidence. Their purpose is to protect the integrity of the evidence, maintain a record of the examination and to protect the rights of the original owner of the evidence. The principles are:

1. No action taken by police or their agents should change data held on a computer or other media that may subsequently be relied upon in court.
2. In exceptional circumstances where a person finds it necessary to access original data held on a target computer that person must be competent to do so and to give evidence explaining the relevance and the implications of their actions.

3. An audit trail or other record of all processes applied to computer-based evidence should be created and preserved. An independent third party should be able to examine those processes and achieve the same result.

4. The officer in charge of the case is responsible for ensuring that the law and these principles are adhered to. This applies to the possession of and access to, information contained in a computer. They must be satisfied that anyone accessing the computer, or any use of a copying device, complies with these laws and principles.

Combined, these principles mean that examinations of computer-based evidence should only be performed by suitably qualified individuals, in a forensically sound manner, with sufficient records maintained to ensure that another examiner can reconstruct their actions.

9.2.2 Imaging

Principle 1 for the examination of computer-based evidence states that 'No action taken by police or their agents should change data held on a computer or other media that may subsequently be relied upon in court'. An essential step in the examination of computer-based evidence is therefore, where possible, to create a forensically sound and accurate copy or 'image' of the media under examination. This step is commonly referred to as 'imaging'.

The purpose of imaging is to enable the forensic examiner to copy the original media in a forensically sound manner that will not alter the integrity of its contents. Any subsequent examination can then be performed on the copy, leaving the original intact and unchanged. Imaging or 'bitstream imaging' is the process of replicating each individual bit of data from a storage device such as a hard drive, USB memory stick, a floppy drive, a digital camera card or any other.

An image can be made of either a physical disk or a logical volume. It is always preferable to image the physical disk where possible. Unused areas of the disk can and do exist outside those normally visible to the user for a variety of reasons. They may not always have been unused and may have previously contained data that we are interested in. For this reason unused areas should be included in an examination.

9.2.3 Examinations

The actions taken by the forensic examiner after imaging are largely determined by the type of investigations they carry out and the forensic software used. Commonly the 'image' of the media exists as a series of files produced by the imaging software which contains all of the data from the imaged media. Dependent on the software used, the image can then either be examined directly within the forensic software as a 'virtual drive', or a 'clone' of the original media made for the purposes of the examination. In this way the examiner is then working on an exact replica of the imaged media, and the original remains intact and unchanged by the examination process.

The procedures to be followed throughout the forensic examination of computer-based evidence are not discussed here, as there is a myriad of software applications available to the forensic examiner. They range from utilities specifically designed for the imaging and analysis of computer-based evidence to utilities designed for other purposes, *e.g.* recovery of lost or accidentally deleted data, which lend themselves to the needs of forensic examiners. The procedures to be followed for each piece of software, or hardware for that matter, are generally unique to that equipment.

An examination process can be set out technically for the use of any piece of software or hardware, but there is more to the examination than 'ticking the boxes'. The methodology differs for each individual case and depends on many factors: the type of investigation, what is required to be proven, the physical hardware or software applications encountered, the timescale for the examination. Computer forensic examiners rarely exist as examiners alone. They are investigators with specialist technical skills who play an integral part in the entire investigation process.

9.3 STORAGE DEVICES

9.3.1 Ones and Zeroes, Bits and Bytes

Information is stored and processed by computers in terms of binary data, *i.e.* 1s or 0s. Each 1 or 0 is referred to as a 'bit' of information or data. Eight bits of information together make up a 'byte'. 1024 bytes make up a kilobyte (kB). Similarly, 1024 kB (1,048,576 bytes) make up a megabyte (MB), 1024 MB (1,073,741824 bytes) make up a gigabyte (GB) and so on for terabytes (TB), *etc*.

That is not to say that computer hard drives, CD-ROMs, and disks are full of 1s and 0s. The 1s and 0s are a mathematical representation of a two-state or binary system, *i.e.* 'on' or 'off'. This is how digital data is physically stored on hard drives or removable media such as CD-ROMs or floppy disks. On hard drives or floppy disks changing magnetic orientations encoded on a magnetically responsive material represent the on/off states. CD-ROMs and DVDs represent them by the inter-ference patterns of a beam of laser light bounced off the surface of a disk caused by pits on the disk surface. All computer media, whatever their physical form, store data in two states. This two-state information is interpreted physically by the hardware, and logically by the operating system, as 1s and 0s.

Storage devices are generally based on magnetic media, optical media or a combination of both technologies. I intend to describe the most commonly encountered storage devices for each category. The examples are by no means exhaustive.

9.3.2 Magnetic Media

9.3.2.1 Hard Disks. Hard disk drives are the heart of data storage in modern computer systems. The computer operating system resides on the hard drive. Every program installed by the user is stored there. Details of the system configuration are found in files saved on the hard disk by the operating system or particular programs. Most importantly, it is the most commonly used location for the user to store files they create. From an evidential point of view the hard disk drive is an extremely rich source of information. It may contain not only the evi-dence of offences that have been committed, but also how they were committed, which may reveal the intentions of the user.

Physically the hard disk drive (HDD) is a sealed unit usually roughly 3.5 inches (9 cm) in diameter (5.25 inches or 13 cm for older drives), or in the case of a laptop drive 2.5 inches (6.35 cm). The only visible com-ponent is the printed circuit board attached to the outside of the sealed unit. In modern hard dives this contains the hard disk controller which handles conversations between the HDD and the processor, interpreting instructions from the processor and converting information from the HDD into a form the processor can understand and *vice versa*. Con-tained within the sealed unit is one or more discs mounted centrally on a rotating spindle, each of which has two read/write heads sliding over their surface on arms, much like the stylus on a record player.

Each rotating disk is made up of a thin layer of magnetically responsive material laid or 'spun' on to a rigid substrate of glass or aluminium. It is within this magnetically responsive layer that the data is stored. In modern hard drives, to increase capacity this may actually be a number of layers each capable of storing data independently. Each disk is referred to as a 'platter'. The magnetic media layer is coated with an extremely thin layer of carbon to protect it from the read/write heads and a lubricating layer to help the heads slide over the surface.

During operation the platters spin at extremely high speeds, 3600–12 000 rpm depending on the HDD used. As they do so the read/write heads glide fractionally (millionths of an inch) above the surface, held up by the pressure of the airflow from the spinning disk. The spinning motion combined with the side-to-side movement of the heads on their arms allows any area of the platter to be accessed very quickly.

Data is written to and read from the disk by the read/write heads. Each head is an electromagnetic induction coil. The read/write head works in two modes:

- In the 'write' mode data is encoded into electrical pulses that are passed through the coil in the head causing a change in the local magnetic field. Because the head is only a microscopic distance above the platter this change in magnetic field is sufficient to reorient the magnetic alignment of a tiny area of the surface. The affected area retains this orientation until a subsequent write operation occurs. These areas are of a finite size.
- In the 'read' mode the head passively passes over the surface of the platter 'sensing' the changes in the magnetic alignment. These changes induce tiny electrical currents in the head that are converted back into data by the appropriate encoding algorithm.

We can therefore see that the basic principle is to write the data, not in 1s or 0s, but in changing magnetic orientations encoded on the surface of the platter which we can then interpret as a two-state system of 1s or 0s.

As the disk rotates, the position of an individual head describes a circular 'track'. There are at least two heads in a hard drive and their combined motions describe a 'cylinder'. If we then consider that there are only so many areas in a track that can be individually written to before the magnetic orientations become so compressed as to interfere with each other, we can segregate the track into 'sectors'. By describing the physical structure of the hard drive in this way we can 'address' the location of any data storage area in coordinates of cylinder, head and sector. This is known as CHS addressing. As part of the physical preparation of disks by the manufacturer they are physically formatted, *i.e.* arranged into discrete sectors for data storage. The individual sectors have a structure within themselves containing a 'header' noting their physical location (CHS) followed by error correction, the data area and finally more error correction information.

Modern hard drives use logical block addressing (LBA) where each sector is assigned a sequential number, starting at zero, and the hard disk controller remembers its physical location on the disk. Modern drives may still report the capacity of the drive in terms of the CHS but this is a legacy restriction imposed by the architecture and operating systems of older computers to ensure compatibility. Each time the hard disk controller receives a request to read or write data to or from a particular location on the drive it simply performs the appropriate translation to access that data.

9.3.2.2 Floppy Disks. Floppy discs were developed to enable data to be transferred or provide a back-up medium and generally provide flexibility. Although virtually obsolete now they are still encountered in casework.

As the hard disk drive is named for the rigidity of its platters, so the floppy disk is named for its flexibility. Instead of a rigid substrate, the magnetic compound is layered on to a single flexible plastic disk and encased within a rigid plastic case to provide protection. In the centre of the base of the disk is a circular piece of aluminium that provides for alignment of the disk in the floppy drive. The sliding aluminium shutter on the top of the disk provides access for the read/write heads. As with hard disk drives, the data is written on both sides of the disk. Rather than floating above the disk, the heads contact with the magnetic surface to read the information. This necessitates a reduction in the speed at which the disk can spin without the heads stripping the magnetic material from the disk. Floppy disks commonly spin at 300–360 rpm. The heads are also fairly crude devices compared with those on the hard disk drive, and so fewer tracks and sectors can be fitted onto the surface of the platter. This results in a much lower data capacity than a hard disk, 1.44–2 MB, but greater mobility and tolerance to damage.

A physical write-protection switch is built into the casing of 3.5 inch floppy disks. If this switch is in the closed position the floppy drive cannot write to the disk, and works in read-only mode. The only safe way to examine a floppy disk in a forensically sound manner is to ensure all disks are write-protected prior to examination. If disks are not write-protected it is possible that the examination process itself may interfere with the forensic integrity of the data on the disk by updating file 'Last Accessed' dates. This is discussed later in section 9.4.

9.3.2.3 Zip Disks. Zip disks look like a slightly larger chunkier version of a floppy disk. They contain a single magnetically responsive disk mounted on a spindle within a rigid plastic case. Access to the platter is via a small slot with a sliding aluminium cover on the top edge of the case. They have a much greater storage capacity than floppy disks, ranging from 100 MB to 750 MB. They require a proprietary hardware drive that can be internally mounted on the computer or attached externally.

Zip disks have no hardware write protection facility. Therefore precautions must be taken during the examination to prevent alteration of the media, *e.g.* disabling antivirus software, or imaging in a read-only operating system environment. As with floppy discs these are now virtually obsolete but still encountered in casework.

9.3.2.4 Digital Magnetic Tape.   Magnetic tape is commonly used as a backup medium, for archive purposes or for system recovery after failure. Access to the data is linear, *i.e.* the tape must be read from the start to the finish, and is therefore relatively slow compared to disk-based media. Difficulties arise in the examination of digital tape as it exists in many proprietary physical and logical formats and often requires the use of proprietary hardware or software to access the data.

Digital tape media come in all sizes and capacities, from the large 'dinner plate' tape spools to the DDS tape, similar in appearance to a small 8 mm video or DV tape. Modern operating systems sometimes provide their own native tape backup utility.

9.3.2.5 Solid-State Storage.   Solid-state storage is based on semiconductor technology. There are no moving parts. These media are rewritable, and the materials used retain the ability to reorient through hundreds of thousands of write cycles. There are currently a wide variety of solid-state media available for data storage. They range in shape, size and capacity depending on their use. They are commonly used in digital cameras and camcorders, MP3 players, as an alternative to floppy disks/CDs in the form of small USB 'thumb' drives, and as solid-state disk drives.

9.3.3 Optical Media

9.3.3.1 Compact Disc (CD).   The most common optical format examined in casework is still the CD, which in itself comes in numerous varieties—CD-ROM, CD-R, and CD-RW are the most relevant for our purposes. Physically a CD consists of an internal data storage layer or layers, sandwiched between two protective plastic layers in a 12 cm diameter disk. Data is stored in a series of pits arranged in a spiral around the centre of the disk. Data is read from the disk by a low-power laser passed over the surface and reflected back to a photosensor. When the laser beam encounters a pit on the surface an interference pattern is created in the beam and is interpreted by the photosensor and drive hardware. CDs have a physical capacity ranging from 650 Mb to 700 MB.

9.3.3.2 CD-ROM. CD-ROMs are commercially produced read-only media. The recording layer is physically pressed by the manufacturer before being coated with the reflective metal layer that gives them their distinctive silver appearance. Given their origins, the data contained within the disk is generally not of forensic interest other than in cases of suspected software counterfeiting on a commercial scale. The presence of disks containing applications relevant to the offence in question may, however, be of evidential interest to an investigator.

9.3.3.3 CD-R. CD-Rs are write-once media. They enable any user with a CD writer to create their own CDs. A laser in the CD writer 'burns' the pits into the recording layer, which in this case is an organic dye. The green, blue, or gold colour of the dye characterises CD-Rs. As the data is recorded on to the CD the appearance of the dye changes and the recorded area is visible around the centre when the CD is viewed at a slight angle.

9.3.3.4 CD-RW. CD-RWs are write–erase–write or rewritable media. They require a CD rewrite-capable drive. The recording layer is a metal alloy into which pits are burned by a relatively high-powered laser. The data is read as previously but there is the added functionality to 'melt' the recording layer back into its original form and delete any data present. This data is irrecoverable. The read–write process may be repeated many times (up to 1000) before the CD starts to fail. CD-RWs are similar in appearance to CD-ROMs, but with a smoky glaze over the silver layer and usually an absence of commercial artwork on the top of the CD.

9.3.3.5 Data Layout. Data on a CD is stored in sectors, normally containing 2048 bytes of data, arranged in a spiral track starting at the centre. The user can write to the CD in a single 'track' or a series of tracks up to the capacity of the CD. A 'session' is a collection of one or more tracks. When a session is completed, it may be left 'open', *i.e.* available for further data storage or 'closed', preventing further writing to that session. The CD itself may also be left open for future use or closed preventing further writing. Data may therefore be recorded to the CD over a period of time in a number of sessions, each session containing one or more tracks. Each session appends data to the CD, rather than replacing the data already present. It is quite common to find CD-Rs/CD-RWs containing data placed there on different dates.

A further method of writing to a CD is 'packet writing'. Using appropriate software the CD can be used similarly to a large floppy disk. The CD is formatted before use and files can then be 'dragged and dropped' on to it. If a CD-RW is used, data can even be erased from the CD and the space reused. The user adds data to the CD in sessions as previously. After adding data the user may again leave the session open or closed, or even close the entire CD.

9.3.3.6 Forensic Implications.   Difficulties can and do arise in the examination of CD-Rs and CD-RWs. The major difficulty is incompatibility between the CD and drive formats in general, and also in differences in quality of CDs and drives. CD-Rs and CD-RWs are potentially volatile media and thus the safest environment for their examination is a read-only CD-ROM drive.

Modern CD-ROM drives will read multisession CD-Rs, but older CD-ROM drives may not 'see' all of the sessions. CD-ROM drives will also not read CD-Rs created with the UDF packet writing applications unless the particular session has been closed. A session left open for further use is only visible to CD writer or CD rewriter drives with the appropriate UDF software installed. In this environment there is the potential for changes in the content of the media, and great care must be taken during the examination.

CD-RWs may be viewed in some modern CD-ROM drives, but not necessarily all. UDF packet CD-RWs may be read with some modern CD-ROM drives, but only if UDF software is available. Similarly, some CD writer drives will view the contents of CD-RWs. All formats of CD-RW will be readable in a CD rewriter drive, but again this is an environment where changes could potentially be made to the media. In these circumstances the examiner should be guided by Principle 2 (Section 9.2.1) for the handling of computer-based evidence.

The contents of CD-Rs/CD-RWs can be examined thoroughly with the aid of dedicated forensic software that allows examination of the track and session structure.

9.3.3.7 DVD Formats.   DVDs are the successors to CD-ROMs and are almost identical in appearance. Their data storage capacity is much greater and this provides for their use as entertainment media, *e.g.* for films. The technology behind DVDs is the same as that of CDs. Data is arranged in a spiral track and recorded in a series of pits that are interpreted by reflected laser light. Reduced spacing between pits allows greater data density and provides their increased data capacity. Further capacity can be gained by using both sides of the DVD and multiple data storage layers. As for CDs there are a number of types of DVD.

9.3.3.8 DVD-ROM. DVD-ROMs are commercially produced read-only media. Like CD-ROMs these are of limited forensic interest unless their content relates directly to an offence. Their capacity ran⁻ges from 4.7 GB to 17 GB.

9.3.3.9 DVD-R. DVD-Rs are write-once media with a capacity of 4.7 GB per side. They exist in two flavours: DVD-R(A) for professional DVD authoring, and DVD-R(G) for consumer use.

9.3.3.10 DVD-RW. DVD-RWs have read–write capability. DVD-RWs have a capacity of 4.7 GB per side, and similarly to CD-RWs can be rewritten approximately 1000 times.

9.3.3.11 DVD-RAM. DVD-RAMs are rewritable media using the same phase change technology as CD-RWs. DVD-RAM operates on the same principle as a very large floppy, allowing data to be dragged and dropped to a DVD, and even erased with up to 100 000 rewrite cycles. Their capacity ranges from 2.7 GB to double-sided media with capacity of 9.4 GB.

9.3.3.12 DVD + RW/DVD + R. DVD + RW is yet another rewritable format of the DVD with a capacity of 4.7 GB per side. DVD + R is a write-once implementation of the same format with a capacity of 4.7 GB.

9.3.3.13 Forensic Implications. Again the difficulties arise through incompatibility. Any modern DVD-ROM drive can read most DVD media with the exception of DVD-RAM (DVD-R + RW is fully compatible with all DVD-ROM drives, DVD-RW may not be). They are therefore relatively easy to examine in a forensically sound manner.

DVD-RAMs require the use of a DVD-RAM drive, which obviously presents the danger of writing to the DVD. As for all other volatile media, adequate precautions should be taken in their examination to ensure forensic integrity.

9.3.4 Magneto-Optical Media

9.3.4.1 Magneto-Optical Discs. Magneto-optical discs are a combination of the technologies used in magnetic and optical storage devices. These discs and the technology that support them are more expensive than other storage options and tend only to be used by larger commercial concerns for the storage and backup of important information. The concept is based on magnetically responsive discs and strong magnetic fields are required to affect the orientation of the media, thus providing protection against accidental corruption or loss, together with a long data life (up to 390 years is claimed). In order to write to the disc, a laser beam is used to heat the magnetically responsive material of the disc while it is exposed to a small magnetic field. This allows the relatively weak magnetic field to reorient the heated area of the disc, which retains this orientation after cooling.

Magneto-optical discs are available in both rewritable and write-once read many times (WORM) forms. They exist in a number of physical formats and capacities: 5.25 inch, 12 inch, 1.2 GB to 30 GB. The discs can only be read in a magneto optical drive matching their physical characteristics.

9.3.4.2 'Super' Floppies. A cheaper form of magneto-optical disc is the LS120 super floppy technology. Based on a 3.5 inch floppy disk, this uses normal floppy disk read/write technology with the head alignment accuracy improved by the use of a laser beam. The disc requires the use of a proprietary hardware drive and has a capacity of 120 MB. The super floppy incorporates a write protection switch similar to a normal floppy.

9.4 LOGICAL STRUCTURE

9.4.1 Partitions and Logical Drives

We have seen how data is physically stored on media, but how does this relate to the operating system and its 'logical' data structure? Logically, we talk in terms of 'drives' rather than physical disks. A physical hard disk drive, as illustrated in Figure 9.1, may be 'partitioned' into logically

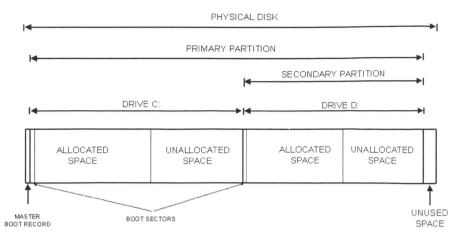

Figure 9.1 Physical and logical disk structure.

discrete data storage areas within the hard disk drive, each containing its own independent file structure and format. Partitions can contain one or more logical drives, or even another partition. Removable media such as floppy disks, CDs, *etc.* do not normally contain more than one partition.

At the start of the physical hard disk is the master boot record which contains information relating to the number of partitions on the disk, where they start and finish, which partition the operating system resides in and the type of file system they use—*e.g.* file allocation table (FAT) or new technology file system (NTFS). The logical volumes within these partitions are referred to as 'drives' in common usage and are assigned letters by the operating system to identify them. Letters A: and B: are historically reserved for the floppy drives. The initial boot drive is normally the C: drive. At the start of each logical drive is another boot sector containing the partition table for that partition. Further hard disks (if any) are assigned drive letters after C: followed by any further logical volumes. CD-ROMs, DVDs, Zip drives, *etc.*, are normally assigned letters following the drives contained on the physical hard disk, *e.g.* D:, E:, *etc.*

9.4.2   Directory Structure

In Windows operating systems the data in a drive is arranged in a tree or directory structure, with a 'root' folder off which branch folders or files. The 'branch' folders themselves contain further folders or files. As an analogy we can think of a drive as a filing cabinet. The filing cabinet is the root folder. The drawers in the cabinet represent the folders branching off the root folder. Contained within the drawers are further folders and files. A hard disk drive has a huge data storage capacity and, as mentioned, is often partitioned into smaller discrete areas or drives. To further our analogy, a hard disk drive can be thought of as a room containing one or more filing cabinets. Each cabinet then represents a drive on the hard disk and has its own directory structure of drawers with their folders and subfolders.

9.4.3 File Allocation Table and Master File Table

To keep track of the data within a drive the operating system makes use of an 'index', usually found at the start of the drive. In older systems such as Windows 98 this is known as the file allocation table (FAT). In more modern systems such as Windows XP or Vista the FAT is still used (FAT32). Each Windows directory contains a record (directory entry) for each file or folder in that directory detailing their name and file extension, their starting location, their size, their relevant dates and times, and the file attributes, *i.e.* system, hidden, archive or read-only. The space physically occupied by the file is recorded within the FAT. The FAT has an entry for each cluster on the drive. If the cluster has been allocated for storage the FAT entry indicates whether the file allocated that cluster resides there solely and if not, points to the next cluster occupied by that file and so on. The concept of file structure and clusters is discussed later.

In NTFS operating systems such as Windows NT or Windows 2000, the situation is more complex. Each file or folder (as folders are essentially just another type of file) has its own separate entry in the 'master file table' which records all relevant information relating to the file. In fact, an entire file can exist solely within the master file table, which is in itself a file. A further file known as the 'cluster bitmap', a representation of the volume showing which clusters are in use, keeps a record of which clusters have been allocated for data storage.

9.4.4 Allocated and Unallocated Space

In discussing the logical structure of the disk the concept of space being 'allocated' to a file was introduced. This is a convenient way to think of how data exists on the disk or in fact on any media. Any file that is stored on a disk has an area that it is allocated for storage. The combined allocated areas of all the files on the logical drive are often referred to collectively as 'allocated space'. Allocated space is therefore the areas of the disk currently assigned by the operating system for the storage of files. Obviously, unless the logical drive is full, not all of the available space is allocated for the storage of data. The remainder is known as 'unallocated space' (see Figure 9.1).

The amounts of allocated and unallocated space are dynamic and change, as files are created and deleted, but always add up to the size of the logical drive. The physical disk may be larger than the combined sizes of the logical drive or drives contained within it, leaving 'unused space' at the end of the drive. There are a number of possible reasons for this, including the fact that the sizes of logical drives are entirely user definable whereas the manufacturer defines the physical disk size at the time of production. It is possible and conceivable that the user may wish to define their logical drives or partitions to be smaller than the physical constraints of the disk. Unused space can also be created by mathematical constraints on the way space is allocated.

9.4.5   File Structure

The structure of a file is roughly illustrated in Figure 9.2.

9.4.5.1   Header.   Included in the directory entry for a file is its file extension, *e.g.* 'txt' for a text file, 'jpg' for a JPG picture file, *etc.*, and these identify the file by type to the operating system and any software application. Additionally, within the body of the file itself is a sequence of bytes characteristic to the particular file type. This is known as the 'header'. Each type of file has its own header signature that uniquely identifies it. This signature is contained in the first few bytes of the file and it is this code that the operating system or application recognises. These signatures can easily be seen with a hex viewer application. For example, Figure 9.3 shows the header signature for a simple JPG picture file.

The first two bytes read hexadecimal 'FF D8'. This is the characteristic header signature of a JPG file. Other information is clearly visible within the file, *e.g.* 'Photoshop 3.0' suggesting this application was used

Figure 9.2   File structure.

```
            0  1  2  3  4  5  6  7   8  9  A  B  C  D  E  F
00000000   FF D8 FF ED 00 10 4A 46   49 46 00 01 02 01 00 F0    yøya..JFIF.....6
00000010   00 F0 00 00 FF ED 0B 1C   50 68 6F 74 6F 73 68 6F    .6..yí..Photosho
00000020   70 20 33 2E 30 00 38 42   49 4D 04 04 00 00 00 00    p 3.0.8BIM......
00000030   00 43 1C 02 00 00 00 02   00 02 1C 02 78 00 0C 59 6F .C.........x..Yo
00000040   73 65 6D 69 74 65 2C 20   43 41 1C 02 74 00 26 4A    semite, CA..t.&J
00000050   61 79 20 54 6F 72 62 6F   72 67 20 31 39 39 39 20    ay Torborg 1999
00000060   28 77 77 77 2E 74 6F 72   62 6F 72 67 2E 63 6F 6D    (www.torborg.com
00000070   2F 6A 61 79 29 00 38 42   49 4D 03 ED 00 00 00 00    /jay).8BIM.í....
00000080   00 10 00 F0 00 00 00 01   00 01 00 F0 00 00 00 01    ...6........6....
00000090   00 01 38 42 49 4D 04 0D   00 00 00 00 00 04 00 00    ..8BIM..........
000000A0   00 78 38 42 49 4D 03 F3   00 00 00 00 00 08 00 00    .x8BIM.ó........
000000B0   00 00 00 00 00 00 38 42   49 4D 04 0A 00 00 00 00    ......8BIM......
000000C0   00 01 01 00 38 42 49 4D   27 10 00 00 00 00 00 0A    ....8BIM'.......
000000D0   00 01 00                                             ...
```

Figure 9.3   JPG header signature.

to create the image. Although not forming part of the file structure this sort of information is commonly included in files.

9.4.5.2   Data Block.   The layout of the data block depends on the type of file. Some file formats are universal standards common to numerous applications, *e.g.* JPG files, whilst others, such as ART picture files, are proprietary. The data contents of some types of files, *e.g.* Microsoft Word files, can be clearly seen in plain text. The layout of the document is contained in formatting information that surrounds the plain text, and so the actual appearance of the document cannot be immediately ascertained. As will be discussed later it is possible to search allocated and unallocated space for this plain text. It is this that allows us to recover even partial fragments of deleted files from unallocated space. An example of such data content is shown in Figure 9.4, again as viewed with a hexadecimal viewer.

The content of the document is clearly visible in plain text. As well as formatting information, details such as the author and the full path of where the document was stored are also included within the file, as shown in Figure 9.5.

The value of such information as author details must not be overstated. The author details are most likely those that were input when the software was installed and registered by the user. They are not necessarily those of the user who created the document. Information such as file paths may be of great evidential and investigative importance, particularly if they point towards removable media, which may contain the complete file. The example given is specific to MS Word documents but the principle may apply equally to other file formats, but others more may appear incomprehensible when examined.

```
          0  1  2  3  4  5  6  7   8  9 10 11 12 13 14 15

00001456  00 00 6D 05 00 00 00 00  00 00 AA 01 00 00 00 00   ..m.......a.....
00001472  00 00 6C 01 00 00 00 00  00 00 BC 00 00 00 12 00   ..l.......X.....
00001488  00 00 CE 00 00 00 0E 00  00 00 A8 00 00 00 00 00   ..Î.......".....
00001504  00 00 A8 00 00 00 00 00  00 00 A8 00 00 00 00 00   ...".......".....
00001520  00 00 A8 00 00 00 00 00  00 00 02 00 D9 00 00 00   ...".........Ù...
00001536  0D 0D 48 65 6C 6C 6F 2E  20 20 54 68 69 73 20 69   ..Hello.  This i
00001552  73 20 61 20 65 78 61 6D  70 6C 65 20 64 6F 63 75   s a example docu
00001568  6D 65 6E 74 20 74 6F 20  69 6C 6C 75 73 74 72 61   ment to illustra
00001584  74 65 20 66 69 6C 65 20  73 74 72 75 63 74 75 72   te file structur
00001600  65 2E 0D 00 00 00 00 00  00 00 00 00 00 00 00 00   e...............
00001616  00 00 00 00 00 00 00 00  00 00 00 00 00 00 00 00   ................
00001632  00 00 00 00 00 00 00 00  00 00 00 00 00 00 00 00   ................
00001648  00 00 00 00 00 00 00 00  00 00 00 00 00 00 00 00   ................
00001664  00 00 00 00 00 00 00 00  00 00 00 00 00 00 00 00   ................
00001680  00 00 00 00 00 00 00 00  00 00 00 00 00 00 00 00   ................
```

Figure 9.4   Data content within file.

```
          0  1  2  3  4  5  6  7   8  9  A  B  C  D  E  F

00001330  0E 00 4A 00 6F 00 6E 00  61 00 74 00 68 00 61 00   ..J.o.n. . . . .
00001340  6E 00 20 00 48 00 65 00  6E 00 72 00 79 00 1C 00   . .H.e.n.r.y...
00001350  43 00 3A 00 5C 00 4D 00  79 00 20 00 44 00 6F 00   C.:.\.M.y. .D.o.
00001360  63 00 75 00 6D 00 65 00  6E 00 74 00 73 00 5C 00   c.u.m.e.n.t.s.\.
00001370  74 00 65 00 73 00 74 00  20 00 64 00 6F 00 63 00   t.e.s.t. .d.o.c.
00001380  2E 00 64 00 6F 00 63 00  FF 40 03 80 01 00 18 00   ..d.o.c.ÿ@.□....
00001390  00 00 18 00 00 00 30 98  8F 02 01 00 01 00 18 00   ......0~□.......
000013A0  00 00 00 00 00 00 17 00  00 00 00 00 00 00 02 10   ................
000013B0  00 00 00 00 00 00 00 19  00 00 00 90 00 00 08 00   ..........□....
000013C0  40 00 00 FF FF 01 00 00  00 07 00 55 00 6E 00 6B   @..ÿÿ......U.n.k
000013D0  00 6E 00 6F 00 77 00 6E  00 FF FF 01 00 08 00 00   .n.o.w.n.ÿÿ.....
000013E0  00 00 00 00 00 00 00 00  00 FF FF 01 00 00 00 00   .........ÿÿ.....
000013F0  00 FF FF 00 00 02 00 FF  FF 00 00 00 00 FF FF 00   .ÿÿ......ÿÿ....ÿÿ.
00001400  00 02 00 FF FF 00 00 00  00 03 00 00 00 47 16 90   ...ÿÿ........G.□
00001410  01 00 00 02 02 06 03 05  04 05 02 03 04 87 3A 00   ............‡:.
00001420  00 00 00 00 00 00 00 00  00 00 00 00 00 FF 00 00   ............ÿ..
00001430  00 00 00 00 00 54 00 69  00 6D 00 65 00 73 00 20   .....T.i.m.e.s.
00001440  00 4E 00 65 00 77 00 20  00 52 00 6F 00 6D 00 61   .N.e.w. .R.o.m.a
00001450  00 6E 00 00 00 35 16 90  01 02 00 05 05 01 02 01   .n...5.□........
00001460  07 06 02 05 07 00 00 00  00 00 00 00 10 00 00 00   ................
00001470  00 00 00 00 00 00 00 00  80 00 00 00 00 53 00 79   .........□....Σ.y
00001480  00 6D 00 62 00 6F 00 6C  00 00 00 33 26 90 01 00   .m.b.o.l...3&□..
00001490  00 02 0B 06 04 02 02 02  02 02 04 87 3A 00 00 00   ............‡:..
000014A0  00 00 00 00 00 00 00 00  00 00 00 FF 00 00 00 00   ............ÿ....
000014B0  00 00 00 41 00 72 00 69  00 61 00 6C 00 00 00 22   ...A.r.i.a.l...™
```

Figure 9.5   Typical information within MS Word file.

9.4.5.3   Footer.   The file footer is the signature that indicates the end of the file to the software application, which then knows to read no further data for this file. Returning to our example of the JPG picture file, Figure 9.6 shows the end of the file.

The last two bytes contain the hexadecimal characters 'FF D9'. This is the footer signature for a JPG file. It should be noted that not all files have a common footer signature.

Above we have noted the information that can be found within the file. What cannot normally be determined from examining the file only is information such as the file name and its relevant date and time stamps.

```
            0  1  2  3  4  5  6  7    8  9  A  B  C  D  E  F
0001E910    F4 EB 40 60 18 D0 FD 0D   54 10 0D 33 B4 DF CA 80    ôë@`.Ðý.T..3´ßÊ□
0001E920    81 3E 07 C3 4A A1 82 47   43 7F 2A C6 EB F8 1A 08    □>.ÃJ¡¬GC□*Æëø..
0001E930    82 46 87 E9 57 41 4F 91   A2 38 DE D0 68 04 CF 81    ¬F‡éWAO'¢8ÞÐh.Ï□
0001E940    FA 1A 8A 0B 12 FA C1 83   24 88 D6 A2 3F FF D9       ú.o..ûÁf#¢Ö¢?ÿÙ
```

Figure 9.6   Footer signature for JPG file.

This information is not normally contained within the file itself but in its directory entry (though there are exceptions). Such information may, however, be inferred from the contents of the file, *e.g.* where a date is used in a letter.

9.4.6   Dates and Times

Contained within the directory entry for Windows files are two sets of dates and times and one date. These represent respectively the 'file created' date and time, the 'file last modified' date and time and the 'last accessed' date. They are collectively referred to as date and time stamps. For Windows NTFS operating systems the last accessed date and time are recorded, along with a fourth date and time representing the date and time the master file table (MFT) entry for that file was modified. In order to understand the significance of the dates it is first necessary to have an understanding of what they represent.

9.4.6.1   File Created Date and Time.   This is set when the file initially comes into existence on a particular logical drive, regardless of its source, *e.g.* created using an application or downloaded from the Internet. It is not a permanent record of the date and time a file was created, only when it initially existed on this logical drive. Within a logical drive, files may be 'moved' between directories without changing the file created date and time. The term 'move' means using drag and drop or cut and paste for single or multiple files. However, if a file is 'copied' from one directory to another using copy and paste, or drag and drop with the Ctrl key depressed, then the file created date and time will be updated. This makes sense, as effectively the copy command brings into existence a second version of the original that was created at the time of copying. If a file is moved or copied between two logical drives, normally a second copy of the file is created on the target drive leaving the original on the source drive and the file created date and time of the new file will therefore be updated. To further complicate matters, if a file is dragged and dropped between two logical drives with the shift key depressed or using cut and paste then the file is removed from the source drive and placed on the target drive and the file created date and time remain unchanged.

9.4.6.2 File Last Modified Date and Time. The file last modified date and time represents the date and time that the system last modified the contents of the file and wrote the changes to the disk. Logically this is initially when the file was created and changes every time the file is modified and the changes saved. Viewing the contents of a file or changing its name does not affect the last modified date and time, as these are changes to the directory entry and not the data contained within the file itself. When a file is moved within or between logical drives the last modified date and time is not affected. The contents of the file have not after all changed, only its location.

9.4.6.3 File Last Accessed Date. The last accessed date for a file is simply the date on which the user last accessed it. It may be thought of as a 'hair trigger' date as it is updated each time the file is accessed and each access overwrites the previous date. No record exists of previous access, only the most recent.

9.4.6.4 Forensic Implications. Interpretation of file dates can provide information of what the user may have known or done with files. For example, an apparent anomaly often arises where examination shows that the created date and time for a file is after the file was last modified by the system. Interpreted literally, this suggests that the file was somehow changed before it ever existed. This can obviously never be the case. However, an understanding of how the dates and times work suggests that the file may have been initially created on an unknown date, prior to its last modified date, and copied to its present location on the date now stated as its file created date. This does not tell us much about the origin of the file, it may have been simply copied from another directory or even extracted from a zip archive file, and the result would be similar. What it may imply, however, is knowledge on the part of the user of the file's existence in that they have performed actions in relation to it.

Individual software applications may affect date and time stamps differently. Compression software is a good example of this. Winzip, a very common application, maintains the last modified date only when it compresses a file. When uncompressed, the file created date and time is exactly that of the last modified date and time. Other compression programs, however, *e.g.* EnZip, similarly maintain only the last modified date on compression, but on decompression give the file created date and time as per the current date and time, but maintain the last modified date and time as per the original file.

The dates and times used by the operating system are obtained from an electronic clock on the motherboard with its own small independent power supply. As the computer starts up, the operating system reads the current date and time from this clock and applies this to itself. Similarly, if the user changes the date and time in the operating system environment, this updates the system clock. The impact of this is that the dates and times associated with files are dependent on the accuracy of the system clock and are subject to manipulation by the user. It is an essential part of any examination of a computer to establish the date and time as recorded in the system BIOS and note any discrepancy.

9.4.7 Sectors and Clusters

Previously we have noted that the sector is the smallest physically addressable area on a disk. Logically a sector corresponds to 512 bytes of storage capacity (with the exception of CDs with a sector size normally of 2048 bytes). Historically, the origin of hard disks and modern storage media is the floppy disk. For small-capacity media such as the floppy disk, data can be effectively and efficiently written to the disk sector by sector. The capacity of storage devices has increased rapidly since the days of floppy disks, and to access huge numbers of sectors individually is now neither effective nor efficient. To overcome this difficulty the operating system may allocate data not to individual sectors but to blocks or 'clusters' of sectors. The number of sectors in a cluster is determined by the size of the disk and how it is logically formatted, *i.e.* FAT12, FAT16, FAT32 or NTFS. Whatever the logical size of a file, it is allocated disk space in a single cluster or multiples of clusters. If a file is larger that a single cluster, a second cluster is assigned for the excess, and so on. The sectors within a cluster are contiguous, that is to say they occupy areas of the disk that are adjacent to each other. Multiples of clusters forming the same file however may not be contiguous but 'fragmented', *i.e.* occupying areas throughout the disk.

9.4.7.1 Slack Space. Although data is allocated to files in cluster-sized blocks, it is written to the disk in sector-sized chunks of 512 bytes at a time. A whole cluster, or multiple of clusters, is assigned for the storage of a particular file and that file only, regardless of how little of the cluster it occupies. The forensically interesting thing about this is contained within the final cluster (or at the end of its only cluster) for each file. The whole cluster is allocated to the file but the file may not be large enough to 'fill' it completely.

Consider the theoretical situation as illustrated in Figure 9.7. The operating system is allocating space to files in four sector/cluster blocks, *i.e.* 2048 bytes. The file 'Hello.txt', a simple text file, is 800 bytes in size, but is allocated one cluster (2048 bytes) of storage space. The operating system can only write to the disk in sector-sized 512-byte chunks, so it writes 512 bytes to each of the first two sectors in the cluster and leaves the remainder untouched, but still allocated to 'Hello.txt'. This raises two interesting situations:

- The file is only 800 bytes in size, but the operating system has written 1024 bytes of data to the first two sectors. Where did the extra 224 bytes of data come from? The answer is from whatever surplus data the operating system has lying around in its data buffers or registers from previous operations it was performing. Thus, data that the user has been working on may be found in the 'slack space' between the end of the file and the end of the last sector the file was written to. This will obviously only be small amounts of data, but potentially could include user names or fragments of documents. This type of slack space on the disk is referred to as 'buffer' or 'file slack'.

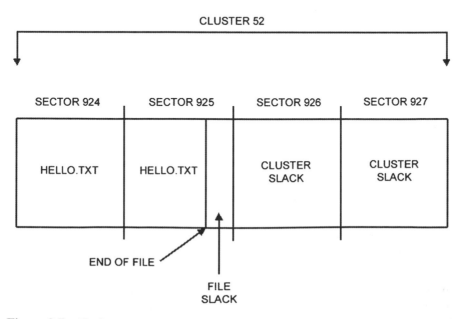

Figure 9.7   Slack space.

- The remaining two sectors in the cluster are untouched by the file being written to the disk. Their contents are quite simply whatever was last written to those sectors by the operating system. This is due to the manner in which files are deleted, which will be discussed later. Therefore the sectors may contain the remnants of whatever file was last allocated to this cluster but not overwritten by 'Hello.txt'. This type of slack space is referred to as 'cluster slack'.

The example used of 4 sectors of cluster is artificial; ratios of 32 or 64 sectors/cluster are more likely, giving us a possible cluster slack space approaching 32 kB in size. This is a considerable amount of storage space and could contain a great deal of evidential information. As slack space forms part of the area assigned to a file it will remain untouched unless the file 'grows' larger, when it will start to overwrite its own slack space. To the normal user of the system, slack space is invisible. Slack space is part of what we have considered to be allocated space, even though it contains the remnants of deleted files that once formed part of unallocated space. It should be noted that when a file is moved or copied from one logical drive to another it does not retain the slack space of the original file, either file slack or cluster slack.

9.5   CONTENTS OF ALLOCATED SPACE

All files that are 'live' exist in allocated space, *i.e.* the space assigned by the operating system for storage of those files. If a file exists on a storage device it must have been created by a user or by an application as a part of its operating processes, or received from an outside source, *e.g.* the Internet or email. Although the content of any file in allocated space may be of an evidential nature, the following are examples of files of particular forensic interest. Details of Internet activity as discussed in section 9.7 may also fall within the boundaries of allocated space.

9.5.1   Link Files

Link files are 'shortcuts' created by the Windows operating system for ease of use by computer operators. They are the shortcut icons on the Windows desktop, the list of recently opened documents we see when we click on 'Start, Documents', the contents of the 'Start Menu' and the list of 'Send To' options presented when we right-click on a file. Link files have the file extension '.lnk'.

9.5.1.1 Desktop. The desktop icons we look at each time we use Windows are contained within the 'Windows\Desktop' folder. Their purpose is to provide quick and easy access to installed programs or a favourite file or folder. The presence of link files on the desktop can indicate which applications or files the user wishes to access frequently or quickly, *e.g.* an encrypted disk area that the user can simply drag and drop files into. Further, they may indicate a degree of knowledge on the part of the user that a relevant application was installed or a file present on their computer.

9.5.1.2 Recent Documents. The list we see in the 'Start, Documents' menu is of link files contained in the 'Windows\Recent' folder relating to the most recent files accessed by the user. These are particularly valuable links. They may point to files now deleted, files the user accessed over a network and never saved locally, files stored on removable media or even files in encrypted volumes. The presence of the link file proves that someone using the computer has accessed these files. It is not enough that the user has right-clicked, moved or deleted a file; in order to create a link file in the 'Recent' folder, they must have opened the file.

9.5.1.3 Start Menu. The contents of the start menu are link files within the 'Windows\Start Menu' folder. As previously, they may prove knowledge of the user that applications relevant to particular offences were installed on the system.

9.5.1.4 Send To. When we right-click on a file or folder, we are presented with a series of options including 'send to'. The link files within the 'Windows\Send To' folder provide these links to applications or physical locations. These links can provide very valuable indications to the examiner that the suspect is using removable media for data storage, file encryption software or even secure deletion applications.

Link files are extremely valuable sources of information about the files they relate to. A link file is just like any other file, with date and time information of its own. However, contained within the link file itself is not only the name and location of the target file, but also the date and time stamps of the target file. This information would for example, assist an examiner in connecting a file found on removable media to a link file found on the computer, by examination of date and time stamps, thus proving access on the suspect computer. The format of the information stored within the link files is such that all the data content is not viewable in plain text. Forensic utilities are available that allow examiners to extract information from these types of file.

9.5.2   System Swap File

When running applications, computers make use of their RAM to hold short-term data necessary to perform tasks. Only a finite amount of RAM is available to use, so the operating system also uses temporary space on the hard disk as memory. This operation is known as 'paging'. The area used for paging is referred to commonly as the 'swap' file. In Windows operating systems the swap file is a single system file whose size is dynamic and controlled by the operating system (by default), or set to a fixed size by the user. In Windows 95/98 systems this file is called 'win386.swp' and resides in the root folder of the operating system partition. In Windows NT and 2000 the swap file is called 'pagefile.sys' and again resides in the root folder of the operating system partition.

9.5.2.1   Forensic Implications.   The contents of the swap file can literally be any part or all of any file—a document, image or other—that was created, accessed or edited, either locally, *i.e.* on this machine, or remotely via a network. Obviously this material is of great interest in any investigation. The swap file is a hidden system file and its operation is invisible to the user. Its contents can, however, be examined with any file viewing utility, *e.g.* Quick View Plus.

As noted, the size of the swap file can be fixed or dynamic, defaulting to dynamic unless fixed by the user. A dynamic swap file increases and decreases in size automatically as required by the operating system and can reach a considerable size (hundreds of megabytes). A great deal of information may therefore be contained within it. Its contents are continually written and overwritten, and as its size decreases the excess becomes part of unallocated space. Unallocated space is further discussed in section 9.8. A fixed swap file will not change size, and its contents, as they are written and overwritten, will remain in allocated apace.

9.5.3   Digital Cameras

The use of digital cameras by commercial and private photographers is now widespread. An integral part of digital photography is the use of computers for image enhancement or printing.

9.5.3.1   Storage Formats.   Digital cameras store their images in internal solid-state memory, or removable media, *e.g.* solid-state 'cards' or microdrives. Most modern cameras use the latter. There are a number of different physical solid-state formats available including Smartmedia, compact flash and secure digital (SD) cards, and the Sony Memory Stick. A further high-capacity storage method, the 'microdrive' employing a miniaturised hard drive, is available for some cameras with up to 1 GB of capacity currently available. Data format on the drive is in a FAT format with a sector size of 512 bytes. Images are stored in any of a number of different graphical formats (JPEG, TIFF, *etc.*) depending on the size of the images and the quality required by the user.

These media are rewritable and the data contained on them is therefore volatile. Write-protection is available on most of the media formats and should be used where present to ensure integrity of evidence. Users either access the images directly from the camera using a proprietary cable and software provided by the manufacturer, or make use of a hardware card reading device. The card or microdrive is simply treated by the operating system as a removable disk.

9.5.3.2   Exif Data.   All images on camera storage media have the date and time information associated with any normal file recorded within their directory entry. In addition, when an image is taken with a digital camera it is very common for technical data about the image, known as 'exif' data, to be recorded. This data is recorded within the body of the image by the camera. Exif data normally includes information about exposure, aperture, ISO speed, *etc.* However, it may also contain information about the date and time the image was originally taken, the make, model, and possibly even the serial number of the camera used. These may be preserved when the image is transferred to the hard disk drive or CD-ROM for storage or archiving, providing very valuable information about its source.

Whether or not the exif data is preserved depends on number of factors including the software used to transfer the image from the camera, the format chosen by the user for storage, *e.g.* JPEG or TIFF, and the use of image manipulation software. The exif data is visible to many utilities including the more common image manipulation packages, or utilities designed specifically for reading exif data such as 'Exifread'. Note that the dates and times recorded in exif information are set by the camera clock and are dependent on these settings being correct or even having been set at all. They are therefore subject to manipulation by the user.

9.5.3.3   Deleted Images.   The deletion process is the same as for any other file. The unallocated space on the disk will therefore contain the data from any deleted images. Because the space available on the storage media is limited, it is likely that images will be overwritten very quickly after deletion if the camera is in constant use.

9.6   CONTENTS OF UNALLOCATED SPACE

The contents of unallocated space (if anything at all) are the remnants of data that must once have existed on the system as 'live' files, and have subsequently been deleted. The files could have come to be on the system in any of the ways listed for the contents of allocated space. What is important is that the files must have existed as live files on the media in order for them to now reside in unallocated space. Unallocated space exists on any storage device that is not filled to capacity. Examination of the contents of unallocated space can provide valuable evidence or suggest further directions of enquiry to the examiner.

The contents of unallocated space are invisible to the user accessing the storage device through their operating system. However, forensic tools, from complex imaging and examination tools to simple hexadecimal viewers, allow examiners to see the physical data layout, *i.e.* to look beyond the logical data structure as shown by the operating system and examine what data lies outside the boundaries of live files.

The following is a description of the general process of deletion, and examples of the type of processes that normally take place in the general use of a computer that may leave traces of this activity in unallocated space. They are only a few examples; the list of potential processes leaving traces in unallocated space is endless.

9.6.1   Deleted Files

As explained earlier, files are allocated storage space on the logical drive by the operating system and their locations are tracked by means of their directory entry and FAT, or *via* the master file table. But what happens when a file is deleted? Contrary to what is commonly believed, data is not removed from the drive.

9.6.1.1   The Deletion Process.   The detailed mechanics of the deletion process are not vital to understanding the forensic implications. Windows FAT and NTFS systems handle file deletion in different ways. They both make use of a 'recycle bin'. For FAT systems there is only one recycle bin, whereas for NTFS systems a folder called 'recycler' exists containing a recycle bin for each user profile. When the user initially deletes a file, the Windows operating system effectively 'places' the file in the recycle bin from where it can either be deleted (by emptying the recycle bin) or restored. When the recycle bin is emptied the operating system marks the area previously occupied by the deleted file as available for reuse, so that the file is no longer 'addressed' by the operating system. The data remains in its original location and form until the operating system uses that same space again to store another file. What it is important to understand is that regardless of the method of deletion used by the operating system, the data is in no way affected by the process.

9.6.1.2 The Recycle Bin. Examination of the recycle bin entries can provide valuable information for the examiner. The naming of files in the recycle bin is significant, as it adheres to a strict protocol. Files are named according to the convention 'D<Original drive letter>-<number>.original file extension' *e.g.* 'DC1.doc'. This may point the examiner or investigator towards files in removable media of which they were previously unaware. A special system file called the INFO2 file also resides in the recycle bin. This file is hidden from the normal user but can be seen on an imaged disk just as any other file. Its purpose is to record the names and locations of files that have been deleted, and the date and time of their deletion to allow them to be restored if necessary. The exact format of this file is dependent on the variant of Windows installed, but the names and original location of files are in plain text, and the dates and times recorded in Windows date and time format. When the recycle bin is emptied, the file entries and the INFO2 file are deleted.

9.6.1.3 Forensic Implications. The essentials of the above are that on deletion, the data contained in a file is not 'gone', it is merely 'hidden' from the operating system and the space it occupies made available for reuse. While the data is deleted, it still resides in the space previously allocated to it unless overwritten. Thus deleted data may be overwritten, either completely or partially, and become corrupted. Large files often exist in multiple fragmented clusters and after deletion become individual fragments of the file, each of which may be overwritten or remain separately intact in unallocated space.

It is possible to 'undelete' a file even after the recycle bin has been emptied. This can be done manually, but realistically is best left to software designed for this purpose. When files are undeleted the examiner has all of the information associated with the original file, such as dates and times, available to them. The only information that cannot be recovered is the first letter of the file name, which unfortunately is overwritten in the deletion process.

Even though a file cannot be successfully undeleted, it may still be intact in unallocated space. By searching unallocated space for the header and footer signatures of particular file types we can effectively 'recover' its contents. Again, it is possible to perform this task manually, but the facility is built into many modern forensic examination tools. Even if the file has been partially overwritten, some of its contents may still be visible, as shown in Section 9.4.5.2. Although these do not form complete files they may still contain valuable evidence.

There are evidential difficulties with files recovered from unallocated space. We cannot state the date and time attributes of even a complete file found in unallocated space, as we have no directory entry or master file table entry relating to the file. It is possible that there may be information contained within the file from which we may infer an approximate date of creation, *e.g.* the contents of a letter may contain a date of writing or refer to incidents whose timing is known.

We have previously referred to slack space as part of allocated space, which it correctly is. However, the contents of the cluster slack were in unallocated space until the cluster was reallocated, so we are really looking at a piece of unallocated space. It is possible to infer date and time information for data contained in cluster slack not only from content but also from the date and time information of the file that overwrote it, *i.e.* the data in cluster slack must have existed prior to the file that overwrote it.

9.6.1.4 Wiping. Deletion is often confused with other actions such as wiping or formatting the drive. Wiping is better described as secure deletion. It differs from deletion in that wiped files have their directory entries and their allocated space physically overwritten by a random or even user-definable series of characters. The overwriting is often repeated to allow for deviations in the movement of the disk read/write heads. Free space or unallocated space wiping is also often available where the contents of unallocated space are overwritten as above. If files are properly wiped they are irrecoverable by normal forensic means. That is not to say there will be no evidence of their having been present at some time, *e.g.* link files.

9.6.1.5 Formatting. Formatting a drive or disk is normally done initially to prepare the drive for use. Reformatting a drive or disk that has been in use will destroy the FAT/MFT and mark all space on the drive as available. This effectively deletes all the pointers that mark the start of files but leaves the data that formed the files intact in their original locations.

9.6.2 Word-Processed Documents

9.6.2.1 Temporary Files. Word processing applications are essential parts of most business and personal computer software libraries. Users may create or edit their own documents, or those created by others, or simply view the contents of a document previously created on another computer. In all of these situations it is possible that all or part of the document may be left in unallocated space long after the user has viewed, printed, deleted or saved the document to removable media. This is because of the way in which applications of this type operate.

Currently, the most commonly used word processing application is Microsoft Word. Word makes extensive use of temporary files to allow normal operation of the computer to continue unhindered while the user works on documents. Temporary files are also created to ensure that data is not lost when the application performs certain tasks, *e.g.* when saving a file. When Word has completed the task for which the temporary file was created, it is deleted. Word uses a large number of temporary files, and their interactions can be complex. However, a few deserve particular mention such as those created during the normal save process, and as automatic saves performed by the application to recover the file if the application is unexpectedly terminated, *e.g.* by a system crash. These temporary files are a copy of the original file and contain all the information normally associated with that file: content, author details, full path, *etc.* When the application has completed the task for which the temporary file was created, it is deleted and enters unallocated space.

9.6.2.2 Locations of Documents. The locations of user-created files in general can be of significant evidential value. If a user creates a new folder to store documents or images they have created, Windows will by default name that folder 'New Folder'. A second new folder in the same directory will be named 'New Folder 2', and so on. If the user renames the new folder 'Young' and stores indecent images of children in it, for example, the implication is that the user created the folder for that specific purpose, and was aware of the presence of such images and their contents. The principle is equally relevant to documents.

The default save location for Word files is the 'My Documents' folder, though the user can reconfigure this. If a Word document is saved elsewhere it must have been placed there by the user. For Windows NTFS systems (*i.e.* all recent versions of Windows), a separate profile exists for each user with their own 'My Documents' folder.

9.6.2.3 Naming of Documents. Microsoft Word not only defines a default location for saving documents but also automatically suggests a name for the document to be saved, based on the contents of the first line or sentence of the document.

9.6.2.4 Summary. The above details are specific to Word, but the general principles may be equally applicable to any word-processing application. The details are dependent on the individual application used. In relation to the production of documents in general, it is possible that while a document was being produced a portion of the document, particularly if it is a large document, may have been paged in the swap file and thus entered unallocated space. Although it is maybe not feasible to search unallocated space for all coherent text fragments, it is certainly possible to link a document suspected of being produced on a computer to that computer by means of a keyword search for relevant or unusual terms contained within the text. This would identify any of the text fragments left behind by auto saves, deleted copies of the document, or contents of the swap file.

9.6.3 Printed Documents

The printing method employed by the various Windows operating system variants has differences in detail, but the concept is the same throughout. In order to print the contents of a file, whether it is a document or an image, Windows makes use of two temporary files.

The purpose of the first temporary file is to convert the data contained within the file to a format that the printer can understand. To do this Windows creates a temporary file in either Windows Enhanced Metafile (EMF), or in a RAW graphical format. These are both graphical formats; in simple terms, the printer receives a 'picture' of the document to be printed.

In addition to the 'picture' of the document, Windows also creates another temporary file containing the information needed to complete the print job, *e.g.* the name and path of the file to be printed, the printer to use, *etc.* These files are only temporary files. After printing is complete they are deleted by the operating system. Their contents then become part of unallocated space.

9.6.4 Summary

As we can see, the contents of unallocated space can be a very rich source of information. Files that the user has deleted and thought gone may still reside there. Searching unallocated space with suitable forensic tools for the content of documents, names, or even particular types of files can produce valuable evidence. As previously noted in relation to the swap file, literally any file accessed on the computer can leave traces of its passing in unallocated space. The rate at which the contents of unallocated space are overwritten is determined by a number of factors including the size of the hard disk, the size of the area currently unallocated, and the amount of use of the computer. Obviously data will be overwritten at a slower rate on an infrequently used computer equipped with a large hard drive and only a small amount of allocated space.

9.7 INTERNET ACTIVITY

9.7.1 The Internet

First, what is the Internet? It would be an oversimplification to describe the Internet as a huge network of computers, but it does consist of literally millions of computers, either standalone or in networks of their own, with the ability to communicate with each other by means of a common protocol known as the TCP/IP protocol.

These millions of computers are not directly connected to each other all the time. When a computer or 'host' is connected to the Internet it is assigned an 'address' in the form of an IP number. The IP number system uniquely addresses each computer and allows the host to send and receive data in small parts called 'packets' to and from another host. Error checking and acknowledgement facilities are included to ensure the data arrives intact at the correct location. The packets of data travel from host to host independently of each other by whatever is the most viable route available to them at that time.

The parts of the Internet that we are familiar with such as email, web browsing, *etc.* are utilities that make use of this connectivity between computers.

9.7.1.1 Connecting to the Internet. The majority of personal users access the Internet via a connection, most commonly broadband now, to their Internet service provider (ISP) whereby the user connects to their ISPs internet server via their telephone line, are assigned an IP address by the ISP, and then proceed however they choose. Larger organisations such as large businesses or academic organisations connect to the Internet via their own servers with direct connections via 'leased lines', *i.e.* high-speed telephone lines. The physical links that provide the connections between computers are leased lines, satellite links, microwave transmitters and receivers. In its travels from one computer to another, a packet of data may well make use of all of these types of connections.

9.7.2 Internet Protocol (IP) Numbers

An IP number takes the form of four numbers each separated by a period (.), each number in the range from 0 to 255. This number forms a unique address or identifier for each host. However, the situation is slightly more complex. Large organisations are assigned a 'block' of IP addresses for their exclusive use. What normally happens is that they have more customers than IP numbers, so each customer cannot be assigned an IP number for their exclusive use. The ISP overcomes this by dynamically allocating IP addresses to each customer as they go online. When the customer goes offline and no longer requires the IP number it is placed back into the pool and can then be reallocated to another customer. The implication of dynamic IP allocation is that an IP number uniquely identifies a customer for the date and time that IP number was assigned to them. Forensically, IP numbers are the fingerprints of users when they are online.

9.7.2.1 Domain Name System. In order to make the Internet more human friendly, a system of converting the IP numbers of hosts to names was established. This is known as the 'domain name system' (DNS). Domain names are registered to particular companies or organisations. When a user attempts to connect to a host, *e.g.* a web page, or send an email, they enter the domain name of the host they wish to connect to, rather than the IP number. This is then converted to an IP number by reference to one of a number of worldwide servers that maintain details of the registered domain names and their corresponding IP numbers. The connection can then proceed based on the IP numbers of both parties. Similarly, given an IP number it is possible to 'look up' the various registries and establish to which domain this IP number belongs.

The allocation of IP numbers and domain names is handled by ICANN, the Internet Corporation for Assigned Names and Numbers.

Regional Internet registries detailing the allocation of IP number blocks and domain names are maintained by a number of regional organisations worldwide including, for example, Reseaux IP Europeans (RIPE), American Registry for Internet Numbers (ARIN), Asia Pacific Network Information Centre (APNIC).

9.7.3 World Wide Web

The World Wide Web is a vast collection of 'web pages', produced using a common language (hypertext mark-up language, or HTML) with multimedia-based content. The pages are graphical in nature and highly interactive. Each web page has a unique address known as a uniform resource locator (URL), *e.g.* 'http://www.microsoft.com/downloads.html'. The URL consists of the protocol used to access the page ('http://' or HyperText Transfer Protocol), the domain (www.microsoft.com), and the filename (downloads.html). The URL may not point directly to a filename but merely the domain of the site, in which case a default index page for the site is automatically loaded.

To give an idea of scale, Google, one of the most popular search engines for the World Wide Web, currently gets 1.5 billion hits per day (May 2010).

9.7.3.1 Web Browser Applications. The World Wide Web is accessed using web browsing software applications. There are a large number of web browser applications available, from commercial products (MS Internet Explorer, Firefox, Opera *etc.*) to free shareware utilities developed by users for their own needs. Browsers interpret the contents of web pages and present it to the user in visual and even audio format. Regardless of the particular browser used, the concept of how they work is as follows.

9.7.3.2 Web Page Structure. The structure of a web page is contained within its HTML code. This records the layout of the page and where its contents, *e.g.* the images on the page, are to be found on its host server. The user views the contents of a web page by copying it to their own computer. This is known as 'downloading'. The user's web browser downloads the HTML code for the page requested by the user along with any content, *e.g.* pictures, as indicated within the HTML code. The browser then interprets the data and displays it to the user. The web page is not simply 'viewed' on the computer monitor; a copy of the web page is made on the hard drive of the user's computer.

9.7.3.3 Internet Cache. The copy files stored by the browser are generally known as 'Internet cache' files or 'temporary Internet files'. Their location on the hard drive is dependent on the web browser used. Most browsers have a default location that may be user configurable. Unless otherwise instructed by the user, the web browser will store the cached files in case the user wishes to visit the page again, in which case the browser will refer to its cached copy of the web page to speed up access. The browser normally retains these files for a set period of time that can be reconfigured by the user.

To keep track of which files relate to which web page, when that page was visited, *etc.*, the browser may create further files for its own use. Microsoft Internet Explorer is perhaps the most popular web browsing application in use today. Internet Explorer stores information such as this in a number of files named 'index.dat' files. These are found in default locations and record information including the URLs of every site visited, along with the date and time of the visit, files accessed, *etc.* The format of the index.dat files, and indeed such files for other browsers, is proprietary to the application. As such they can only be interpreted fully either by their parent application or by the use of specialised software designed specifically for this purpose.

9.7.3.4 Forensic Implications. The content of these files can be extremely valuable. Included within the URLs will be the name of any file accessed, *i.e.* downloaded, using the browser. The URLs also contain information regarding the use of common search engines. When the user submits a query to the search engine, the results are displayed in a web page whose URL may contain the search terms. For example, a search on Google for 'computer forensics' will return a search page whose URL is:

'http://www.google.com/search?hl=enandlr=andie=UTF-8andoe= UTF-8andq=computer + forensics'.

Information such as this can obviously be indicative of a suspect's intentions when browsing.

9.7.4 Email

Email is undoubtedly the most popular and widely used Internet application. Put simply, it is a means of communicating electronically with a single person, or group of people. Email messages take the form of a text message with the facility to send any other type of file with the message, including graphics, multimedia, executable files, *etc.*, as an attachment. The email message consists of two parts—the message 'body' and the message 'headers'. Email addresses also contain two parts, the domain name of the recipient, and the recipient's user account in that domain, *e.g.* 'jon.smith@fireserve.com', where 'fireserve.com' is the domain name for the recipient, and 'jon.smith' is the particular user in that domain.

9.7.4.1 Headers. The headers are the information that helps the message to reach its destination and allow the recipient to reply. The fields visible at the top of every email message, *i.e.* the 'From:', 'To:', 'Date:' and 'Subject:' lines, are part of the headers but do not show the full information. These are normally referred to as the simple headers. The full header information is much more detailed and can be viewed by accessing the properties of an individual message. The message headers contain a record of each stage of the journey from sender to recipient, tracking its origin and the location of the computers its passes through *en route* by means of IP numbers.

In the first stage in the journey, the sender composes the message using their email application software (e.g. Outlook Express, *etc.*), connects to the Internet *via* their ISP, and sends the message. The email application adds the first set of headers to the message, the simple 'From:', 'To:', 'Date:' and 'Subject:' lines, and passes the message to an server belonging to their ISP dedicated to dealing with email, known as a 'mail server'. The ISP mail server determines the location of the intended recipient's mail server by examining the simple headers and converting the domain name portion of the recipient's email address into an IP number using the DNS. It then directs the message onwards to the mail server of the intended recipient.

The message does not usually pass directly from the sender's mail server to the recipient's. Normally it is routed through one or more servers on the way. Each time the message is received by another server another line of information is added to the top of the headers in the form of a 'Received From:' field. This line details:

1. The domain name and IP number of the computer the message was received from.
2. The domain name of the receiving computer.
3. The date and time (including time zone or deviation from GMT) of the transfer.
4. A unique ID number assigned by the receiving computer for its own handling purposes.
5. The intended recipient of the message (by the recipient's mail server on the final transfer only).

For example:

Received from: mail.fireserve.com (mail.fireserve.com [198.212.34.56])
by server8.messlabs.com with SMTP; id 183uhG-0004vH-00 12 Oct
2009 08:47:29–0000 *i.e.* a message received at 08:47:29 on 12th October
2009 (–0000 hours from GMT) by 'server3.messlabs.com' and assigned
ID '183uhG-0004vH-00', from 'mail.fireserve.com' whose IP address is
'198.212.34.56'

The recipient ISP's mail server determines which of their customers
the email is for and places it in their mailbox, awaiting collection. When
the recipient next checks their mailbox the message is passed to their
computer and removed from the mail server.

9.7.4.2   Forensic Implications.   The exact content of the email headers will vary from message to message. Some applications or servers
may add additional lines to the headers for their own purposes. It is
normal to trace the route of a message by reading from the top of the
headers to the bottom, as a line or layer is added to the top of the
header each time a transfer takes place. The IP address of the sender
is normally the last IP number before the message body.

Examination of the contents of email message headers is obviously a
valuable source of information about the identity of the sender.
Unfortunately, some of the information can be suspect and may not be
reliable. All fields attached to the message by the sender's computers are
suspect in that the sender can change them using simple tools or techniques in an attempt to mask their true identity. This could include
information such as the sender's email address or date and time of the
message. However, once the message is in transit it is beyond the sender's control. This means that the information contained within the
'Received From' lines is independent and reliable. Tools used for mass
mailing or spamming often insert spurious 'Received From' lines into
the message header, but careful examination and verification of the
details should identify these types of entry.

9.7.4.3   Body.   The message body contains the 'data' of the email—
the text message (if there is one) and any attached files. The protocol
that handles email (Simple Mail Transfer Protocol or SMTP) is text
based, therefore any graphics, *etc.,* attached to the message are 'encoded' by the sender and 'decoded' by the recipient. This is done automatically by the mail handling software and is invisible to the user.

9.7.4.4   Storage.   Depending on the email application, the user will
normally have an 'Inbox' folder for incoming/received messages, a

'Sent Items' folder for messages sent by them, a 'Deleted Items' folder much like a recycle bin, a 'Drafts' folder for messages written but not yet sent, and any user-created and -named folders.

As previously stated, the protocol used by email messages is text based. For the forensic examiner this has the benefit that the textual content of the messages, including all of the date and time information in the message headers, may be visible using a file viewer. The matter is complicated by the fact that messages may not be stored individually, but grouped together in an 'archive' file. The content of the messages in an archive file may be visible with a file viewing application, but this is dependent on the application. For, example, Outlook Express messages in their archive files are clearly visible in clear text when examined with a file viewer, but Outlook messages, which are all stored in a single archive file, are not. Regardless, unless the messages are viewed with the email application that created the archive file, generally they cannot be seen as they appeared to the sender or recipient. In particular, any files such as pictures sent with the message are text-encoded, and must be decoded by the email application to be properly viewed. The location and format of the archive files is dependent on the application used.

As a final note, the protocol used by email applications to send messages is independent of the individual application.

9.7.5 Webmail

Webmail can be thought of as where email meets the World Wide Web. Whereas email requires the use of an application such as Outlook Express to send and receive messages, webmail allows the user to perform the same tasks using their normal web browsing software. The user simply accesses the website providing the webmail service, *e.g.* Hotmail, Yahoo, *etc.*, and after entering their username and password, can compose and send messages in email format, or read any email messages sent to them. Messages are still sent and received using the SMTP protocol and so are compatible with ordinary email applications. The difference is in how the messages are 'retrieved' from the server. Users browse their email messages and view their contents without downloading them to their own computer. Services such as this are also provided by most ISPs, thereby allowing their subscribers to access their email account from computers other than their own.

9.7.5.1 Forensic Implications. Email messages received via webmail are not stored on the user's computer unless deliberately saved. The messages remain on a server belonging to the webmail or service provider, where the user can view, delete or save them. The messages are simply 'viewed' as the contents of a web page. They may therefore be found in the Internet cache, in unallocated space, or in print spool files if printed. Note that the Internet cache stores web pages downloaded to the computer (*i.e.* received messages), but not information sent to the server by the user (*i.e.* sent messages). Therefore it may be possible to recover messages the user has read, but possibly not those they have sent. However, different webmail providers operate in their own way, and it is not uncommon for the user to see a preview web page showing a message they have typed and asking for confirmation to send the message. In this way messages the user has sent may be cached on their computer and be recoverable.

Webmail and email operate under the same protocols and the information recorded in the headers of an email message is similarly recorded in the headers of a webmail message. Again, normally only simple headers are shown at the top of the message, but options are available to show the complete header information. Some webmail providers now include an extra line in the header information by default, clearly stating the 'Originating IP' of the sender in an attempt to discourage abuse of the system.

9.7.6   File Transfer Protocol and Peer-to-Peer Applications

The file transfer protocol (FTP) is a set of common rules that allows computers to connect to each other directly and transfer data. Modern peer-to-peer applications are a new incarnation of this existing Internet technology. The first well-publicised peer-to-peer application was Napster, which allowed users to connect to each other and share MP3 music files.

Napster has been succeeded by many others. The idea behind the applications is that individual users choose to share certain files on their computers with other persons generally unknown to them. The application software handles the communication and transfer of files between users. Users choose which files they wish to share (if any), generally by placing them in a default 'shared' folder. The location of the shared folder is dependent on the application used, but is normally within the program folder created on installation of the application.

Downloading files using peer-to-peer applications differs from downloading from a web site in that the file transfer is directly between two or more users. There is no central repository of all the files on a server as with a web site. Each user is in effect making their own computer available as a server for others to download files from. To download from other persons, the user can choose to view all available content of a certain type, *e.g.* music, video or still pictures, or enter search criteria based on filename, file type or content. The application software then establishes which other users are online currently sharing files which meet the requested criteria. The user is then presented with a list of files available for download including details such as the file name, the user making the file available, the size of the file, the speed of the Internet connection available for that user, *etc*. The user chooses which files they wish to download and the download then commences direct from one or more users who possess that file and have chosen to share it. It is even possible to view the contents of another user's shared folder directly and choose what material to download. As the user is downloading files, other users may be simultaneously downloading that file or other files from the user's computer. The application software manages the entire process. The content of the files downloaded is generally not visible to the user until the download is complete.

9.7.6.1 Forensic Implications. Certain applications maintain a database file logging files that have been downloaded including information such as file names, date and time information, or even the user the file was downloaded from. The format of these files is proprietary to the application. They may contain information in plain text, but the full content is visible only with specialist forensic software.

The applications mentioned are all highly configurable by the user. Users can set their own preferences for any number of the application features: the location of the shared/download folder, whether or not other users can download from them, whether or not files are to be viewed immediately on completion of the download, and many more. The preferences of the individual user may be recorded within a configuration file forming part of the program installation or within the Windows Registry entry for the application, again usually in a format proprietary to the application. The contents of these configuration files can obviously be of benefit in determining how the user obtained material, whether or not it was shared, viewed immediately, *etc*.

9.7.7 Newsgroups

9.7.7.1 Introduction. Newsgroups are the precursors of chat rooms but do not provide 'real-time' conversations. News servers throughout the world contain thousands of newsgroups, each created for discussion on a particular topic. Users wishing to share files, information, or just their opinion can 'post', *i.e.* upload, a message to a newsgroup. Similarly, any user viewing the newsgroup can read (*i.e.* download) the contents of a posted message, which may include attached files such as pictures.

9.7.7.2 Newsgroup Applications. Newsgroup reading applications are many and varied, as are the records they maintain. Normally, however, a user must do two things before viewing a newsgroup. Firstly, they must download a list of the newsgroups available on that server. Secondly, they must 'subscribe' to the newsgroup whose contents they wish to view. Contents of particular postings in that newsgroup can then be downloaded. Both of these actions involve downloading of data from the news server. Obviously, it would be very cumbersome for the application to download the same information each time the user wished to access a newsgroup, so a record is usually maintained by the application of the newsgroups available on that server, and the newsgroups to which the user has subscribed, possibly including the current postings the user has downloaded to enable offline viewing of the material. The location and format of these records is dependent on the application used.

The user cannot see the contents of a newsgroup posting until it has been downloaded to their computer. The user normally sees a list of messages posted to that group showing a subject line indicating content, the date of posting, the name of the sender and an indication that a file is attached to the posting. Applications are available to automatically download all postings to particular newsgroups without the user being present. The user defines the newsgroups whose contents are to be downloaded and the application takes over. The user can then view the postings at a later time and decide what to keep and what to discard. Such applications are commonly used to download large numbers of picture files. The records stored are again dependent on the application used, but may include details of the newsgroups marked for download by the user.

9.7.8   Chat Rooms and Applications

9.7.8.1   Chat Rooms.   Chat rooms are 'virtual rooms' where computer users can engage in real-time 'conversation' in text, audio, or even video. Text-based chat rooms are by far the most common. As a user types a message it is visible to all persons in the room at that time. Other users can reply to the message in a similar fashion. New users entering the room can view the conversation from the time of entry onwards.

There are a plethora of chat rooms on the Internet, each dedicated to a particular subject or group of people, *e.g.* 'scuba diving', or 'Thirtysomethings'.

Chat rooms are generally hosted on a server, and communication between all persons in the chat room takes place via that server. It is also possible for users in the room to have a 'private' conversation with another user, and possibly even exchange files with them. In this case the conversation takes place directly between the two users and is not visible to other persons in the room.

9.7.8.2 Chat Applications. Access to chat rooms requires either the downloading of a chat program (e.g. Microsoft Internet Relay Chat, MIRC) or is by access to a World Wide Web based chat room.

Other chat applications are available that allow conversations to take place solely between two people privately, *e.g.* Instant Messenger, Skype or ICQ. Users log into the service when they go online and are notified when any 'friends' also come online. A chat session can then take place directly between the two users, with no server involved. Some applications *e.g.*, Facebook also allow users to exchange files.

9.7.8.3 Records. There are many chat applications. The records maintained on the users' computers are dependant on the application and the settings applied by the user. For example, Microsoft Internet Relay Chat (MIRC) is a long-standing and still popular application that allows users to 'text chat' in 'channels' hosted in servers worldwide, engage in direct conversation privately, and exchange files of any type. Logs of MIRC chat sessions are created at the user's discretion in the form of plain text files of the chat session conversations. The location of the log files is within the program group for the application or another location as determined by the user.

For all applications, even if 'logging' is not switched on or simply not available, it may be possible to recover fragments or entire conversations from the swap file or unallocated space using keyword searches, or from print spool files if a printed record of the session has been made. Similarly, when webcams are used for video links, the video is made up of a series of still images, some of which it may be possible to recover, again from unallocated space or the swap file. For web-based chat rooms, Internet cache files may indicate visits to the web site, and unallocated space may contain parts of chat sessions as mentioned previously.

In some chat applications it is possible to use other utilities from within the chat room, *e.g.* to identify the IP number of other persons in the room, or to set up a file server to allow automated file transfers (using a file server utility such as fserve). Logs may be maintained by these utilities that record the details of other persons in the room or files transferred. The location, format, and content of these logs will be dependent on the utility used.

9.8   CONCLUSION

Computing is a rapidly developing technology in terms of both hardware and software. Advances in storage technology are providing users with ever increasing amounts of storage capacity in their computer systems, forcing the development of more powerful forensic tools with faster imaging speeds and greater analytical functions. Development of software operating systems and applications is constant, with each new variant requiring further knowledge on the part of the examiner.

BIBLIOGRAPHY

B. Carrier, *File System Forensic Analysis*, Pearson, Englewoods Cliffs, NJ, 2005.

H. Carvey, *Windows Forensic Analysis*, Elsevier, San Diego, 2007.

F. Clarke and K. Dilberto, *Investigating Computer Crime*, CRC Press, Boca Raton, 1996.

T. A. Johnson (ed), *Forensic Computer Crime Investigation*, CRC Press, Boca Raton, 2005.

D. G. Parker, *Fighting Computer Crime*, John Wiley & Sons, New York, 1998.

P. Stephenson, *Investigating Computer-Related Crime*, CRC Press, Boca Raton, 1999.

CHAPTER 10

Fire Investigation

DAVID HALLIDAY

Fire Investigation Unit, Forensic Science Service, Lambeth Road, London
SE1 7LP

10.1 INTRODUCTION

Fires are part of our everyday experience. Without fire, the things we take
for granted today would not exist and our lives would be very different.
Fire also exerts a fascination over people, and its destructive potential is
well known, so it is not surprising that fire is used to attack both people
and property. In UK law a deliberate act that damages property without
the permission of the owner is an offence against the Criminal Damage
Act 1971, and if the destruction is by fire then the offence is classed as
arson. (In Scotland this offence is called 'wilful fire raising'.)

Arson is one of the few crimes that by its very nature destroys forensic
evidence; not simply the evidence that might be used to determine who
was responsible, but also the evidence of how the crime was committed.
Since many fires are the results of accidents, the investigation of a
potential arson is unlike many other areas of forensic science because the
fire investigator will be asked to prove that a crime has been committed
before the question of who was responsible can be addressed. It is this
double burden of proof that makes prosecution in cases of criminal
damage by fire one of the most difficult tasks the courts have to deal
with, and it places considerable demands upon the knowledge and
experience of the scientist giving evidence on the fire's cause.

Crime Scene to Court: The Essentials of Forensic Science, 3rd Edition
Edited by P. C. White
© Royal Society of Chemistry 2010
Published by the Royal Society of Chemistry, www.rsc.org

Fires are not only investigated for forensic reasons: there may be issues of public safety or regulatory compliance that require an in-depth study of ignition or propagation mechanisms at some incidents. Fire investigation is not, therefore, the sole province of forensic scientists in the public and private sector, but may be carried out by fire brigade personnel, who have a statutory duty to fulfil in reporting the cause of fires, or fire engineers from organisations such as the Fire Research Station. In recent years the emergency services have developed a partnership approach to fire investigation, with teams including scenes of crime examiners and fire brigade officers taking responsibility for most of the incidents that are reported to the police as suspicious. In technically complex cases, however, or those that rely on a combination of evidence drawn from work both at the scene and in the laboratory, it may be desirable to call upon the dual skills of a trained and experienced forensic scientist.

Forensic scientists who specialise in fire investigation must possess a basic knowledge of chemical and instrumental analysis, combustion chemistry, fire dynamics and fire engineering to augment the vocational competence that they acquire during training at scenes and in the laboratory. Interpersonal and communication skills are also important, given that they will meet and manage people from all walks of life at a fire, and may eventually be required to present their findings in court.

10.1.1 What is a Fire?

In the real world fires are extremely complex chemical processes, but in simple terms a fire is an exothermic oxidation involving a fuel and an oxygen supply, usually air. Air is all around us but even highly flammable materials do not burst into flames at random because the oxidation must be initiated by the application of a heat source. All fires depend upon the presence of these three interdependent components—air, fuel and heat—for their existence. If any one of these is missing then the fire will not start, and if you take away one or more of them while the fire is burning it will go out.

Air, fuel and heat are the three sides of what is commonly referred to as the 'fire triangle', although combustion scientists will by preference use the term 'fire tetrahedron' because a self-sustaining fire will require, as a fourth apex, a propagation mechanism if it is to keep burning. Considering the simpler fire triangle, in real life the air around us is one side of the triangle and the other two sides are formed by the combustible items in our homes or offices and heat from sources such as flames, smouldering cigarettes and other incandescent objects. The mechanism that produces the heat that starts the fire—the ignition source—is the aspect of the fire that is of most interest to forensic fire investigators, and a large part of the work that they carry out at the scene and in the laboratory focuses on answering the question, 'What started the fire?'

10.1.2 Ignition of Gases, Liquids and Solids

Flammable gases and vapours will only ignite if they are mixed with air in the correct proportions. Above a certain concentration the mixture will be too rich and below a certain concentration it is too lean; in either case the application of a flame or spark will not cause the mixture to catch fire. These concentrations, the upper and lower flammability limits, are different for each type of fuel.

When a fuel–air mixture is within its flammability range a small volume of it must be heated to a certain minimum temperature, called the autoignition temperature of the gas or vapour, before it can catch fire. In most cases a flame burning at 900–1200 °C or an electric spark at a temperature in excess of 3000 °C will be sufficient. Studies of ignition by static electricity, however, have revealed that temperature is not the sole criterion: for fuel gases such as natural gas or propane the ignition source must also have a minimum energy of approximately 0.3 mJ.

Combustion can be a simple process: in an everyday context a lit Bunsen burner will burn natural gas (methane) to produce carbon dioxide, water vapour and heat:

$$CH_4 + 2O_2(air) \rightarrow CO_2 + 2H_2O + \Delta H_c$$

where ΔH_c is the heat released by combustion. The stoichiometric nature of this reaction is only true for a premixed flame, where the correct amount of oxygen is mixed with the fuel before it burns. If the amount of available oxygen is severely reduced by adjusting the air inlet at the base of the burner then the fuel will burn with a luminous diffusion flame; the fuel–air mixture will be too rich and it will only burn at its interface with the surrounding air. Incomplete combustion can occur in a diffusion flame; its luminosity is due to the presence of unburnt carbonaceous (soot) particles that have been heated to incandescence. Analysis of a sample of the gases in the flame would reveal a more complex reaction taking place:

$$CH_4(excess) + O_2(air) \rightarrow CO_2 + 2H_2O + CO + C + \Delta H_c$$

This equation has deliberately not been balanced because the amounts of carbon dioxide, carbon monoxide and carbon produced depend upon the amount of oxygen available. Real-life fires are generally oxygen deficient, hence their propensity to produce large amounts of soot and unburnt volatiles as well as heat and flames.

At normal room temperatures the vapour pressures of many liquid fuels are too low to produce concentrations of vapour in air that are above the lower flammability limit. These liquids need to be heated to enrich the vapour–air mixture directly above the liquid before the concentration can reach the lower flammability limit. The lowest temperature at which sufficient vapour is present that it can be ignited by a small flame is known as the liquid's flashpoint. There are a number of different methods by which flashpoints are measured, and figures quoted in the literature for any particular liquid will vary depending on the method used. A number of common gaseous and liquid fuels with their flammability limits, flashpoints and autoignition temperatures are listed in Table 10.1.

Liquids with a flashpoint above room temperature can be ignited by a small flame without the bulk of the liquid being pre-heated, but only if they are soaked into an absorbent material that is capable of acting as a wick. In such circumstances the application of the flame to the wick will quickly heat the small volume of liquid soaked on to it, vaporising it and producing a locally enriched fuel–air mixture that will ignite if it is within the flammability limits. The heat released by the combustion of the fuel–air mixture will continue to vaporise more of the liquid from the wick to sustain the flame, and the fuel that is burnt will be replaced as the wick draws more of it from the reservoir by capillary action.

Common combustible organic solids include natural materials (such as wood, cotton, leather or wool) and synthetic polymers that are primarily hydrocarbon-based. The chemical composition of these materials consists principally of carbon and hydrogen, with varying amounts of

Table 10.1 Flammability limits, flashpoints and autoignition temperatures for some common fuels and solvents

Fuel	Flammability limits (vol% in air)		Flashpoint (°C)	Autoignition temperature (°C)
	Upper	Lower		
Methane	5.0	15	−188	640
Propane	2.1	9.5	−104	500
Ethanol	3.3	19	13	365
Petrol	1.3	7.1	−45 to −42	440
White spirit	1.1	6	38–60	149–232
Diesel fuel	0.6	7.0	>55	>250

Data collated from a number of sources—see bibliography.

oxygen, nitrogen and other elements such as sulphur and chlorine. When heat is applied to an object made from these materials its surface temperature will rise; thermoplastic polymers will melt and other materials will scorch and char. If the temperature is high enough (150–500 °C for wood and many common polymers) thermal decomposition will take place. The molten thermoplastic, or the charring object, will start to pyrolyse, with the weakest chemical bonds in its molecular structure breaking down to produce smaller molecules called pyrolysis products. Low molecular weight pyrolysis products are released as gases or vapours, many of which are flammable when mixed with air. Once the concentration of the pyrolysis products reaches the lower flammability limit, and assuming that the heat source is sufficient to raise them above their autoignition temperature, they will catch fire. The combustion of the pyrolysis products then generates more heat, which will help to maintain the decomposition process. Provided the heat released by the combustion of a given quantity of pyrolysis products feeds back into the fuel more energy than is required to produce them then the fire will continue to burn for as long as additional fuel and fresh air are available.

10.1.3 Self-Heating and Spontaneous Ignition

Many organic materials undergo slow surface oxidation reactions even at room temperatures, but normally over a short period of time these will have no perceptible effect on the physical properties of the material (although they may have a distinct effect on the flavour of foodstuffs, for example). If the oxidation takes place on the surface of a solid or liquid then the small amount of heat that is generated will usually be lost to the environment by convection. If the material has a large surface area to volume ratio, however, then heat production is much greater and heat loss will be greatly reduced because of the insulating effect of its bulk. As a result the internal temperature of the material will rise and, provided the oxygen supply is sufficient, the rate of reaction will increase until there is a balance between heat production and heat loss. If no balance is achieved the reaction will undergo thermal runaway and the material may ignite. The process of self-heating, thermal runaway and ignition is called spontaneous combustion as the material involved can catch fire in the absence of an external ignition source.

In all cases of spontaneous ignition resulting from prolonged self-heating, the process starts within the bulk of the stacked or packed material rather than on its surface. Provided that the scene has not been too badly damaged by fire fighting, the investigator will look for signs of this 'inside-out' burning when considering spontaneous combustion as a cause of the fire.

A wide variety of materials and products can undergo self-heating, including hay, coal, cotton waste and grains, and also unsaturated oils and fats if they are soaked on to fabric or a fibrous substrate with a large surface area. Spontaneous ignition is a problem frequently encountered in the food processing industry if heat-dried powders or granular materials are not cooled before bulk storage, and it can also occur in commercial laundries if freshly washed and dried clothing is stacked while still hot and the fabric retains traces of oil or fat. The risks of spontaneous combustion are well documented in product and process safety literature, and the fire investigator will draw upon this information when considering whether materials at the scene are capable of self-heating.

10.1.4   Flaming and Smouldering Fires

A fire that is supplied with adequate fuel and air will usually continue to burn until it is extinguished or until the fuel supply runs out. Most of the combustion processes take place in the gas phase, producing the flames we commonly associate with a fire. Smouldering fires, by contrast, do not produce flames because the combustion process takes place directly on the surface of the fuel, which glows. Smouldering fires propagate directly through the fuel they are burning on: the combustion front preheats the unburnt material ahead of it until it ignites and starts to glow. Since the minimum ignition temperatures necessary to initiate glowing combustion are lower than the autoignition temperatures of many liquid and gaseous fuels, smouldering fires can be started by heat sources that are incapable of igniting flammable liquids. Thus a lit cigarette can easily start a smoulder on many common fabrics and padding materials but it cannot set light to petrol (despite the many apparent instances of this happening in action movies).

In general, only materials that are capable of forming a rigid char will smoulder. The temperatures necessary to initiate and maintain smouldering would melt and break down thermoplastic polymers. Since a polymer melt will not undergo glowing combustion, fires will only start on thermoplastic material if the temperature of the heat source applied to it is higher than the autoignition temperature of any vapours or pyrolysis products that are generated, and these conditions of course produce flames. Placement of a smouldering ignition source such as a lit cigarette on the surface of a thermoplastic material will therefore only result in a hole being melted in it.

If the air supply sustaining a flaming fire is restricted then the reduction in available oxygen will lead to a drop in the burning rate and heat output. There may come a point at which the heat fed back into the fuel is not capable of generating sufficient quantities of pyrolysis products to sustain the flames. In such circumstances the fire may die down and go out, or it may smoulder and continue to burn at a rate determined by the air flow.

Fires can also make the transition from smouldering to flaming. As a smoulder grows its total heat output will rise, a greater volume of unburnt material will be pre-heated and increasing quantities of gaseous pyrolysis products will be released. At some point the concentration of these products in the air above the smouldering material may exceed the lower flammability limit and, if the heat output from the smoulder is sufficient to raise them above their autoignition temperature, they will start to flame. The time taken for a smoulder to grow and undergo the transition to a flaming fire is highly sensitive to the nature of the fuel and the environmental conditions. With a lit cigarette, and given the right circumstances, paper tissue can take as little as 12 minutes to start flaming and bulk cotton fabrics 20 minutes; however it is generally accepted that an induction period of 30 minutes or more is required for foam-padded furniture that has not been treated with fire retardants. Some smouldering fires never make the transition, and since the outcome for any particular combination of materials and environmental conditions is difficult to predict it is best determined by testing.

As a smouldering fire propagates it almost completely consumes the fuel that it is burning on, leaving an area of severe localised damage that may be surrounded by completely intact materials. In contrast, flaming fires tend to cause relatively superficial and widespread damage as they grow. The appearance of the damage after the fire is one of the clues that the fire investigator will look for when trying to determine how the fire started, as it may indicate that the fire has either flamed or smouldered for much of the time it was alight.

10.1.5 Fire Growth and Propagation

If a fire is to spread then heat needs to be transferred to nearby combustible surfaces that have not yet ignited. This can happen in three ways: by radiation, convection or more rarely conduction. As might be expected, radiative heat transfer from flames has a strong influence on the growth of a flaming fire in its early stages as the unburnt materials around the perimeter of the burning area are heated to ignition temperatures. As the fire grows, however, it will produce increasing quantities of hot gases and smoke. These are less dense than the surrounding air and will rise in a plume above the burning fuel, being replaced by incoming cold air at the base of the fire. Convective flows will elongate the flames, and if the fire is burning against a vertical combustible surface then contact with, and radiation from, the flames and hot gases will pre-heat that surface a greater distance vertically than the fire heats surrounding surfaces horizontally. This will result in the fire propagating upwards faster than it spreads outwards at its base, as shown in Figure 10.1.

Within a compartment the plume of hot gases and smoke from a fire at low level will be confined and deflected by the ceiling, rapidly forming a hot smoky layer across the entire room. Deflected flames can spread across the underside of a ceiling up to ten times further than they would extend vertically if unconfined. As a consequence, combustible materials within the hot gas and smoke layer will rapidly ignite some distance away from the base of the fire, and may fall into areas that the fire has not yet spread to at low level. These pilot ignition sources can start additional small, and apparently separate, fires that may be misinterpreted by inexperienced fire investigators. One of the most commonly encountered examples of pilot ignition involves curtains catching

Figure 10.1   From low level on the seat of the settee the fire had propagated upwards faster than it has spread outwards. Hot gases spreading under the ceiling have ignited the curtain to the right of the window and it has dropped to the floor.

fire and falling to the floor (as can be seen on the right of Figure 10.1). If the fire is put out quickly then the patterns of damage may give the appearance of more than one fire having been started in the room.

As the hot gas and smoke layer gets deeper, and its temperature rises, the amount of heat radiation it emits increases. Exposed combustible surfaces throughout the room start to heat up, pyrolysis is initiated and within a short time the gases and vapours that are produced can ignite. This period of extremely rapid fire growth is referred to as flashover when, as fire fighters say, 'the fire really starts to motor'. In a domestic fire, a room can go from ignition to flashover in less than 5 minutes depending upon its contents.

Once the materials in a compartment have become fully involved in a fire the rate at which they burn will depend upon how much air can get into the room through the existing openings. The fire at this stage is said to be ventilation limited. A long period of burning under such conditions may result in greater damage in areas where the air enters and the fire burns more vigorously.

If a fire in a building is not put out it can spread a considerable distance from the compartment of origin. Hot gases will flow at high level through openings such as doors and windows and can ignite materials in other rooms. In multistorey structures, if vertical shafts such as stairwells are accessible to the fire then it will be drawn upwards in a 'chimney effect' rather than spreading outwards at the lowest level. This may eventually leave the level on which the fire started much less damaged than the floors above. In extreme cases the entire building may become fully involved in the fire, although in urban areas it is more likely that fire fighters will intervene before this occurs.

10.1.6 Interpreting the Physical Evidence Resulting from Fire Behaviour

The mechanisms by which a fire grows and spreads leave characteristic patterns of damage that can be interpreted by the fire investigator. This has led to the development of some general rules of thumb that investigators find useful when assessing fire damage.

Fire propagation mechanisms result in the fire spreading faster vertically than it does horizontally, and the fire gases themselves flow outwards from the fire at high level. When comparing two fire-damaged compartments connected by a doorway the fire investigator will look for evidence of the direction of this flow—damage at lower level in the originating compartment—to determine where the fire was located. If the room of origin has been identified, and if the damage is not too severe, then the tendency for flaming fires to spread upwards and out-wards may allow the investigator to locate the area of origin of the fire at the lowest point of burning.

The rate at which a combustible material will burn is dependent on both the nature of that material and the fire environment to which it is exposed. In general, however, the longer a fire burns on an object the greater the degree of damage it will suffer, so provided you compare like with like then the worst damage will be found closest to where the fire started.

Many simple fires can be successfully investigated using these two guidelines; however, as with all rules of thumb there are a great many exceptions: pilot ignition may cause additional areas of low level burning; the degree of damage sustained by objects in the scene can be strongly influenced by the way the fire is fought; and restricted ventilation may smother the fire at its point of origin while it continues to burn where an air supply is still available. Recognising when the rules of thumb do not apply is an essential skill for the competent fire investigator

10.2   FIRE SCENE INVESTIGATION

Although every fire scene is different, the investigation of fires benefits from a methodical approach as this ensures that the work is carried out efficiently and that nothing is omitted. However, any general investigative strategy may require modification before or after the work starts if the situation requires it, and the experienced fire investigator will always be prepared to adapt to changing circumstances.

A robust scene investigation process includes most or all of the following components:

1. Safety assessment.
2. Collation of information.
3. Inspection and strategy decisions.
4. Clearance and reconstruction of relevant areas.
5. Recording and retrieval of evidence.
6. Evaluation of findings.

10.2.1   Scene Safety Assessment

Burnt structures probably contain a greater number of hazards than any other type of forensic scene. If the building is to be examined safely, the investigator must ensure that these hazards are mitigated sufficiently to allow work to be carried out.

Structural damage may leave walls, floors and roofs in an unstable condition if the fire has destroyed supporting elements. As a result, parts of the building will be prone to failure if they are disturbed by post-fire settlement or by the weight of a person walking on them. Before entering any part of the premises the investigator will assess the structure's safety; cracked or leaning walls and wooden floors that have been severely damaged from below will be treated with caution as they could fail without warning. Stonework may be a particular problem as it can crack internally due to thermal stresses generated during the fire and then break up and collapse as it cools and contracts. Sharp items are almost always present at fire scenes, in the form of broken glass, protruding nails, splintered wood and torn metal.

Exposed electrical conductors may pose an electrocution risk if the supply has not been cut off, so the building service inlets will be checked to make sure that gas and electricity supplies have been isolated.

Older buildings may contain asbestos that can be released as airborne fibres after a fire, and industrial sites may contain specific process hazards. If necessary the investigator will seek information about the building's history and the nature of the work carried out in it before deciding that it will be safe to work there.

Hazards arising from the nature of the building's occupancy may not simply be confined to industrial premises. Squats and semi-derelict houses are frequently used by vagrants and drug addicts, and may contain biohazards such as used syringes and needles, faecal material and parasite infestations.

Anyone entering the premises will need appropriate personal protective equipment; at a minimum boots with reinforced insoles and ankle protection, sturdy gloves, a helmet and overalls will be worn when examining a fire scene. In cold conditions the protective clothing will include garments chosen for warmth, as prolonged exposure to low temperatures inevitably leads to physical and mental impairment.

Before starting any work inside the scene the fire investigator will take into consideration all the potential hazards before deciding whether it is safe to proceed (see Figure 10.2). Advice may be sought from other better-qualified persons if there are specific structural or process safety issues that need to be resolved. Only if all of the potential risks have been reduced to an acceptable level will the site investigation begin.

Safety assessment does not end once the decision is made to start work inside the premises; it is a continuing process in which the risk to persons working in the building will constantly be re-evaluated to ensure that their activities, or changing circumstances, do not expose them to new hazards.

Figure 10.2   Scene safety assessment: obvious hazards include unsupported walls and RSJs, loose masonry, torn steelwork and falls from a height. Given these hazards, the scene investigator would not enter the area.

10.2.2   Information Gathering

Before the fire investigator's arrival a considerable amount of information may have been gathered about the incident. Some of this will have an immediate effect on the scientist's overall investigative strategy, including professional advice on safety issues or allegations that other serious crimes having taken place.

Other information, such as the state of security of the premises when the firefighters arrived and the scope of their subsequent firefighting activities, may be relevant both to the surrounding circumstances of the incident and the interpretation of the physical evidence at the scene. Firefighters often need to break into locked premises that are on fire, so evidence of forced entry found after a fire is not always a sign of a criminal act. Forensically aware fire crews will take pains to ensure that any of their activities that might impact upon a crime scene examination are made known to investigators and logged.

Many modern alarm systems are computer-based and can store activation and fault logs in their electronic memories. If this information is retrieved by an engineer with the right equipment it can offer an insight into the time, area of origin and pattern of spread of the fire. Because of the finite life of battery backup power supplies, and restrictions on memory capacity that may lead to data being overwritten, it is a matter of some importance that alarm systems are interrogated as soon as possible after the fire is put out. Event logs from combined intruder and fire detection systems may offer an additional insight into the cause of the fire if they have recorded evidence of a break-in or, as has sometimes been observed, if a keypad has been used to deactivate sensors to allow someone to enter the premises just before the incident.

Any information about what happened during the incident is useful material for the investigator, and eyewitness accounts can be particularly valuable if they cover the events leading up to the fire. Normally witness information will be taken into consideration before deciding upon a strategy, but there is one exception to this: many investigators prefer not to be informed of any admissions that have been made by a suspect until after they have completed their work at the scene. This avoids any suggestion that their findings have been biased towards one particular outcome because of prior knowledge, and it adds weight to the evidence if their findings independently corroborate the suspect's account of events.

10.2.3 Visual Inspection and Strategy

Before anything is disturbed at the scene the investigator will take the time to carry out a visual inspection of the entire premises. During this inspection the general patterns of fire damage will be evaluated to establish the directions in which the fire has spread through the structure. Hot gases and smoke are buoyant; if they vent through an opening from a compartment that is on fire they will predominantly cause damage at a higher level in the area they spread into. A pattern of decreasing damage at high level indicates to the investigator that they are moving away from the fire and conversely, if the damage increases and is found at lower levels, that they may be approaching the area of origin. These simple observations are often sufficient to allow the investigator to place the origin of the fire within a single compartment that can then be prioritised for detailed examination.

As smoke spreads through a property it leaves sooty or tar-like deposits on exposed surfaces. After a fire the pattern of smoke staining that is left can be a valuable indicator of the positions of doors, window latches and free-standing objects, offering the investigator an insight into the condition of the premises before the fire and indicating whether items have been moved after the incident as can be observed in Figure 10.3.

During the walk-through the scientist will formulate a forensic strategy, identify any vulnerable evidence that will benefit from early retrieval and decide what records need to be made and what

Figure 10.3 The patterns of smoke staining show that papers had been scattered across the floor before the fire, evidence that the premises had been ransacked.

photographs need to be taken before any of the evidence is disturbed. In some instances the visual inspection will immediately reveal strong evidence of a malicious act, such as multiple areas of fire damage that are entirely separate or signs that a flammable liquid has been used. Early advice to that effect may have a significant influence on the successful outcome of any criminal investigation, as it might result in the timely seizure of items from suspects before trace evidence is lost.

Modern firefighting techniques are usually highly effective, and fire crews are increasingly aware of the importance of preserving scenes for forensic examination. If a fire is discovered early and there is a timely response by firefighters then normally domestic premises will be left in such a condition after a fire that only one room, or a very small number of rooms, will need to be examined in detail. Although in most cases a small amount of fire debris will still need to be cleared or excavated, the evidence of interest to the fire investigator may survive almost intact.

10.2.4 Detailed Examination, Clearance and Reconstruction

During a fire the contents of the structure will burn and break up; ceiling and wall linings may fail and in extreme cases walls and roofs may collapse. All of this debris must be removed as it may conceal evidence of interest to the fire investigator. Excavation is not simply about shovelling the debris away: the stratigraphy or layering of the materials that have been deposited may well yield useful information on events leading up to, or during, the fire. Thus evidence that the contents of a room had been strewn across the floor before any fire-related collapse of the linings might indicate that the premises had been searched or vandalised.

In extreme cases large amounts of collapsed building materials may need to be removed mechanically, but once the layers of material from the compartment of interest are uncovered the fire investigator will start to remove the debris using a shovel or a trowel, or even by hand. The positions of significant items that are uncovered will be recorded before they are moved, with a view to replacing them later after the debris below has been searched and cleared away. Any evidence of potential ignition sources, such as electrical wiring and appliances, will be retrieved for later examination if necessary.

The layers of debris covering the floor may also act to protect materials of evidential value from the full effects of the fire. Flammable liquid residues that might otherwise vaporise and burn away will be preserved and after excavation they can be recovered for analysis. In many instances the quantity of flammable liquid that survives the fire is large enough that it can readily be detected by smell, but there are other aids that can be used at the scene such as portable electronic instruments incorporating flame ionisation detectors (FIDs) or photo-ionisation detectors (PIDs). These devices react to the presence of organic vapours in air, but they must be used with caution as they are not specifically detectors for flammable liquids but respond to any vapours including pyrolysis products. They are perhaps most useful when guiding the investigator in the selection of suitable samples for laboratory analysis at scenes where large quantities of flammable liquid have been distributed and olfactory fatigue has set in. Dogs trained to detect flammable liquid residues are being deployed increasingly at fire scenes, usually by fire brigades. Properly trained and handled these dogs can be of great assistance to the fire investigator, but they are not entirely reliable and any positive indications must be confirmed by laboratory analysis if the investigator is to say with confidence that a flammable liquid is present.

Removal of fire debris may also uncover protected areas on surfaces to show where items were standing while the fire was alight. These patterns of protection may have characteristic shapes or dimensions that uniquely identify the object that made them, allowing the replacement of that object as part of a scene reconstruction. In some incidents the contents of a room could have been entirely cleared during fire

(a)

(b)

Figure 10.4   Although the entire contents of the bedroom had been removed during
fire fighting, the contents of the room can be reconstructed.

fighting and yet floor coverings, furniture and personal effects can all be
replaced with certainty in their original positions by the investigator; see
Figure 10.4.

10.2.5   Recording and Retrieval of Evidence

During the course of work at the scene the investigator may uncover
items that need detailed examination in the laboratory. These will be

retrieved, appropriately packaged to prevent further damage or cross-contamination and labelled for continuity purposes. Most samples recovered from fire scenes are taken for subsequent analysis for flammable liquid residues, but items may also be seized for inspection or testing.

As the results of a fire investigation may have to be presented in court many months or even years later, the investigator must make a permanent record of the evidence at the scene. This can be done in a variety of ways, depending on the circumstances of the incident and the resources available. Most commonly the investigator will take photographs and draw plans of the relevant areas. They will make written or dictated notes about the construction, internal layout and building services and their interpretation of the patterns of fire damage to identify the point of origin of the fire and its cause. Any handwritten notes and plans that are prepared at the scene will be signed and dated to show that they are a contemporaneous record. Similarly, dictated notes will be transcribed and the transcript signed as a true record as soon as possible after the work is done.

The ready availability of robust digital cameras with preview screens means that fire scene photography is no longer exclusively the province of the professional photographer, but correctly exposed high-quality images are still an essential part of the evidence package. Since high-capacity memory cards are inexpensive, many investigators include large numbers of scene images in their case notes. Given the questionable permanence of digital storage media, and the need to retain some case records almost indefinitely, all forensic practitioners have storage protocols for digital material that are designed to ensure that data is not lost.

The scene record must be as comprehensive as possible because it will be the primary source of information if the investigator is asked to prepare a statement of findings. Before or during court proceedings the scene notes may be scrutinised by other experts under the rules of disclosure, and if any important information is missing then this may result in the investigator's competence being called into question.

10.2.6 Interpretation and Evaluation

Before communicating any findings to interested parties the fire investigator must interpret the evidence that has been recovered during the scene examination. Ideally the evidence will be sufficient to identify the area or areas where the fire started and the ignition sources that are present. If flammable liquid has been distributed around the scene, or if a number of fires have been started in entirely separate areas, then the investigator's task may be very simple as the presence of a flammable liquid or multiple seats of fire is generally a good indicator of arson. In the absence of obvious signs of arson, however, the investigator may have to consider a variety of explanations for the fire. To determine which is the most probable a series of hypotheses will be formulated based upon viable ignition scenarios and these will be tested against the physical evidence at the scene and other reliable information that may be available. Some of the hypotheses will not withstand scrutiny and testing—they will be discarded as unworkable, thus ruling out certain ignition mechanisms as in the following examples:

- A flaming fire is discovered in a room that had been occupied only minutes before. After excavation and reconstruction the investigator considers the hypothesis that a lit cigarette was the ignition source. The surviving materials found at the seat of the fire cannot support a smouldering fire, so the hypothesis that a cigarette started the fire appears unlikely. Further support for this conclusion comes from the witness evidence. A smoulder will produce copious quantities of acrid smoke as it develops, and this smoke should be readily perceptible to anyone in the room shortly before the transition to flaming. Since the room was occupied just before the flames were seen, the presence of a well-developed smoulder at that time would appear to be highly improbable if the witness evidence is to be trusted.
- Electrical wiring is present within a room where a fire has started. The investigator considers the hypothesis that the fire was the result of an electrical fault. Examination of the evidence from the reconstruction of the scene reveals, however, that no electrical wiring or appliances were located within the localised area of damage apart from cables running under the floor. There is a hole in the floor, but the patterns of damage indicate that it has burnt through from above and the insulation on the cables is intact as seen in Figure 10.5. The available evidence is not consistent with an electrical cause of fire and consequently the investigator discards it as a working hypothesis.

In many cases even after hypothesis testing more than one viable explanation for the fire may remain, and the investigator will use their technical knowledge and experience to assign a likelihood to each possible cause based upon the circumstances of the incident.

Throughout the evaluation process the investigator will also test the factual evidence from the scene against information from witnesses to

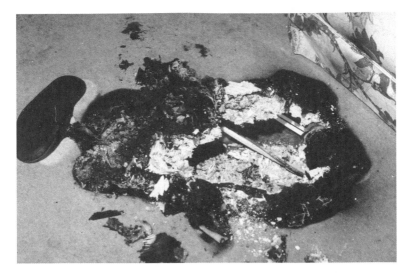

Figure 10.5 A smouldering fire has burnt a hole in the floor but insulation on the electrical cable in the void below is intact, indicating that the fire did not start in the void as a result of a fault to the cable.

assess the reliability of the latter. Although witnesses can be genuinely mistaken, in some instances there may be such large discrepancies between the evidence at the scene and the account given by a witness that the witness may become a suspect.

10.3 FATAL FIRES

In general the methodology employed at fatal fires is the same as that used to determine the area of origin and cause at other fires. If the victim is still *in situ*, however, the investigator will need to examine the body.

10.3.1 Causes of Death by Fire

As a fire burns it will produce heat, smoke and toxic gases that can pose a potentially lethal threat to anyone in the immediate vicinity. Although some people die of thermal injuries in flash fires, the majority of fire victims do not succumb to flames but are overcome by smoke inhalation. In many accidental fires the body will be found in an area that is smoke stained but unburnt. In sufficient quantities smoke is an asphyxiant, but it is poisoning by carbon monoxide (a product of the incomplete combustion of carbon-containing fuels) that statistically leads to most fatalities. Carbon monoxide binds strongly with haemoglobin, having an affinity almost 200 times greater than oxygen. As a consequence carbon monoxide will progressively accumulate in the blood of a victim breathing it, inducing a variety of symptoms that culminate in unconsciousness and eventually death. Hydrogen cyanide can also be found in fire gases, but it is rare for a fire victim to die solely of cyanide poisoning because toxic cyanide levels are usually accompanied by potentially fatal carbon monoxide concentrations as well.

10.3.2 Examination of Fire Victims

Any fatal fire is potentially a murder scene, and the fire investigator will take appropriate precautions when dealing with the body to avoid compromising other forensic evidence. If the body is extensively burnt then signs of other non-accidental injuries may well be concealed; however, the presence of a large amount of blood under or around a fire victim is grounds for suspicion as thermal injuries alone do not normally lead to blood loss.

After a fatal fire it is generally desirable that the victim is removed as soon as possible to a mortuary where a proper examination can be carried out, but the investigator will record the condition and position of the body before it is moved and supervise the process of removal to ensure that fire-related evidence is not lost. Any personal effects that might aid in the identification of the victim will be retrieved at this time.

When the body is lifted, the presence or absence of fire debris or smoke staining under it may offer some insight into the stage of the fire at which collapse occurred. A clean contact area between the victim and the surface they are resting on may indicate that the victim was prone prior to, or at an early stage of, the fire. Conversely the presence of fire debris or smoke deposits trapped between the contact surfaces indicates that the victim must have been upright and mobile after the fire started. If the fire started as a result of the victim accidentally igniting their clothing then the garments trapped beneath the body may show signs of burning, and in cases of self-immolation these burnt materials might also retain the only traces of flammable liquid to survive the ensuing fire.

10.3.3 Lifestyle and Fire Fatalities

Because most fatal fires are accidental, the habits and activities of the victim may be of considerable importance when looking for a cause. A combination of elevated blood alcohol levels and smoking is frequently reported at inquest, and evidence of the victim's drinking and smoking habits may be readily apparent to the investigator at the scene.

Advanced age, infirmity, poor living conditions and some social or religious customs are also associated with enhanced fire risk. There is a possibility that the fire in which the victim dies is the culmination of a series of similar incidents; evidence of previous accidents with smoking materials, or misuse of candles or heating appliances may be apparent to the investigator when examining the premises. Relatives, friends, neighbours and, in the case of elderly or disabled people, the social services may be valuable sources of information on the victim's lifestyle and activities. However, when considering this information the investigator will always bear in mind that the witnesses may not be in possession of the true facts; fire victims are known to have successfully concealed the fact that they smoked even from close relatives. When there is a conflict between the witness information and the facts established at the scene by examination, the investigator will always rely on the latter when deciding what started the fire.

10.3.4 Post-mortem Evidence

When interpreting the scene findings and reconstructing the events leading up to the fire the investigator will also draw upon information obtained from post-mortem and toxicological examinations. This may provide unambiguous evidence to indicate that the victim died after the fire started. The presence of soot in the airways and elevated levels of carbon monoxide in the blood are indicators that a person has inhaled fire gases and must, therefore, have been breathing while the fire was burning. The converse is not always true, as victims of flash fires may not have time to inhale significant quantities of smoke before they die.

10.3.5 'Spontaneous' Human Combustion

In some fires the investigator may find that the body of the victim displays exceptionally severe localised damage while other combustible materials in the immediate vicinity appear to be unaffected. Often the victim's torso is partly or completely consumed and large bones reduced to a crumbly grey ash. Many observers of this phenomenon, in trying to reconcile the conditions they know are required to destroy a body with those that must have been present at the scene, explain the damage in terms of a supernatural 'fire from within'. Spontaneous human combustion (SHC) is the popular term for such fires and it has also been used as an explanation for fatal fires where no ignition source has been found.

Alleged instances of SHC involving severe damage to a body arise from a mistaken belief that a process akin to cremation is the only way that a fire can completely to destroy flesh and bone. Cremation involves the destruction of a body using a furnace fitted with gas burners operating at high temperatures; equipment designed to make this process as quick and efficient as possible. In contrast, most domestic fires that cause partial or complete destruction of a body are neither quick nor particularly efficient, but even a relatively small flame is capable of causing massive damage if it is allowed enough time to burn. All that is required is that the fire around the body is provided with a source of fuel that will sustain it for the time needed to burn away the flesh and calcine the exposed bone. This fuel is supplied by the body itself in the form of liquefied fat rendered out from the tissues as they are heated. Such fires tend to be more common with victims who are obese, as they can provide a greater quantity of fat to fuel the fire, allowing it to burn for longer. As long as there are absorbent materials nearby which are capable of acting as a wick, such as the victim's clothing, then the fat will burn as if in an oil lamp and the process of rendering and tissue destruction will continue, sometimes leaving only the extremities around a hole in the floor. As they involve the burning of fat, these fires tend to leave very greasy smoke deposits within the premises.

Close examination of the scene by the fire investigator will almost always reveal evidence of a viable ignition source in the vicinity of the victim. In some cases, however, the evidence can be complicated by the actions of the victim who may have moved away from or turned off the ignition source before they collapsed.

10.3.6   Multiple Fatalities and Major Disasters

At fire scenes where there has been a major loss of life there may be problems identifying the victims because of the destruction of clothing and personal effects. Severe damage to the body itself may also render it impossible or inadvisable for relatives to provide evidence of identification. DNA profiling is a valuable tool for confirming the identity of fire victims, but it is only effective if there is a sample of known origin available for comparison. At the early stages of the investigation of a major disaster, therefore, preserving any evidence of identity associated with the victims is extremely important. Because of the large-scale nature of such incidents victim recovery will be a team effort, and interaction with the site teams will be factored into the investigator's strategy. Depending on the degree to which the investigator is involved with victim recovery, work to determine the origin and cause of the fire may be deferred or will be carried out in such a way as to ensure that any bodies and associated materials remain undisturbed until they can be removed.

Inevitably at major incident investigations there are a great many interested parties, some of whom have an interest in the cause of the fire. The scene investigation process will take longer under such circumstances, even though the work carried out will follow the same principles as with much smaller fires.

10.4 ENDANGERMENT OF LIFE

Section 1(2) of the Criminal Damage Act 1971 states in essence that a person who damages any property with intent to, or being reckless as to whether they might, endanger the life of another is guilty of an offence. Although a charge of endangerment could be applied to any type of criminal damage, it is almost exclusively used by the criminal justice system in cases of arson. To support such charges fire investigators may be asked to comment on the degree of risk posed by the fire they are investigating.

Endangerment carries with it the suggestion that death or serious injury is a likely outcome of events. The principal life hazards in any fire are exposure to hot gases, soot and toxic combustion products; and injuries sustained while attempting to escape. In evaluating life threat the fire investigator will consider whether there is a risk that occupants of the building would be exposed to fire gases for the length of time it takes to produce impairment, or whether there would be sufficient thermal contact to produce burns.

One of the most obvious ways in which endangerment can occur is if potential victims are unable to avoid exposure because their means of escape is blocked. The way the fire has been started, its location and its potential to develop further will all be assessed by the investigator together with the fire precautions that are present and the available means of escape from the premises. On the basis of this assessment the investigator will comment on the likelihood that persons in the premises might sustain injuries or die.

10.5 LABORATORY INVESTIGATIONS

Within forensic laboratories at least 80% of all fire-related examinations are directed towards the recovery and identification of flammable liquid residues. The presence of such residues in a sample recovered from a fire scene provides strong support for the proposition that the fire was deliberately started, although the accidental spillage or leakage of liquids that were legitimately present at the scene must be ruled out before the scientist can say with confidence that the fire was arson.

Other examinations that the fire investigator may be asked to carry out include the inspection of clothing for flash burns, ignition and burning tests on materials recovered from the scene, the identification of improvised and manufactured pyrotechnic or incendiary devices and the inspection and testing of small domestic appliances for evidence of fault or misuse.

10.5.1 Flammable Liquid Recovery and Analysis

In most instances the flammable liquids used to start fires are the ones that are readily available to the general public. These tend for the most part to be hydrocarbon liquids such as petrol, white spirit and kerosene-based liquids (paraffin and turpentine substitutes). Of these petrol is the most frequently used, probably because it is easy to purchase in reasonably large quantities. The laboratory methods for the recovery and analysis of flammable liquids are capable, however, of identifying most hydrocarbon and non-hydrocarbon liquids that can be used to start fires.

Liquid samples recovered from containers need no preparation before analysis, but the majority of samples taken from scenes or suspects consist of absorbent substrates such as debris or clothing. Any liquid residues present on these items need to be recovered before they are analysed. Non-volatile liquids such as mineral and edible oils can be recovered by solvent washing, and hydrocarbon liquids can be extracted by steam distillation, but nowadays volatile flammable liquids are usually recovered using headspace techniques.

Most flammable liquids are volatile to a greater or lesser degree. In sealed forensic samples the vapours that accumulate in the headspace inside the packaging can be recovered and concentrated on a variety of adsorbent media. If necessary the concentration of these vapours can be increased by heating the item. Passive headspace techniques usually involve placing the adsorbent within the package and incubating it, while dynamic sampling involves the withdrawal of a measured volume of the headspace that is passed over the adsorbent. The adsorbent is then either treated with a solvent, or thermally desorbed, and the resulting sample is introduced into an analytical instrument for identification.

Gas chromatography–mass spectrometry (GC-MS) is the generally accepted method of choice for the forensic analysis of volatile organic compounds such as flammable liquids. Within the gas chromatograph the complex mixture of chemical compounds that makes up the flammable liquid is driven by a carrier gas into a long capillary tube coated with an adsorbent. During transit through the capillary these compounds will interact with the coating to an extent that depends upon their physical and chemical properties. This has the effect of slowing the passage of some compounds along the capillary with respect to others, separating them so that they emerge into the detector at different times.

As the chemical compounds enter the detector, a mass spectrometer, they are ionised and break down into fragmentation patterns that are characteristic of the molecules involved. A mass detector then records these fragmentation patterns. In its simplest form the mass spectrometer's integrated output or total ion count (TIC) can then be reproduced on a chart as a series of peaks. These peaks may have the characteristic pattern of a flammable liquid (Figure 10.6), but even if they do not the MS data will provide details of the molecular structure of the compound that each peak represents (Figure 10.7), normally allowing the liquid to be characterised. This is an extremely powerful method for identifying flammable liquids and their residues.

10.5.2 Examination of Clothing for Flash Burns

When a volatile liquid such as petrol is spilled or poured it will start to evaporate, and the vapour that is released will mix with air to form a diffuse flammable cloud above the liquid. If the cloud of vapour is ignited, the resulting burst of flame can cause heat damage to vulnerable surfaces nearby, including the clothing of a person applying the ignition source.

The response of garment fabrics to this brief application of intense heat will vary depending upon the nature of the fibres they are made from. Thermoplastic materials will melt and natural fibres will scorch or char. The tips of fibres and threads protruding from the surface of the fabric are most vulnerable to heat, and at high enough temperatures they will either melt back or discolour. This damage is not always visible to the naked eye, so microscopic examination of the surface of the garment may be necessary to identify it. The distribution of the damage across various garments may be evidentially significant, depending on the circumstances of the incident—petrol vapour is denser than air, so if a pool of petrol is ignited then the burning vapour is likely to cause greatest damage to garments at low level such as shoes and the legs of trousers rather than upper items of clothing.

Some fabrics and garments are heat-treated or flamed to remove loose threads during manufacture, and there are other ways in which clothing can be thermally damaged by heat after purchase (over-hot tumble drying or ironing are two examples). As a result the identification and

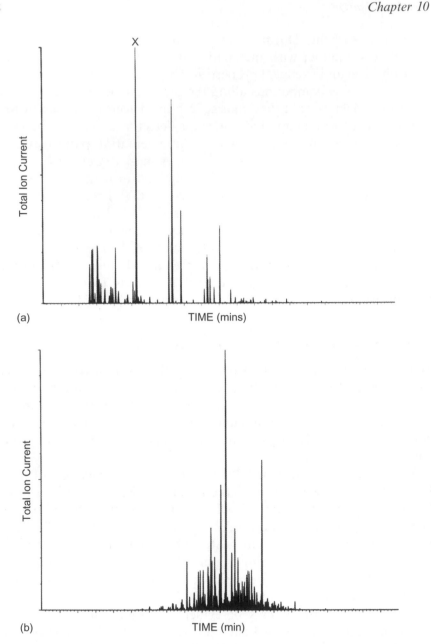

Figure 10.6 GC-MS total ion current (TIC) plots of (a) petrol and (b) white spirit. Note the peak marked X on the petrol plot and refer to Figure 10.7.

interpretation of flash burning requires considerable expertise on the part of the scientist.

The presence of flash burning on a suspect's clothing is only eviden-tially significant if it can be shown that a volatile flammable liquid was

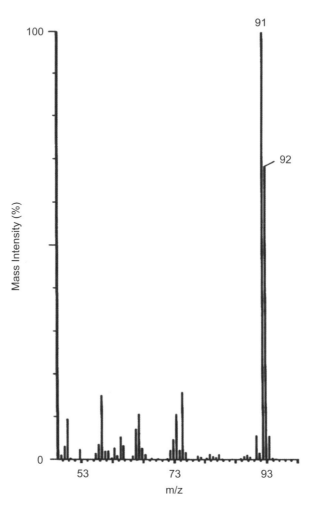

Figure 10.7 Mass spectrum of peak marked X in Figure 10.6. The mass response at m/z 92 is the molecular ion of toluene.

used to start the fire. In such circumstances, and provided that alternative damage mechanisms can be ruled out, then the existence of flash burns offers considerable support for the suggestion that the suspect was close to the fire when it was ignited.

10.5.3 Ignition and Burning Tests

The process of hypothesis testing during the evaluation of a scene investigation might indicate that there is more than one possible explanation for the fire. If the investigator has been able to identify the material that was first ignited then tests can be used to establish which candidate ignition sources are viable, and which can be ruled out, as causes of the fire. Equally, if there are unanswered questions about certain aspects of the fire's subsequent development, such as the time needed for a transition from smouldering to flaming, then a burning test might provide the answers.

Ignition and burning tests need to be designed carefully if they are to reproduce as closely as possible the circumstances at the scene. The more a test differs from the situation at the scene the more likely it is that the results will be challenged in court. Forensic burning tests are not therefore like the national and international standard tests that are used to assess ignitability or fire performance.

If possible the investigator will use materials recovered from the scene to carry out the tests and will try to replicate as closely as possible the conditions at the time the fire started. This is particularly important if the size of the reconstruction, or the restricted availability of items to burn, means that only one test will be carried out. If a completely intact test object is not available then a mock-up may be constructed using whatever materials can be salvaged from the original item at the scene.

For small-scale burns, replicate testing is desirable as it allows the investigator to explore the effect of changes to the starting conditions. Ignition sources such as smouldering cigarettes are particularly sensitive to small changes in orientation and ventilation, to the extent that apparently identical tests can have different outcomes. In such circumstances the investigator may have to carry out a large number of tests to determine the probability of ignition so that they can offer an opinion on the likelihood of it taking place.

For safety reasons, anything other than simple ignition tests are only carried out in purpose-built laboratory facilities fitted with smoke extraction equipment and fire extinguishers.

10.5.4 Examination of Incendiary Devices

Improvised incendiary devices are sometimes used to start fires or to provide a time delay so that the person responsible can arrange an alibi. Since such devices are often made up from commonly available household materials, it is only the combination of items that allows the scientist to say that the recovered item is an incendiary device.

Simple time delays can be constructed from cigarettes and matches, or employ electrical time switches connected to heating elements or other incandescent sources. Such devices are often deployed with a flammable liquid that will assist the growth of the small fire that is produced, and if petrol is used the delay may allow sufficient flammable vapours to build that there is the risk of an explosion.

Laboratory testing of time-delay devices is normally carried out to determine how long a delay they produce and how reliable they are. As with other burning tests, it is important that the reconstructed device reproduces as closely as possible the one found at the scene if the results are to be relied upon in court.

Incendiary devices that do not incorporate a time delay are generally intended to start a large fire quickly and efficiently. They will almost always include flammable liquid, and they tend to be variations on the design of a petrol bomb. Although all such devices may have similar effects when used, there are specific legal rulings in case law that apply only to genuine petrol bombs, which are classified as explosive substances. Manufacture or possession of a petrol bomb is therefore subject to the provisions of, and the penalties imposed by, the Explosive Substances Act 1883. For the purposes of the Act a petrol bomb is strictly defined as a glass bottle, containing petrol and with an absorbent wick fitted through its neck. Possession of the components needed to make a petrol bomb is not sufficient; the device must be made up as prescribed if the law is to apply. The forensic scientist will be asked to certify that a particular device conforms strictly to the definition of a petrol bomb if charges are brought against an individual alleging manufacture or possession of an explosive substance.

Occasionally, manufactured items such as fireworks, distress flares or military pyrotechnics will be submitted to the laboratory for identification or testing. The casings of civilian or military pyrotechnic devices often have batch numbers on them, and in some instances it may be possible to trace them back through manufacturers' and suppliers' records to a particular retailer or armoury. Depending upon the charges faced by a defendant, the court may want the scientist to test a similar intact device and make a comment on its effect.

10.5.5 Appliance Examinations

Electrical and gas appliances can start fires through fault or misuse. When examining such items often the most important consideration is to establish whether the appliance was turned on or not, and in the case of a gas appliances whether gas was supplied to it. If the appliance was switched off at the time of the fire then it is unlikely to have been the ignition source, although if the scientist is to say this then the possibility that the controls were interfered with after the fire must be ruled out.

Normally, any appliance recovered from the area of origin of a fire will be damaged as a result. Plastic casings and light alloy components will be melted and metal parts will be oxidised and distorted. The investigator will use non-destructive techniques such as visual inspection and x-radiography to establish the condition of external and internal components before disassembling and testing the appliance. If the fire damage is not too severe then internal fuses and electrical switches can be checked for continuity, and gas valves tested for flow. After the damaged casing is removed inspection may reveal that internal components are intact, indicating that the fire has not originated inside the appliance.

If an appliance is not too severely damaged then it can be repaired and refurbished if tests are required; otherwise, the scientist will try to obtain an undamaged example to carry out the work.

BIBLIOGRAPHY

V. Babrauskas, *Ignition Handbook*, Fire Science Publishers, Issaquah, WA, 2003.

P. C. Bowes, *Self-heating: Evaluating and Controlling the Hazards*, HMSO, London, 1984.

J. D. DeHaan, *Kirk's Fire Investigation*, 6th edition, Prentice Hall, Englewood Cliffs, NJ, 2007.

D. Drysdale, *An Introduction to Fire Dynamics*, John Wiley & Sons, Chichester, 1985.

D. J. Icove and J. D. DeHaan, *Forensic Fire Scene Reconstruction*, 2nd edition, Prentice Hall, Englewood Cliffs, NJ, 2009.

N. NicDaeid (ed.), *Fire Investigation*, CRC Press, Boca Raton, 2004.

E. Stauffer, J. A. Dolan and R. Newman, *Fire Debris Analysis*, Elsevier, New York, 2008.

The SFPE Handbook of Fire Protection Engineering, 4th edition, National Fire Protection Association, Quincy, MA, 2008.

CHAPTER 11

Explosions

CLIFF TODD, LINDA JONES AND MAURICE MARSHALL

DSTL, Forensic Explosives Laboratory, Fort Halstead, Sevenoaks, Kent
TN14 7BP

11.1 INTRODUCTION

The investigation of accidental or illegal explosions, and scientific analysis of their causes, has a long history which in the UK started formally
in 1871. An explosion at a factory in Stowmarket making guncotton led
the then Home Secretary to instigate an inquiry. A Royal Engineer,
Captain Vivian Majendie, who was an expert on explosives, led the
investigation. He recruited a chemist, Dr A. Dupre, to assist. The
arrangement proved so successful that the Home Office decided to
continue it, leading to the present-day Forensic Explosives Laboratory
(FEL). One of the first fruits of their collaboration was the 1875
Explosives Act which embraces various aspects of explosives including
their manufacture and storage. Subsequent bombing outrages led Par-
liament to enact the 1883 Explosive Substances Act which deals with the
criminal use of explosives and devices; this Act was intended specifically
to deal only with the most serious offences, hence the special provision
made requiring the fiat of the Attorney General for any prosecution
under it. More recently, these Acts have been amended by the
Manufacture and Storage of Explosives Regulations (MSER) 2005.

Crime Scene to Court: The Essentials of Forensic Science, 3rd Edition
Edited by P. C. White
© Crown Copyright 2010
Published by the Royal Society of Chemistry, www.rsc.org

Many of the questions which Captain Majendie and Dr Dupre were asked to address are still relevant to today's forensic explosives investigators, including the following:

1. Was it an explosion?
2. Was it an accident or a bomb?
3. Is this an explosive?
4. Was this a viable device, or a hoax?
5. Are these items or materials intended for making explosives or bombs?
6. Has this person been in contact with explosives?
7. Have explosives been stored in this place?
8. Are there similarities between these items or incidents that link them together?
9. Could the items have an innocent use?

Actual cases involve a variety of circumstances and the questions which are appropriate will vary accordingly.

In order to address the above points, the forensic explosives scientist requires a sound grounding in the requirements of the judicial system and the ethical principles underlying it, together with a detailed knowledge of the science and technology of explosives, practical experience in the construction of all types of explosive devices, and a clear understanding of their effects.

11.2 EXPLOSIVES TECHNOLOGY

11.2.1 What is an Explosion?

A convenient working definition is 'a sudden and violent release of physical or chemical energy, often accompanied by the emission of light, heat and sound'.

To the human observer an explosion seems instantaneous; however, the chemical reaction actually proceeds at a finite, albeit high, speed, progressing through the material as a definite 'front' or 'wave'. Two types of event may be defined: a 'deflagration' is an event where the decomposition within the explosive occurs at a speed equal to or less than the velocity of sound within the material, and a 'detonation' is an event where decomposition occurs at a speed greater than the velocity of sound within the material. Sometimes deflagrations are referred to as 'low-order explosions', and there is a mistaken tendency to assume that they are in some way less serious than detonations. This is an error: many of the most devastating accidents with explosives have involved deflagrations rather than detonations.

Typically deflagrations occur with velocities of less than 2000 m/s while detonations may reach velocities of 6000–8000 m/s for certain high-performance military explosives.

11.2.2 Types of Explosion

Explosions may be characterised by the source of their energy, *i.e.* physical or chemical, and also by their locus, *i.e.* whether dispersed or condensed phase.

Examples of the various classes are:

- **Physical.** An exploding pressure vessel, *e.g.* an overheated gas cylinder.
- **Chemical.** Explosion of a mass of black powder, or gunpowder as it is more commonly known.
- **Dispersed.** Detonation of a cloud of flour in air.
- **Condensed.** Detonation of a stick of dynamite.

Each of these types can be further classified according to the speed and duration of the explosive event, and this can be linked to the type of practical effect observed, enabling the skilled investigator to draw some meaningful conclusions from examination of the types of macroscopic and microscopic damage found at the scene of an explosion. Thus high-velocity detonations cause shattering and cutting of metal, whereas low-velocity events result in tearing and heaving.

11.2.3 Types of Explosives

Although dispersed-phase explosions can be of great power, as can physical explosions, the vast majority of practical explosives used for either commercial or military purposes are condensed-phase chemical explosives. These energetic materials are best considered according to their function; thus we have:

- **Pyrotechnics.** Used for the production of heat, light, sound or smoke, *e.g.* in fireworks or signalling flares.
- **Blasting explosives.** Used *e.g.* to break up rock in quarrying operations.
- **Initiatories.** Used to transform a small mechanical or thermal impetus into a violent shock wave capable of causing detonation of less sensitive energetic materials such as blasting explosives.

Initiatories are also referred to as primary explosives, indicating their role at the start of a chain of explosive events. Analogously, the less sensitive explosives used for the main charge are referred to as secondary explosives.

- **Propellants.** Energetic materials which deflagrate in a controlled fashion to allow their energy to be used *e.g.* in propelling rockets or projectiles from guns.

Apart from the above scientific definitions, the forensic scientist needs also to be familiar with the legal definitions; e.g. section 3 of the UK's Explosives Act 1875 states:

The term 'explosive' in this Act—

1. Means gunpowder, nitro-glycerine, dynamite, guncotton, blasting powders, fulminate of mercury or of other metals, coloured fires, and every other substance, whether similar to those above mentioned or not, used or manufactured with a view to produce a practical effect by explosion or a pyrotechnic effect; and
2. Includes fog-signals, fireworks, fuses, rockets, percussion caps, detonators, cartridges, ammunition of all descriptions, and every adaptation or preparation of an explosive as above defined.

This Act is updated and further expanded upon, under the Manufacture and Storage of Explosives Regulations 2005. Other countries have their own legal definitions.

Some examples of military and commercial explosives are shown in Figure 11.1.

11.2.4 Chemistry of Explosives

The requirements are simple: the ingredients of the explosive need to undergo some very rapid chemical reaction liberating large amounts of energy. In practice the most difficult part is achieving control of this process so that the explosive only reacts when required, and not otherwise.

The earliest known explosive was gunpowder, more correctly known as black powder, first developed by the Chinese over 1000 years ago. This is an intimate mixture of charcoal and sulphur (the fuel) and potassium nitrate (the oxidant). The chemical reactions which occur are in fact highly complex, resulting in formation of oxides of carbon, sulphur and potassium, together with potassium/sulphur compounds, all in a range of oxidation states. Black powder illustrates the way in which

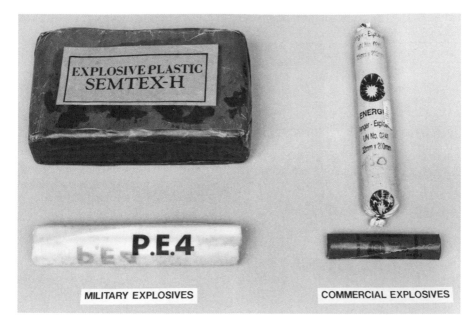

Figure 11.1 Some commercial and military explosives.

the behaviour of explosives can be tailored to a particular application. If spread out in a thin layer in the open air it will merely burn violently, but if confined in a pressure-tight container it can be made to detonate if ignited. Under normal conditions black powder deflagrates and can be usefully employed as a propellant, *e.g.* in the guns of Lord Nelson's *HMS Victory* at Trafalgar, or in modern-day display rockets on Guy Fawkes night. This versatile material is also very widely used in current military munitions as an igniter powder; the fiery particles of molten inorganic slag which form on its decomposition make it particularly suitable for this purpose.

In broad chemical terms black powder is an intimate mixture of fuels and oxidants. Other inorganic and organic materials can also be combined to produce explosive mixtures, the commonest example being the use of ammonium nitrate/fuel oil mixtures (ANFO) in quarrying. Inorganic compounds are also used widely in pyrotechnic mixtures, *e.g.* chlorates, perchlorates, and nitrates of alkali metals are common oxidants.

The other approach, which can yield even higher explosive performance, is to combine the fuel and oxidant in the same molecule. Although some organofluorine explosives are known, these have not been generally adopted, and the majority of organic explosives are

Table 11.1 The common and chemical names of organic explosives with their chemical structures and properties.

Common name	Chemical name	Properties
TNT	2,4,6-Trinitrotoluene	Low-melting solid; can be conveniently cast into bombs and shells; dissolves in organic solvents
Nitroglycerine	Glyceryl trinitrate	Viscous liquid; toxic vapour; the neat liquid is sensitive and unstable
PETN	Pentaerythritol tetranitrate	White crystalline solid
RDX	Cyclotrimethylenetrinitramine	White crystalline solid; high density; high detonation velocity; moderately soluble in acetone
HMX	Cyclotetramethylenetetranitramine	White crystalline solid; high density; high detonation velocity; high melting point (275 °C); very stable; moderately soluble in acetone
Tetryl	2,4,6-Trinitrophenylmethylnitramine	Yellow solid

organic nitro compounds. The earliest example was nitroglycerine (glyceryl trinitrate), commercialised by Nobel as 'Dynamite' or 'Gelignite'; others include the nitroaromatics such as 2,4,6-trinitrotoluene (TNT), or the nitramines, *e.g.* RDX. Some commercial and military explosives together with their names and properties, are listed in Table 11.1.

11.2.5 Initiation and Detonation of Explosives

As mentioned above, the most important practical problem in explosives technology is the safe control of initiation. This has led to the widespread use of the concept of the 'explosives train'; a small quantity of a very sensitive explosive is used to receive an initial stimulus in the form of thermal, mechanical or electrical energy and amplify this so as to start a reaction in a larger mass of less sensitive material.

This leads to the concept of the 'detonator' invented by Nobel. Commercially, two main types are produced: 'plain detonators' and 'electric detonators'. Figure 11.2 shows the essentials of the latter type, together with some examples of the wide variety available.

A sealed pencil-shaped metal tube contains a metal wire filament embedded in a heat-sensitive match composition; when an electric

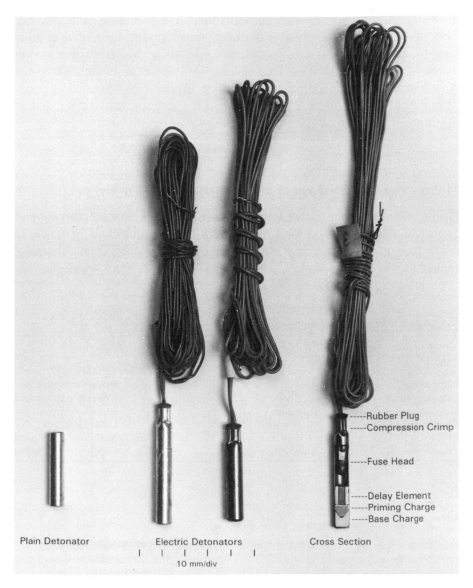

Rubber Plug
Compression Crimp

Fuse Head

Delay Element
Priming Charge
Base Charge

Plain Detonator   Electric Detonators   Cross Section

10 mm/div

Figure 11.2 Detonators.

current is passed through the wire the resulting heat from the filament ignites the match composition(the 'fuse head'). This in turn ignites the next element in the detonator. In practice this may be a delay composition (the 'delay element') used to provide greater control in the sequencing of industrial blasting operations, or the ignition may pass directly to the small charge of primary or initiatory explosive (the 'priming charge') which once ignited burns to detonation. The detonation of the primary explosive in turn causes detonation of a slightly larger quantity of a secondary explosive (the 'base charge') which acts as a booster, amplifying the effect several-fold. In a plain detonator the wire filament is replaced by an igniferous fuse.

Table 11.2 lists some examples of the energetic materials used in detonators for each of the functions shown above.

11.2.6   Essential Elements of an Improvised Explosive Device

Great ingenuity has been displayed in the creation of improvised explosive devices, but common elements can be identified in terms of basic functions. Figure 11.3 shows the essential elements in schematic form.

Table 11.2   Some explosive materials used in detonators.

Common name	Function
Mercury fulminate	Primary explosive
Lead azide	Primary explosive
Lead styphnate	Primary explosive
Potassium chlorate/lead thiocyanate	Match composition
Barium peroxide/selenium	Delay element
PETN	Secondary explosive
Tetryl	Secondary explosive

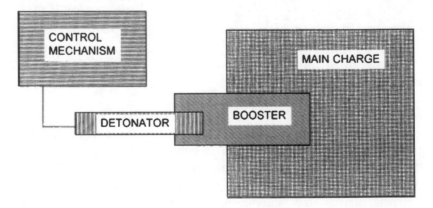

Figure 11.3   Elements of an improvised explosive device.

11.3 FACILITIES REQUIRED FOR FORENSIC EXPLOSIVES EXAMINATIONS

In the UK, strict regulations apply to the keeping and use of explosives, and subject to a few exceptions such as for some fireworks, no laboratory or individual may possess explosives without appropriate licences. Similar regulations exist in many other countries. Apart from these legal aspects, safety considerations also need to be taken into account and explosives should not be handled without adequate training and facilities for the purpose.

11.3.1 Safety

Explosives explode. This simple rule needs to be borne in mind at all times when handling energetic materials of any kind; all processes and equipment should be designed on the basis that at some time an unintended explosion will in fact occur and therefore the safety procedures must be such as to minimise the effect. Most nations have detailed explosives regulations covering the safe storage, transport and use of commercial and military explosives. However, these often cannot be applied to the case of illegal materials, and scientists investigating such matters must be careful to minimise hazards both to themselves and to others.

A particular hazard, in recent years, has been a significant increase in the number of incidents involving the use of the very sensitive improvised primary explosives TATP (triacetone triperoxide) and HMTD (hexamethylene triperoxide diamine). These have been produced both for serious criminal purposes and by amateur experimenters, but in all cases are extremely hazardous, to the people making them and all who come within their immediate vicinity. Death and serious injuries have resulted from the manufacture of these materials in the UK, irrespective of the intended reason for making them. In addition, the possibility that improvised explosive devices may incorporate concealed anti-handling features should always be addressed.

Explosive devices should always be examined at the scene by properly qualified bomb disposal personnel before being disturbed in any way whatsoever; the bomb disposal personnel will also be able to provide the necessary advice on safe packaging and transport arrangements. Practical experience shows that the most dangerous devices are often those made by amateur experimenters, *e.g.* schoolchildren.

Specialist facilities must allow for four activities: receipt, storage, examination and disposal.

11.3.2   Receipt

The arrangements for receipt of items have some features in common with those general in forensic work, *e.g.* the possibility of cross-contamination between submitted items must be rigorously and demonstrably precluded. The general requirements for preserving the integrity of trace and contact evidence are discussed in previous Chapters. In addition, the system must always preserve the chain of custody of evidence. In addition, explosives laboratories need pre-planned procedures for dealing with devices which may deteriorate into an unsafe condition or are suspected of incorporating anti-handling features. Thus immediate access to bomb disposal equipment and personnel is essential, together with radiographic facilities designed to allow safe examination of explosive devices and provide details of the internal construction of such items before disassembly.

The mechanics of receipt have to cater for a range of items which can cover anything from a small sealed package of hand swabs for trace examination to a blown-up motor vehicle or several tonnes of bomb scene debris. In practice it is convenient to have different physical arrangements for the various streams of evidence types whether bulk or trace, explosive or non-explosive, clean or dirty.

11.3.3   Storage

Detailed safety guidelines and regulations cover the design of explosives storehouses (magazines). These require *e.g.* separation of detonators and bulk explosives; they also include limits on the mass of explosives which may be kept in particular designs of store, and also specify security features to prevent unauthorised access.

A typical laboratory magazine is strongly constructed of reinforced concrete, provided with spark-proof electrical lighting and lightning protection, and fitted with high-security doors and locks. The whole is enclosed within a substantial barricade intended (1) to protect the magazine from an explosion in any other adjacent building, and (2) to deflect upwards the blast wave that would occur if the contents of the magazine itself were to explode.

The laboratory will also need to store non-explosive exhibits ranging from items for explosives trace analysis to bulk debris in a way that meets all the requirements for preservation of the chain of custody and protection of exhibit integrity.

11.3.4   Examination

The items to be examined can be categorised as bulk explosives, invisible traces of explosives and non-explosive materials.

11.3.4.1 Examination of Bulk Explosives. The design of facilities for bulk explosives is governed by both the type and mass of explosive being handled. In general, different work areas need to be provided for examination of detonators and primary explosives, pyrotechnics and incendiaries, and secondary explosives. Where practicable it is also wise to segregate activities involving large and small quantities. For example, it is often convenient to break down devices containing large explosive masses in a separate building, take small samples and then examine these in more detail in the main laboratory. Buildings for explosives work require appropriate electrical fittings to prevent spark and dust hazards, together with lightning protection and anti-static precautions; particular attention needs to be paid to fire prevention and escape in the building design.

In a laboratory for examining detonators and primary explosives the floor and benches are normally covered in conducting rubber sheet which is earthed to reduce hazards from sparks due to static electricity, as is all the laboratory equipment. All personnel entering the laboratory have to check themselves on a special meter to ensure that they are properly earthed. In dry climates it is wise to humidify the laboratory atmosphere to reduce hazards from static electricity. The detonators and primary explosives are manipulated behind stout armoured screens which need to be of a design that has actually been type-tested to prove that it really does withstand the effects of an explosion of the quantity likely to be involved. It is not sufficient to rely on normal laboratory safety screens, which are most unlikely to contain the explosive effects of a typical commercial detonator. Similarly, fume cupboards used for chemical treatment of primary explosives (which are generally toxic) should be suitably armoured with arrangements to vent blast pressure waves due to any detonation.

Special machinery is essential if detonators are to be opened to allow examination: lathes, cutters, manipulators and screens must all be type-tested to prove their suitability and safety before use. The superficial simplicity of the process conceals many potential hazards and it most definitely should not be attempted without proper facilities and training.

11.3.4.2 Detection of Invisible Traces of Explosives. This can include swabs or other samples taken from people's hands or property, or samples obtained from items that have been involved in an explosion. As with all other contact trace evidence where the objective is to detect very small quantities of the analytical species, it is essential to carry out the examination in a controlled environment where it can be demonstrated by objective measurement that the results obtained are meaningful and not the consequence of contamination either from the general environment or other items at the laboratory.

In general this will require precautions such as the use of clean disposable overclothing, separate work areas for different trace operations, regular monitoring and cleaning, and control of access to ensure that people and items which enter the trace area are free from explosives. It should also be borne in mind that such contamination can occur at the crime scene from which the samples or items are recovered, and similar precautions must be taken there, as at the laboratory. In particular, samples for trace analysis should not be taken by people who are likely to be contaminated with explosives in their normal work environment, such as bomb disposal operators or other military personnel.

11.3.4.3 Non-Explosive Items. Bomb Components. Live explosives will often be accompanied by non-explosive items, either as part of a device or in a collection of suspected bomb making equipment. These may include containers, timing devices, power supplies, arming mechanisms and switchgear. Facilities for mechanical examination and disassembly, measurement, photography and microscopy will be needed.

Post-Explosion Debris. This could include the remains of an exploded firework or more generally an exploded device, a bomb-damaged motor car or tonnes of debris from a bombed building. Each of these evidence types requires a different approach and different facilities. Conventional well-lit, clean laboratory benches covered with fresh disposable paper will suffice for examination of the remains of a small exploded device. Handling bomb-damaged motor vehicles requires special trolleys and lifting gear, particularly since such vehicles are likely to be structurally unsafe and may easily collapse on people nearby if lifted without properly designed equipment operated by trained personnel. Examination of bulk debris from bombed buildings requires a system for the mechanical handling of large quantities of material, together with facilities for drying, sieving and searching.

Debris from bomb scenes can present a range of hazards: there may be materials such as asbestos present in building debris, which is also liable to be contaminated with sewage; damaged vehicles may contain flammable liquids; and all are liable to be contaminated with biohazard materials, particularly from events where people have been injured or killed. Thus laboratory facilities for this type of work need appropriate mechanical ventilation for dust and vapour extraction, supplemented as necessary by personal breathing apparatus. Vaccination of staff against a range of diseases is also advisable.

11.3.5 Disposal

A major consideration in explosives work of any kind is the need to provide for the safe and environmentally acceptable disposal of all waste and unwanted material. It is not permissible simply to mix explosives waste with normal waste, and most particularly not with waste chemicals. Waste explosives and explosives removed from improvised devices are likely to be less stable than newly manufactured material as a result of rough handling, inadequate storage and contamination and should therefore be treated with especial care. Material may be destroyed chemically or more commonly by burning at an appropriately licensed facility.

11.3.6 Reference Collections and Databases

Reference collections are commonly used in forensic science; they are of great value in explosives work. Thus samples of compositions, and examples of packaging and labelling are all most useful in the identification of individual commercial explosives. Although the identification of an intact cartridge of explosive bearing the manufacturer's label may appear to be a trivial task, identification of the same item from a few small damaged fragments of wrapper and a milligram of undetonated composition is a more challenging problem. Likewise, specimens of items such as detonators and blasting accessories are useful for physical comparison. Such collections of physical hardware can advantageously be supplemented with a library of commercial literature which can prove especially helpful in identifying unusual or old items.

A set of well-characterised samples of known explosives are essential as comparison and calibration standards for chemical analysis by techniques such as GC or HPLC. Collections of reference spectra are also helpful, but the possibility of errors in commercially published spectral libraries must be borne in mind, and it is wise to establish the authenticity of spectra used in critical work.

A carefully indexed database of devices should be maintained to facilitate correlation between incidents, yielding pointers to individuals and organisations involved in series of outrages.

11.4 FORENSIC QUESTIONS

11.4.1 Was it an Explosion?

This may at first sight appear to be a fairly straightforward question to answer. Generally it will involve examination of the damage at the scene, either at first hand or, less satisfactorily, by viewing photographs, reviewing witnesses' accounts of the event and laboratory examination of debris from the scene. The occurrence of sudden loud reports, flashes, violent projection of debris, smashing, tearing, or rupturing of structural materials and the formation of craters at the seat of a supposed explosion are all useful indicators. However, one instance where it can be a very difficult question to answer is in the case of air crashes, where there will be a huge amount of damaged material, only a very small proportion of which is likely to show any characteristic explosive damage. This situation is dealt with in more detail in the next section.

11.4.2   Was it an Accident or a Bomb?

This has to be a combined process both of elimination of possible accidental causes, such as leaks of flammable gases or the presence of flammable dust clouds; and wherever possible, the positive identification of physical or chemical evidence resulting from the bomb.

Physical evidence can involve recovering fragments of the bomb from the debris: contrary to popular misconception, explosions do not always vaporise everything in the immediate vicinity. Rather it is more accurate to say that 'bombs shatter and scatter'. Thus teams of trained and diligent searchers were able to recover evidentially significant fragments of the device in many of the cases that have occurred in the UK between 1989 and 2009. Such fragments may include pieces of the bomb container, the waterproof plug from the detonator, parts of any timing mechanism such as clockwork, electronic circuitry, batteries, and wires. Fragments of the device are likely to be found lodged in comparatively soft objects near the seat of the explosion, *e.g.* in vehicle tyres and seats in the case of under-car booby traps. The bodies of victims are also potentially good receptors for bomb fragments; medical staff should routinely be asked to x-ray victims and pass on any fragments recovered for scientific examination.

A small victim-actuated bomb in a book is shown in Figure 11.4, together with the accompanying bag in which it was transported. Such an item would be intended to explode when the book was opened.

Although containing only some 25 g of plastic explosive, such a device would be likely to maim or kill anyone opening it. The fragments recovered after explosion can be seen in Figure 11.5; information of considerable investigational and evidential value might potentially be obtained from such material. The metal fragments, for example, might yield chemical residues of the explosive, the battery fragments type and batch numbers; the watch might have tool marks or manufacturer's

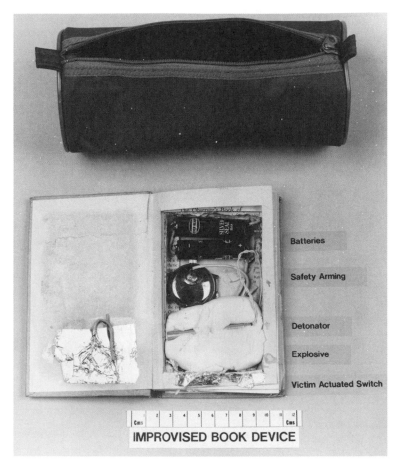

Figure 11.4 A small improvised bomb in a book.

marks, and it might be possible to link the wires with similar wires from a suspect's premises. Likewise, the book and the bag might provide links to a suspect.

Incendiary devices that have partially functioned often produce a fused mass of charred and burnt material, particularly if the device was in a plastic container. In such cases radiography can often be used to reveal details of the construction of the device from otherwise intractable evidence.

In addition, explosions cause characteristic damage to many receptor surfaces as a result of the unique combination of high-velocity shock and transient high temperatures. Surfaces near to the seat of an explosion, at a range of a few centimetres to some metres depending on the size of the explosion, are likely to be exposed to the effects of very hot

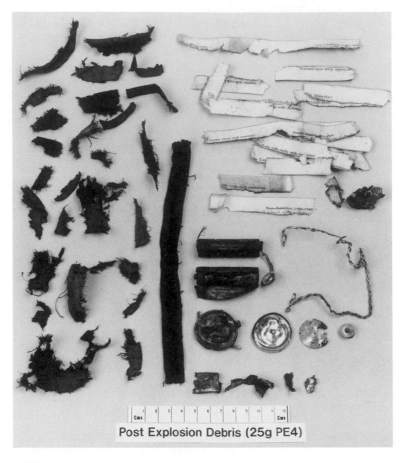

Post Explosion Debris (25g PE4)

Figure 11.5 Fragments from a small bomb in a book.

gases, and tiny particles, ejected at high velocity. Hot gases can give rise to the phenomenon referred to as 'gas-wash', which gives rise to a mottled irregular surface effect due to the partial melting and erosion of the target material. Fast moving particles can result in 'microcratering' on nearby surfaces, particularly metal. These tiny craters, characteristically with a rolled lip, range in size from fractions of a millimetre to 1–2 mm diameter. An example of a microcrater on a piece of aluminium is shown in Figure 11.6.

Many explosives also leave deposits of soot on cooler adjacent surfaces where partially combusted material and reaction products condense. Large metal fittings and items such as metal window frames and railings often act as receptors for soot in this way.

Examination of the surface damage with a scanning electron microscope (SEM) can be highly informative. For example, nylon textiles may display

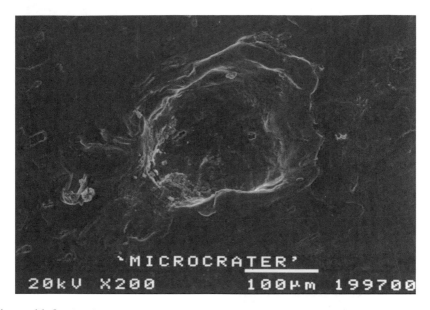

Figure 11.6 A microcrater in aluminium revealed by SEM.

characteristic 'toffee apple' or 'clubbing' damage; the momentary high temperature causes the polymer to melt and then re-solidify to produce a small spheroid on the fibre end which is readily visible in the SEM and an example of this is shown in Figure 11.7. Other characteristic effects also occur on metals; ferrous alloys may display changes in crystal structure, and explosive cladding of one metal upon another can also occur.

Chemical residues of explosives or their decomposition products may also be sought and if found provide useful evidence. In the case of uncommon chemicals with no non-explosive uses (*e.g.* tetryl or RDX), the detection of residual traces is strong evidence. This is especially true in the case of improvised explosives such as TATP and HMTD, since they are not used or produced commercially at all, owing to their sensitivity and poor storage qualities.

The gross physical damage at the scene of a suspected bombing is also informative, since careful examination will enable the scientist to distinguish between a dispersed-phase explosion and a condensed-phase explosion and in the latter case to locate the seat of the explosion. Structural damage, crater depth, and breakage of window glazing are all useful in assessing the mass of explosive involved; however, it must be realised that calculations of the size of bombs from observations of scene damage are subject to large margins of error.

Particular difficulty can be experienced in determining whether an aircraft crash is due to the explosion of a bomb, or some other cause. In

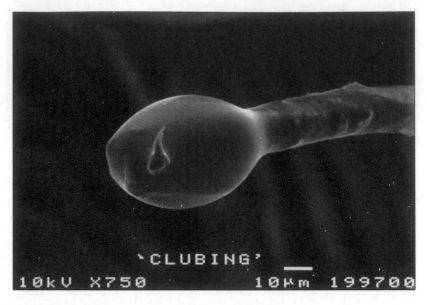

Figure 11.7   Clubbing on a nylon fibre as seen by SEM.

this type of case it is necessary to distinguish between what may be very extensive mechanical damage due to the crash and damage that could only be due to an explosion. Where reconstruction of the crashed aircraft is feasible, this can be extremely helpful in locating the position of an explosion. Tracing of the path of shrapnel from a bomb can sometimes be accomplished by examination of the aircraft seats, *e.g.* by threading stiff wires through the fragment penetration holes, thereby enabling location of the site of the bomb. Where a bomb has been placed in hold baggage, careful examination of the recovered baggage can sometimes enable identification of which items of baggage were adjacent to the bomb and which specific item of baggage contained the bomb. This can provide exceptionally valuable assistance to the investigator, enabling possible links back to the individual responsible for checking the baggage onto the aircraft.

11.4.3   Is This an Explosive?

The forensic scientist is likely to encounter explosives samples in five main ways:

1. As a bulk of an unknown substance, whether in isolation or together with components suspected of being related to bomb-making activities.

2. From a suspect device that has been rendered safe by bomb disposal personnel. Apart from suspect explosive, other components of the device such as batteries, timepieces and igniters are also likely to be submitted.
3. As traces on items of post-explosion debris.
4. As traces on items suspected of having been in contact with explosives.
5. As traces on people suspected of having handled explosives.

11.4.3.1 Examination of Bulk Explosives. Although some organic compounds are well-known explosives, *e.g.* TNT, there are many mixtures of ingredients which may be explosives depending on the precise formulation and circumstances. Thus it is necessary to make use of both chemical analysis to identify suspect materials and also a variety of physical tests to demonstrate both the energetic nature and detonability of samples.

The very first step is of course to look at the sample: if it happens to bear a manufacturer's label identifying it, this can be quite helpful! However it is still necessary to confirm the true identity of labelled items since the label may not be accurate. Visual examination with a low-power microscope is of considerable value, particularly for recognising materials such as home-made black powder.

The simplest test is to observe the behaviour of a small (tens of milligrammes) quantity of the material when it is ignited. This test must be done behind a substantial armoured glass screen, preferably within an armoured fume cupboard since some explosives produce toxic vapours on decomposition, *e.g.* lead salts. Points to note are:

1. Does it burn? Smoulder? Melt? Char?
2. What colour is the flame? Are there flashes or sparks?
3. What is the appearance of any residue?
4. Did the burning material emit a characteristic odour?

A more elaborate test is the Lidstone cartridge test, which assesses the detonability of small quantities of suspect materials. A 2 g sample of the suspect material is packed in a .303 brass cartridge case, a standard detonator fitted in a sleeve in the open neck of the case and the assembly fired in a tank filled with wood pellets. The wood pellets are then sieved and the remnants of the cartridge case recovered. Comparison with firings of known explosives allows a rough assessment of the sample's explosive performance to be made.

Since some explosives, particularly mixtures, react differently when initiated in very small quantities compared with how they behave in larger quantities, a larger-scale version of this test, known as a gap test, is sometimes used. This test uses up to 200 g of the questioned material, in standard-size steel tubes, initiated with a standard-size pellet of commercial high explosive. This, however, requires a properly licensed bomb chamber in which to conduct the test.

Thereafter a whole range of chemical analyses can be performed, including wet chemical techniques such as spot tests, thin-layer chromatography (TLC), or solvent extraction and gravimetric analysis, and instrumental methods such as FTIR, ion chromatography (IC), SEM-EDX, GC–TEA, GC–FID, GC–MS, LC–MS, and LC–MS–MS.

11.4.3.2   Examination of Items for Traces of Explosives. Recovery and Sampling.   It is necessary to recover samples for detection of invisible traces of explosives from a wide variety of types of evidence, and sampling techniques need to be chosen accordingly. Because of the enormous range of possible explosives that may be encountered it is necessary to devise an analytical approach for a given sample which covers the range of most likely candidate explosives whilst providing timely and reliable results. In order to do this the scientist will need to make use of any available information which may provide guidance about the nature of substances likely to have been used.

The following sampling methods are often used:

- **Swabbing, either dry or with solvent.** A cotton wool swab is moistened with the selected solvent and gently rubbed over the test surface; when the swab becomes dirty or dry a further fresh swab is moistened with the solvent and the process continued in this fashion until the entire area has been sampled. The resulting swabs from any one test area are combined for later analysis.
- **Solvent washing.** Small items are placed in a beaker, covered with the chosen solvent, and then the whole agitated in an ultrasonic bath to aid recovery of soluble materials.
- **Contact heater for adsorbed and absorbed volatile explosives.** This is a specially designed device with an electrically heated, ventilated, metal platen which is placed on the surface to be sampled. Air is drawn in across the test surface, out through the centre of the platen and passed through a glass tube containing an adsorbent such as Tenax-GC which traps any vapours evolved from the sampled surface.

- **Vacuum sampling to trap particulates and vapours.** A glass syringe barrel is fitted with a filter support and a microporous filter and then a length of disposable tubing connected between the needle fitting and a vacuum pump to apply suction, via a glass tube containing an adsorbent. The open end of the syringe barrel is rubbed over the sample surface, so that any mobile particles are trapped on the filter for subsequent examination. Any vapours, *e.g.* TATP, that may pass through the filter, are trapped by the adsorbent tube.

Solvents for swabbing or washing need to be chosen to suit the type of explosive being sought and the analytical method which is to be used. Table 11.3 lists some commonly used solvents and Table 11.4 gives some examples of the use of these different methods in the collection of samples for trace explosives analysis from a motor vehicle.

Table 11.3 Solvents for recovering traces of explosive.

Solvent	Type of explosive	Comments
Water	Inorganic, *e.g.* nitrates, chlorates, perchlorates	Slow to evaporate
Methanol	Organic, *e.g.* nitrate esters, nitroaromatics, nitramines	Toxic, poor volatility
Ethanol	Organic, *e.g.* nitrate esters, nitroaromatics, nitramines	Poor volatility
Acetone	Good solvent for organics, especially nitrocellulose	Good volatility, but leaves aqueous residue unless dried thoroughly
Iso-propanol	Organic	Moderate volatility
Ethyl acetate	Good solvent for a wide range of explosives	Available in high purity
Methyl t-butyl ether	Organic	Good volatility
Diethyl ether	Organic	Good volatility, but serious fire hazard
Acetonitrile	Peroxides, *e.g.* TATP, HMTD	Toxic, but good for use with LC–MS techniques

Table 11.4 Sampling of the surfaces and materials in a motor vehicle.

Substrate	Method
Glass and painted metal	Solvent swabbing
Impermeable plastic surfaces	Solvent swabbing or contact heater
Porous surfaces, woven materials	Vacuum sampling, contact heater
Small loose items, *e.g.* keys, coins	Solvent washing

Controls. With any of the methods outlined above, no results can be considered to have forensic significance unless proper environmental control measures are taken before the work of trace analysis starts. These need to be designed to take account of the sensitivity of the analytical method used, and the potential exposure to contamination of the exhibits and sample solutions, throughout the process of sampling and analysis. As more sensitive detection methods are used, more stringent environmental control measures are needed. When nanogram quantities are being sought, operators should, as a minimum, wear fresh disposable overgarments and gloves to preclude transfer of suspect materials either between casework items or between samples and the general laboratory environment or the operator. Control swabs should always be taken from the actual work surface and also the operator's gloves and oversuit, and these control swabs should be prepared and analysed in the same way and at the same time as the actual trace analysis samples from the case under examination.

Sampling kits. A number of forensic laboratories prepare kits of materials ready for use in sampling at scenes of crime. Kits for the collection of explosives traces should contain items such as swabs and solvent which, by means of a rigorous quality control process, have been shown to be free of explosives traces. Three official kits are issued to police in Great Britain for recovery of explosives traces. The general swabbing kit is shown in Figure 11.8; a similar but smaller kit, with fewer components, is used for swabbing suspects' hands, and a portable vacuum sampling kit is supplied for porous and woven surfaces.

Sample pretreatment and concentration. This will depend again on both the likely candidate explosives and the analytical method chosen. For example, particles of sugar and sodium chlorate mixture recovered by vacuum sampling can be dissolved in water and analysed by IC or capillary electrophoresis (CE) to identify anions, cations and sugars. The same basic technique can be applied to aqueous extracts from vacuum sampling tubes, water wetted swabs or washings. The extracts can be concentrated by partial evaporation to increase analytical sensitivity.

Liquid chromatography and gas chromatography are used in conjunction with different selective detectors for determination of organic explosives traces. For organic explosives of moderate to high volatility such as nitroglycerine or RDX, capillary gas chromatography with a selective chemiluminescence (thermal energy analyser) detector or a mass spectrometer is suitable. For involatile explosives, or for less thermally stable explosives such as TATP and HMTD, liquid chromatography, coupled with single or tandem mass spectrometry, provides an

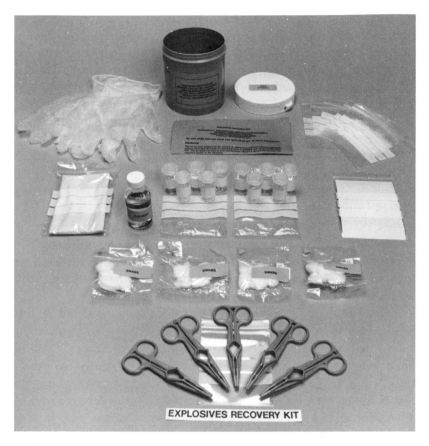

Figure 11.8 Explosives traces recovery kit.

alternative approach. Although these techniques are extremely sensitive, being able to detect picogram amounts of some explosives in pure solutions, they are generally liable to be seriously affected by the large amounts of non-explosive material present in real samples such as oils, fats, soot and soil. A good technique is to elute the initial sample extract through a simple solid-phase chromatographic column which adsorbs the explosives selectively for subsequent elution with a different solvent. The Forensics Explosive Laboratory has taken this approach with a sample pre-treatment method that allows a single sample, be it swab, wash or vacuum, to be separated into different fractions for analysis by all of the above techniques.

Interpretation of explosives trace evidence. A surface such as the seat of a car or a person's hand may acquire traces of explosives either by direct contact with explosive, or by secondary transfer from another item contaminated with explosive. The quantity of explosives traces found will depend an a whole range of factors—the amount of explosive present on the donor, the duration and intimacy of contact between the donor and recipient, the nature of the respective surfaces, the elapsed time between contact and sampling and whether the recipient surface has been cleaned since the contact. Experiments done on transfer when a person directly handled a stick of gelignite showed they had micrograms to milligrams of nitroglycerine on their hands shortly afterwards. Persons who shook hands with them acquired much smaller traces, with transfer efficiencies of a few per cent. When a third person shook hands with the second, higher transfer efficiencies of up to 35% between the second and third person were observed for short time intervals. It was suggested that this enhanced transfer efficiency was a function of the distribution of the explosive traces on the second person's hand. Some very limited experiments with the transfer of penataerythritol tetranitrate (PETN) between individuals suggested that it behaved in a broadly similar fashion to nitroglycerine. Other experiments have suggested decay rates for traces of nitroglycerine and RDX on hands, of three to five orders of magnitude in the first day, for normal handwashing regimes. Much lower decay rates might be expected for traces on inanimate surfaces subject to infrequent cleaning or otherwise undisturbed. Decay rates for solid explosives such as RDX and PETN, having relatively low vapour pressures, would also generally be lower than, for instance, the relatively volatile nitroglycerine.

Consideration must also be given to the likelihood that an explosives trace might have been acquired in the course of legitimate activity, *e.g.* quarry workers or military personnel might come into contact with explosives in the course of their normal duties. Nonetheless, the finding of traces of organic high explosives can be quite significant. Studies of casework samples received in our laboratory show that only a few per cent of samples taken from suspect's hands, clothing or premises exhibit detectable explosives traces, with a somewhat higher proportion of suspect's vehicles yielding positive results. In addition, surveys of traces of organic high explosives in the general public environment in Great Britain have shown that such traces are rare. However, higher levels of inorganic species of possible explosives significance, such as ammonium, nitrate or chlorate ions are likely to be widely encountered in the general environment, so this needs to be taken into account when interpreting the significance of inorganic trace findings. Hence, 50 ng of RDX, for example, is likely to be significant evidence, but a similar finding of nitrate ions would be entirely insignificant. The quantity of nitrate ions would certainly need to be one, and maybe two, orders of magnitude higher before any kind of evidential significance might be attached, depending on the circumstances of the case.

11.5 PHOTOGRAPHY

Because of the hazardous nature of explosives and bombs, it is not possible to take most explosives exhibits to court. As a consequence it is essential to produce a high-quality photographic record of exhibits, particularly when devices are being taken apart, in order to show all the relevant details to the court. Previously, large negatives were used to give greater detail in the final print, which were more suitable than 35 mm film for production of courtroom-quality photographs. Nowadays, this has been superseded by digital photography, but this similarly requires specialist high-resolution equipment. It also requires dedicated recording equipment, to demonstrate to an evidential standard that the original image has not been altered or tampered with in any way.

11.6 LINKS WITH OTHER FORENSIC DISCIPLINES

Explosives examinations generally take precedence over other forensic tests for two fundamental reasons: (1) safety, and (2) because unless there are items of explosives significance then it is unlikely that a case can be established under explosives-related legislation.

As in all complex forensic cases, discussion of requirements between scientists from different disciplines and the investigating officer is essential if the best evidence is to be protected and realised. Practical compromises between the requirements of explosives trace analysis, DNA, fingerprints, fibres and tool marks need to be agreed and the work planned accordingly since each aspect can make a vital contribution in many cases.

11.7 CASE STUDY

As in all forensic science, correct interpretation of the results is crucial. A delicate balance has to be struck, ensuring that all the relevant facts are made known, that the appropriate weight is given to each aspect of the evidence, and that the work is recorded in a manner which allows meaningful review by others skilled in the art so that, for example, other possible interpretations of the results can be explored by different experts assisting the various parties in a case. That said, how are the various scientific and technical threads pulled together? Consider the following hypothetical case.

11.7.1 The Scenario

A violent sharp report is heard just after 3 a.m., apparently from the grounds of a large country house, whose alarmed inhabitants summon the local police. Upon arrival the police make a search of the grounds and find that the base of one of the large (and valuable) statues in the ornamental gardens has been damaged; they decide to secure the scene and wait until daylight before making a detailed search.

Meanwhile other officers have been called to an accident just a few miles away involving a motor car and a small lorry. The behaviour of the lorry driver and his passenger causes them to make further inquiries and their suspicions are increased when the passenger is reported to have been convicted of burglary some years before.

Further inquiries are made, and a link to the attempted theft of the statue suspected. By now daylight has broken and examination of the scene around the statue has yielded a bootprint from a patch of soft ground, and revealed a pattern of damage on the stonework of the statue's base. Further searching results in the discovery of a pair of long thin wires, various fragments of broken stonework, and an assortment of general litter and debris of uncertain relevance; in addition a tyre print is found in mud alongside the driveway. Scene photographs are taken, the recovered items carefully bagged and identified, and the damaged area of the statue's base is swabbed using an explosives trace recovery kit; the whole is then submitted to the laboratory for examination.

Chemical analysis of the swabs rapidly indicates the presence of PETN residues, and this, coupled with examination of the damage to the statue, suggests an unsuccessful attempt to cut the statue from its base using detonating cord. After discussion with the scientist the police decide to have the suspects' lorry, their clothing and their hands tested for explosives traces. It is agreed that in this instance the scientist will give priority to the trace explosives analysis and leave examination of the various items from the scene of the explosion until later; this simplifies matters since it neatly allows clearly separate examination of the 'clean' and 'dirty' exhibits and avoids any possible suggestion of contamination between them.

Most of the results from the trace analysis are negative. However, positive results for PETN are found for a pair of rubber gloves found in the cab of the lorry, from a penknife found in the pocket of the passenger's jacket, from the area of the pocket that held the penknife and from the passenger's trousers. The swab samples taken from both men's hands are negative.

A search of both men's homes reveals nothing relevant; however, a lock-up garage rented by the passenger yields a bag containing what initially seems to be a length of plastic clothes line; however, it has the words 'Danger—Explosive' printed along its length. This sample is therefore examined by a bomb disposal operator who packages it safely and advises forensic examination to confirm if it is in fact detonating cord. The rubbish bin from outside the garage is also searched; some discarded packaging appears significant and is also submitted to the laboratory.

These new items are identified at the laboratory as a particular brand of commercial detonating cord containing PETN, and the remnants of a box which once contained a particular type of commercial electric detonator widely used in quarrying.

Meanwhile the inhabitants of the house where the crime occurred have found another pair of rubber gloves discarded in their garden at some distance from the statue; unfortunately there have been several nights of heavy rain before the gloves were found. Examination of the gloves for explosives traces does indeed prove negative.

Searching of the debris from the explosion scene reveals a small burnt and deformed rubber plug, which the scientist is nonetheless able to identify as similar to the plug from detonators made by a particular company; there are also the wires, which prove to be a long length of a standard twin-core cable attached to two short lengths of plastic insulated single-strand wire whose colour and diameter match that from a particular range of commercial detonators. In fact the packaging from the lock-up garage also matches and the three items: wires, plug and packaging all belong to the same product range from the same company.

Further examination of the lorry reveals a partially discharged battery lying in a corner of the loading area and a torn packet matching the brand of rubber gloves found both in the lorry's cab and at the scene. One of the tyres on the lorry is found to match the tyre print in the garden.

11.7.2 The Prosecution Case

It was suggested that the facts were consistent with a bungled attempt to steal a valuable statue; detonating cord had been wrapped around the base in an effort to cut the statue loose so that it could be removed. In fact all that had resulted was noise and damage. The frustrated thieves had then fled, leaving behind the rubber gloves worn by one of them; the second pair of gloves was found in the cab. No fingerprints were found at the scene, and the accuseds' hands were free of explosive traces because they had worn gloves. However, the penknife that was found in the passenger's jacket pocket had been used to cut the detonating cord and hence had contaminated the pocket with the traces of PETN found. Likewise, the gloves found in the cab had been worn when handling the detonating cord, hence the traces of PETN on them. The absence of traces on the gloves found later in the garden could be due to their having been washed away by the heavy rain.

The detonating cord found in the garage contained PETN; the packaging found in the rubbish bin at the garage matched the detonator plug and wires found at the scene.

11.7.3   The Passenger's Defence

The lock-up garage had been rented by a group of five men, including himself, each of whom had a key. The bag found in the garage belonged to one of the others.

While working in the garage on the previous day he had needed some rope to secure a bundle of wood; when he looked in the bag he saw what seemed to be a coil of clothes line and so he cut a length off with his penknife. However, a white powder then fell out of the line; he didn't know what it was, but realised it wasn't suitable for his purpose so he put it back in the bag, wiped the knife blade on his trousers and put the knife back in his pocket. He did all this while kneeling on the garage floor.

The reason he was in the lorry was that he had been out poaching and had been walking home when his friend drove by in the lorry and gave him a lift.

11.7.4   The Lorry Driver's Defence

He had been employed recently to remove heavy rubbish from the gardens of the house and had legitimately been near the statue in the course of his work, hence the bootprint. Likewise he had parked his lorry in the driveway, and this must be the explanation for the tyreprint.

11.7.5   What Really Happened?

The passenger was telling the truth.

The lorry driver was in league with the owner of the bag, who was one of the group renting the garage. The bag-owner had left the crime scene on foot when he realised that the noise of the explosion had alerted the inhabitants of the house. In his haste he had discarded the gloves found later in the garden. The lorry driver had had no choice but to leave via the driveway unless he wished to abandon his vehicle at the scene.

This case study illustrates the fact that there are often a number of explanations for a given set of circumstances. The forensic scientist's task is to conduct impartial examinations and formulate unbiased conclusions within the remit of their specific discipline, in order to help the court reach an informed judgement. In most cases the scientific findings will form only a part of the jigsaw, the other pieces of which may radically affect the interpretation of the science.

ACKNOWLEDGMENTS

We wish to thank the Defence Science and Technology Laboratory for permission to publish the photographs.

BIBLIOGRAPHY

J. Akhavan, *The Chemistry of Explosives*, Royal Society of Chemistry, Cambridge, 1998.

Anon, *The Manufacture and Storage of Explosives Regulations 2005*, TSO, Norwich, 2005.

A. Bailey and S. G. Murray, *Explosives, Propellants and Pyrotechnics*, Brassey's (UK), London, 1989.

A. Beveridge (Ed.), *Forensic Investigation of Explosives*, Taylor & Francis, London, 1998.

T. L. Davis, *The Chemistry of Powder and Explosives*, Angriff Press, Hollywood, CA, 1943.

S. Fordham, *High Explosives and Propellants*, 2nd edition, Pergamon Press, Oxford, 1980.

Guide to Explosives Acts 1875 and 1923, HMSO, London, 1992.

J. Kohler and R. Meyer, *Explosives*, 4th edition, VCH, Weinheim, 1993.

T. Urbanski, *Chemistry and Technology of Explosives (3 volumes)*, Pergamon Press, Oxford, 1964.

H. J. Yallop, *Explosion Investigation*, Forensic Science Society, Harrogate, 1980.

CHAPTER 12

Firearms

MARK MASTAGLIO AND ANGELA SHAW

Forensic Science Service, 109 Lambeth Road, London SE1 7LP

12.1 INTRODUCTION

The practitioner in the discipline of the forensic firearms examination, sometimes called forensic ballistics, can be required to attend the scene of a shooting and the post-mortem examination, carry out subsequent examination of items at the laboratory and, ultimately, to present expert testimony at court.

The earliest references to what we would recognise as a firearm date from the 14th century. In simple terms a firearm incorporates a mechanism, mounted in an ergonomically designed frame, which discharges a projectile through a tube. This is where the phrase 'lock, stock and barrel' comes from. Dictionary definitions of the noun 'firearm' usually refer 'to a portable weapon that discharges a projectile by means of an explosion; a rifle, gun, pistol, *etc.*' A more useful definition of firearm for the forensic firearms examiner is provided by the legislation. In the UK a firearm is defined by Section 57(1) of the Firearms Act 1968 as being:

A lethal barrelled weapon of any description from which any shot, bullet or other missile can be discharged and includes—

(a) Any prohibited weapon, whether it is such a lethal weapon as aforesaid or not.

Crime Scene to Court: The Essentials of Forensic Science, 3rd Edition
Edited by P. C. White
© Royal Society of Chemistry 2010
Published by the Royal Society of Chemistry, www.rsc.org

(b) Any component part of such a lethal or prohibited weapon.

(c) Any accessory to any such weapon designed or adapted to diminish the noise or flash caused by firing the weapon; . . .

This definition obviously applies to military-style weapons, *i.e.* those designed to kill; target rifles and pistols, which clearly can be used as weapons; air guns (including those powered by compressed carbon dioxide); and shotguns, together with less obvious items such as lachrymator canisters and electronic stun guns. Component parts of firearms such as barrels and the firing mechanism themselves also constitute firearms for the purposes of the legislation, as do so-called silencers, more correctly called sound moderators, and flash suppressors.

Ammunition in the 1968 Act means 'ammunition for any firearm and includes grenades, bombs and other like missiles, whether capable of use with a firearm or not, and also includes prohibited ammunition.' This definition clearly applies to conventional bulleted cartridges and shotgun cartridges but can also be applied to airgun pellets, blank cartridges and cartridges that contain a lachrymatory agent such as CS.

The forensic firearms examiner has to be competent in a wide range of examinations. The most commonly asked questions include:

1. Is this item a firearm?
2. What part of the firearms legislation is applicable?
3. Did this gun discharge this cartridge case or bullet?
4. How many guns and what type were used in the incident?
5. Could the gun have gone off accidentally?
6. Is this a bullet hole?
7. Where was the shooter standing in relation to the damage or victim?

The application of what can be referred to as firearms chemistry, a branch of analytical chemistry, is also an important tool in the forensic investigation of shooting incidents. This includes the study of gunshot residue (GSR), its origin, formation, identification and persistence. The aim of this chapter is to provide the reader with a basic grounding and overview of the subject and stimulate further interest.

12.2 HISTORICAL DEVELOPMENT

The beginnings of forensic firearms examination can be dated back to the 19th century; however, it was during the first 30 or so years of the

20th century that a few notable pioneers took the discipline further, entrenching it as one of the essential tools of forensic science.

In 1900 Dr Albert L. Hall of Buffalo, USA, published the article 'The missile and the weapon', which detailed his experience as a clinician dealing with gunshot injuries and the identification of bullets. Work continued in Europe and the USA, and the subject progressed through the work of Professor Balthazar in France who published papers in 1912 and 1922 and Major Calvin Goddard who worked in the Bureau of Forensic Ballistics in New York. During the 1920s and 1930s a plethora of important papers were published, further establishing the discipline. In the UK, pioneers included the gunsmith Robert Churchill, and Major Gerald Burrard who went on to write *The Identification of Firearms and Forensic Ballistics* in 1934. During this period expert testimony was given in a number of high-profile cases including the St Valentine's Day massacre in the USA and in the trial that followed the murder of an Essex policeman, William Gutteridge, who was shot dead on 27 September 1927. In the latter case the application of the relatively new discipline of forensic firearms examination led directly to the conviction of the two accused.

12.3   FIREARMS AND AMMUNITION TERMINOLOGY

As in all specialist subjects, technical terms and jargon abound. Firearms terms used by the layman can be confusing, and sometimes just plain wrong. Arcane units are also encountered. At this point it is best to define some of the more frequently used terms and units and explain their significance.

12.3.1   Firearms Terminology

12.3.1.1   Action.   This refers to the firearm's working mechanism.

12.3.1.2   Barrel.   The barrel of a firearm is a tube which directs the projectile, or projectiles, under the influence of the propelling force, which may arise from powder gases or compressed air. In the UK legislation the length of the barrel is taken from the point of ignition to the barrel's end or muzzle.

12.3.1.3   Breech.   This is the part of the firearm at the end of the barrel where the cartridge or propellant is loaded. The breech face is that part of the breech bolt or breech block that abuts against the head of the loaded cartridge.

12.3.1.4 Rifling. This consists of a series of helical lands and grooves machined in the surface of the inside of the barrel. James Bond film fans will recall that during the opening credits, the silhouette of 007 is pictured with the viewer looking down the barrel of a gun; the helical lands and grooves are visible in the animation. The lands are the raised portions and the grooves are the valleys that run between them. Rifling imparts rotary motion to the projectile, stabilising the projectile in flight. The ballistic stability, known as gyroscopic stability, imparted to a spinning bullet is the same as that given to an American or rugby football when it is thrown, the ball is spun around its long axis when it is launched, ensuring that the ball has a tendency to fly nose first. The ball does not tumble and so it can travel a prodigious distance. A ball thrown with no spin would not travel as far or be as accurate as one launched with spin and with the same force. When the relatively soft bullet travels down the steel bore of a rifled firearm, the rifling engraves marks on to the bullet which can identify the type and the specific firearm from which the bullet was discharged.

12.3.1.5 Calibre. One of the most misunderstood and potentially confusing terms is calibre. The calibre of a firearm can be defined as the diameter of the barrel's bore. In rifled firearms this is interpreted as the diameter of a circle whose circumference would be tangential to the tops of the lands. The calibre of the firearm's ammunition is a nominal description, which includes a numeric reference to the bore diameter. It can be alphanumeric, such as 9 mm Parabellum, the 9 mm referring to the nominal bore diameter and the word 'Parabellum' being a trade name derived from the Latin phrase '*si vis pacem, para bellum*', which translates 'if you wish for peace, prepare for war'. The calibre designation can also include the dimensions of the cartridge, for instance, the current NATO rifle cartridge is 5.56×45 mm, where the 5.56 mm refers to the bore diameter and the 45 mm to the cartridge case length. The calibre designation .30-06, a rifle cartridge, includes a reference to the nominal 0.3 inch bore diameter and the 06 is refers to 1906, when the calibre was introduced into US military service.

As can be seen calibre can be expressed in millimetres or inches: 9 mm Short is also known as .380 Auto. Each calibre can have synonyms, *e.g.* 9 mm Parabellum is also known as 9×19 mm or 9 mm Luger. The last synonym illustrates that a calibre designation can also include the name of a firearms designer or manufacturer, or the type of gun the calibre was designed for, such as a 40 Smith & Wesson. To further complicate the issue some calibres can be discharged in firearms of a different calibre, *e.g.* the 9 mm Short or .380 Auto cartridge can be discharged in a 9 mm Parabellum pistol.

Shotgun calibre designation is more arcane and dates from the days of muzzle loading, when the calibre was described by the number of lead balls of the same diameter as the gun's bore which would weigh 1 lb. This means, in the case of 12-bore, the most common shotgun calibre, 12 spheres of the same diameter as the shotgun's bore would weigh 1 lb. All this may seem confusing and a little daunting to the student firearms examiner, especially when one realises that there are thousands of different calibres. Fortunately, a relatively small number of calibres are encountered on a day-to-day basis. In casework in the UK the most commonly encountered calibres are .22 Long Rifle, .25 Auto (6.35 mm Auto), .32 Auto (7.65 mm Browning), .38 Special, .357 Magnum, 9 mm Parabellum, 9 mm Short, .45 Automatic Colt Pistol, .455 Revolver, 5.56×45, 7.62×51, 7.62×39, .410 and 12-bore shotgun cartridges. Modified blank calibre cartridges, principally 8 mm Knall, 9 mm PAK and 9 mm Rimmed are also encountered.

12.3.2 Cartridge Terminology

A cartridge is a whole round of ammunition. In modern terms it consists of a cartridge case, loaded with a primer, propellant and a projectile, or projectiles. Figure 12.1 illustrates the different kinds of cartridge that are commonly encountered.

A cartridge is sometimes wrongly called a bullet; the term bullet refers specifically to the projectile. Shotgun cartridges usually contain multiple projectiles together with cartridge wadding. When a cartridge is discharged, marks can be transferred from the firearm on to the relatively soft cartridge case and primer. These marks, as identified in Figure 12.2, can enable the make and specific firearm to be identified and can originate from any part of the firearm that has come into contact with the cartridge such as the breech face, firing pin, extractor, ejector and chamber.

12.3.2.1 The Cartridge Case. Bulleted rounds normally use cartridge cases made from brass, although materials such as aluminium and steel are also used. Shotgun cartridge cases are usually made from plastic or paper, with a brass or steel base.

12.3.2.2 Propellant. Propellant is a mixture of chemicals which when ignited burns in a predictable manner producing large quantities of hot expanding gases which propel a projectile. The fundamental

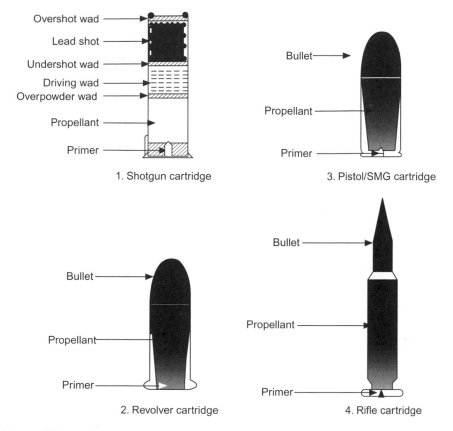

Figure 12.1 Schematic representations of commonly encountered cartridge types.

principle is to transform a portion of the propellant's stored chemical energy into the kinetic energy of the projectile. The first type of propellant, gunpowder, also known as black powder, still used today for vintage design firearms, consists of an intimate mechanical mixture of potassium nitrate, sulphur and charcoal. Modern so-called 'smokeless' propellants contain nitrocellulose (single-based propellant), a substance that has the oxidiser and fuel components in the same molecule, and sometimes mixed with nitroglycerine (double-based propellant). The discharge of a propellant in a firearm can result in trace evidence, *i.e.* propellant grains and combustion products, being deposited on substrates close to the discharge.

12.3.2.3 Primer Cup. This is the component of the cartridge that contains a sensitive explosive mixture which detonates when impacted with sufficient force by the firing pin or striker and then ignites the

Figure 12.2   The base of a spent cartridge case showing the position and shape of the firing pin impression in the centre of the primer, the extractor mark at 3 o'clock and ejector mark at between 6 and 7 o'clock.

propellant. For centre-fire cartridges the primer sits in a recess in the centre of the cartridge base. This explosive mixture is housed in a brass, nickel or gilding metal cup.

There are two main types of centre fire primer design, Berdan and Boxer. The former consists of the explosive mixture covered with a foil or paper disk, the primer being detonated by the impact of the firing pin forcing the explosive mixture against an anvil integral to the cartridge case. In the case of the Boxer primer the anvil is integral to the primer. The latter design is more commonly encountered in casework. In rimfire cartridges the primer composition is present inside the cartridge rim. The compositions and analysis of both propellant and primer trace evidence are discussed in section 12.11.

12.3.2.4   Bullets.   A bullet is the projectile loaded into the cartridge case. Bullets for revolvers are normally made of lead, but those for self-loading pistols and rifles are encased in a hard metal covering, known as a bullet jacket. Bullet jackets are normally made from cupronickel, brass or steel. The jacket is present to prevent the lead being stripped off in the bore and to prevent misfeeds in self-loading guns. Lead used for bullets and shotgun pellets is normally hardened by the addition of small quantities of a hardening agent such as anti-mony. Other metals and alloys, such as brass, bronze, aluminium and steel are also used to make bullets.

The most commonly encountered bullet shape is the round-nosed bullet. However, there are a range of shapes, the design being suited to their purpose. For example, some bullets have a flat nose; they are essentially cylindrical in shape and when they strike a card target they punch a circular hole in it. This style of bullet is known as a 'wadcutter' and is designed to be used by target shooters. Some bullets have a hollow nose or a partially exposed core; their function is to expand on entering a quarry, be it animal or human, and increase the wounding potential. Military-purpose rifle bullets can incorporate steel penetrators, be designed to pierce armour, have an incendiary capability or have a pyrotechnic component so that the discharged bullet's trajectory is visible.

12.3.2.5 Shotgun cartridges. These too are loaded with propellant and a primer, but usually contain multiple shot pellets, together with card, fibre or plastic wadding. The number of pellets that can be loaded into a shotgun cartridge depends on the size of the pellets and the calibre of the cartridge. For example a 12-bore cartridge loaded with a 32 g load of number 6 size shot will contain approximately 320 pellets. The largest shot encountered in a British loaded cartridge has diameter of approximately 9.1 mm. Some shotgun cartridges are loaded with a single projectile, known as a slug, but these are rarely used in crime in the UK. In recent years concern over environmental pollution has increased and there has been a move away from the traditional lead-based shot. Use of other types of shot, such as tungsten or bismuth–tin alloy shot, has become more common.

12.3.3 Ballistics

Ballistics is the science of projectiles and the word is derived from *'ballista'*, a Roman military weapon in the form of a large mounted crossbow. There are two main areas of study, internal and external ballistics.

12.3.3.1 Internal Ballistics. This is the study of the motion of the projectile while it is still in the firearm, from the point of initial discharge to when the projectile leaves the barrel. The time taken from pulling the trigger of a firearm to when the projectile leaves the barrel can be as short as a few milliseconds, but what happens during this short period profoundly influences the design of both the firearm and the ammunition. What happens to the cartridge case and projectile during this short period of time is of vital importance to the forensic firearms examiner.

12.3.3.2   Exterior Ballistics.   This is the study of the motion of the projectile from when it leaves the muzzle to when it comes to rest. The forensic firearms examiner is sometimes asked to determine various parameters concerned with the exterior ballistics of a projectile. Depending on the type of firearm and the range over which it is fired, the time can be a fraction of a second to many seconds. A rifle bullet discharged vertically upwards could take a minute or so to return to earth. What happens to the projectile when it is close to and emerging from the muzzle influences the projectile's stability, is also of interest; this is sometimes referred to as intermediate ballistics. The terminal ballistics, *i.e.* how the projectile interacts with a target, is also of interest when it comes to interpreting damage or wounds.

12.3.4   Units

In the UK the grain is the unit of mass which is most often used to measure propellant and bullet weights. There are 7000 grains to the pound and 1 g is equivalent to 15.432 grains. The foot pound is a unit of energy frequently encountered in the UK and USA when referring to the kinetic energy of a projectile. In Europe the equivalent SI unit, the joule, is used: 1 ft lb is equivalent to 1.356 J.

12.4   FIREARMS

Firearms can be categorised in a number of ways according to their form and function, as outlined below.

12.4.1   Long Arms

These are intended to be fired from the shoulder. They have a butt stock which can be drawn into the shoulder to accommodate the recoil experienced when a cartridge is discharged. Rifles are so called because of the presence of rifling in the bore. Some types of short barrelled rifle may be referred to as a carbine, a word derived from 'carabineer', a reference to French cavalry troops who carried a short-barrelled musket.

The design of a rifle depends on whether it is intended to be used in a sporting context, such as target shooting or hunting, or whether it has a military purpose. They have two things in common; they are designed to be accurate over several hundred metres and to be fired from the shoulder.

Most rifles incorporate a magazine capable of holding multiple cartridges, although single-shot rifles are also infrequently encountered in casework. The firing mechanism can be bolt-action, like the old British military .303 Short Magazine Lee Enfield rifle; self-loading; lever action; pump action; or have a full automatic capability such as modern military so-called assault rifles which can fire in the self-loading or automatic modes.

Another important type of long arm which is frequently encountered in firearms-related crime is the shotgun. These have smooth-bore barrels and are designed to discharge pellets also known as 'shot'. Typically they are designed for game or clay pigeon shooting. When the shot and cartridge wadding emerge from the barrel the shot disperses as a function of distance. Most shotguns have a constriction of a few thousandths of an inch at the muzzle, known as choke; this serves the function of controlling the pellet dispersion. It is unusual to encounter full-length shotguns in crime, other than in domestic incidents; the shortened, so-called sawn-off, shotgun is more common.

The usual barrel length of a full-length shotgun is some 26–30 inches (66–76 cm), whereas a typical barrel length for a sawn-off shotgun is 12 inches (30 cm). In extreme cases even shorter barrels are seen. In addition to shortening the barrel the butt stock is usually also cut off, leaving a pistol-like grip. The resultant shortened weapon is easier to conceal, but a significant amount of the gun's mass has been lost, so when a cartridge is discharged the recoil is all the more fierce. Shotguns can be single-barrelled, double-barrelled with the barrels side by side or over and under, pump action or self-loading.

12.4.2 Handguns

These are usually much smaller than long arms, in general use less powerful cartridges, are designed to be used over shorter ranges and can be fired with one hand. In common with long arms, early examples were smooth bore and are infrequently encountered in casework. In general handguns can be split into a further two categories, revolvers and pistols.

12.4.2.1 Revolvers. Revolvers were developed in the early part of the 19th century and became increasingly common from the 1850s onwards. These are instantly recognisable to the layman from their depiction in cowboy movies. They all have one thing in common: they use a rotating cylinder, a kind of magazine, which holds the cartridges. On pulling the trigger the cylinder rotates to bring the next chamber in line with the barrel. The most common cylinder capacity is six cartridges, although some can hold twelve or so cartridges.

12.4.2.2   Pistols.   The type of handgun most frequently encountered in case work is the self-loading pistol, also known as the semi-automatic pistol. Like the revolver, this is a repeating firearm. It is not uncommon for this type of gun to be referred to as an automatic; this term is incorrect and its use in this context should be avoided. Single-shot pistols are occasionally encountered in casework. Self-loading pistols use a detachable magazine which is usually inserted in a recess in the pistol grip. Self-loading pistols are so called because the firing action automatically removes the spent cartridge case from the pistol's chamber and ejects it from the gun; a new unfired cartridge is then stripped from the magazine and loaded into the chamber. A subsequent pull on the trigger will discharge the newly loaded cartridge and the cycle begins afresh. It is worth noting that since the introduction of the UK ban on small firearms in the wake of the Dunblane tragedy of 1996 many novel designs have emerged increasing the size of traditional handguns so that they do not come under the ban.

In the last decade the numbers of handguns used in the commission of crime in the UK has increased. A significant proportion of these handguns have been converted from guns which were designed and sold as blank cartridge firing guns. The level of conversion sophistication varies hugely: some have a piece of plumber's copper pipe for a barrel, others a highly engineered rifled barrel. Most of the converted and reactivated guns encountered in the UK are capable of causing lethal injury at close range.

12.4.3   Automatic Weapons

This category of firearm is intended for military and, in certain countries, police use. They include machine guns, submachine guns, assault rifles and some handguns. They all have one thing in common: they continue to fire as long as the trigger is pulled and there is ammunition in the magazine.

Machine guns are large and tend to use rifle ammunition. They are very rarely encountered in crime in the UK, and the same is true of military assault rifles. The majority of such weapons tend to find their way into criminal hands from war zones.

12.4.4   Air Guns

Most air guns can cause lethal injury and are classified as firearms. They can be handguns or rifles; they all employ a mechanism whereby compressed air propels a small pellet or dart from the gun. The commonest type uses a spring and piston to compress a column of air which is pushed through a transfer port. Other types include pre-charged mechanisms, gas piston mechanisms, and pump-up so-called pneumatic mechanisms. Although millions of such guns are in circulation they are rarely used in fatal shootings. When they do occur, such incidents frequently involve juveniles and claims of accidental discharge. A more common crime where the firearms examiner may be asked to help is in the investigation of criminal damage cases involving air guns.

In the last decade there has been a large increase in the availability of so-called soft air guns which have been a designed to look like conventional firearms but are usually made from plastic and discharge small plastic pellets at low velocities. Conventional firearms manufacturers, such as Beretta and Walther, have also allowed other manufacturers to make pellet-firing copies of their conventional self-loading pistols. Air guns using self-contained compressed air cartridges, which look like conventional rifles, revolvers and self-loading pistols, have also been developed and successfully marketed. Soft air guns, carbon dioxide guns and converted self-contained compressed air are frequently used in crime, as are cartridge guns and the firearms examiner needs to know how they work and how to classify them.

12.4.5 Miscellaneous Firearms

The firearms examiner will also need to recognise and examine a range of weapons that do not readily fall into the above categories, *e.g.* electric shock devices such as stun guns, lachrymator devices, and disguised firearms like pen guns, walking stick guns, key fob guns and guns disguised as mobile phones. All of these are classified as firearms and prohibited under UK firearms legislation. The firearms examiner may also be asked to examine and classify old designs of firearm which may be viewed as being antiques. It is therefore essential that the examiner has a thorough grounding in the history and development of firearms and the legislation that governs their use and possession.

12.5 FIRING MECHANISMS

There is a relatively small number of firing mechanisms or action designs. The most frequently encountered actions are described below:

12.5.1 Revolver

This is usually a handgun which uses a cylinder with multiple chambers which rotates around an axis, bringing the chambers in line with the bore when the firing mechanism is activated, allowing successive cartridges to be discharged. Most revolvers can be operated in one of two modes, either single or double action. Single action involves cocking the mechanism by manually pulling back the exposed hammer until it engages on an internal sear, pulling the trigger allows the hammer to fall and the cartridge is discharged. The double action mode allows the cartridge to be discharged by a single pull on the trigger, the hammer is retracted and the cylinder rotated by action of pulling the trigger. Most double action revolvers can also operate in the single action mode. In both cases the cylinder revolves and brings the next cartridge in the adjacent chamber in line with the barrel. The spent cartridge cases remain in the gun and have to be manually removed from the cylinder's chambers.

12.5.2   Self-Loading

This is used in repeating weapons, allowing cartridges to be successively discharged by using some of the energy of the discharge to activate the removal of fired cartridge cases and the loading of unfired cartridges. A separate pull on the trigger is required for each discharge.

12.5.3   Automatic

This allows successive cartridges to be discharged with a single pull of the trigger; the gun will fire until the cartridge supply is exhausted. Spent cartridge cases are extracted and ejected automatically from the gun.

12.5.4   Bolt Action

The action uses a breech bolt. Unlocking the breech and pulling the bolt to the rear and then forwards will cock the weapon and remove any spent cartridge from the chamber and load the subsequent cartridge ready for firing. A bolt action is used in rifles and also some relatively low-quality shotguns.

12.5.5   Pump Action

This utilises a manually activated slide mechanism situated underneath and parallel to the barrel. When pulled to the rear the slide opens the breech, extracting any cartridge or cartridge case; pushing the sliding mechanism forwards loads a cartridge into the chamber and closes the breech. The gun, typically a rifle or shotgun, is then cocked and ready to fire.

12.5.6 Lever Action

This design uses a manually activated lever, usually situated on the underside of the gun, which when pulled down opens the breech, extracting and ejecting any cartridge case or cartridge. Returning the lever loads a cartridge and closes the breech with the gun cocked and ready to fire.

12.5.7 Side Lock

This action describes the position of the firing mechanism, which is situated in removable side plates as opposed to being integral to the gun. This action is most commonly seen in good-quality side-by-side shotguns.

12.5.8 Box Lock

This action is generally found in double-barrelled shotguns; the firing mechanism is situated inside the gun's frame.

12.5.9 Hinged Barrel

The barrel or barrels are pivoted to the gun's frame. Opening of the action is normally facilitated by a catch. The manual act of opening and closing the barrels cocks the firing mechanism.

12.5.10 Repeating Actions

12.5.10.1 Blow Back. This is the simplest of these designs. It relies on the recoil of the gun, the force of which is transmitted to the breech face from the base of the cartridge case. The breech block will move to the rear once its inertia has been overcome, and this results in the extraction and ejection from the gun of the spent cartridge. At the rearmost point of travel, the gun's recoil spring will force the breech block forward, a cartridge will be stripped from the magazine and loaded into the breech. The gun is now cocked and ready to be fired; the cycle will then begin again.

12.5.10.2 Locked Breech System. For more powerful cartridges used in self-loading pistols, simple blow back designs do not keep the breech closed long enough for the pressure to reduce to an acceptable level. In these cases the breech must be locked to the barrel until the pressure diminishes. Various solutions have been designed to achieve this with self-loading pistols, such as short recoil, rotating barrel and roller locking systems.

12.5.10.3   Gas Operated.   Many high-velocity rifles such as military assault rifles, together with some large-calibre pistols and self-loading shotguns, use a gas-operated system to cycle the gun's action and unlock the breech. In this case some of the gas resulting from the discharge of the propellant charge is tapped off to cycle the action.

12.6   MANUFACTURING PROCESSES

One of the most frequent tasks the forensic firearms examiner is asked to perform is to establish whether or not a given gun can be associated with recovered bullets or cartridge cases. In order to do this the examiner must know what kind of marks a firearm can transfer to the ammunition. Additionally, an assessment of the significance of these marks is required. Firearms are tools made up of a number of machined and finished components. Each one of these components can act as a tool transferring tool marks to the cartridge, the cartridge case, bullet, air gun pellet or plastic shotgun wad. Knowledge of how these components are manufactured is needed to enable the examiner to assess whether the transferred marks are unique to the firearm, so-called 'individual characteristics' or whether they arise from design features which could be common to a group of firearms, so-called 'class marks'.

Individual characteristics arise from the irregularities or imperfections of the tool surface, and the profile of the tool may also change during the manufacturing process due to wear or damage. The latter point also applies to the firearm's internal components, which as has already been noted are themselves tools, and hence also subject to wear and damage. These individual characteristics can result in characteristic marks being transferred to the ammunition component enabling the firearm to be identified.

12.6.1   Rifling Process

In general there are two ways in which the barrel's bore can be rifled: by either removing metal or displacing it. The principal metal-removing technique is by cutting, either by using a hook, broach (both single and gang) or scrape method. All involve a tool with one or more cutting surfaces.

After the rifling process the bore is crowned, the muzzle is chamfered to protect the rifling, and the bore is lapped or burnished. Some firearms, such as military rifles, also have a thin chromium layer added by an electrical process to protect the bore from erosion. These processes can also result in individual characteristics being transferred to the firearm.

12.6.2 Firearms Components

Firearms components such as the breech block/bolt, firing pin, extractor, ejector and magazine are made by a variety of engineering processes. These processes, such as milling, casting, turning, drilling, reaming, grinding, stamping together with the finishing processes, will also lead to the components exhibiting individual characteristics.

12.7 ATTENDANCE AT CRIME SCENE AND POST-MORTEM EXAMINATION

The firearms examiner is frequently required to attend the scene of a shooting incident and, in the case of a fatal shooting, the post-mortem examination. This is a crucial part of the investigation where vital information can be determined and passed on to the investigating officer.

12.7.1 Crime Scene

It is only highly experienced and demonstrably competent personnel who are qualified to attend scenes in the lead firearms examiner role. As in all scene attendance, precautions must be taken to minimise the risk of contaminating or compromising any trace evidence. This usually involves taking full DNA and fingerprint precautions by wearing a bodysuit, double gloves and face mask. Special consideration must be given to the risk of GSR contamination.

Safety is paramount when dealing with firearms, and the forensic firearms examiner may be asked to 'make safe' a gun. In any event it is important that the location and status of the gun is recorded before it is handled. Specific notes should be taken of whether or not the gun is cocked, the position of any safety catch or mode of fire selector, and whether or not the gun is loaded.

The most obvious questions asked of the firearms examiner at the scene involve the assessment of possible firearms-related damage; what type of projectile might have caused it; the possible location of cartridge cases, bullets, *etc.* and their identification; and the possible position of the shooter. In cases where the evidence allows a full scene reconstruction, the bullet's flight path can be determined. The scene, including all damage sites, firearms and spent ammunition components, should be photographed with a scale placed next to each item when photographed. The use of digital imaging and when required, computer-aided three-dimensional scanning techniques can greatly assist subsequent interpretation.

12.7.1.1 Possible Bullet Damage.   The question of whether a bullet caused some given damage is easy to answer if a bullet can be recovered from the damage site. If no bullet is present, then a close examination of the damage and the use of two simple chemical spot tests can help. The position and size of the damage site should be recorded. The appearance of the bullet damage will depend on the substrate material, and the shape, composition, velocity and orientation of the bullet relative to the substrate. Since Locard's principle implies that there will be a reciprocal exchange of material when two objects come into contact, it is highly likely that the bullet will leave material, such as lead, copper, or GSR at the impact site and the substrate will transfer material to the bullet. Spot tests for lead and copper and bullet jacket material can be used to assist in the determination of whether any bullet material is present at an impact site. Microscopic lead particle deposition arising from GSR and vaporised bullet lead can also be detected. An assessment of the possibility of ricochet damage can also be made by careful examination and application of the spot tests.

12.7.1.2 Projectile Flight Path.   The shape of the bullet damage can indicate the angle of impact. For example, a circular bullet hole and wipe may indicate an orthogonal or near orthogonal impact, whereas an elliptical hole with or without an associated bullet wipe can indicate an angled impact. The shape of the soot or powder deposition can also assist in angle of impact determination. The extent of any such deposition should be measured as this can assist in determining the range of fire.

If the substrate is reasonably thick, such as a wooden fence post, it may be possible to determine the bullet's path by the angle of the bullet track. Also, the appearance of the damage on both sides of the damaged object can assist in determining the projectile's entry and exit sites. If the material is glass, a larger crater is often seen on the opposite side to impact side. Over short ranges bullets tend to travel in straight lines. Consequently, if

Figure 12.3 Bullet damage to a car. The bullets' path is visualised with the use of trajectory rods.

there is more than one damage site attributable to a single bullet, a straight line joining up these damage sites will give the bullet's flight path. The use of string or thin cylindrical rods and/or small lasers can be very helpful when extrapolating or interpolating a bullet's flight path and Figure 12.3 shows the use of such rods. This type of reconstruction often leads to an accurate assessment of the shooter's position.

12.7.1.3 Cartridge Case Location. The position of one or more cartridge cases can help to determine the position of the shooter; their locations should be noted so that they can be plotted on a scene plan. A number of cartridge cases grouped together could help to indicate the position of a self-loading firearm when it was discharged. Caution is required in such situations as ejected cartridge cases, being almost cylindrical in shape, can bounce and roll several metres on a hard surface. Nevertheless the extent of any cartridge case groupings should be measured.

12.7.1.4 Shotgun Discharge Damage. Shotgun cartridges contain multiple pellets, so when discharged the pellets can cause multiple damage sites. At close range, within 1 m or so, the pellets can act as a single projectile, they then start to disperse. Consequently, the resultant damage to the target can be in the form of a single hole, possibly with scalloped edges or a pellet pattern. The size of this pattern should be measured as this will assist in the estimation of the range of fire. Large quantities of pellets can be present at the scene, together with shotgun cartridge wadding. Plastic buffer material, which is used in shotgun cartridges containing large pellets, may also be present. At close range this buffer material and cartridge wadding can cause damage in their own right.

12.7.2 Post-Mortem Examination

The forensic firearms examiner can play a vital role in assisting the pathologist during the post-mortem examination. The principal questions asked are similar to the ones asked at the scene of a shooting: what was the gun used and how far away was it from the victim when it was discharged? The scientist works with the pathologist to identify the entry and exit wounds. This is not always straightforward, as it is a myth that the exit site is always large and the entry site is small. From establishing the bullet's path through the deceased it is sometimes possible to indicate where the shooter may have been relative to the victim.

If projectiles, be they bullets, shotgun wadding or shot, are recovered from the deceased, the calibre of firearm can be determined. The shape and extent of soot or powder deposition around an entry wound can indicate the orientation and distance the gun's muzzle was from the deceased when it was discharged. Similarly, the size of a shotgun pellet pattern together with the presence of cartridge wadding injuries can give an insight into the range of fire. Where there is doubt as to whether the death was accidental, suicide or homicide the determination of the range at which the firearm was discharged is of crucial importance. When assessing a potential hypothesis concerning the mode of death, such as whether it was it suicide or homicide, the evidence from the scene examination is brought together with the pathological findings. Clothing from the deceased and any firearms-related items recovered at the scene and post-mortem examination should be submitted to the laboratory for examination.

12.8 EXAMINATION AT THE LABORATORY

12.8.1 Firearms

Safety is paramount and all firearms must be checked to see if they are loaded when they are submitted to the laboratory. If required, full trace evidence anti-contamination precautions should be taken. During this safety check the condition of the gun should be noted. Protocols should be in place to ensure that live or spent ammunition is not submitted in the firearm or in the same exhibit packaging. Sometimes guns are submitted in a condition where the safety has not been assessed, such as when the breech is jammed closed, *e.g.* through corrosion. This can happen if a gun has been buried or has been recovered from a river. The use of an X-ray scanner can be useful in such situations to see if there is a cartridge in the gun's chamber.

Once the gun has been made safe, the full examination will usually take place after all other trace evidence work has been carried out. Trace evidence recovery may occur during a joint examination with scientists from other disciplines, or the firearms examiner may be cross-trained. When the firearm is first examined it should be photographed. The nature and positions of any foreign material adhering to it should be noted. Notes including a full description of make, model and type of firearm should be taken. The overall length, barrel length and magazine capacity should be recorded, as these parameters may have a bearing on the legal classification of the firearm.

A low-power stereo microscope can be used to examine the condition of the breech and bore. An endoscope is also a useful tool to assist in bore examination. If the bore is rifled, the direction of twist and the number of lands and grooves should be noted. A bore wipe, using a cleaning patch, filter paper or swab, should be taken as this will remove some of the GSR particles present which can be analysed. The position and nature of any markings stamped on to the firearm should also be noted; these can include the serial number, proof marks and date codes. Proof marks are stamped on to the firearm when the firearm is sent by the manufacturer to a proof house. At this testing facility the firearm will have been test fired so that it can be authenticated as being safe to be released into the marketplace. The serial number may have been partially or totally removed in an attempt to conceal the firearm's provenance. It is sometimes possible to discern the serial number by use of a stereo microscope, otherwise the examiner can use an etching technique to recover it.

One of the propositions frequently put forward is that the firearm discharged accidentally or inadvertently. To address this, it is essential to determine if the safety catches and mechanisms are working. It is also important to determine the firearm's trigger pull by carrying out a quantitative assessment of the force required to pull the trigger. To see whether or not the firearm is prone to accidental discharge, a series of drop and impact tests are performed.

Once these tests and observations have been made an assessment should be carried out to see whether the firearm is safe to test fire. This is

particularly important with modified, converted, reactivated or improvised firearms. If the firearm is deemed unsafe, it may be appropriate to
discharge a primed cartridge case, *i.e.* one that does not have a bullet or
propellant loaded into it. A remote test firing device should be used with
suspect firearms.

Firearms should be test fired using a range of ammunition with different bullet and primer materials. Varying the nature of the bullet
material will help to maximise the range of characteristic markings
transferred to the ammunition from the gun. Undistorted bullets can be
recovered by firing into a water tank or box containing cotton wool or
bespoke fibrous material. The recovered bullets and cartridge cases
should be recovered and placed into an appropriately labelled bag.

In order to assess the lethal potential of a discharged projectile it may
be necessary to measure its velocity using a chronograph. Once the
velocity has been obtained the kinetic energy of the projectile can be
calculated. A chronograph can also be used to carry out other ballistic
calculations such as extreme range of the weapon.

Comparative penetration tests using ballistic soap, gelatine or other
homogeneous media can be carried out when assessing the wounding
potential of a projectile. Such tests are used to compare how far the
projectiles penetrate from a firearm which is just capable of causing a
lethal injury, and a submitted gun. In order to classify air guns with
respect to the UK firearms legislation it is necessary to determine the
muzzle kinetic energy of the discharged pellets. Air pistols have a limit of
6 ft lbs, other types of air gun 12 ft lbs; if the gun is capable of discharging pellets in excess of these limits further legal restrictions apply to
their possession. Figure 12.4 shows the effect of a high-velocity bullet
passing through ballistic gelatine.

The firearm can also be test fired at various ranges in order to
duplicate the soot, propellant or shotgun pellet pattern observed at the
scene, at the post-mortem examination or on recovered clothing. To
obtain precise range estimations the recovered gun and ammunition
should be used. Sometimes this is not possible and approximations have
to be made. Figure 12.5 shows the soot and powder deposition for four
different types of .357 Magnum calibre ammunition discharged in a
Smith & Wesson revolver as a function of distance.

12.8.2 Ammunition

Unfired and spent ammunition should be examined to determine its
calibre and make. The calibre is usually readily identifiable by the
experienced examiner; however, if it is an unusual cartridge then its

Figure 12.4 The effect of a 7.62 NATO bullet on ballistic gelatine. Note the back-spatter and temporary cavitation.

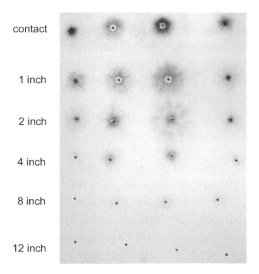

Figure 12.5 Soot and powder deposition for four different brands of .357 Magnum calibre cartridges as a function of distance of the gun from the target.

dimensions should be determined and reference made to specialist books and databases. The viability of unfired ammunition should be determined, either by dismantling the cartridge and examining the primer and propellant or by test firing the cartridge. Spent bullets are weighed and any rifling characteristics present determined by identifying the number of lands and grooves and their direction noted. The dimensions of the

Chapter 12

land and groove widths can also be measured. These rifling characteristics can be used with suitable databases, such as those created by the
FBI and the Bundeskriminalamt, to determine what make and model of
firearm may have been used.

The presence of any trace evidence, such as powdered glass or paint,
can help to determine what the bullet may have struck: damage to the
surfaces of the bullet other than the nose can indicate a ricochet, and the
angle of impact can also be assessed by the angle of the plane of damage
to the bullet. Similarly, any gun-dependent class marks present on spent
and unfired cartridge cases or cartridges are assessed. The principal
marks as identified previously in Figure 12.2 are the shape, position of
the ejector and extractor, firing pin impression, breech face and chamber
marks. These can all help to identify the type, make and sometimes
model of gun used.

Any shotgun pellets submitted should be examined to determine if
they are conventional lead pellets, and a representative sample should be
weighed so that their nominal size can be determined. The size of
shotgun cartridge wadding will give the calibre of the parent cartridge
and may also have pellet impressions pressed into its upper surface. The
size of these impressions will also enable the shot size to be determined.
Plastic shotgun wadding should be closely examined, since the shotgun's
barrel, especially if sawn-off, can transfer individual characteristics on to
the wad.

12.8.3   Clothing

Any garment with suspected gunshot damage should be photographed
and the positions of any damage noted and the extent of the damage
determined. The use of low-power stereo microscopy is helpful in
looking at the nature of the damage sites. The spot tests for lead and
copper, previously discussed, should also be used to assess entry and exit
holes. The extent of soot and propellant deposition can assist in range of
fire determination. The use of a mannequin can be helpful in visualising
the bullet's path though the body and clothing.

12.9   COMPARISON MICROSCOPY

Individual characteristic gun marks present on cartridge cases and
projectiles, such as bullets, air gun pellets and plastic shotgun cartridge
wads can be compared to one another by using a comparison microscope. This piece of equipment, in effect two microscopes joined by an

optical bridge, uses a prism to enable two objects to be viewed simultaneously, side by side, at the same magnification.

The rifling or bore marks present on the bullet, air gun pellet or plastic shotgun cartridge wad which have arisen from the passage into and down the barrel are compared to other recovered samples or test firings using the comparison microscope. Similarly firing pin, extractor, ejector, chamber, or breech face marks transferred by the firearm to the cartridge case or cartridge can also be compared to other submitted samples and test fires. Figures 12.6 and 12.7 show a bullet and a cartridge case comparison, respectively.

Live cartridges can also be examined to see if there are any individual characteristic marks on them which have resulted from being worked through the action or magazine of a self-loading firearm. The principal skill required by the examiner, together with years of experience, is pattern recognition. In order to come to a positive conclusion the examiner must be satisfied that the level of agreement exceeds that of any known best non-match. If the level of agreement is sufficient, one can conclude that for all practical purposes a single firearm was used. Any critical comparison microscopy conclusion should be subject to peer review. Recent research has built on work carried out in the 1950s to give a statistical foundation to the theory of tool-mark identification.

Figure 12.6 Use of a comparison microscope to match individual characteristics on the land impressions of two bullets.

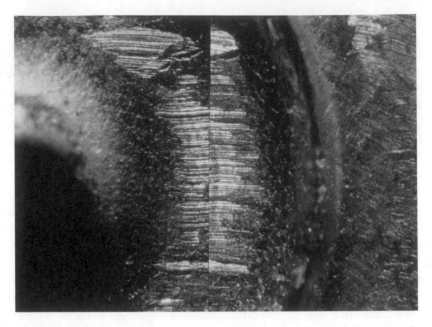

Figure 12.7   Use of a comparison microscope to match individual characteristics on the primers of two cartridge cases.

Most government or police forensic firearms laboratories have what is known as an open case file (OCF). This consists of a collection of ammunition components, such as bullets, cartridge cases and plastic shotgun wads which have been recovered from the scenes of unsolved crimes. The collection is subdivided into calibre, and in the case of bullets, rifling characteristics. When appropriate spent or unfired ammunition and its components are submitted to the laboratory they are screened through the OCF to see if there is any previous use of the same gun. Similarly when a firearm is submitted, its test-fired bullets and cartridge cases are screened through the OCF. Depending on the size of the OCF, this can be a long process. In recent years there have been huge developments in the automation of the comparison microscopy process using hardware and software to capture the individual characteristic marks and search them against the uploaded OCF database. However, it is still the case that any potential match must be manually looked at by the experienced examiner.

12.10   FIREARMS CHEMISTRY

12.10.1   Gunshot Residue (GSR)

This section illustrates the value of GSR as trace evidence and the factors affecting its detection and identification in crime. When a gun is fired thousands of microscopic particles, called gunshot residue, are produced by the ammunition. They are emitted from the end of the muzzel of a gun and from any gaps or openings in the gun's action and can be deposited on the firer, any persons or surfaces sufficiently close to the firer, and the gun itself.

The recovery and identification of GSR particles can help address questions such as, 'Has the suspect recently fired a gun?' That the particles originate from a firearm is rarely contested. Of far more interest to the court is how these particles came to be on a suspect's clothing or hands. GSR evidence is one of the most heavily scrutinised types of trace evidence in criminal investigations.

12.10.2   Origin and Chemistry of GSR

Modern percussion primers are the products of more than 100 years of research. The earliest primers contained potassium chlorate as the main ingredient, but mercuric fulminate mixtures were introduced in 1831 and used widely thereafter. Mercury caused cartridge embrittlement and barrel corrosion, so the first non-mercuric, non-corrosive primer was introduced in 1928. This 'Sinoxyde' primer was composed of lead styphnate (an initiator), barium nitrate (an oxidant) and antimony sulphide (a fuel). This primer composition continues to be used alongside lead-free primers which were introduced in the 1980s.

When the primer is detonated by the gun's firing pin the primer composition decomposes exothermically and drives hot gases and molten material into the cartridge case, igniting the propellant. The burning propellant decomposes, producing large volumes of hot gaseous products which dramatically increases the pressure within the cartridge. The cartridge case expands tightly against the chamber walls (preventing the rearward escape of gas), while the high pressure forces the bullet out of the cartridge, into the barrel and out of the gun.

The high temperature inside the cartridge creates conditions in which the individual components of the primer can fuse together, in a matter of ten thousandths of a second. This very short time frame accounts for the variability and inhomogeneity that is observed in the composition and numbers of GSR particles produced. After the material has cooled discrete particulate residue remains, containing combinations of the elements lead, barium and antimony and any other components present in the primer. The number of variations in the components is limited and therefore the number of different types of residue is small. Other variations can arise from components used in the construction of the primer cap, the firearm and the bullet, *e.g.* the use of tin foil to seal the primer cup. Nevertheless, different types of GSR particle can be identified and used to eliminate a particular gun or spent cartridge case as the source of residue found in a case.

In any population of residue particles there will be a variation of the proportions and combinations of the chemical components. The compositions in most cases will not be in fixed or consistent proportions and can even vary from shot to shot. This can be understood by considering the very short time available for the components of the primer to mix and fuse together during the detonation and subsequent ignition of the propellant. Particles containing all three elements *i.e.* lead, barium and antimony were previously considered unique to the discharge of a firearm, however they have been subsequently observed elsewhere as will be discussed later (section 12.11.4). For this reason GSR is now widely referred to as being characteristic of originating from a firearm. Particles without all three components are referred to as 'indicative' or 'consistent' with a firearms origin.

The particles of GSR may also contain contributing elements from the cartridge case, bullet, gun barrel and propellant. Particles from the bullet can also often be observed alongside particles originating from the primer. High levels of small spherical lead particles often with a trace of antimony, used to harden bullets, are produced by unjacketed leaded ammunition and lead shot.

12.10.3   Detection of GSR

The earliest methods for the detection of residue on the hands of a suspect were based on simple colour tests. These tests enabled the detection of lead, antimony and barium from primers and other elements, such as copper and zinc, from the cartridge case. These tests suffered from a distinct disadvantage in that they were not source specific and many sources of these elements other than firearms exist. Today these tests can only be used for screening or as presumptive tests, but they can be very useful in demonstrating the presence of lead or copper around a bullet entry hole.

In the early 1970s work began independently at the Metropolitan Police Forensic Science Laboratory in London and the Aerospace Corporation in California on a new method of identifying individual particles of gunshot residue using the scanning electron microscope equipped with an x-ray analyser (SEM–EDX). Importantly, this work established that discrete particles composed of lead, barium and antimony appeared to occur only in percussion primer residue. There was no known source other than firearms, and hence the first conclusive method for identifying GSR was established. This particle analysis method is still employed today.

12.11 ANALYSIS AND IDENTIFICATION OF GSR USING SEM-EDX

Briefly, the SEM functions by scanning a finely focused beam of electrons over a sample. The impinging electrons interact with the sample and produce an image of a particle. Furthermore, the electron beam results in a displacement of electrons and emission of characteristic x-rays from the particle. These x-rays are then used to identify the elemental composition of the particle. Overall, the SEM locates particles of interest, generates a characteristic X-ray spectrum and software classifies the particle.

In general GSR particles will be rounded because they have been generated during a fusion process, although an irregular appearance does not necessarily result in a particle being ruled out. The final decision on whether a particle constitutes GSR will depend on the morphology, size, chemical composition, homogeneity and the nature of the sample as a whole and must be done manually. In terms of size, the largest particles will generally be lost first and so most particles found on clothing are of the order of 1–10 µm in size (1 µm is one thousandth of a millimetre). In terms of numbers of particles, a finding of six particles or more would constitute a good result. Figure 12.8 shows a typical SEM image of a GSR particle.

Over the past 20 years or so the original acceptance criteria for the chemical components constituting a particle of gunshot residue have been refined and widened. As a general rule, the following five compositions of GSR particles are commonly seen in casework and can be used to distinguish between different sources of residue:

1. Lead–barium–antimony.
2. Lead–barium–antimony–aluminium.
3. Lead–barium–antimony–tin.
4. Lead–barium–calcium–silicon.
5. Lead–barium–calcium–silicon–tin.

Particles containing barium–lead where the barium peak is larger than the peak for lead are unusual in the general environment and are indicative of GSR originating from .22 calibre rim fire ammunition. The particles can be considered to be characteristic if the spent cartridge case or gun from the crime is available for comparison of the residue type.

Figure 12.8   A scanning electron micrograph of a GSR particle.

Ammunition containing the older primer formulations based on mercury is still available, particularly in criminal circles. Such ammunition can produce residue particles where the mercury itself is not always present or detectable since is it so volatile, leaving, for example, only antimony–tin particles. This can mean detection and conclusive identification of the particles as originating from a firearms source is not always straightforward.

More recently, lead-free primers have been developed. The first and most common is Sintox, which is based on the chemical elements titanium and zinc. Newer lead-free formulations now exist based on potassium and boron, elements that are more difficult to detect in the SEM. Ammunition containing these components has so far rarely been seen in crime in the UK.

12.11.1   Deposition and Transfer of GSR

As indicated earlier, GSR is emitted from the end of the muzzle of a gun and from any openings such as the breech face or cylinder gaps in a revolver on to the firer and anyone standing next to the firer. The majority of GSR particles are deposited within approximately 3 m from the end of the barrel and approximately 1 m on either side of a gun. The amount of residue produced will depend on the type and condition of the gun, the type of ammunition and calibre of the firearm. Greater quantities of residue may be deposited if a shot is fired inside rather than outside, since external climatic conditions such as wind, rain and temperature can influence the amount deposited.

After the bullet emerges from the gun's muzzle and strikes a substrate it can transfer gunshot residue from its surfaces, in particular the base, to the perimeter of the entrance hole in the target. This appears to occur irrespective of the range of fire. On passing through the target surface most of the residue will be removed from the bullet's surface, such that little or none will be present on the exit hole. The presence or absence of gunshot residue therefore allows differentiation between the entry and exit holes. Examination for gunshot residue on the clothing of a shot victim, or *e.g.* a car window which has been shot at, can help establish whether the shot was fired close to or from afar and also the composition of the residue.

Handling a recently fired gun with residue on its surfaces can also result in a transfer of residue to the person's hands. This is termed secondary transfer. It is generally accepted that only a small proportion of the residue will be transferred at each contact. Consequently, it is often not usually possible to distinguish between residue that has been deposited as a result of firing a gun from that which has originated from handling a gun with residue on its surfaces or from standing next to someone firing a gun.

12.11.2 Persistence of GSR

Although thousands of particles can initially be produced upon firing a gun, in reality only a fraction of these ultimately remain on a suspect's clothing by the time they have been arrested and their clothing seized. Primer residue is chemically stable and does not evaporate or dissolve. It is lost by physical disturbance, with washing and wiping considerably accelerating the loss. Up to 90% of GSR particles can be lost in the first hour following deposition. On the hands of a firer, residue may only remain for a few hours at most and may be lost much more quickly. The face and hair of a suspect usually suffer from less disturbance and persistence can be a few hours longer. On clothing, GSRs can remain for up to a day on the outer surfaces if the same clothing is continuously worn. Walking, running or brushing the surfaces of the clothing will increase the rate of loss. However, residue may be transferred to a pocket from the hand or from a gun and can remain there indefinitely if not subject to any disturbance.

12.11.3 Recovery of GSR

A number of methods of collecting primer and organic residue are employed in forensic science laboratories. The optimum recovery of primer residue is by a taping method. Small adhesive strips or carbon tape mounted on small metal stubs are pressed repeatedly over the surface of a suspect's hands, face, hair or clothing. Any particles of gunshot residue will adhere to the tape which can then be analysed in the SEM.

The forensically significant components of organic residue, namely nitroglycerine and 2,4-dintrotoluene, are best collected using either a swabbing or vacuuming technique. Any primer or organic residue can be extracted from the resultant swabs or filters and the organic residue analysed using chromatographic techniques such as GC–TEA or HPLC. These methods are not employed routinely in most laboratories since examination for primer residue alone is more discriminating and sufficient to establish exposure to a firearms source. The chromatographic methods can also be used to detect the components of explosives, and such organic analysis is still undertaken in Northern Ireland.

Generally, it is outer items of clothing that can be expected to offer the best opportunity of detecting GSR following a firearms offence since residue will remain for longer on clothing than on skin and hair. Other items which are also routinely examined include gloves, balaclavas, bags which may have been used to convey firearms before or after an incident and vehicles from or at which shots may have been fired.

12.11.4 GSR-Similar Particles

A variety of materials have been reported in the scientific literature as potential alternative sources of particles that may be mistaken for GSR. In general terms the particles are generated during a hot process, thus conferring a spherical often molten appearance.

Fireworks are often mentioned as being a possible source of particles similar to GSR. The main ingredient is black powder and to this is added an oxidiser, metal fuel and binders. To produce the colour observed when a firework burns, a number of different chemical compounds are employed such as potassium chlorate, strontium nitrate, barium nitrate and aluminium or magnesium powder. Antimony is often used to produce a glittering effect and lead produces an audible crackling sound. The potential can be seen for the production of pyrotechnic residue containing similar combinations of elements to GSR. One report of finding isolated lead–barium–antimony particles in the population of firework residue has been published in the scientific literature, but this has not been reproduced in other studies.

A profusion of barium–aluminium–magnesium particles has been observed in casework samples, particularly around the month of November. However, the presence of other particles commonly containing elements such as potassium, chlorine, copper and most notably magnesium would alert an experienced scientist that they were not observing GSR.

Lead–barium–antimony particles have been observed in samples taken from certain car brake linings and hence could possibly be present on people employed in occupations such as car mechanics. However, most of these particles also contained high levels of iron, magnesium, sulphur, silicon and titanium. The particle's appearance is usually one of a compacted mass which lacks the features of particles produced at high temperature and therefore not typical of GSR.

Cartridge-operated tools such as nail guns use blank cartridges as the energy source for providing the driving power. Blank cartridges contain all of the components of a round of ammunition except the projectile. In the UK most cartridge-operated tools employ rim fire ammunition and the resulting residue can, from the coating on the nails, contain higher levels of zinc than GSR.

12.11.5 Contamination Avoidance

The largest single factor that must be considered when interpreting GSR findings is the possibility of contamination. In general, small amounts of GSR particles are anticipated so the opportunities for cross-contamination must be kept to a minimum and the risk must be assessed within the overall context of the case.

At a crime scene, as with any evidence type, appropriate protective clothing must be worn by anyone entering the scene. Clothing and samples from suspects' skin and hair surfaces need to be recovered as soon as possible after the incident. This will depend on the circumstances of the case and the time interval between the incident and the apprehension of a suspect. Generally, the suspect will be taken back to a police station where the samples will be taken from them along with their clothing. It is very important that any suspect is not conveyed in a vehicle used by armed police officers, or taken to a police station that has a firing range, or sampled by someone involved with firearms, as this may compromise the evidence.

At the scene it is critical that the investigator does not handle any guns or spent cartridge cases and then come into contact with a suspect or other item that is to be examined for GSR. Separate individuals should be involved in the collection of evidence that may be highly contaminated with GSR and that which may only contain low levels. Any other items recovered at the scene, such as abandoned items of clothing, bags, gloves or masks, should as with most forensic evidence be packaged, sealed and exhibited as soon as is practical to ensure its integrity.

At incidents involving firearms it is common practice for armed police officers to attend to secure the safety of the general public. In the most severe incidents this can mean incapacitating a suspect to prevent them from discharging a firearm that may injure themselves or others. The officers will also be involved in searching suspects thought to be in possession of a firearm and addresses where firearms may be located. There is a distinct possibility that firearms officers will have GSR on their hands and operational clothing, and that this may be transferred through physical contact to a suspect. The officers may also deploy a distraction device such as a flash-bang either in training or when operational. These devices produce smoke and a loud bang and produce distinctive particles containing zirconium, barium, aluminium, tungsten and chromium among other chemical elements. If found on a suspect's clothing they provide further information in relation to the arrest of the suspect. All details of arrests of suspects involving firearms officers need to be considered when interpreting the GSR findings in a case.

Stringent procedures should be in place to ensure that the risk of any cross-contamination at the laboratory is kept to an absolute minimum. Quality control samples are generally taken from the examination bench and the outside of the packaging of an exhibit. Full personal protective equipment should be worn when conducting examinations and gloves should be changed frequently. Most laboratories undertake routine environmental monitoring by taking samples from the areas where items are examined and in the area where the SEMs are utilised. This ensures that the environment does not present a contamination risk to the integrity of the exhibits being examined.

12.11.6 Interpretation of GSR Findings

As can be seen, deposition of GSR is a highly variable event and subject to a large number of factors that determine what is eventually found. This represents the challenge in interpreting the findings in a case to determine if someone may have been in possession of or fired a gun.

A brief outline of the circumstances will normally be available, but the practitioner must establish the full case circumstances as far as is practicable if they are to come to a reliable interpretation. Key factors include the timings between the incident and the recovery of the items, the recovery of a weapon or spent cartridge cases, the number of any shots fired, whether armed police officers were involved and what the suspect has said in response to the allegations. This last point is important in establishing if the suspect has any other explanation for the presence of GSR on their clothing, *e.g.* being a legitimate holder of a firearms licence or a member of a gun club. However, this tends not to be the case in most cases involving firearms and the suspect may admit to being present but not to discharging or handling the weapon. It is then for the firearms chemist to consider the relative likelihood of the GSR evidence in terms of the questions asked by the investigator compared to any reasons given by the suspect.

It is often alleged that a suspect used a particular hand or drew a gun from a pocket or the waistband of their trousers. Finding residue in pockets suggests that a gun or a hand with residue on it has been placed inside the pocket at some time. Residue in pockets is less likely to be the subject of accidental contamination since it requires an additional step in the transfer process. This is important in cases where it is alleged that a suspect may have been in possession of a firearm. It is a common misconception that the finding of residue on a particular hand of a firer proves that this was the firing hand. Residue is redistributed and lost almost immediately after it is deposited. However, finding residue on the hands, face or hair of a suspect supports recent deposition and can be very helpful in an investigation. Finding residue on the back of a garment rather than on the front or sleeves can indicate a proximity to firing rather than being the firer. Significant amounts of residue are rarely found inside the waistband as this area is subject to much disturbance.

It must always be considered that if the time interval between an incident and recovery of a garment is longer than a few days then the finding of GSR is more likely to be attributable to a more recent event. Conversely, the absence of GSR does not exclude a firearms involvement as residue may have been lost in the intervening time.

Although the presence of GSR in the environment is considered to be rare, in recent years there has been an increase in the criminal use of firearms and a rise in the number and use of armed police officers, so the risk of accidental contamination has risen. People who associate with firearms users might unknowingly get a small amount of residue transferred to them. For these reasons, very little can be said about the finding of single particles.

Residue is also important in helping to establish range of fire. It is by no means as precise an indication as that given by the firearms examiner following laboratory tests, but it is useful in instances where no visible blackening or powder patterns have been left or where the crime weapon has not been recovered. It can also be used when differentiating between entry and exit holes.

BIBLIOGRAPHY

Gunshot Residue Detection, Aerospace Report ATR-75 (7915)-1, Aerospace Corporation, El Segundo, CA, 1975..

L. C. Haag, *Shooting Incident Reconstruction*, Elsevier Academic Press, San Diego, 2006.

A. J. Schwoeble, *Current Methods in Forensic Gunshot Residue Analysis*, CRC Press, Boca Raton, 2000.

J. S. Wallace, *Chemical Analysis of Firearms, Ammunition and Gunshot Residue*, CRC Press, Boca Raton, 2008.

T. A. Warlow, *Firearms, The Law, and Forensic Ballistics*, 2nd edition, CRC Press, Boca Raton, 2005.

CHAPTER 13

Drugs of Abuse

MICHAEL COLE

Anglia Ruskin University, Department of Life Sciences, East Road, Cambridge CB1 1PT

13.1 INTRODUCTION

The problem of drug abuse continues to increase and the drug analyst is always being presented with new and interesting challenges. Drug use is a dynamic problem which changes with, amongst other factors, preference amongst drug users, socio-economic climate and changes in legislation. The drug analyst must therefore be at least as versatile and dynamic as the drugs market. For example, in 2007, 15 new compounds were notified across Europe to the early warning system of the European Monitoring Centre for Drugs and Drugs Addiction (EMCDDA) and each requires a method of unambiguous identification. Drug analysis is one area of forensic science where a positive identification of a material must be made and as a consequence there is also a requirement for the development of novel and improved analytical methods, data interpretation and reporting of the evidence to court either in written and/or verbal form.

The legislation in the UK prescribes different charges and sentences for different drugs and for differing amounts of those drugs. In addition to the identification of any controlled substance that might be present, the analyst may also be required to quantify any drugs of abuse in the sample. Depending on the drugs thought to be involved, some techniques are

Crime Scene to Court: The Essentials of Forensic Science, 3rd Edition
Edited by P. C. White
© Royal Society of Chemistry 2010
Published by the Royal Society of Chemistry, www.rsc.org

more appropriate than others for these tasks and it is a knowledge of the drugs involved and which techniques to apply, and when, that form the major part of a drug analyst's expertise and activity. There is also increasing interest, both nationally and internationally, in the comparison of drug samples with each other. This provides information concerning the manufacture, distribution and supply networks and movement of drugs at the street (user) level. These comparisons require different methodologies depending on the drug in question. If such data is to be shared and used in a meaningful way, however, then the same instruments and methods should be used in the participating laboratories who should all train their staff to a given set of standards and engage in the same quality control and management procedures and protocols. Even when this is the case, the success of such endeavours depends on the drug class in question and the methods to be used. Understanding these processes is an important part of the drug chemist's work.

13.2   DRUG CONTROL LEGISLATION IN THE UK

A number of principal pieces of legislation are employed to control the drugs of abuse, namely the Misuse of Drugs Act 1971, the Misuse of Drugs Act (Regulations) 1985 and the Misuse of Drugs Act (Regulations) 2001. There have also been a number of other legislative instruments amending these principal documents. The 1971 Act lists the drugs in one of three classes as part of Schedule 2 to the Act. Examples of such drugs and penalties associated with them are given in Table 13.1.

A number of classes and individual drugs have subsequently been added and deleted, as detailed in the subsequent Amendment Orders, the Misuse of Drugs Regulations 1985 which controlled benzodiazepines and barbiturates for the first time, and the Misuse of Drugs Regulations 2001.

Importation and exportation of controlled substances are prohibited by Section 3 of the Act, production, supply or offer to supply a controlled substance are prohibited by Section 4. Possession is controlled by Section 5. Cultivation of any plant of the genus Cannabis is controlled by Section 6 of the Act, while Section 8 states that

> A person commits an offence if, being the occupier or concerned in the management of any premises, he knowingly permits or suffers any of the following activities to take place on those premises . . .

and the list includes producing or attempting to produce a controlled drug, supplying or attempting or offering to supply a controlled drug, preparing opium for smoking or smoking cannabis, cannabis resin or

Table 13.1 Examples of different drugs classified under Classes A, B and C of the Misuse of Drugs Act 1971 and penalties imposed.

Classification under the Misuse of Drugs Act 1971	Penalty for possession	Penalty for supply	Examples
Class A	Up to 7 years in prison or unlimited fine or both	Up to life in prison or unlimited fine or both	Heroin, cocaine, LSD, methylamphetamine
Class B	Up to 5 years in prison or unlimited fine or both	Up to 14 years in prison or unlimited fine or both	Cannabis, barbiturates, amphetamines
Class C	Up to 2 years in prison or unlimited fine or both	Up to 14 years in prison or unlimited fine or both	Ketamine, benzodiazepines

prepared opium. Section 9 of the Misuse of Drugs Act is interesting in that it relates specifically to opium. The section makes provision for the control of smoking or other use of prepared opium, the frequenting of such a place where this occurs and also being in possession of pipes or other utensils made or adapted for use in connection with the smoking of opium or preparation of opium for smoking.

The legislative materials also make provision for the control of any salt, ester or stereoisomer of the compounds listed in the legislation. A controlled substance is therefore defined as one which falls under the umbrella of being controlled by one or other, or both (the Act or the Regulations), of these pieces of legislation. In short, The Misuse of Drugs Act prescribes what cannot be done and the regulations prescribe what may be done, with the correct authority to do so.

Other legislative tools, *e.g.* the Traffic Offences Act 1986, make provision for the confiscation of the proceeds in trafficking in drugs of abuse. Recent Amendments to the Misuse of Drugs Act add specific compounds or small groups of compounds to the list of controlled substances and, rarely, remove a substance from the list. Sometimes drugs are reclassified and so it is important to ensure an awareness of all modifications. Compounds added under the modifications include some steroids, phenethylamines and most recently the piperazines in addition to individually named compounds.

13.3 DRUGS OF ABUSE AND THEIR SOURCES

During the course of a drug analyst's career, they are likely to encounter a wide variety of drugs. What follows is a description of the principal classes of drugs likely to be encountered.

13.3.1 Cannabis and its Products

The products of *Cannabis sativa* L. form the major part of the drug materials currently submitted to forensic science laboratories. The materials come in three major forms: herbal material, resin and oil. All contain the active constituent, Δ^9-tetrahydrocannabinol (Δ^9-THC), sometimes with a small amount of its isomer, Δ^8-tetrahydrocannabinol. The precursor of this is cannabidiol and in older samples in which decomposition of the sample has started, cannabinol will also be observed (Figure 13.1).

Herbal material of *C. sativa* comes in a variety of forms, from whole plants to small amounts of leaves, stems and flowering tops, and can also be encountered fresh or, more commonly, dried. The fresh material has a very characteristic smell, like spearmint, although not exactly the same. The leaves are palmate, with serrated edges. The stems are square and four-cornered. The flowering tops are recognised by the presence of buds, flowers and sometimes seeds. Three principal morphological characteristics allow the recognition of cannabis plant material. The first is single-celled trichomes, which are colourless and, unusually, point unidirectionally up the stem. The second feature is the cystolithic trichomes, with crystalline materials at their base. The third feature is the presence of glandular trichomes, which consist of a glandular structure on a thin stalk of cells. It is within the glandular trichomes that the resin is stored. If all three of these morphological features are observed, the

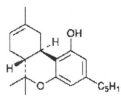
Δ^9 - Tetrahydrocannabinol

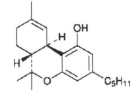
Δ^8 - Tetrahydrocannabinol

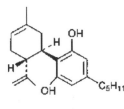
Cannibidiol

Cannabinol

Figure 13.1   Chemical structures of some cannabinoids present in *Cannabis sativa* L.

plant is almost certainly *C. sativa*. Even the closest relative of *C. sativa*, the hop (*Humulus lupulus*) does not have these morphological features, although it does have glandular trichomes at the base of the leaves which may, by the inexperienced eye, be confused with the trichomes of *C. sativa*. The herbal materials may be packaged in a number of different ways and contain differing amounts of leaf, stem, flowering tops and seed material. These materials contain about 2% (w/w) of the active component Δ^9-THC The form of the packaging and its contents may be used to identify the country or region of origin of the plant materials.

Resin comprises the material obtained when the glandular trichomes are collected, by a variety of means. These include passing large hessian nets through the crop, on to which the resin becomes stuck, as is done in North Africa, and rubbing of the plant materials between the palms of the hands, as occurs in Pakistan. Either way, the resin is compressed into blocks and shipped. The resin may contain other cellular debris and it is frequently possible to observe parts of both the glandular hairs and the cystolithic trichomes under microscopic examination. This material typically contains 5–7%(w/w) of Δ^9-THC.

The oil, commonly referred to as 'hash oil', represents the material that is extracted using solvent extraction of the plant materials. The concentration of the active constituents and other cannabinoids is much higher and contains about 10–12% (w/w) Δ^9-THC.

In addition to bulk samples of the drug a number of different types of items might be encountered by the forensic analyst. These include wrapping materials which should be compared to determine whether there is a relationship between samples being analysed. In addition, there may be trace items, such as scales, knives, cutting implements and stamps, all of which may be the subject of investigation to determine whether they have been in contact with controlled substances and whether they can be used to determine any commonality between samples. Further, the remains of cigarettes used to smoke cannabis products, smoking utensils (*e.g.* roach stones) and even the contents of ashtrays can all be analysed for the presence of cannabis materials.

13.3.2 Heroin

Heroin is derived from the opiate alkaloids produced by the field poppy *Papaver somniferum* L. It is not a pure compound, but a mixture of the impurities, reaction intermediates and breakdown products manufactured during the synthesis of diamorphine; compounds carried through from extraction of the morphine from the plant and subsequent processing; excipients, *e.g.* sugars; and adulterants, *e.g.* other drugs, used to dilute, or cut, the drug. Diamorphine is derived from morphine, one of the alkaloids from opium resin. The seed pods of the plant are lacerated and latex exudes which then dries. From the dried latex, morphine is extracted using a series of solvent/solvent extractions and precipitations. The morphine is then acetylated, producing the mono-acetylmorphines and diamorphine, which are substances controlled under the Misuse of Drugs Act 1971. Since codeine is also extracted from the opium resin using these methods, acetylcodeine will be present in the final product. The chemical structures of alkaloids commonly found in heroin are shown in Figure 13.2.

Thebaine is also present because it is extracted from the opium resin in the same way as the morphine. Two other alkaloids, of different classes, are also co-extracted, namely noscapine and papaverine. The amounts of all of these alkaloids present in the final products will depend upon a number of factors, including geographical origin of the drug, quality of

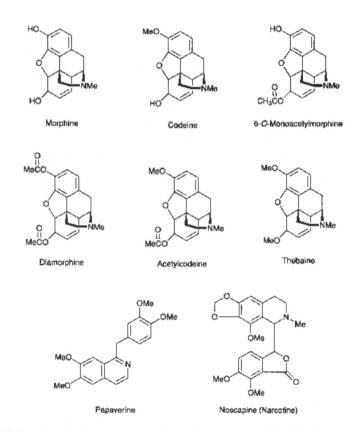

Figure 13.2   Chemical structures of alkaloids commonly found in heroin.

the extraction chemistry, quality of the acetylation and quality of the purification processes. Since heroin is produced in a batchwise manner, no two samples will ever be the same. However, heroin samples from different geographical regions may exhibit certain chemical characteristics, as outlined in Table 13.2.

It is due to this variation in appearance and chemical nature that the origins and relationships of drugs can be identified. This is examined in greater depth later in this chapter.

Cutting agents are used to dilute the drug and hide its impure nature and/or lack of active ingredient. Examples of common cutting agents include sugars, caffeine, benzodiazepines and barbiturates. These too can be used to compare drug samples. Interestingly, the benzodiazepines are added as anxiolytics to control anxiety in heroin users and barbiturates are found in the heroin as a drug added to control pseudoepileptic fits which are can occur as the diamorphine is cleared from the body.

Although bulk or trace samples of the powders may be encountered it is equally likely that utensils will be found. These include, for example,

Table 13.2 Examples of how heroin samples can vary both chemically and in appearance, within and between regional sources.

| Origin | Drugs Present | | | |
	Acetyl codeine	6-O-monoacetyl morphine	Noscapine	Papaverine
S.W. Asia				
Type I	5%	3%	10%	4%
Beige–brown, powder with small aggregates 60% diamorphine, free base				
Type II	3%	2%	ND	ND
White–off white 80% diamorphine, hydrochloride salt				
Middle East				
Type I	3%	2%	ND	ND
Beige–light brown powder, aggregates rare 50–70% diamorphine, hydrochloride salt				
Type II	2–3%	2%	NA	
White–off white, fine powder 70–80% diamorphine, hydrochloride salt				

NA, data not provided; ND, drug not detected.

spoon surfaces. Spoons are often used to mix the heroin with a water-soluble organic acid so that the diamorphine itself becomes water soluble. This solution can then be drawn through cotton wool, which acts as a crude filter to remove solid materials, and injected. As a consequence the cotton wool and syringes used may also be encountered. It should be noted that such items represent a considerable health and safety risk and great care should be exercised in their handling to prevent 'needlestick injuries'. The spoons themselves are often interesting—the lower side will be black where heat from a flame has been applied, while the spoon itself may contain the residue from the extracted drug. The handle may be covered in small white flecks of rather purer heroin and breakdown products, which can serve to indicate that heroin was once heated in the spoon.

13.3.3 Cocaine

Cocaine is produced from the tropane alkaloids produced by the trees *Erythroxylon coca* Lam, *E. novogranatense* (Morris) Hieron and *E. novogranatense* var. *truxillense* (Rusby) Plowman. Its chemical structure is shown in Figure 13.3.

   The species examined depends on the region in South America from which the sample originates. As with heroin, the tropane alkaloids are extracted from the leaves by a series of solvent/solvent extractions and precipitations. A hydrolysis step follows, so that all of the ecgonine-based tropane alkaloids are hydrolysed to ecgonine. From this ecgonine methyl ester is manufactured, from which the final product, cocaine, is synthesised. Since the material, like heroin, is produced in a batch process, no two samples are the same and it contains impurities from the natural products in the leaves, including other tropane alkaloids and esters with plant acids, *e.g.* the cinnamoylcocaines (Figure 13.4).

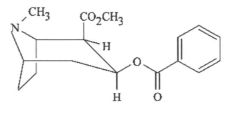

Cocaine

Figure 13.3   Chemical structure of cocaine.

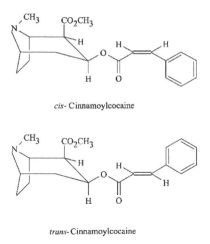

cis- Cinnamoylcocaine

trans- Cinnamoylcocaine

Figure 13.4   Chemical structures of cinnamoylcocaines.

Again, it is the composition of the product that allows samples to be compared and geographical origin to be determined. Two principal forms of cocaine may be encountered. The first is cocaine hydrochloride. This is a white powder, which has a fluffy, snow-like appearance, hence one of its street names, 'snow'. If the free base is then re-prepared from the hydrochloride salt, hard, granular lumps of cocaine are formed. This form is known as 'crack'.

As with heroin, cocaine may also be cut. The cutting agent that is employed depends on the region in which the drug is encountered. In the United States, the drug is commonly adulterated with sugars, *e.g.* mannitol or sucrose, but in the UK local anaestetics (*e.g.* lignocaine, procaine) are frequently used.

As with cannabis, trace samples may be encountered on materials used to administer the drug. This includes traces of drug on surfaces used to mix the drug. Any small pieces of rolled paper or banknotes should also be analysed for the presence of cocaine although it should be noted that bank notes in many currencies are highly contaminated with drugs, particularly cocaine.

13.3.4   Amphetamines

In the UK, under the Misuse of Drugs Regulations 1985, Schedule 1, Part (c), any compound structurally derived from phenethylamine is controlled. These compounds include amphetamine, methylamphetamine

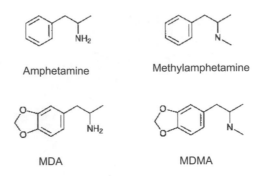

Figure 13.5 Chemical structures of common amphetamines.

(MA), methylenedioxyamphetamine (MDA) and methylenedioxymeth-ylamphetamine (MDMA); their chemical structures are shown in Figure 13.5.

These compounds can be produced *via* a number of different synthetic routes, using as starting materials either natural products derived from plants (such as ephedrine, safrole and isosafrole), or chemicals available from chemical companies (*e.g.* benzenemethylketone, BMK). It is unlikely that the natural products will ever come under control, but certain purified precursor chemicals are now only supplied to organisations who have the requisite licence. Examples include BMK and ephedrine, both precursors of amphetamine.

These phenythylamines can be synthesised *via* a number of different routes, including, for amphetamine, the Leuckart synthesis, the nitrostyrene route and the reductive amination of BMK. The most popular routes depend on the country in which the drug is being synthesised. For example, in western Europe, the synthesis of amphetamine is principally *via* the Leuckart synthesis, but in parts of northern Europe and the United States the drug is manufactured principally *via* the reductive amination of BMK, and in Japan ephedrine is currently favoured as the starting material. Since different synthetic routes are employed and the drugs are made in batches, the impurities and reaction by-products will also vary between batches. Some will be 'route specific', indicating which of the synthetic routes has been employed. Further, the overall composition of the drugs and impurities will vary between batches, allowing differentiation between samples from different batches and identifying those street samples that have come from the same parent batch.

In addition to traditional chemical methods of analysis and comparison, amphetamines are interesting because physical comparison can provide a wealth of information about the origin of the drug. The tablets that are manufactured are often coloured, and analysis of the colourant can occasionally lead to the identification of links between the samples. The logos which are stamped into the tablets can sometimes be used to link samples, particularly if the logo is rare or unusual. It should not, however, be assumed that because a tablet is the same colour and has the same logo that the drugs contained within the tablets are the same. This can only be determined through chemical analysis. One interesting facet of this drug type is the use of 'ballistics' to make comparisons between drug samples. The tablets are produced in large numbers in tabletting machines. Should one of the dies in the instrument become flawed, the flaw is imparted to every tablet that is made in that die. This can be used to link samples together as having been pressed in the same machine even when the drug content of the tablet is different.

13.3.5 *Psilocybe* Mushrooms

Occasionally, material from mushrooms with hallucinogenic properties may be encountered by the forensic chemist. The hallucinogens in the fungi are based on an indole-amine nucleus. In the genus *Psilocybe*, of the approximately 140 species, 80 are known to contain hallucinogens. The three most important species are *Psilocybe semilanceata* (Fr.) Quel., *P. cubensis* (Earle) Singer and *P. mexicana* (Heim). These contain psilocybin (4-phosphoryloxy-*N*,*N*-dimethyltryptamine), its biochemical precursor, baeocystin (4-phosphoryloxy-*N*-methyltryptamine, norpsilocybin), and psilocin, the dephosphorylated compound, which is also a hallucinogenic metabolite. The chemical structures of the major constituents found in *Psilocybe semilanceata* are shown in Figure 13.6.

The effects of baeocystin are not known. The materials may be encountered either as dried mushrooms or as powdered materials. As with many mushrooms, identification of these species using morphology alone is a difficult area in which to work and ideally a qualified mycologist should be engaged to positively identify the species of mushroom.

13.3.6 Mescal Buttons

The peyote cactus *Lophophora williamsii* (Lem. ex Salm-Dyck) Coult. is known to produce the hallucinogen mescaline (Figure 13.7). The material is usually encountered as disc-like slices of the body of the cactus which have been dried. Occasionally (particularly in the USA), seizures are encountered where the active constituent has been synthesised from 3,4,5-trimethoxybenzaldehyde.

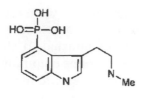

Baeocystin

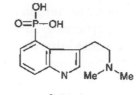

Psilocybin

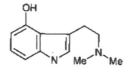

Psilocin

Figure 13.6 Chemical structures of major constituents found in *Psilocybe semilanceata.*

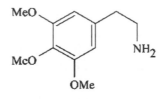

Mescaline

Figure 13.7 Chemical structure of mescaline.

13.3.7 Lysergic Acid Diethylamide (LSD)

Lysergic acid diethylamide, also known as lysergide or LSD (Figure 13.8), is one of the most potent hallucinogens known and is usually encountered in one of four forms: blotter acids, gelatine blocks, small tablets and microdots. It can be synthesised following the isolation of lysergamide from morning glory, Hawaiian baby woodrose seeds, or the ergot alkaloids from *Claviceps purpurea* or *Aspergillus clavatus*. No single route dominates over the others.

LSD can be encountered in gelatin blocks, into which the LSD has been incorporated while the gelatin is liquid. It can also be encountered as small, highly coloured tablets, known as microdots. Both of these carrier media present problems in terms of the homogeneity of dosage units: one unit may contain a very much larger or smaller dose than another. It is for this reason that the most frequently encountered dosage form is the blotter acid. Adsorbent paper is passed through a

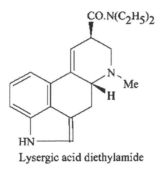

Lysergic acid diethylamide

Figure 13.8 Chemical structures of lysergic acid diethylamide (lysergide, LSD).

solution of LSD and then allowed to dry, leaving the drug impregnated in the paper. The papers are often decorated with emblems, patterns or symbols, that may pass over the whole sheet, or individual dose units. The dosing on LSD blotter acids is far more uniform, giving doses of 30–50 µg per dose unit.

13.3.8 Barbiturates and Benzodiazepines

In addition to synthetic drugs and drugs derived from plant products, some drugs are 'diverted' from licit sources. This is particularly true of the benzodiazepines and barbiturates. The principal dosage forms of both of these drug classes are tablets and capsules. However, a few of the drugs are prescribed as solutions for injection and the drug chemist ought to be aware of the possibility of encountering these drugs in this form.

13.3.9 Tryptamines

The tryptamines are a chemically diverse group of drugs based on the indole nucleus. The class includes LSD, psilocybin and psilocin, discussed above, and also the natural products ibogaine, bufotenin and harmaline in addition to the rarer dimethyltryptamine (DMT) and 5-methoxy-DMT. The method of ingestion of the drug depends upon the drug being administered. Doses and methods of administration are given in Table 13.3.

The drugs may be used as plant powders, extracts from the plants which are ingested or injectable preparations of the pure compounds. As a consequence, it may be necessary to identify the plant material, the compounds in the extracts and compounds on paraphernalia associated with use of drugs of this type.

Table 13.3 Common tryptamines with details of dose and administration route.

Drug	Dose (mg)	Administration route
5-methoxy-*N*,*N*-dimethyl-	2–3	Intravenous
tryptamine (5-MeO-DMT)	6–20	Oral
Bufotenine	8–16	Intravenous
Harmaline	150–300	Oral
Ibogaine	Up to gram amounts	Oral
N,*N*-Dimethyltrypatamine	4–30	Intravenous
(DMT)	60–100	Smoked
	>350	Oral

13.3.10 Piperazines

Of the drugs of abuse that are currently found on the illicit market in Europe, synthetic drugs represent the second most widely abused group after cannabis products. In order to avoid the legislation controlling these drugs, drug users are increasingly turning to another drug group— the piperazines (Figure 13.9). These are amongst the newest of the classes of drugs to be controlled across the globe and for this reason are detailed here.

There are a number of piperazines likely to be found in illicit drug samples, namely *N*-benzylpiperazine (BZP), 1-(3-trifluoromethylphenyl) piperazine (TFMPP), *ortho*-methoxyphenylpiperazine (oMeOPP), *para*-methoxyphenyl piperazine (pMeOPP), and 1-(4-methylphenyl)piperazine (pMePP).

N-Benzylpiperazine (BZP) [1] was synthesised as an antiparasitic drug by Wellcome Research Laboratories in the UK in 1944 but found not to be effective. There was further interest in BZP when scientists discovered it might have antidepressant activity. Benzylpiperazine is a synthetic drug and is easily synthesised from benzyl chloride and piperazine hexahydrate.

The related compound trifluoromethylphenylpiperazine (TFMPP) [2] is also found as a street drug. Introduction of a methoxy moiety is a common structural variation of designer drugs of the amphetamine type. The analogous piperazine derivatives are *ortho*- and *para*- methoxyphenylpiperazines (oMeOPP [3] and pMeOPP [4]). Methylphenylpiperazine [5] has also been encountered in seizures of drugs.

In general, piperazines are sold via the Internet and non-Internet-based herbal shops in tabletted dose formats. BZP is commonly sold by itself as either 'BZP' or 'A2'. The typical dose of BZP is 75–150 mg taken orally, that of TFMPP is 25–100 mg. The most popular forms of pills are those that contain a blend of both BZP and TFMPP but there are

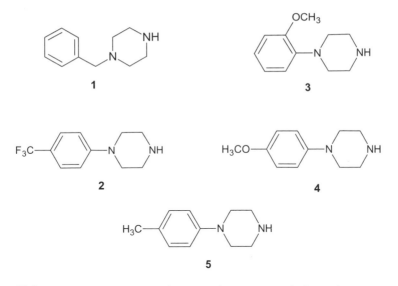

Figure 13.9 Chemical structures of commonly encountered piperazines.

'blends' that contain the other substituted phenylpiperazines. BZP was first reported to EMCDDA and Europol in 1999 through their early warning system on potential drugs. By March 2008, 15 EU member states and non-member Norway had reported seizures of BZP in tablet, capsule or powder form to either Europol or EMCDDA.

In terms of legislative control BZP and TFMPP are the most widely used piperazines and are currently the only drugs of this type placed under legislative control in various countries around the world at this time of writing. From March 2008 EU member states have been given a deadline of 1 year to take necessary measures to submit BZP to 'control measures proportionate to the risks of the substance' and 'criminal penalties' in line with their national laws. All states have now complied with the EU legislation although Belgium, Denmark, Estonia, France, Germany, Greece, Italy, Lithuania, Malta and Sweden already control BZP under drug legislation. Two Member States—the Netherlands and Spain—had also applied control measures to BZP under their medicines legislation.

13.4 IDENTIFICATION OF DRUGS OF ABUSE

There are a number of materials which may be encountered during the course of a drug analyst's career. These include drugs in bulk form (anything that can be seen), trace samples, wrapping materials and paraphernalia that may be associated with drugs of abuse. Whichever of these is encountered, it is usual to use presumptive tests, sometimes thin-layer chromatography (TLC) and then a chromatographic method linked to a spectrometer to identify and quantify drugs in a bulk sample. Trace samples are usually described and the analysis then proceeds directly to a chromatographic method which is linked to a spectroscopic technique. In addition, other methods have been developed particularly for drug comparison and profiling including the analysis of the inorganic components of drugs, the use of stable isotopes and the use of DNA.

Although by no means perfect for eliminating drug classes, the visual appearance of a sample may well indicate the class of drug that may be expected in the sample. When examining an item, every precaution should be taken to prevent contamination between samples. Trace samples should not be examined in the same room as bulk samples. Trace samples can be considered to be materials in, or on, which drugs are likely to be found but cannot be seen. The potential for contamination is enormous and great care should be taken to eliminate it. This impinges on consideration of lab design, throughput of samples, siting of equipment, the movement of personnel and even the design of the fabric of the building, *e.g.* the air conditioning system. All should be designed to minimise the risk of contamination.

In addition to drug samples themselves, there are a number of items of paraphernalia that may well be encountered on a regular basis by the drug chemist. These include glass surfaces (*e.g.* oven doors) on which drug samples may be cut or mixed; scales or balances on which drugs are weighed and measured; knives and cutting implements that have been used to divide and apportion the drug materials; and items that have been used to administer or take the drug. These include spoons, syringes and items for smoking cannabis or methamphetamine-based products. Further, drug debris may also be present. For example, there may be residues from heating of a heroin sample on a spoon or smoked material from cannabis.

Provided that there is sufficient sample, the analysis should proceed through a logical sequence of events leading to drug identification and if necessary, quantification. If only a tiny amount of sample is present, the non-specific non-instrumental methods are usually omitted and the analysis proceeds straight to the instrumental methods that provide definitive information concerning the identity of the drug. The analysis proceeds as shown in Figure 13.10.

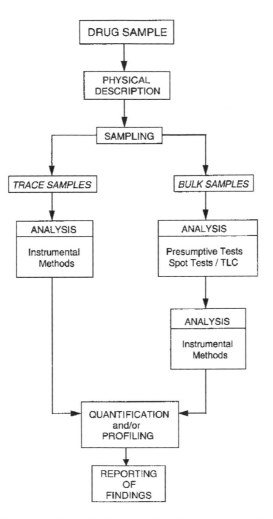

Figure 13.10 Scheme for the definitive identification and quantification of drugs of abuse.

13.4.1 Physical Descriptions and Sampling

Initially a full physical description of the item should be made, including a note of the integrity of the packaging. Physical comparison of the packaging can be useful in linking samples together, especially if one or more of the samples has changed chemically as happens for some drugs, *e.g.* cannabis products. In addition, with extruded polymers, *e.g.* cling film, it is possible to examine by polarized light the striations which are a result of the film manufacturing process. Under such conditions it is

possible to observe striae not visible under normal daylight and sometimes it is possible to match striation patterns.

Following this, decisions should be made as to how the item(s) should
be sampled. Each item that appears different should be considered to be
a member of a separate group. Where there are a large number of
packages in a group (*i.e.* many identical items) there are a number of
different models that can be adopted (Table 13.4).

The items to be analysed should be chosen at random from the group.
The easiest way to do this is to assign each member of the group a
number and then choose the subsample using a random number gen-
erator or random number tables.

Powders are analysed after thorough homogenisation, *e.g.* by placing
the sample in a polythene bag or Waring blender to simply mix it up, or
using the 'cone and quarter' method where the sample is mixed thoroughly and divided into four equal parts. The process is repeated for two
opposite quarters. This process is repeated until a sample suitable for
analysis is obtained. Where very large samples are to be analysed, *e.g.* a
dustbin liner full of drug, a 'core' sample should be taken. Small samples
are easily removed from liquid samples. Stains and marks can be sampled using a swab soaked in an appropriate solvent. Rinsing is often
used to remove drug samples from some items of paraphernalia, *e.g.*
syringe barrels. The solvent should fulfil a number of criteria, including
not reacting with the drug in question, not causing or catalysing the
breakdown of the drug, dissolving the drug quantitatively and being
compatible with subsequent analyses. For example, methanol is suitable
for most very polar drug classes, whereas ethanol is preferred for

Table 13.4 Numbers of items to be analysed from a group of materials
thought to contain drugs.

Number to be sampled	Advantages	Disadvantages
All ($n=N$)	Certainty about composition	Excessive sample sizes
$N=0.05\ N$, $0.1\ N$, *etc.*	Simple	Excessive sample sizes when N is large
$n=20+10\%\ (N{-}20)$ where $N>20$	Different populations likely to be discovered in large sample if present	Excessive sample sizes when N is large
$N<x$ then $n=N$ $x\leq N\leq y$ then $n=z$ $N>y$ then $n=\sqrt{N}$	UNDCP methods ($x=10$, $y=100$, $z=10$)	Excessive sample sizes when N is large
$n=3$ or 4 regardless of N	Mathematically proven to work	Only used by a few law enforcement agencies

surfaces thought to have been in contact with cannabis products, since these are most stable in ethanol. Chloroform is the best solvent for heroin and cocaine samples, which are easily hydrolysed in methanol because of the presence of water. Whichever sampling technique is employed for a single item, sufficient should be left for a re-examination by any other interested party. This is particularly important in any adversarial legal system. If leaving a sample for re-analysis is not possible, the value of the data should be very carefully considered before the analysis is undertaken.

13.4.2 Presumptive Tests

If the sample is a trace sample, such as a swabbing, then there may only be enough to perform instrumental tests. If there is sufficient sample, however, the analyst should proceed to the next stage, the presumptive tests. These are cost effective, easy, simple tests, which provide accurate and reproducible colour changes on the addition of certain reagents to certain classes of drug. The principal importance of these tests is that they provide the analyst with an idea of the class of the drug to which the seizure is thought to belong. Additionally, because of their simplicity, they are ideal for use by police officers and customs officers, who are not necessarily trained laboratory scientists, working at airports, seaports and other duty stations. It is normal to apply a battery of common tests to discriminate between drug classes and because there may be more than one drug class in the sample, as exemplified by the opiates, benzodiazepines and barbiturates that are all commonly found in heroin samples.

Presumptive tests have several disadvantages, including the fact that they do not definitively identify which member of a particular class of drug is present. They also react with substances that are not under control, thus providing what are somewhat inaccurately described as 'false positives'. This term should be avoided if at all possible: the reaction is just as positive, only with a compound that is different from the one for which the analyst may be testing.

13.4.3 Thin Layer Chromatography (TLC)

Following presumptive tests, the next stage in the analysis is the use of TLC. The stationary phase, mobile phase and developing reagent selected will have all been determined on the basis of the class of drug present, identified through the presumptive tests. Aliquots of the sample(s), together with both positive and negative controls, are spotted separately onto the plate which is then developed with the mobile phase. On completion of this stage the plate is removed from the developing tank and dried. Next it is examined visually under ultraviolet light (long and short wavelengths) and any compounds observed are marked. The plate is then sprayed with a developing reagent. The distances moved by the drug relative to the distance moved by the mobile phase (R_f values) and colour reactions of any compounds observed are recorded in the laboratory notes. TLC is relatively rapid, enables several samples to be run on one plate and separates many of the major components of different classes of drugs of abuse. Compounds are identified on the basis of two physico-chemical parameters: R_f value and colour reaction with a developing reagent.

The positive controls provide R_f and colour reaction data against which the sample can be compared, while the negative control demonstrates that the compounds observed are due to the sample applied and not to contaminants of the solvent. These controls are essential on every plate since the R_f values obtained will not be identical between operators or replicates of the same analysis, owing to a number of physico-chemical properties of the analytical system.

Because of the nature of the chromatographic system, two principal limitations are inherent to the technique. First, the separation between compounds (resolution) is limited, and secondly, even with these data generated there is no definitive proof of the identity of the drug. The power of the technique lies in the ability to determine which members of a particular class of drug are present, and on some occasions, which adulterants are present too, in a short time interval. Further, TLC finds particular application in drug comparison where the concentrations of the materials being compared are relatively high, *e.g.* in cannabis products and in heroin. The power of the method lies in the rapid way it can be used to differentiate samples which are clearly different (both for cannabis and heroin samples) and those in which decomposition products, *e.g.* cannabinol in cannabis, are present and hence further profiling would not be valid.

13.4.4 Instrumental Techniques

Following the identification of the members of the class of drug present, instrumental methods are employed to identify the drug. The techniques used fall into three principal categories: chromatographic, spectroscopic and hyphenated techniques.

13.4.4.1 Chromatographic Techniques. The techniques that find common application are high-performance liquid chromatography (HPLC) and gas chromatography (GC). Other techniques, such as supercritical fluid chromatography, are not used routinely and hence not considered any further in this chapter although on occasion they do have certain advantages. HPLC separates analytes on the basis of chromatographic mobility in a stream of liquid flowing through a matrix of materials on which the compounds are retained to different extents and hence are separated. The time a compound is retained on the column (retention time) is recorded. The choice of stationary and mobile phase is again dictated by the drug class in question, illustrating once again the central importance of the correct interpretation of the presumptive tests. GC separates the analytes in the gas phase, again providing a retention time. The principal means by which compounds are identified through these methods is the comparison of the retention time of the components of the sample with those of a standard compound, chromatographed on the same instrument under identical conditions.

HPLC requires that the compound exhibits good liquid chromatographic properties and is readily detected *e.g.* through the absorbance of ultraviolet light or by fluorescence. It represents an improvement over TLC because of its greater resolving power of components in a mixture. However, alone, it does not definitively prove the identity of the drug. GC has similar advantages and disadvantages. The latter technique also presents problems, in that some drugs, *e.g.* certain benzodiazepines, are thermally labile (decompose when heated), or others, which are polar, acidic or basic, require modification (derivatisation) of their chemical structure before analysis. This requires the analyst to ensure that each compound is fully derivatised before analysis.

13.4.1.2 Spectroscopic Techniques. The principal spectroscopic technique that is employed is infrared (IR) spectroscopy. The IR spectrum is obtained by measuring the interaction of IR radiation with a drug. The spectrum interpreted in four ways. One method is to compare the spectrum of the sample with that obtained for a standard of the pure drug. Another is method is by comparison of the fingerprint region of the spectrum (frequency range 450–1500 cm^{-1}) of the sample to the standard or to literature values. The fingerprint region is of particular value because it represents the low energy vibrations of the molecule due to absorbance of the IR radiation and is unique to any one analyte, and therefore provides definitive identification of the drug. Alternatively, the principal peaks (strongest absorptions) can be listed and the compound identified through examination of the literature values. In the final method the sample is compared with a library of data prepared in the laboratory, preferably by computer. This has a particular attraction because the spectra will have been prepared on the same instrument, eliminating the differences observed due to any inherent variation between instruments. This should also be borne in mind when comparing data to literature values—operator-dependent and instrument-dependent differences will always be observed. In addition to the identification of the active drug, other drugs and excipients in a sample may also be identified by IR spectroscopy.

An HPLC system using a diode array multiwavelength ultraviolet/visible detector not only provides the retention time of a compound, but also its ultraviolet, or ultraviolet/visible absorption spectrum as it elutes from the end of the chromatographic column. This provides another means of compound identification. The spectrum should always be compared to a sample analysed in the same batch of solvent since variation in solvent strength and pH may lead to variation in the observed ultraviolet spectrum. HPLC coupled to diode array detection provides stronger evidence for the identity of the compound—the degree of proof that this represents is arguable and the discussion as to whether or not this technique provides definitive proof of the identity of a drug is beyond the scope of this chapter.

13.4.4.3 Hyphenated Techniques. Techniques which can provide definitive identification of drugs of abuse include the so-called hyphenated techniques, *i.e.* gas chromatography–mass spectroscopy (GC–MS) and gas chromatography–Fourier transform infrared spectroscopy (GC–FTIR). The GC separates the analytes and the spectrometers provide unique signatures which can be used to identify a compound. The degree to which such techniques are successful depends on a number of factors. The more concentrated the sample, the better the signature for any given compound. The better the instrument's tuning, the more accurate the signature. Thus in these cases, it is the unique nature of the mass spectrum for GC–MS or IR spectrum for GC–FTIR, that provides data for identification of a compound. If the sample is very weak, or the detector is dirty, then interferences in the spectra will be observed and it will be difficult, if not impossible, to call a match between samples analysed under identical conditions to the standards, or between the samples and the literature values. Literature values should always be treated with caution, since the spectra obtained by both MS and IR are dependent on the conditions under which they were obtained. The instrument and operator in the laboratory will be different, in most cases, from those by which the literature values were obtained and this should always be borne in mind when data interpretation is undertaken.

13.5 QUANTIFICATION OF DRUGS OF ABUSE

Apart from forensic scientists wishing to identify drugs, diluents and other components in samples, a knowledge of the quality and quantity of the drug present is often required. Under the Misuse of Drugs Act 1971, there is also a requirement, for certain controlled drugs, to quantify the amount of drug in a sample, since the length of imprisonment or level of a fine is dependent on the quantity involved in a case. Furthermore, by determining the quantity of drugs and other components, comparisons between samples can be made which can aid drug profiling and intelligence studies.

To quantify the amount of drugs present in a sample, a number of techniques, including UV, HPLC, GC and GC–MS, may be employed. All are based on a measure of a change in a physico-chemical property measured by the detector in response to different amounts of analyte passing through the detector.

A few points are pertinent to drug quantification. Whether single-point, two-point or full linear regression methods are used, linearity of response must be established prior to use of the method as a quantitative technique. This is essential when the instrument is installed, but also whenever a component of the instrument is changed, and is vital if the data generated is to be accepted in court. Further, data should be generated from the lowest to the highest expected concentration, thus reducing the risk of column priming and carry-over between samples. The samples should always be separated by blanks, *i.e.* the solution or solvent in which the sample has been prepared. This best demonstrates the cleanliness of the instrument, equipment and operator. With the instrumental chromatographic systems described, the output from these instruments is a graph (chromatogram) of time versus detector responses. Hence, peaks are detected for components as they elute from a column, but for quantification it is important that there is baseline resolution between the peaks of interest and any others, and the response for the compound being quantified is due solely to that compound. If these general rules are observed, then valid quantification data will be obtained.

13.6   PROFILING OF DRUGS OF ABUSE

In addition to quantification of drugs of abuse, as indicated earlier it may be necessary to compare two or more samples of drugs to determine whether or not they came from the same parent batch. The term 'common origin' should be avoided since it is plainly obvious, for example, that all cannabis resin has a common origin: the plant *Cannabis sativa*. What is really meant is, 'Do the resin samples form part of the same batch?' The same question can be asked of other plant-derived drugs, such as heroin and cocaine, or synthetic drugs, such as the amphetamines. It is now recognised by the European Network of Forensic Science Institutes Working Group on Drugs that there are four levels of drug comparison which can be used to determine whether the drug samples came from more than one batch:

1. Drug identification.
2. Drug quantification.
3. Identification of the cutting agents.
4. Chemical impurity profiling.

Each of these will yield different information about the drugs and their relationships. In addition, it is possible to determine any relationship between wrapping materials—the drugs may be different but the wrapper might be linked.

Steps 1–3 are carried out using all of the principles described above. The methodologies and controls are identical and represent the first stage in comparison. However, the information that such analyses reveal about sample relationships is variable. If a drug is identified and found to have been mixed with a very uncommon cutting agent then a link may be inferred, but it may also be that the same batch of drug was cut with two different agents and a match may be missed. It is vital, therefore, to examine the chemical impurity profile, the signature of impurities themselves, in the drug samples if meaningful data is to be obtained. This is because unless the cutting agent is itself one of these impurities, the proportions of these in the sample will remain unchanged on cutting the drug.

The chemical profile may take many forms. It may be a profile of the drugs themselves as in cannabis products or heroin. It may be the profile of the chemical impurities, *e.g.* those examined when amphetamines are compared. A concentration step for the impurities is usually involved. Either solid phase extraction or liquid/liquid extraction steps can be used. However, the essential point is that they must be rapid, cheap, quantitative and extract all of the compounds relevant to the profiling with virtually 100% efficiency, without decomposition or production of artefactual impurities. Newer methods which have been suggested include the examination of inorganic ions found in drug samples, the analysis of stable isotopes in samples and the analysis of DNA in drug samples of plant or fungal origin. Having created the profile the analyst must then place an interpretation on the data. How the data is interpreted and reported depends upon the materials examined, the methods used, the question being answered and the statistics used to assist in interpretation. Examples of each are given below.

13.6.1 Profiling of Cannabis Products

Cannabis resin and oil (herbal material does not give meaningful chemical profiling results but is especially useful as a source of DNA) contain tetrahydrocannabinolic acids. These are thermally labile and hence GC is not a good means of drug profiling because the drugs decompose on heating in the injection block of the instrument. GC is used in some laboratories to quantify total Δ^9-tetrahydrocannabinol content, but HPLC is preferable since no decomposition occurs and both the acidic and neutral cannabinoids can be detected and quantified directly. It is also for these reasons that HPLC can be used or profiling these samples.

The profiling of cannabis samples is performed by extracting the herbal material, resin or oil with ethanol; an aliquot of the centrifuged solution is then analysed by HPLC. The chromatograms obtained are then either compared directly as shown in the example in Figure 13.11, where three cannabis samples have been analysed, or normalised (scaled with respect to a component in the mixture) and overlaid if computing facilities are available.

From these results the profiles for samples (a) and (b) cannot be discriminated, but are different from the profile for sample (c). Hence this example illustrates how profiling can be used to determine if materials are related to each other. Such data does, however, illustrate the difficulty in drug profiling when HPLC is used because of its lack of resolution when compared to GC.

13.6.2 Profiling of Heroin

In recent times, with advances in instrumentation, both HPLC and GC have been applied to heroin profiling. The sample requires a simple re-treatment to extract the opiate drugs noscapine and papaverine, which are then subjected to either HPLC or GC analysis. If GC is used, it is common to derivatise the sample prior to analysis to eliminate the

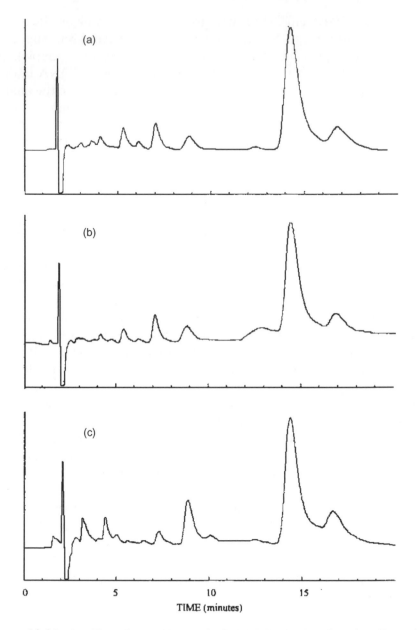

Figure 13.11 Profiling of cannabis samples by HPLC using UV detection. From these profiles it impossible to discriminate between samples (a) and (b) but these are different from sample (c).

problem of artefact formation. For example, it is known that transacetylation of diamorphine (the 3,6-diacetylated product of morphine) and morphine can occur if both are present in an injection block of a GC and certain solvents are used at the point of injection, with the production of both 3- and 6-monoacetylmorphine. Typical profiles obtained from two related heroin samples are shown in Figure 13.12.

A great deal of work has been carried out on data processing for heroin comparisons. This ranges from comparison of paper copies of chromatograms and examination of the ratios of different compounds, through to cluster analysis and the use of euclidean distances for retrospective comparisons of large datasets. Such methods have been used in, for example, determination of country of origin of heroin samples and also to examine the relationships of samples at the street level. However, there are still problems in terms of data exchange, comparison and the setting up of large databases for heroin comparison. Problems

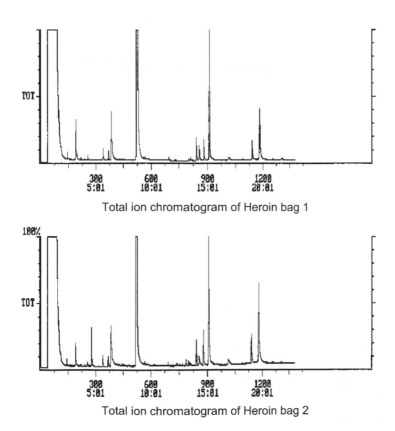

Figure 13.12 Profiles of two related heroin samples following derivatisation with *N,O*-bistrimethylsilyl acetamide and analysis by GC–MS.

with the methods include intra- and inter-laboratory variation and obtaining the same result from identical samples. It has been argued that, for heroin comparisons at least, there is some argument for using a centralised laboratory.

13.6.3 Profiling of Amphetamine

Although there have been various studies of how to carry out amphetamine comparisons there are a number of problems associated with these methods because they have been formulated around amphetamine synthesised by the Leuckart route. A number of different synthetic routes are now used (including variations on the Leuckart, reductive amination and nitrostyrene routes). The route of preference depends on where the amphetamine is being manufactured—as mentioned earlier, the Leuckart route is popular in north-east Europe, the reductive amination of BMK in parts of Scandinavia and the nitrostyrene route in parts of the Far East. The situation is further complicated by the fact that the impurities occur at extremely low concentrations, can potentially decompose in the solvent in which they are dissolved following extraction, require a complex extraction and recovery step, are relatively polar and reactive in nature; and profiles often contain a very large number of compounds including stereo and regioisomers. However, following extensive collaborative work it is now possible to reliably store amphetamine, extract the impurities, analyse them using GC–MS using a collaboratively tested and harmonized method and compare the data to that obtained in other laboratories. This work highlights the considerable amount of work that is required before such a position can be reached and it is only for very few drugs that such detail is known about the profiling process.

13.6.4 Drug Profiling Using Metal Ion Content

In addition to examination of the drug impurity profile obtained from the organic content of the drug, it is also possible to examine the inorganic content. A number of studies have shown that, particularly where there is chemical modification of the drug, it is possible to use metal ion content to determine relationships between samples. The metals themselves can occur in the drugs for a number of reasons. They may be there because of the container in which the chemical reaction is being carried out, *e.g.* aluminium from a saucepan. They may be present as a consequence of the catalyst in the reaction being used, particularly when a reduction is being carried out, *e.g.* aluminium, platinum or tin. Sodium, potassium and silicon may be present as components of the additives added to the street sample after manufacture, *e.g.* if an additive is a sodium or potassium salt then the levels of these metals may be changed.

A number of techniques are available to provide quantitative data for the analysis of metals in drug samples. These include flame atomic absorption spectroscopy (FAAS), electrothermal atomic absorption spectrometry (ET–AAS) and inductively coupled plasma mass spectrometry (ICP–MS). The latter has the capability of multielement analysis but is expensive to run and maintain.

There are a number of considerations to take into account before using such analysis methods. Appropriate blanks should be employed to ensure that the metals chosen for discrimination come from the drug samples and are not leached from the vessels used to prepare the samples or from the reagents employed to prepare samples. Appropriate techniques should be used for metal dissolution. Studies have shown that dissolving samples in nitric acid and microwaving may be more appropriate than vortexing or sonicating the same extract. The analytical method should be optimised for the metals under consideration and appropriate linearity of detector response demonstrated prior to any sample analysis. Under such circumstances the data so generated can be used in complex algorithms used to determine sample relationships.

The use of metal ions has a number of advantages over the traditional analysis of organic impurities. These latter may be volatile and may be subject to degradation processes. The consequence of this is that the organic content of the samples may not be stable. With metal ions, unless there is an addition to or deletion from the sample the metal content will remain the same. It is worth considering the way in which such data is reported. Good practice would suggest that the best way to report this data is for the metal content to be reported for each metal in its uncharged state. This is because the methods used determine the metal content in terms of concentration but do not, in general, provide information on the valency of the ions present. This too may be a function of storage conditions of the drugs.

13.6.5 Drug Profiling Using Stable Isotopes

In addition to examining the organic chemical impurities of the drug sample it is also possible to examine the composition of the drugs at the atomic level. Of particular significance are the stable isotopes of carbon and nitrogen although it is possible to use hydrogen, oxygen and sulfur in such work. Work has been undertaken on the profiling of cannabis products, heroin, cocaine, MDMA and amphetamine. The basic premise is that the isotopic ratios of carbon and nitrogen reflect either the area in which the drugs were grown and/or the chemicals which were used to prepare the drugs.

By way of example, considerable work has been undertaken on cannabis. It is possible to relate herbal material to the geographic area in which is was grown on the basis of the $^{12}C/^{13}C$ and $^{14}N/^{15}N$ ratios. It is known that the different ratios of the isotopes are particularly influenced by water availability and in studies where cannabis of known origins has come from areas of different rainfall it has been possible to discriminate between the samples. This is easily observed in a plot of the difference in the nitrogen ratios against the difference in carbon ratios. Groups of drugs from different areas can easily be differentiated using principal component analysis. This work has further been developed to the extent that once a number of samples have been analysed linear discriminant analysis can be used to make predictions about where a sample may have come from and under experimental conditions it has been found that such predictions can be accurate.

The use of stable isotopes is particularly advantageous because the drugs maintain, in general, their original elemental and isotopic profiles. This is in contrast to organic composition or occluded solvent content which can vary as time progresses as the drugs decompose or dry out.

13.6.6 Plant Drug Identification and Profiling Using DNA

On occasion it is necessary to identify drug materials from plant material but the necessary microscopic characteristics have been lost because of physical damage to the sample or the chemical characteristics required have been lost, *e.g.* because of the age of the sample or poor storage conditions. On such occasions the analyst may be able to use DNA analysis as a means of identifying a drug sample, or indeed making comparisons between drug samples. This is particular relevance and importance where, for example, genetically identical plants have been found to exhibit phenotypic plasticity and hence different chemical profiles because they have been grown under different environmental conditions. One such drug where this occurs is cannabis.

Early work on DNA analysis focused on the use of random amplification of polymorphic DNA (RAPD) using non-specific primers. However, this has the disadvantage of problems with reproducibility. It is now possible to use RAPD with species-specific primers. This can be used to identify the species of plant—in drug chemistry this is generally cannabis—but cannot be used to relate two samples to each other. It is also possible to use the analysis of chloroplast DNA. Using optimised primers it is possible to amplify conserved DNA and then digest with a series of restriction endonucleases. Each of these will produce a series of DNA fragments and it is possible, using the correct suite of enzymes, to produce fragments that indicate that the sample is indeed cannabis. Again, profiling is not appropriate. However, using STR analysis it is possible to generate profiles, specific to cannabis, which can be used to profile and compare samples. Indeed, such approaches have already been used in casework for some time.

It is also possible to use DNA technologies to identify fungal materials, which are notoriously difficult to identify on the basis of morphology alone. Studies have also extended to the study of drug materials used in traditional and herbal remedies which have been used to perpetrate crime, including murder.

13.6.7 Numerical Methods

Detailed discussion of the methodologies used for comparison of drug samples using statistical methods is beyond the scope of this text. However, a few general principles should be outlined.

Before employing any numerical method it is essential that the correct methodologies have been followed. Additionally, for instrumental methods, it is important to determine that that the instrument is functioning correctly and that a linear detector response is being obtained for the concentration range of the analyte in question. Appropriate blanks and check standards should be analysed between samples.

Once the numerical data is obtained, a number of different methods can be used to compare samples. How closely samples are related can be determined using euclidean distances and illustrated using dendrograms. However, one assumption, often ignored, is that the analytes considered should be independent of each other in terms of their concentrations. This is true for some, but not for others, *e.g.* the concentrations of some opiates in heroin samples are not mutually independent. Under such circumstances the use of these techniques is not appropriate.

The quotient method, Canberra distances, principal component analysis (PCA), hierarchical cluster analysis (HCA), and Fisher's linear discriminant analysis are among the methods that have been used for the comparison of samples and the explanation of relatedness. However, at the end of the analysis the only truly reliable method for determination of whether two samples are related or not is for the analyst to compare the chromatograms if chromatographic methods have been used, or some method of directly comparing numerical data for multivariate analysis of components of the sample.

13.7   QUALITY ASSURANCE IN DRUG ANALYSIS

In addition to simply analysing the drugs of abuse, it is important that quality assurance is considered. Every necessary control should be taken, and these have been considered above. Further, the instruments should be demonstrably functional, both qualitatively and quantitatively. This requires that standard test samples providing defined responses are analysed at the start and end of each analysis, and if the analysis is long, within the analytical sequence too.

Computerised data handling should be backed up by analogue output. In the UK this is an important requirement. This is because there is a myth among the lay population (and unfortunately among some scientists too) that if the answer comes out of a computer, it must be right. With numerical data handling and quantitative analyses, this is especially important.

Recent developments include the introduction of internationally accepted quality assurance practices, *e.g.* ILAC 17025. These require all the above-mentioned precautions to be taken and, in addition, good laboratory note-taking. Everything that is done, right down to changing a pair of protective gloves, should be recorded in writing. Glove changes are necessary to prevent contamination between samples and it is unlikely that in a court room 6 or 12 months after the analysis was performed, you will remember whether or not you changed gloves between samples. If all of the above points are considered, the drug analysis will be valid and successful.

BIBLIOGRAPHY

M. Cole, *The Analysis of Controlled Substance—A Systematic Approach*, John Wiley & Sons, Chichester, 2003.

M. D. Cole and B. Caddy, *The Analysis of Drugs of Abuse—An Instruction Manual*, Ellis Horwood, New York, 1995.

T. A. Gough, *The Analysis of Drugs of Abuse*, John Wiley & Sons, Chichester, 1991.

Guidelines on representative drug sampling, *European Network of Forensic Science Institutes Drugs Working Group*, 2002. www.ENFSI.org.

L. A. King, *The Misuse of Drugs Act—A Guide for Forensic Scientists*, Royal Society of Chemistry, Cambridge, 2003.

L. A. King, *Forensic Chemistry of Substance Misuse: A Guide to Drug Control*, Royal Society of Chemistry, Cambridge, 2009.

E. Lock, L. Aalberg, K. Andersson, J. Dahlén, M. D. Cole, Y. Finnon, H. Huizer, K. Jalava, E. Kaa, A. Lopes, A. Poortman and

E. Sippola, Development of a harmonised method for the profiling of amphetamine V. Determination of the variability of the optimised method, *Forens. Sci. Int.*, 2007, **169**, 77–85 (and references contained therein).

Rapid Testing Methods of Drugs of Abuse, United Nations Drug Control Programme Handbook ST/NAR113, United Nations, Vienna, 1988.

CHAPTER 14

Forensic Toxicology

ROBERT ANDERSON AND HAZEL TORRANCE

Forensic Medicine and Science Division of Cancer Sciences and Molecular Pathology, Faculty of Medicine, University of Glasgow, Glasgow G12 8QQ

14.1 INTRODUCTION

14.1.1 What is Toxicology?

Toxicology is the study of poisons, including their origins and properties, and their effects on living organisms. The meaning of the term poison is dealt with later. A toxicologist is primarily concerned with the chemical analysis of specimens such as blood and urine for the presence of poisons and subsequently with interpreting the significance of the results obtained. In some countries outside the United Kingdom, the toxicologist is also directly involved in treatment of the poisoned patient. The remit of the toxicologist requires expertise not just in analytical chemistry but also in pharmacology and aspects of biology and medicine.

Appending the word 'forensic' to the name 'toxicology' simply means the use of toxicological investigations in court. Any type of toxicological work, in its widest sense, may appear in criminal or civil court proceedings and a list of examples is given below. However, because most court cases currently have been concerned with the effects of poisons on humans, 'forensic toxicology' has traditionally been considered to be the study of the effects of drugs and poisons on human beings and the

Crime Scene to Court: The Essentials of Forensic Science, 3rd Edition
Edited by P. C. White
© Royal Society of Chemistry 2010
Published by the Royal Society of Chemistry, www.rsc.org

investigation of fatal intoxications for the purpose of some sort of medico-legal enquiry, the exact nature of which depends on the country involved. Typically, such deaths are caused by suicidal or accidental overdose of drugs. Alternatively, the death may have occurred under the influence of a drug or poison but was in fact a consequence of another event, such as a car accident or a fall from a height. Very occasionally homicidal poisonings occur involving drugs, such as insulin, but also other substances such as carbon monoxide gas, cyanide or pesticides. In these investigations, the toxicologist works in close collaboration with the forensic pathologist, who carries out the medico-legal autopsy and furnishes samples of blood, urine and other tissues for analysis in the laboratory. The toxicologist is rarely required to attend the locus of a death or even the autopsy room, and specimens for analysis are normally delivered to the laboratory by an appropriate method, to preserve their integrity (preventing any tampering) and legal validity (preserving a chain of custody). In some jurisdictions, suppliers or traffickers of drugs which are subsequently implicated in a drug-related death are prosecuted on a charge of culpable homicide, and it is under these circumstances that the toxicologist is most commonly involved in homicide cases.

At present, however, post-mortem forensic toxicology accounts for only a minor percentage of the field as a whole and most work involves living cases, such as performance testing of employees (workplace drug testing, when either legislation in force or else the employment contract prohibits employees from using or working under the influence of a proscribed panel of substances) or drivers of vehicles allegedly driving under the influence of alcohol and or drugs. The forensic toxicologist is also employed in other types of living case, notably the clients of drug abuse clinics and inmates of prisons, who are regularly tested to determine whether they are using controlled drugs, complainers in cases of alleged drug facilitated assault, including sexual assault ('date rape'), child abuse cases, in which a parent or guardian is accused of administering drugs to the child in their care, and child custody cases, in which a parent or guardian must demonstrate abstinence from drug use before the child is allowed back into their care.

Other types of toxicological investigation that may appear in court proceedings are:

1. **Clinical toxicology**. The investigation of poisoning in a supervised medical setting, usually a hospital.
2. **Environmental toxicology**. The investigation of the harmful effects of poisonous substances in the environment.

3. **Regulatory/animal toxicology**. The investigation of the toxic
   properties of substances to fulfil the legal requirements for the
   licensing of commercial products.

The remainder of this chapter deals primarily with post-mortem
forensic toxicology but includes sections on cases involving living
subjects.

14.1.2   Forensic Toxicology in the United Kingdom

The systematic toxicological investigation of human organs for the
presence of poisons is widely considered to have been initiated by
Mathieu Orfila, a physician working in Paris whose treatise on toxicology was the first substantial work on the subject. Several students of
Orfila returned to the United Kingdom in the 19th century and established the subject here in their home universities within the wider discipline of forensic medicine.

Toxicology services have therefore been provided traditionally by
university forensic medicine departments, and this is still true in some
cases today. However, other organisations now largely provide these
services including the Forensic Science Service in England and Wales,
the Scottish Police Services Authority Forensic Services in Scotland and
Forensic Science Northern Ireland in Belfast. In other countries,
laboratories are operated by law enforcement agencies, by government
or by the medical services. Increasingly, commercial organisations have
developed to provide scientific support including forensic toxicology for
forensic investigations and this is almost universally true for high-
throughput performance testing toxicology. Defence counsel has free
access to these commercial suppliers, unlike the state-owned laboratories, either because these work only for the law enforcement agencies,
or else, having been engaged on a case by the prosecution, they are
unable to work simultaneously for the defence. Many laboratories are
accredited by their national accreditation body (UKAS in the United
Kingdom) for some or all of their scientific procedures, such as the
analysis of blood samples, to the relevant quality standard ISO/IEC
17025:2005.

Qualifications accepted by the courts for forensic toxicologists are
degrees in medicine or science in an appropriate subject (chemistry,
biochemistry or pharmacology) plus experience in a routine forensic
toxicology laboratory. In common with other fields of forensic science,
most toxicologists are trained in the laboratory in which they are
employed. In the United Kingdom, the Home Office Forensic Science

Regulator is considering options for certification of forensic toxicologists but in other countries, notably the USA, systems are already in place for certification of individual scientists.

14.2 POISONS AND POISONING

14.2.1 Definition

A poison is 'any substance which, taken into or formed in the body, destroys life or impairs health'. This definition does not restrict poisons to particular classes of chemical. Almost all substances are capable of causing harm if administered at a high enough dose, including such essentials for life as salt, water and even oxygen. As examples, consider the use of salt and water as an emetic to induce vomiting when a poison has been swallowed. If care is not taken to restrict the quantity of salt to a spoonful or two, a dangerously high amount of salt can be ingested and this can have its own toxicological consequences. Also, at least one teenager, recognising the dangers of dehydration when using ecstasy at a dance club, has almost died by drinking too much water (about 6 L) too quickly. A number of factors are involved in determining the dose at which toxic effects are produced and these are considered below.

14.2.2 Factors Affecting the Toxic Dose of a Substance

Most substances can act as poisons at a high enough dose and the crucial question of whether or not a substance is a poison relates to the dose administered rather than to the substance involved. Figure 14.1 may be useful in visualising the relationship between the size of the dose administered and the effects on the subject.

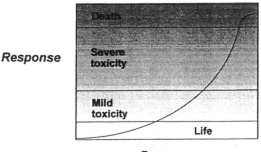

Figure 14.1 Relationship between the dose of a substance administered and the effects or response on the organism.

The graph indicates that low doses of a substance are consistent with life; *e.g.* if a drug is administered at the normal dose (the therapeutic dose), then life will continue and the organism or person may benefit from the substance. As the dose increases, however, mild toxicity may occur, shown by the appearance of adverse or side effects. At this stage, the beneficial effects of a drug may be outweighed by the appearance of these undesirable effects. As the dose continues to rise, the adverse effects become more severe and may start to become life-threatening. Severe toxicity occurs at too high a dose and is seen, *e.g.*, as depression of respiration: the breathing becomes more shallow and the rate slows down until it finally ceases altogether. At this point the threshold for fatal poisoning is passed. Note that there is no indication of timescale in the diagram. The dose may be administered at one time or in small amounts over a period of time, and similarly the response may appear in the short term (acute poisoning) or long term (chronic poisoning).

It should also be noted that, for some substances, the lower end of the dose curve starts to rise again as the dose approaches zero, giving a U-shaped curve. Such substances are often trace elements and vitamins which must be present in the diet. Both too much and too little is harmful: *e.g.* copper, iron and zinc are essential trace metals. If the dietary intake is too low, the term used is deficiency rather than toxicity. In the context of forensic toxicology, an analogous situation can arise in which a patient may take too low a dose of medication and suffer as a result. Common examples are underdosing on insulin (resulting in a diabetic coma), antiepileptic drugs (resulting in an epileptic fit, which can be fatal) or antipsychotic drugs (resulting in a psychotic episode in which the individual can be harmed or can hurt those in the vicinity).

What has been described qualitatively above is a dose–response curve and in an ideal situation the curve is predictable and the thresholds for each transition stage are clear-cut. This does not happen in the real world, and hence problems of interpretation arise, and these are the province of the toxicologist. Some of the factors that upset the simple dose–response curve are examined below. It is important to note here, as well as later, that there is no specific 'fatal dose' or 'fatal threshold' for a drug or other substance.

14.2.2.1 Carcinogenic and Mutagenic Substances. These include, *e.g.* asbestos and ionising radiation, which cause cancer or mutations, respectively. There is no simple dose–response relationship for them because there is no safe exposure level. Any exposure can cause a catastrophic injury to the living organism and, once initiated, it is essentially irreversible. In the case of ionising radiation this can cause damage to genetic material in living cells, *i.e.* a gene mutation—an alternation to the normal DNA sequence. This mutation, however, may not be apparent in the generation exposed to the radiation, but several generations later.

14.2.2.2 Age and Size. The doses of a substance required to cause mild toxicity, severe toxicity and death vary with the age and size of an individual; *e.g.*, children are often given a lower dose of a drug than adults because their body weight is less. Doses are normally calculated as quantity of substance per unit of body weight, *e.g.* milligrams per kilogram, to compensate for different body sizes. Children also differ from adults in the metabolism and disposition of substances which they ingest. The adult liver, for example, may behave differently from that of a child because it has already encountered substances during its lifetime. As individuals reach old age, their ability to cope with exogenous (foreign) substances is diminished and the threshold for toxicity may be expected to become lower.

14.2.2.3 State of Health. Ill-health can also affect the ability of the body to cope with a toxic insult due to a poison. This is obvious if the liver, the main metabolic organ of the body, is diseased, but many other organs can be involved, including the kidneys which excrete poisonous substances into the urine, and the heart, which may fail at a lower dose of substance than normal if there is already significant heart disease.

14.2.2.4 History of Exposure. The response to a particular dose of a substance is often affected by the previous exposure of the individual to the substance and can take two forms. The most common and better-known effect is the development of tolerance. This means that the individual becomes used to the effects of the drug such that a higher dose or more frequent administration of the same dose is necessary to obtain the original effects. For example, regular heroin users who inject heroin several times a day develop tolerance to the effects of the drug over a period of several weeks and may end up using doses that would have caused death if they had taken them at the beginning of their habit. Equally important in the forensic context is the loss of tolerance. Again considering the case of a heroin abuser who has acquired tolerance to the drug over a period of weeks or months, this tolerance will be lost fairly rapidly during a period of abstinence from the drug, *e.g.* during a custodial prison sentence. If this individual returns to the same heroin habit after release from prison there is a severe risk of accidental heroin overdose and death.

The second way in which previous exposure to a substance can affect the dose–response relationship is through the development of sensitisation. In effect, this means the development of an especially sensitive, idiosyncratic response at doses that may be lower than those normally required to elicit a toxic effect. One example of this type is volatile substance abuse (sometimes called 'glue-sniffing' or 'solvent abuse'). Regular sniffing of butane (the gas used in cigarette lighters) or aerosol propellants containing halogenated hydrocarbons (ozone-damaging propellants such as freons) can cause sensitisation of the heart such that further exposure can cause cardiac arrythmia (irregular heart beat) and death.

14.2.2.5 Paradoxical Reactions. For some substances there exists the possibility of a paradoxical reaction by a small number of individuals. The dose–response curve described above does not apply to these individuals as the effects produced are quite different from the normal ones experienced by the majority of the population. An interesting example is a paradoxical excitement caused by some of the benzodiazepine tranquillisers. These drugs normally cause sedation, relaxation and loss of anxiety, but in a very small number of cases the drugs can cause excitation and aggression. Although this effect is very rare, its occurrence is unpredictable and it is sometimes invoked by the defence to explain bizarre or uncharacteristic behaviour in a client.

14.2.2.6 Genetics. Each individual has a unique set of genes inherited from their parents which provide the genetic codes for enzymes and other proteins used in the body, including those involved in the metabolism/biotransformation of drugs and in the response to drugs. The most important group of enzymes in drug metabolism is the family of cytochrome P450 enzymes, and for each member of the family a number of variations exist. Individuals may have different cytochrome P450 profiles depending on their genetic makeup and these differences lead to some individuals being 'fast' metabolisers or 'slow' metabolisers depending on their cytochrome profiles. The field dealing with the interface between pharmacology and genetics is known as pharmacogenetics and is of significant interest to the pharmaceutical industry because of the possibility of personalizing drug treatment regimes. The corresponding field in toxicology is toxicogenetics, and within forensic toxicology the intention will be to use the genetic profile in the interpretation of toxicology results.

14.2.3 Types and Examples of Poisons

Poisons may be grouped into different classes depending on their chemical composition and properties, and their effects on the human body. Physical classification of poisons might distinguish solids, liquids and gases whereas chemical classification would create acidic, basic and neutral groups of substances. Functional classification creates groups of substances such as drugs, pesticides, metals, *etc.* and pharmacological classification of drugs would subdivide them according to their effects, for example stimulants, analgesics, sedatives, *etc.* An early type of classification was made according to the mode of action.

14.2.3.1 Corrosive Poisons. This group includes acids, alkalis and other substances that cause physical corrosion of body tissues through direct contact. Corrosive poisons such as vitriol (sulphuric acid) and Lysol (a mixture of substances similar to carbolic acid) were more commonly encountered in the early part of the 20th century, when there were fewer drugs to hand. This type of substance is usually taken orally and causes severe and very visible damage to the soft tissues of the mouth, oesophagus, stomach and intestine.

14.2.3.2 Irritant Poisons. These include inorganic and metallic poisons such as lead, mercury and arsenic compounds. They are most often ingested orally and cause severe gastrointestinal irritation and upset, with nausea and vomiting, colic and diarrhoea or constipation depending on the substance involved. They are absorbed from the digestive system and are transported in the blood circulation to the organs of the body. These poisons are compounds of chemical elements which cannot be broken down or metabolised by the body to inactive substances: the body can only excrete them or deposit them where they will do least damage, *e.g.* in the bones of the skeleton. The passage of metals such as lead through the kidneys into the urine causes severe tissue irritation, and kidney damage is a common result.

14.2.3.3 Systemic Poisons. This is the largest group of poisons and it includes those substances that act directly on biochemical processes within cells. The name refers to the blood circulation system, since these poisons must reach their site of action *via* the blood circulation after they are taken into the body by any one or more of a variety of routes, which are considered in more detail below. Within this group are the well-known alkaloid poisons such as strychnine; inorganic poisons such as cyanide and carbon monoxide; pesticides; and drugs. The remainder of this chapter will be concerned primarily with poisons of this type.

14.2.3.4   Toxins. Naturally occurring substances produced in living organisms are usually referred to as toxins rather than poisons. The best-known examples are venoms from poisonous snakes and insects, but the group also includes substances produced by bacteria such as salmonella and cholera and algae such as the blue-green algae which causes problems both in inland water reservoirs and in the sea in the neighbourhood of shellfish beds.

14.2.4   Routes of Administration and Excretion

Administration and excretion are topics of particular importance in the interpretation of analytical toxicology results and are frequently the basis of questions asked in court. Drugs and other substances can be taken into the body by a number of different routes, which affect:

1. The speed with which they act on the body.
2. The fraction of the dose which is absorbed into the bloodstream (the bioavailability) and is able to reach the site of action of the substance.
3. The ultimate concentration of the substance that is achieved in a particular body tissue.
4. The effects experienced by the person receiving the substance.

Similarly, substances are cleared from the body by different routes, which affect:

1. The length of time over which the substance remains active in the body.
2. The methods by which the body inactivates or detoxifies biologically-active materials.
3. The selection of samples for analysis.
4. The length of time over which the substance can be detected.

A generalised diagram illustrating the absorption and excretion of a substance is shown in Figure 14.2. This relates the level of the substance in blood to the time after administration. For the most part, substances must enter the blood circulation before they can exert an effect on the body; and often, but not always, the time when the strongest effects are observed can be related to the maximum blood concentration. The units of concentration currently favoured are *milligrams per litre* (mg/L) of blood (or urine) and *milligrams per kilogram* (mg/kg) of tissue (such as liver). Alternative units are used if the former produce

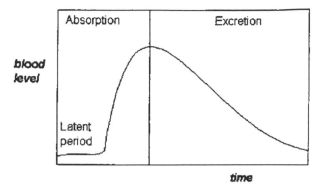

Figure 14.2 Absorption and excretion profile for a substance entering and leaving the human body.

very small numbers, *e.g. nanograms per millilitre* (ng/mL), which is numerically equivalent to *micrograms per litre* (µg/L), is 1000 times smaller than mg/L and is typically used for drugs found in blood at low concentrations, *e.g.* tetrahydrocannabinol, the active constituent of cannabis.

14.2.4.1 Routes of Administration

Oral. Oral administration implies entry to the body through the mouth and digestive tract, whether taken deliberately, accidentally or unknowingly as a result of administration by someone else. To the toxicologist, entry to the body begins not when the substance is swallowed and reaches the stomach or digestive tract, but when it is absorbed from these organs into the bloodstream. In this respect the digestive tract from the mouth to the anus can be considered as a tunnel passing through the body: the space within it is not an integral part of the body itself. Some substances are not well absorbed when taken orally, including liquid mercury and some other toxic metals and their compounds. These mostly pass through the digestive system without entering the body tissues.

The most important questions usually raised in court concern, 'How long did it take for the substance to be absorbed into the body?' and, 'How much of it was actually absorbed?'

The first question depends on the substance involved: as mentioned above some substances are not absorbed at all. However, most drugs are formulated to be taken orally and are usually absorbed within a period of 30 min to 2 h. This also applies to many other poisons when taken orally. The pharmacologist can alter the rate of absorption by changing the physical form (formulation) of the substance. Slow-release capsules contain small granules with different rates of dissolution, although it is interesting to note that drug abusers can increase the rate of absorption of slow-release drugs, *e.g.* oxycodone, by pulverising the granules before administration. Information on rate of absorption can be obtained from the medical literature for the drug in question, contained in pharmacopoeias and, *e.g.*, the Association of the British Pharmaceutical Industry (ABPI) Data Sheet Compendium, also available online.

A related question is, 'Where does the absorption take place?' Most substances are absorbed from the intestine rather than from the earlier part of the digestive system (mouth, oesophagus and stomach) or the end section (large intestine and rectum). There are exceptions, *e.g.* some substances are absorbed through the mucous membranes of the mouth. This is usually considered to be a different route of administration (buccal) and is similar to the intranasal route considered below. Similarly, drugs are absorbed through the rectum but this is also usually considered as a separate route when the substance is introduced as a suppository.

The second question, 'How much of it was actually absorbed?', refers to the pharmacological term *bioavailability*, which is important in determining the dose required for a desired effect. Bioavailability simply gives the percentage of a dose that reaches the blood circulation and is thereby available to exert a biological effect. If this is small, the dose must be increased proportionately. There are several factors that affect the bioavailability, two of which will be mentioned here.

- The physical and chemical form is important. Thus cannabis resin is poorly absorbed from the digestive tract because it is an oily resin, and some salts of lead are poorly absorbed because they do not dissolve very well in water.
- The effect of the liver must be taken into account. When drugs are absorbed from the intestine, they enter the bloodstream through a network of blood vessels that surrounds the gut. The blood flows into the hepatic portal vein and thence to the liver, where the vein subdivides again into smaller blood vessels that pass through the liver (the portal blood vessels). From the liver, the blood is collected once more into the hepatic veins, which return blood to the heart *via* the inferior vena cava. In this way food absorbed from the gut is taken directly to the liver to be metabolised for energy. However, medicinal drugs are also carried in the same direction and are immediately confronted by the metabolic capabilities of the liver. This is called the first-pass metabolism and explains why many

drugs are rapidly inactivated after they are swallowed and an alternative route of administration is used. Heroin is one example: if it is taken orally, there is almost complete first-pass metabolism and the products formed, morphine and other metabolites (breakdown products) do not give the same euphoric effect as when heroin is administered by injection.

Intravenous. The intravenous route became available in the 19th century, when the hypodermic syringe as we know it was used to administer drugs. This device was almost immediately recognised as an efficient means of introducing abusable drugs such as cocaine and heroin directly into the bloodstream. The bioavailability of substances administered intravenously is almost 100% (some is left in the syringe) and, more importantly, the time taken to reach the blood circulation is essentially zero. This route of administration can give rise to a very rapid rise in the blood concentration of a drug such as heroin or cocaine, causing potentially lethal side effects on the body' physiological systems, notably on the heart rate and blood pressure. Drugs administered intravenously under medical care are usually diluted in a transfusion of glucose or saline (salt) solution or are pumped in slowly using a syringe pump.

Inhalation. Drugs and other substances can readily enter the blood circulation following absorption through the lungs, which are richly supplied with blood vessels. Common instances of this type of administration are the use of an inhalant (such as an antiasthmatic device) and smoking (tobacco and other drugs). Smoke particles and gases enter the lungs and their constituents pass through the thin walls of the air sacs (alveoli) and the surrounding capillary blood vessels. Substances are absorbed rapidly by this route and effects are seen within seconds or minutes of administration. The intense 'high' obtained after smoking crack cocaine contributes greatly to its addictive properties. The bioavailability of inhaled drugs is about 40%.

Through the mucous membranes. The moist surfaces of mucous membranes are found in the nose, mouth, eyes, ears, vagina, rectum and lungs, amongst others, and all of these can and have been used to administer drugs. Perhaps the best known instances are the use of eye and ear drops, nitroglycerine tablets taken sublingually (*i.e.* under the tongue), chewing tobacco and snuff, and also the snorting (a more formal name is nasal insufflation) of drugs such as cocaine and amphetamine. The bioavailability of drugs by snorting is similar to smoking, about 40%.

14.2.4.2   Routes of Excretion.   The excretion phase shown in Figure 14.2 consists of two main components, at least for many substances of interest here. The first consists of the metabolism or breakdown of substances such as drugs and the second involves voiding of the waste products from the body.

The role of the liver in metabolism has already been described. Most other tissues, including the blood itself, also have metabolic capabilities, which are designed to break down food and nutrients to provide energy and sustenance for the body. However, the biochemical reactions involved are not specific to foodstuffs but also act on most exogenous (foreign) substances. The metabolic process in this case often serves to inactivate or detoxify the exogenous materials prior to their excretion from the body. As part of the metabolic sequence (many biochemical reactions may take place one after the other), sugar molecules may be attached to the metabolites to render them more water soluble and easier to discharge into urine. These end products are called conjugates and, as illustrated in Figure 14.3, the metabolism of diamorphine (the controlled drug present in heroin) is *via* monoacetylmorphine, morphine and then finally ending up as the glucuronide conjugate of morphine.

The main routes for actual excretion of substances from the body are: (1) through the kidney into the urine, (2) through the liver into bile and faeces, (3) through the skin in sweat, and (4) through the lungs, exhaled in breath. An interesting consequence of excretion *via* bile into the digestive tract is that substances can be reabsorbed into the bloodstream further along the gut, giving rise to a loop called the enterohepatic circulation, *i.e.* from the gut to the liver, from the liver to the bile, discharge of the bile into the gut, reabsorption from the gut and then back to the liver. The net effect of this circulation can be to retain substances in the body longer than might otherwise be expected and in turn, this can lead to a prolongation of the time over which they can be detected by the toxicologist. A common example of this is the retention of cannabis metabolites in the body; these can be detected for up to several weeks after a single administration, in favourable instances. In one exceptional case, cannabis metabolites were detected for 4 weeks after a single cannabis cigarette was smoked.

Court cases often raise the question, 'How quickly was the substance cleared from the blood?' or, knowing the concentration of a substance in blood at the time of death, 'What was its concentration at a given time prior to death?' These questions can be answered if graphs such as those shown in Figure 14.2 are available for the substance in question. The rate of clearance from blood is usually specified in terms of the

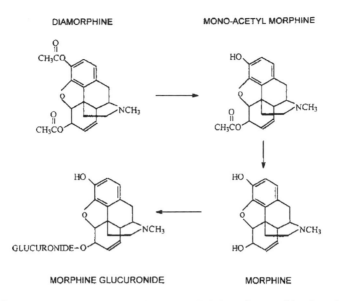

Figure 14.3 Metabolic sequence for the controlled drug diamorphine found in heroin. Monoacetylmorphine and morphine are intermediate metabolites. The final product, morphine glucuronide, is a water-soluble product excreted in urine.

half-life in blood. This is the time it takes for the concentration of the drug in the blood (blood level) to fall to half of its initial value. For example, the average half-life of morphine in blood is about 3 h. This means that a blood level of, say, 1 mg of morphine per litre of blood will fall to 0.5 mg in about 3 h, and this will fall to 0.25 mg after a further 3-h period, and so on. Half-life data for drugs are available from reference tables in pharmacopoeias and other texts.

Unfortunately, the half-life of a substance can vary widely between individuals for reasons similar to those given earlier concerning dose and response. Data which are available from literature sources are usually given in terms of a range and mean (or average) value. For example, morphine has a half-life of 1–5 h with a mean or average value of 3 h. When the toxicologist is asked in court to calculate blood levels at a time prior to death, or prior to the time when a blood sample was taken, the calculation will give the average value and a range, based on data from the literature. For obvious reasons these calculations are referred to as back-calculations. The range can be very wide if a significant time interval is involved, in fact so wide as to be useless for the purposes of the court.

14.2.5 Patterns of Poisoning

14.2.5.1 Time and Place. As indicated earlier, poisoning incidents have tended to involve substances that are readily available in the locality or are familiar to those involved. These have changed during the course of history as the number and variety of chemicals available has increased, but today they also vary according to geographical location around the world. In Europe, the USA and other developed industrial regions, the most common substances involved in fatal poisoning are drugs, both licit and illicit, whereas in most developing countries agricultural chemicals (mostly pesticides) cause the highest proportion of deaths. Interesting regional variations occur even in the types of illicit drugs used, *e.g.* amphetamine is commonly abused in the UK and Europe but in Japan and South-East Asia methamphetamine has always been more popular. Similarly, in South and Central America and the Caribbean, cocaine is the major drug of abuse and heroin is uncommon whereas the pattern is reversed in South-East Asia.

Nevertheless, these patterns are subject to change, sometimes suddenly. The former member countries of the USSR had a very small problem with illicit drugs but were faced with a rapid rise in the abuse of illicit drugs and the attendant rise in drug-related deaths after the collapse of the Soviet Union.

Finally, it should be noted that unexpected incidents can happen, as illustrated by mass poisoning incidents over the years. In Tokyo, the underground system was deliberately poisoned with the chemical warfare gas, sarin, in a politically motivated mass homicide attack. In Switzerland and France, mass suicides by ingestion of cyanide have been encountered in several locations among members of religious organisations.

14.2.5.2 Fatal Poisoning

Suicide. Most cases of fatal poisoning occur as a result of suicide or accident (37% and 17% of drug-related deaths, respectively) and most of these in the UK involve an overdose of medicinal drugs, consistent with the principle that people tend to use what is readily available. In the UK suicidal poisoning with other substances such as cyanide usually involves a laboratory or factory worker who has access to the substance in their place of work. The important medico-legal question in these cases concerns the intent of the deceased, *i.e.* did they intend to commit suicide or not? Circumstantial evidence from the locus of the death can be helpful in this respect, including the presence of a note written by the deceased stating intent, the presence of medicine containers such as empty pill bottles and the absence of any indication of violence or struggle. The medical history of the deceased is also relevant, especially if suicide had been attempted previously. Suicide by deliberate overdose is not necessarily easy to perform: the quantity taken can be too little to cause death or be so large as to cause vomiting of the substance swallowed.

Accident. Accidental overdose cases also occur regularly, although it can be difficult to distinguish between accidental and deliberate overdose on the basis of toxicology alone. The toxicologist is often asked to estimate the quantity of substance taken, to assess if it could have been taken inadvertently. Two points should be taken into consideration when making this estimation:

- The dose cannot be calculated simply and accurately by multiplying the blood concentration of substance by the volume of blood in the average body. This is because the substance does not reside only in the blood but is distributed throughout the body tissues in varying amounts. An accurate estimation of the dose requires the analysis of each of the body organs, including the stomach contents and contents of the digestive tract.
- The minimum dose taken can be estimated very approximately by comparing the concentration of the substance in autopsy blood with the levels obtained therapeutically when it is administered in a known amount. In many overdose cases, however, the blood level is 10 or more times higher than the normal therapeutic level and the likelihood of this being accidental is lower.

It is important to note that the blood level obtained after a given dose of substance varies widely between individuals and that, as for half-life calculations, a range and average value should be calculated. Also, the estimation of dose taken must take into account the length of time the deceased survived before dying and therefore involves a back-calculation to the time when the substance was administered. As described above, these calculations have their own uncertainties which compound the problems of dose estimation.

Industrial deaths. Deaths from poisoning in industry are unusual in the UK; deaths resulting from an accident caused by intoxication with alcohol or drugs are more common. Poisoning cases inevitably involve exposure to substances present in the workplace, and Health and Safety legislation such as COSHH (Control of Substances Hazardous to Health) carefully controls this type of exposure. Deaths are therefore the result of an accidental or unregulated exposure, unless the premises involved are not conforming to legal requirements. Typical problems are exposure to carbon monoxide through faulty ventilation or to cyanide through splashing with cyanide solutions (rapidly absorbed through the skin) and asphyxiation due to entering a closed space in which the oxygen has been displaced by an inert gas such as carbon dioxide (*e.g.* in a brewing tank). If industrial cases of non-fatal poisoning are encountered, relevant data for interpretation purposes can be obtained from tables published by the Health and Safety Executive, among others.

Iatrogenic. These deaths are the result of medical treatment or intervention. They are infrequent and are often the result of errors in calculating the dose of a drug to be administered or else of the use of the wrong route of administration. Deaths due to a general anaesthetic are usually considered separately.

Homicide. Homicidal poisoning is unusual in the UK today, but occurs more frequently in other parts of the world, again often involving agricultural chemicals. Mass deaths due to cyanide poisoning, which may have been at least partly homicidal, have also occurred in different countries. Such cases attract media attention because of their relative rarity. Examples which have occurred in the UK over the last few years include poisoning of spouses with cyanide, paraquat, carbon monoxide, atropine and chloroform. The use of carbon monoxide can involve car exhaust but in one case involved a cylinder of the gas, which was obtained by a technician from his laboratory.

14.3   THE WORK OF THE FORENSIC TOXICOLOGIST

14.3.1   Role of the Forensic Toxicologist in Medico-Legal Investigations

The relationship of the toxicologist to the legal authorities in charge of the case is normally that of a consultant with established expertise in the field of forensic toxicology. The work is usually carried out under contract, either on an individual case basis or as a service provider. In the post-mortem context, specimens for analysis are taken at autopsy by a pathologist authorised by the legal system to do so. Alternatively specimens are taken by a forensic medical examiner (formerly known as a police surgeon) from an alleged offender, typically under the Road Traffic Acts (1988 and 1991), Sections 4 and 5. These specimens remain the property of the legal authorities and are delivered to the toxicologist who is instructed to undertake the analysis. In the wider context of forensic toxicology, urine, oral fluid and hair samples can be collected by any trained personnel.

Decisions on whether or not to have a toxicological analysis carried out, and what types of analysis are to be requested, are usually taken by the pathologist or by law enforcement personnel, in road traffic cases. The factors involved in these decisions are varied but include the facilities available in the toxicology laboratory, the time taken to obtain results and the cost of the analyses. It is recommended that the toxicologist is involved in making these decisions and that background information concerning the case is made available in order to direct the toxicological investigation and subsequent interpretation of results. Relevant information is contained in the pathologist's report, the police report, the medical history and (if involved) the fire service report.

The purpose of involving a toxicologist in an investigation is to establish if drugs or other substances played a significant role in the case, *e.g.* a death, and access to relevant case background can be of great assistance in indicating what substances to look for. The work can then proceed effectively and produce results in the shortest time and at the minimum cost. In most cases, once the immediate objective of the work has been obtained, *e.g.* a cause of death or a cause of impairment has been established, work can cease unless there is a specific need for further information. This might concern the presence of other drugs or substances available to the deceased or the accused. There is certainly no need for blanket coverage of all possible substances which the laboratory is capable of detecting and measuring, as this would take too long and cost too much.

14.3.2 The Forensic Toxicological Investigation

Different scenarios are possible for the toxicology investigation, depending on the nature of the case. The possibilities are as follows:

- **Specific substance.** The simplest situation is when a sample is submitted with a request for analysis of one or more specific substances. Typical examples are requests for analysis of alcohol in blood (almost all cases) or for analysis of alcohol and carbon monoxide in a fire death.
- **Partial unknown.** In this situation, the toxicologist is informed of a number of possible substances that were available to the deceased or the accused, all or none of which might be present in the samples. In this case, each of the possible substances is targeted specifically to determine whether or not it is present.

- **Complete unknown.** Usually, there is no information available concerning what was available to the deceased or accused. The pathologist may not have a cause of death but may suspect that drugs are involved. Alternatively, there may be an established cause of death or impairment and the intention is to rule out the involvement of drugs and other substances in the case. In this situation, the toxicological investigation consists of a systematic application of all of the analytical methods available, beginning with the analysis of drugs and substances commonly found in the locality. The profile of the deceased or accused may be used as a guide: *e.g.* pensioners may be unlikely to abuse illicit drugs. Also, as indicated earlier, if a significant level of a drug or other substance is detected, the toxicologist may stop further work after consultation with the pathologist or police.

Because cases often involve a fatality, there is not the same degree of urgency as in other contexts, with the exception that a homicide involving a poison may require the investigation to be carried out as rapidly as possible. Depending on the complexity of the investigation, the time required to produce a report may vary from less than 24 h (*e.g.* a simple blood alcohol measurement) to several weeks (for a full toxicological investigation of all types of drugs and other possible substances).

14.3.3   General Analytical Approach

Most of the cases handled by the forensic toxicologist involve some type of legal proceedings. In this context, two different analyses are required to provide a high standard of proof that the results are reliable, beyond reasonable doubt. The first test carried out is normally a screening test, which is used to establish if a drug or substance is present. Ideally, this type of test should be quick and inexpensive, but also reliable. If the first test is positive, a second analysis, called a confirmatory test, is carried out. The primary characteristic of this test is that it should be unambiguous and give unequivocal proof of the presence and also the amount of a substance.

14.3.4   Different Types of Specimen

The different types of specimen and their value to the toxicologist are summarised in Table 14.1. Because of recent interest in oral fluid and hair analysis, more detailed accounts are given for these in Section 14.5.

14.3.5 Tools of the Trade—Methods of Analysis

A generalised schematic for methods of analysis is given in Figure 14.4.

14.3.5.1 Sample Pretreatment. This involves physical and chemical manipulation of specimens to render them suitable for further work, *e.g.*, filtration of urine, centrifugation of blood, homogenisation of tissue, adjustment of pH (degree of acidity).

14.3.5.2 Extraction. Drugs and other substances are often present at very low concentrations in specimens, typically comprising less than one part per million of specimen (*i.e.* 1 mg per litre of blood or per kilo of tissue). To detect and quantify these substances, the bulk of the matrix containing the analyte (substance being analysed) must be removed and discarded. This entails a process of extraction and concentration of the analyte(s) using either an organic solvent such as ether and dichloromethane or a solid absorbent, currently often based on silica.

14.3.5.3 Purification of the Extract. Also called the 'clean-up' stage, this purifies the initial crude extract to remove as many interfering substances as possible.

14.3.5.4 End-Step Analytical Procedure. The purified extract is analysed by one or more techniques based on spectrometry, chromatography or immunoassay.

Spectrometric techniques regularly used in toxicology include UV spectroscopy and mass spectrometry. These depend on the isolation of a pure substance from the specimen and the information obtained from the technique is a spectrum. This is valuable information for the identification of unknown substances, and/or confirmation of the identity of a substance thought to be present. Spectrometric methods become more powerful when combined with another technique, especially a chromatographic separation method.

Chromatographic methods in this context should be considered simply as methods of separating out the constituents of extracts of biological specimens. The type of information obtained is a chromatogram or chart, characterised by the presence of peaks, each representing one or more constituents, along a timescale generated by the recording device. The chromatogram gives two pieces of data: the time taken for each constituent to pass through the chromatographic system (the retention time) and the size (height or area) of the peak on the chromatogram. The former is characteristic of the substance concerned and the latter is proportional to the amount. A chromatographic method can therefore be

Table 14.1   Types of specimen and their value to the toxicologist.

Specimen	Attributes
Blood	Available in almost all cases and can be obtained at autopsy without opening the cadaver if it is known to present a risk of infection
	Blood levels of substances can be interpreted using literature data; note, however, that autopsy blood is usually partially clotted and haemolysed (i.e. the red blood cells have burst open) so it is not exactly equivalent to blood taken by a clinician from the living patient
	Window of detection of drugs is typically 24 hours
Urine	A non-invasive specimen which can be collected from living patients by any trained personnel
	Available from only a percentage of autopsy cases
	Contains most substances, including drugs and their metabolites, which have been taken into the body
	The concentrations of drugs and metabolites in urine are often higher than in blood
	Window of detection of drugs and their metabolites typically several days
	Lower infection risk than blood; the major disadvantage of urine is that concentrations of substances in urine cannot be correlated with response in the same way as blood levels and therefore cannot be used to establish a cause of death or a cause of impairment, but only indicate a possible cause
Liver	Available at autopsy from almost all cases
	When available, the liver is a large organ and there is no problem in obtaining an adequate specimen for analysis
	Drug concentrations are usually higher than in blood
	Interpretation of drug levels in liver can be difficult because of lack of reference data
Bile	Bile is obtained from the gallbladder in some autopsy cases
	Drug concentrations can be higher than in blood
	Typically used for opiates such as morphine
	Bile contains many constituents that can cause interference in the analytical methods
	Interpretation of drug levels in bile can be difficult because of lack of reference data
Vitreous humour	Can be obtained from the eyes in most autopsy cases but this may be ethically unacceptable
	Considered to be less prone to artefactual changes in the concentrations of alcohol and drugs
	Few interfering substances present
	Volume available is small
	Difficult to interpret levels of drugs and other substances because of lack of reference data
	Sometimes used for drowning cases, to measure potassium and sodium concentrations
Cerebrospinal fluid	Can be obtained from the brain in most autopsy cases
	Difficult to avoid contamination with blood and brain tissue
	Difficult to interpret levels of drugs and other substances because of lack of reference data

Table 14.1 (*Continued*).

Specimen	Attributes
Brain	Available in most autopsy cases; Difficult to analyse because of the presence of fatty material Difficult to interpret levels of drugs and other substances because of lack of reference data Useful for detection of volatile substance abuse as volatiles are retained in the fatty tissue of the brain for longer periods than in blood
Lung	Used for detection of gases and volatile substances Only qualitative results are obtained, as concentrations are not meaningful
Hair	Useful for detecting both drugs and other substances such as metals and arsenic Provides long window of detection of substances compared to blood and urine Contains a 'profile' of the sample donor's exposure during the life-time of the hair (about 1 month/cm) Only qualitative results are obtained, as concentrations are not meaningful
Nail clippings	Used mainly for assessment of chronic exposure to environmental materials
Oral fluid	Available from living subjects only Easily collected by expectoration or with a swab by any trained personnel Drug concentrations may be related to those in blood plasma and so may provide information on impairment Of interest for roadside testing of drug drivers

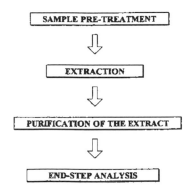

Figure 14.4 Generalised analytical sequence used in most toxicological investigations.

used for identification (based on the retention time) by using a reference table of retention times of drugs and other substances, and also to quantify the amount of substance present (based on peak size). Taken alone, a single chromatographic method cannot give unequivocal identification, as more than one substance may have the same retention time. The reliability of identification can be improved by using more than one system or as indicated above by combining the chromatographic method with a spectrometric method. The most common examples of the latter are gas chromatography–mass spectrometry (GC–MS), high performance liquid chromatography-diode array spectrometry (HPLC–DAD) and liquid chromatography–tandem mass spectrometry (LC–MS/MS). GC–MS and LC–MS/MS are considered to be the most reliable methods for the identification of substances. Taken together, they are applicable to all materials of interest although alternatives can also be used.

Immunoassay tests are based on the use of antibodies, the protein molecules formed in living organisms to control infection or invasion by foreign organisms such as viruses. Each different type of foreign material (antigen) entering the blood causes the formation of an antibody which will 'recognise' and bind to that material forming an antibody–antigen complex. Using a variety of methods, antibodies can be obtained which recognise drug molecules. These can be used to test for the presence of drugs in a specimen if there is some way of detecting the formation of an antibody–drug complex. Many ways of detecting the formation of these complexes have now been developed and these form the basis of different immunoassays, including radioimmunoassay (RIA), enzyme multiplied immunoassay technique (EMIT), fluorescence polarisation immunoassay (FPIA, often known by the commercial trade name TDx) and enzyme-linked immunosorbent assay (ELISA), which is one of the most widely used techniques. These tests are used as screening methods as they are relatively simple to perform and large numbers of specimens can be processed. They are, however, expensive to buy as commercial kits and are not available for all types of drugs—the most commonly available kits are for drugs of abuse. The nature of immunoassays is such that they are usually not specific for a single substance, but for a group of similar substances. This can be advantageous, as a single kit can be used to screen for the presence of several substances, *e.g.* for the opiates. The disadvantage is that the lack of specificity renders the immunoassay test unsuitable for unequivocal identification of a particular substance or for quantification of an analyte in a specimen.

14.3.6   Chemical Classification of Drugs

An area of confusion that sometimes arises, especially during an examination of a toxicology report in court, concerns the difference between the pharmacological and chemical classification of drugs. The former is familiar to most people. Drugs are classified according to their effects on the body, *e.g.* narcotic analgesics, stimulants, hallucinogens, or antidepressants. The chemical classification, however, deals only with the chemical structures and the resulting chemical properties of the drugs, in particular, whether a drug is acidic, basic or neutral. For example, acidic drugs containing carboxylic acid groups include anti-inflammatory drugs (such as ibuprofen and salicylic acid) and the barbiturates. Basic drugs contain an amine structure and constitute the largest group of drugs, including the narcotic analgesics such as morphine, stimulants such as amphetamine and cocaine, major tranquillisers such as chlorpromazine, antidepressants such as amitriptyline, and most other prescribed drugs. Neutral drugs are fewer in number and this group contains *e.g.* anabolic steroids. To the analyst, the chemical properties of drugs can be used to subdivide them into three fractions (acidic, basic and neutral drugs) and toxicology reports will normally refer to these chemical classes as well as to particular drugs (*e.g.* alcohol, paracetamol) and groups of drugs (*e.g.* benzodiazepines).

14.3.7 The Toxicology Report

A toxicology report is issued only by an authorised member of the toxicology personnel after the results of the analyses have been reviewed and checked and, preferably, following consultation with the requester to ensure no further work is required. In current practice, the analytical work is carried out entirely by the reporting scientist or is carried out under the supervision of the scientist with assistance from technicians, or by a combination of these two. Of course, much of the work is actually performed by laboratory instruments under the supervision of the analyst and results are printed by computer. The degree of personal involvement is steadily diminishing with time. In some contexts, the laboratory can be fully automated and all that is required of the personnel is to load the specimens in a rack. The reliability of the results produced must be safeguarded by the instigation of a quality system in the laboratory. More than just quality control, a quality system specifies the procedures used for all aspects of the work of the laboratory.

At its simplest, a toxicology report contains the following information:

1. Date and reference number of the report.
2. The case name (the deceased or specimen donor).
3. The instructing authority *e.g.* the police, coroner or procurator fiscal.

4. The name of the pathologist or police officer from whom the specimen(s) was received.
5. The date on which the specimen(s) was received.
6. The tests that were carried out.
7. The results of the tests.
8. The signature of the analyst.

In addition, according to the practice of the laboratory and legal system concerned, the report may also contain the written interpretation of results and opinion of the toxicologist who signed the report.

Presentation of a toxicology report in court may consist of attesting that the report and its contents are an accurate record of what was done and of the results obtained. This much is the province of the 'professional witness'. If, in addition, the toxicologist is asked to give an opinion on the results, expert testimony is involved. Court appearances by a toxicologist on simple matters of fact are now relatively rare, as legislation in the UK allows a report to be accepted by the court without the verbal testimony of the signatory, if neither prosecution nor defence has any problems with it. It follows that court appearances now occur only in complex or difficult cases in which the opinion of the toxicologist will be carefully examined and cross-examined. In future, the toxicologist will be subject to more searching cross-examination with respect to the analytical methods used. Method validation procedures are largely standardised in toxicology and will give an indication of the statistical uncertainty associated with analytical results.

14.4   INTERPRETATION

As indicated previously, toxicology results are primarily either qualitative, indicating whether a substance is present or absent, or quantitative, indicating the concentration of a substance in a specimen. Given the purpose of the investigation, the interpretation of results concerns their significance with respect to a cause of death or a cause of intoxication.

14.4.1   Qualitative Results

These are derived from either an analytical method that does not produce quantitative results, including many immunoassays, or from analyses carried out on specimens in which the concentration is not of diagnostic significance. The most important examples of the latter are tests carried out on stomach contents or on urine. It is perhaps obvious that the concentration of material in the stomach is less important than what is present or perhaps the total quantity present.

14.4.2 Quantitative Results

These are normally concentrations of substances in blood, urine and other tissues. Their interpretation is intended to relate the concentrations to effects on the sample donors. However, concentrations of substances in urine are more open to misinterpretation and misunderstanding. The problem arises because urinary concentrations vary widely, even more than in blood, and this in turn is because the volume of urine excreted changes according to the state of hydration of the body. The volume is also affected by foodstuffs such as coffee and drugs such as diuretics, which influence the working of the kidneys. Analytical methods used for urine specimens nevertheless often yield quantitative results and the toxicologist is often pressed to use and interpret these concentrations with respect to incapacitation, *e.g.* in road traffic samples.

Although some experts are willing to provide an opinion of this type, it is generally accepted and recommended that such interpretations should not be made except under exceptional circumstances, usually where additional information is available. In these cases, the concentration of drugs and other materials in urine do not prove incapacitation, and it is better to base the case on an examination of the accused by a forensic medical examiner or other physician at the time of the offence. Toxicology can then be used to support this examination.

One final point concerning urine concentrations. In the USA, threshold values (also called 'cut-off' values) have been specified for drugs of abuse in urine and if a drug is detected but its concentration is below the cut-off, then the sample is considered to be negative. These cut-off values are 'administrative' thresholds rather than limits imposed by the sensitivity of the method and were established to deal with problems of inadvertent exposure and analytical methodology. There are no statutory cut-off values in the UK but many laboratories use the US cut-offs which have mostly gained international acceptance.

The medical literature indicates that no concentration–effect relationship exists for many drugs, because the individual response is very variable, as described earlier. This means that, in practice, the interpretation process is limited and may be, at best, an estimate based on the information available. Perhaps the best-known example of this is the wide variation observed in the effects of alcohol on individuals: some people may function quite well at blood alcohol levels that would cause others to be extremely unwell. As a guide, the toxicologist can indicate what would happen to the average person at a given level and point out the possibility of individual variation.

In all branches of forensic science, interpretation is based on a comparative approach. Toxicology results are compared with a database of concentrations found in previous cases. This reference information comes from a variety of sources:

1. The records of the toxicologist's own department.
2. Compilations of records from other departments, *e.g.*, those published by the International Association of Forensic Toxicologists (TIAFT).
3. Reference textbooks and pharmacopoeias.
4. The medical literature, consisting of research papers published in refereed journals.

The information in this database is segregated according to the type of case, including normal-therapy cases, non-fatal poisoning cases and fatal poisonings. The toxicologist can then decide if the results are consistent with therapeutic administration, an overdose or a fatal overdose. What it is often not possible is to give an opinion with certainty that a measured level of a substance in blood would have produced a particular effect.

14.4.3 Specific Problems of Interpretation

A salient problem of interpretation is caused by polydrug use among drug abusers. The effects of drugs administered individually are difficult to predict, but when two or more are taken together then the net combined effect is highly unpredictable. Two sedative drugs such as alcohol and heroin would have a combined effect which is greater than that of the individual drugs, but the combined effect of a stimulant and a sedative such as cocaine and heroin is very uncertain.

Care must also be taken in the interpretation of toxicology results since there could be alternative explanations for the presence of drugs in a sample. Several instances have been recorded in the literature in which drugs were found in test specimens (particularly urine samples) which were derived from legitimate sources rather than from illicit materials. For example, poppy seeds used for baking have sometimes been found to contain traces of opium acquired from the poppy capsule. If enough of these seeds are consumed, morphine can be detected in urine. Similarly, some cases were encountered in the USA in which herbal tea contained coca leaves ('Health Inca Tea'), giving rise to positive cocaine tests in urine. Thirdly, some drugs which can be prescribed or even purchased are metabolised in the body to controlled substances. A wide range of medicines give either amphetamine or methamphetamine, and codeine is metabolised to morphine. Finally, the analyst must take care in testing specimens for amphetamine to avoid false positives due to the presence of similar substances such as ephedrine, a constituent of cold remedies and ginseng tea.

The effects of passive smoking cannot be neglected. Cannabis and crack cocaine are usually smoked, and those present in the same room or car as the smoker will absorb some of the drug due to passive smoking. Recent work has also indicated that rooms used for smoking crack become contaminated by cocaine 'fall-out'—minute crystals of the drug which settle on the furnishings of the room. People, including children, coming in contact with these can give false positive results in a drug screening test. For this reason, cut-off values for urine tests make allowance for the maximum level which could possibly be obtained by passive exposure.

The fact that some drugs are very rapidly metabolised in the body, and cannot be detected a short time after dosage, must not be over-looked. The best-known example is diamorphine, which is present in heroin samples and is converted rapidly to morphine. Heroin abuse is usually inferred from the presence of morphine in the blood and from other circumstantial evidence.

In addition to the problems of interpretation already mentioned, it is also known that drug concentrations can vary according to the source of the blood specimen (*e.g.* from the heart or from a major blood vessel) and can change after death due to passive physical processes of diffusion in the organs. It is generally not possible to predict or allow for these post-mortem changes and redistribution when reporting drug levels, and the part of the body from which blood samples are taken at an autopsy should be chosen with care, to minimise these effects. Guidelines for forensic pathologists recommend that blood samples should be taken from a peripheral vein such as the femoral vein rather than from the heart or aorta. In any event, the source of the blood should be indicated on the submission form which accompanies the blood sample.

14.5 SPECIFIC AREAS OF INTEREST AND CASE STUDIES

The purpose of this section is to provide some additional detail about areas of particular interest, with examples of how toxicology cases are handled and how the results can be used in subsequent legal proceedings.

14.5.1 Alcohol (Ethanol)

14.5.1.1 Chemistry. Ethanol is the most common member of the group of substances known as alcohols, characterized by the presence of a hydroxy group in their chemical structures, and is colloquially referred to simply as 'alcohol'. Ethanol is a colourless flammable liquid, boiling point 78.4 °C, which is used widely in the laboratory as a solvent for organic materials.

14.5.1.2 Alcohol Concentrations and Amounts. Ethanol is completely miscible with water and is a constituent of alcoholic beverages which contain a wide range of alcohol concentrations, usually between 0.5% and 50% by volume. All of these beverages originate in the fermentation of sugars from fruit or cereals, yielding a maximum alcohol concentration of about 15% by volume, after which yeast is deactivated by the alcohol produced. Higher concentrations of alcohol are usually obtained by distillation, up to 94.8% by volume, which is unpalatable, and in beverages such as whisky is usually diluted to approximately 40% by volume or less.

The alcohol content of a beverage is usually expressed as '% by volume', *e.g.* a beer of strength 5% by volume contains 5 mL of alcohol per 100 mL of beer. Alcohol has a density of 0.789 g/mL and so 5 mL alcohol weigh $5 \times 0.789\,g = 3.945\,g$. The weight of alcohol in an alcoholic drink is calculated as

$$\text{volume of drink} \times \text{alcohol content}(\%\text{ by volume}/100) \times 0.789$$

For example, the weight of alcohol in a 125-mL glass of wine containing 13% alcohol by volume is $125 \times 13/100 \times 0.789 = 12.8\,g$.

By contrast, alcohol concentrations in blood and urine are usually expressed as weight per volume, *e.g.* in many countries the prescribed limit for alcohol in blood when driving is 50 mg/100 mL of blood (abbreviated to 50 mg%) which is numerically equivalent to 0.5 g/L. Lower concentration units are used for convenience for alcohol in breath and the corresponding limit for alcohol in breath is 22 μg/100 mL breath (22 μg%).

14.5.1.3 Pharmacology

Effects of Alcohol. Alcohol is a central nervous system (CNS) depressant which acts by (1) augmenting the effects of the inhibitory neurotransmitter gamma-aminobutyric acid (GABA) and (2) decreasing the effects of the excitatory neurotransmitter glutamate. CNS depression can manifest itself in different ways, including sedation, amnesia, unconsciousness and ultimately death. Paradoxical disinhibition occurs when those parts of the brain that govern self control and restraint are depressed by alcohol, which occurs at an early stage of alcohol intoxication.

It can be seen from the calculations given above that even a single drink may contain a substantial dose of alcohol, greatly in excess of any other drug consumed, and, as mentioned earlier, even low levels of alcohol can enhance the effects of other drugs such as sedatives.

Like many drugs, the effects of alcohol are dose-related and, because blood alcohol concentrations are also dose related, the effects can be correlated with blood alcohol level, as summarised in Table 14.2. However, there is a great deal of variation in the response of the individual subject to alcohol due to the factors mentioned earlier plus the effects of mood, expectation and experience with alcohol of the user and the environment in which alcohol is consumed—alone or in a social setting. Tolerance to the effects of alcohol develops with repeated consumption but impairment still occurs, *e.g.* in tasks requiring fine control

Table 14.2 Effects of alcohol versus blood alcohol concentration.

Blood alcohol concentration (mg%)	*Anticipated effects*
0–50	Little effect, except perhaps more talkative
50–100	Tendency to loss of fine motor control, causing perhaps slurred speech. Also, inhibition and self-critical faculty may begin to be lost and there may be mild euphoria
100–200	Generally, both physical and mental functions become significantly impaired as the blood alcohol concentration increases through this range, perhaps resulting in poor balance, staggering when walking, poor driving performance, poor judgement, sedation, emotional instability and nausea. Many fatal cases in forensic medicine occur with blood alcohol concentrations in this range, when the subject performs badly but is not so intoxicated as to be incapable of getting into trouble
200–300	Subject displays typical signs of being 'drunk and incapable': increased impairment of physical and mental abilities, including impaired vision and speech; difficulty in standing; confusion; sedation, possibly coma
300–450	Severe intoxication resulting in unconsciousness, impaired respiration, coma and unresponsiveness in all but very tolerant subjects. Possibly fatal
>450	Probably fatal, due to respiratory paralysis

such as driving a vehicle. Physiologists refer to 'fine motor control', meaning motor control of the small muscles that carry out fine movements, which are also involved in control of a vehicle. Numerous studies have shown that the risk of an accident when driving increases exponentially with blood alcohol concentration. Questions about the effects of alcohol on individuals occur regularly in court, but only a general indication can be given of the range of effects that might be expected at a given blood alcohol level.

Absorption. Absorption of alcohol into the bloodstream starts in the stomach but mostly takes place in the small intestine. Absorption is usually rapid and is complete within 0.5–1.5 h after the last consumption of alcohol. Food in the stomach can delay absorption by slowing down gastric emptying into the intestine.

Distribution. After absorption has occurred, alcohol is distributed uniformly throughout the body water, including all fluids, organs and tissues, and the fetus if the subject is pregnant. It follows that knowledge of the water content of a tissue or fluid allows alcohol concentrations in different body compartments to be compared and this is the basis of the Widmark equation, which relates the blood alcohol concentration to the amount ingested.

Metabolism. Metabolism of alcohol also begins in the stomach but mostly takes place in the liver, where it is oxidized by two different systems. Alcohol dehydrogenase (ADH) is the main metabolic enzyme, producing acetaldehyde which is subsequently oxidized by aldehyde dehydrogenase to acetic acid. This can be assimilated into the normal metabolism of the body and so alcohol can act as a nutrient, at least for some time. The second oxidizing system is the microsomal enzyme oxidizing system (MEOS), which is important in the metabolism of alcohol at high concentrations. The rate of alcohol metabolism is influenced by the genetic profile of the subject, due to polymorphism of the enzymes involved, and varies widely between individuals as well as having a geographical variation between populations.

Alcohol concentrations in the body are high compared to those of most other drugs and saturate the metabolic systems, such that they are working at maximum capacity until the alcohol concentration falls to a low level. Technically this is described as 'zero-order kinetics', in which the rate of alcohol metabolism is independent of the alcohol concentration, *i.e.* a constant amount is lost per hour regardless of how high the alcohol level is in blood. The average rate is 18 mg% per hour within a range of 9–27 mg% per hour, although these figures vary depending on the source of the data. The rates do not, however, vary according to the size of the person as an increase in body weight also means that the liver is proportionately larger. Body size does affect the total amount of water in the body, which in turn affects the alcohol concentration reached for a given dose of alcohol, *i.e.* a small person will reach a higher blood alcohol level than a large person if they both consume equal amounts of alcohol.

Elimination. Alcohol is eliminated from the body mostly by metabolism in the liver but a small percentage (2%) is excreted in urine and breath plus additional small amounts in other secretions such as sweat. Concentrations of alcohol in urine are higher than in blood, reflecting the higher water content of urine compared to blood, during the elimination phase. However, the reverse is true for the absorption phase since there is a lag time between alcohol reaching the blood and then distributing into urine. Alcohol is present in breath at concentrations which are much lower than in blood: the ratio of blood to breath alcohol concentrations is variable but on average it is 2300:1, the figure used in UK road traffic legislation to define the prescribed limit for alcohol in breath.

14.5.1.4 Forensic Toxicology

Prevalence. Alcohol is one of the most commonly encountered drugs in forensic toxicology. During the period 2007–2010, more than 6000 post-mortem cases were examined by Forensic Medicine and Science, University of Glasgow. Alcohol was present in 62% of these and was included in the cause of death as a principal or contributory factor in 16% of the cases.

Alcohol Toxicity. Alcohol causes both acute and chronic effects. Acute intoxication occurs while alcohol is still in the blood with a range of dose-related effects, as described in Table 14.2. Chronic effects of alcohol abuse involve a number of body systems but especially the gastrointestinal, neurological and cardiovascular systems. Alcohol is associated with ulcers in the oesophagus and stomach, and in the liver alcohol causes fat deposition leading to hepatitis, cirrhosis and an increased incidence of cancer. Alcohol also causes pancreatitis, damaging the islets of Langerhans which produce insulin, leading to chronic pain and type 1 diabetes. Alcohol is associated with encephalopathy (disease of the brain) including dementia and with cardiomyopathy (disease of the heart muscle).

Acute Abstinence Syndrome. Sudden withdrawal from alcohol ingestion, especially after a prolonged period of excessive alcohol abuse, can lead to life-threatening withdrawal symptoms, including seizures and delirium tremens. Ethanol is a CNS depressant and prolonged abuse leads to tolerance, in which the brain adapts to the depressant effects of alcohol to restore as much normal activity as possible. If alcohol is removed, the sedative effect disappears and the brain moves into a hyper-excited state in which neurological damage (excitotoxicity) occurs. Management of this condition, *e.g.* in a detainee in a jail cell, involves administration of a sedative.

Artefactual Production of Ethanol. Ethanol can be produced in blood and tissues after death by processes similar to fermentation, especially if the body has been damaged, *e.g.* in an accident, allowing ingress of microbes and the environmental temperature is high enough to permit enzymatic reactions to take place. Concentrations of alcohol produced in this way can be significant, up to 150 mg% or more, although lower concentrations are more likely to be due to post-mortem production. Ethanol can also be produced in stored blood samples but this can largely be prevented by addition of fluoride to inhibit biochemical changes.

One approach to this problem involves the analysis of other samples including urine and vitreous humor, if available, which are considered to be less prone to post-mortem change than blood. The presence of alcohol in blood but not in urine or vitreous may indicate post-mortem production. An alternative approach is to analyse the suspect sample for the presence of metabolites produced during life following ingestion of alcohol or else to look for markers of post-mortem production of alcohol. For the former, alcohol produces glucuronide and sulphate metabolites only in living subjects and so these serve as markers that alcohol was ingested before death. On the other hand, post-mortem formation of alcohol can be indicated by the presence of other volatiles in blood, and these have been used to interpret blood alcohol levels in aircraft accident deaths.

14.5.1.5 Methods of Analysis.   Two methods of analysis of alcohol are of interest in forensic toxicology. One is the laboratory procedure used for samples of blood and urine and the other is the evidential breath test device. Hospital laboratories also use a clinical biochemistry procedure but the methods described below are more appropriate.

Laboratory Method. Blood and urine samples are analysed routinely by gas chromatography with flame ionisation detection. The method is rapid, has good accuracy and precision, better than 2% relative standard deviation, and is selective, since the analysis is carried out in parallel using two gas chromatography columns with different selectivities. It is also a robust and clean procedure as the samples are prepared for headspace analysis rather than being analysed directly, avoiding column contamination with blood proteins and other non-volatile components. In the headspace technique a measured aliquot of blood or urine for analysis is sealed in a partially filled glass vial with a cap containing a rubber septum. A series of calibration standards containing known concentrations of alcohol is prepared at the same time under identical conditions. The accuracy and precision of the measurements are improved by adding a fixed amount of a reference material to each sample as an 'internal standard'. The vials are then placed in a warm oven before being analysed.

Alcohol in the samples equilibrates with the air in the vials, giving concentrations in air which are directly proportional to those in the samples. A small portion of the air is removed with a hypodermic syringe by an automatic sampling device and injected into the gas chromatograph, producing a chromatogram for each sample with peaks for alcohol and the internal standard. The ratio of the area of the alcohol peak to that of the internal standard peak is calculated for each sample. A calibration curve is prepared by plotting the peak area ratios for the standards against their known alcohol concentrations and this curve is used to obtain the alcohol concentrations in the case samples being tested.

Evidential Breath Test. Breath alcohol measurements are made with living subjects under a variety of circumstances including road traffic safety, employment screening and doping control in sport. Instrumental breath test devices are used both at the roadside as a preliminary screening test and indoors, typically in a police station, as an evidential breath test. The portable roadside devices use an electrochemical (fuel) cell whereas the evidential versions use an electrochemical (EC) cell or infrared (IR) spectrometry or both (EC/IR). Current EC/IR instruments are sensitive, specific and automated, requiring little operator involvement in the measurement process. They must be maintained regularly and be located in a stable temperature environment.

The automated test procedure involves a calibration check at the prescribed limit using a gas standard of ethanol in air from a cylinder, analysis of two sequential breath samples from the subject, and a second calibration check. The calibration checks must be within approximately 10% of the target value and the two breath test results must agree within 15%. Additional checks are usually made to detect mouth alcohol (residues from recently ingested materials containing alcohol) and insufficient breath samples. The instrument prints out a record of the test and its results plus any error messages or warnings.

The reliability of these instruments has increased compared to previous versions, and problems caused by interfering substances in breath, such as acetone in the breath of diabetics, have largely been overcome.

14.5.2   Road Traffic Safety

14.5.2.1   Legislation in the United Kingdom and Europe.   The relevant legislation includes the Road Traffic Act 1988 as amended by the Road Traffic Act 1991 and the Railways and Transport Safety Act 2003. The 1988 Act Section 4 deals with *Driving, or being in charge, when under influence of drink or drugs* and Section 5 with *Driving or being in charge of a motor vehicle with alcohol concentration above prescribed limit.* The 2003 Act legitimized the use of oral fluid samples in addition to blood and urine. Samples submitted under Section 4 are usually analysed for the presence of drugs since alcohol is covered by Section 5 and is mostly analysed in breath.

14.5.2.2   Driving Under the Influence of Drugs (DUID).   The roadside test device for alcohol described in section 14.5.1.5 has proved to be an effective and successful tool for law enforcement, creating a demand for a similar device for the detection of DUID. Commercial development of test methods initially evaluated urine 'dipstick' tests in which a urine sample is collected from a driver and tested for drug presence using a disposable device resembling the home pregnancy test and based on an immunoassay technique (ELISA) described earlier. Similar tests were also developed for oral fluid and sweat, along with handheld electronic readers to provide an objective indication of the test result. Some of these systems are now in use in jurisdictions outside the UK and have been found to be effective in deterring drug driving, at least in the initial period after their introduction when public awareness was at its highest.

It is important to note that Section 4 of the UK Road Traffic Act is primarily concerned with the prevention of driving while the driver is impaired due to drink or drugs rather than with driving with drugs present in the blood circulation (a 'zero tolerance' approach). The exception is for alcohol, for which there is a defined limit in blood regardless of whether or not the driver is impaired. Section 4 requires an assessment to be made of whether or not a driver is impaired, which is subjective and difficult to do at the roadside. The preliminary impairment test (also known as the field impairment test) was introduced in the

Railways and Transport Safety Act 2003 to provide the police with a way of assessing impairment, based on familiar tests of physical and mental ability such as balancing on one foot, walking in a straight line and understanding and following instructions. If the preliminary test indicates impairment, a medical examination follows in the police station. Outside the UK, the problem of assessing impairment has been circumvented by the introduction of zero-tolerance legislation, which creates offences based only on the presence of specified controlled drugs in the driver and does not attempt to create limits as for alcohol.

14.5.3 Alternative Specimens

14.5.3.1 Hair. Hair has gained popularity as a matrix for drug testing because it can be used to distinguish between chronic and naive (first time) use. Cranial (head) hair is most common, but axillary or pubic hair can also be used. Drugs circulating in blood diffuse into the hair root and are trapped in the keratinized shaft of the hair as it grows. The mechanism by which drugs are bound to hair is still not fully understood, but it is related to the melanin (pigment) content. Hair with a lower melanin content, *i.e.* blonde hair, has less potential to bind drugs than hair with a higher melanin content, *i.e.* dark hair. Also, basic drugs (a large group which includes most drugs) are positively charged in the blood and bind more strongly to hair than acidic drugs (a smaller group which includes some sedatives, anti-inflammatory drugs and drug metabolites) which are negatively charged. In this way hair incorporates a record of drugs that have been in circulation during its growth period.

A huge advantage of hair is the long time window of detection for drugs. Hair grows at approximately 1 cm per month, therefore a 12 cm length of hair will represent about 12 months of hair growth. Segmental analysis can be performed on the hair; this involves cutting the hair into smaller lengths and analysing each separately. The length of the segments will depend on the reason for the analysis but for longer lengths of hair 3-cm segments are usually used and 1-cm segments for shorter lengths. The number of segments is also limited by the amount of hair available, since approximately 30 mg of hair are required for an analysis.

External contamination of the hair is a major consideration when carrying out hair analysis. Sources of contamination can range from drugs smoked in close proximity to the sample donor to sweat from a parent touching a child's hair. Hair samples must be 'washed' before any analysis is carried out to try to eliminate the possibility of contamination. The samples are typically immersed sequentially in a solvent such as dichloromethane and sodium dodecyl sulphate (SDS) buffer, and ultrasonicated for a short time. The wash liquid is then removed and the process repeated. The washes can be analysed to see if any drugs are present, but if the washes are negative no further washing is required. A compromise needs to be reached between removing as much external contamination as possible without losing any drugs from the inside of the hair. Usually the presence of metabolites in hair confirms that drugs detected have been ingested rather than deposited from outside. However some drugs readily degrade to their metabolites, *e.g.* cocaine hydrolyses to benzoylecgonine, which can make it hard to distinguish between ingestion and external contamination. After the wash procedure, drugs are extracted from the hair by prolonged immersion in a suitable solvent and analysed using conventional methods.

14.5.3.2 Oral Fluid. Testing oral fluid for drugs has increased exponentially over recent years. It has many advantages over other matrices, especially in its ease of collection and representation of recent drug use. Oral fluid can be described as the mixture of fluids and other components sampled from the oral cavity. It is made up mostly of saliva, with other fluids such as oral mucosal transudate and cell and food debris. Drugs in the circulation diffuse through lipid membranes in the salivary glands into saliva. The chemical nature of the drug and the pH of the blood and saliva affect how well the drug moves from blood into saliva. Some drugs are highly protein bound in blood, *e.g.* diazepam, resulting in very low concentrations being found in the saliva.

As in hair analysis, oral contamination is also an issue in oral fluid analysis. In this instance the difficulty lies in distinguishing between drugs which have diffused into saliva from blood and those which have come from contamination of the oral cavity through smoking or ingestion. It is well accepted that the active constituent of cannabis, Δ^9-THC, in oral fluid comes purely from contamination of the oral cavity when cannabis is smoked as opposed to diffusion from the blood.

Expectorate (spitting) samples are unpleasant to collect for the donor and the collector and therefore are not often used. Several commercial companies have developed different collection devices, most of them centred on a sterile fibre pad which is placed in the mouth and left for a period of time to absorb oral fluid. This pad is then placed in a tube, usually with some buffer to preserve the sample before the analysis is done. Oral fluid analysis is comparable to blood analysis with respect to sample preparation; however the concentrations found in oral fluid tend to be much lower than in blood.

Because of its ease of collection and the possibility of detecting recent drug use, oral fluid is now widely used at the roadside to test drivers suspected of being under the influence of drugs. In the UK the Railways and Transport Safety Act 2003 made testing oral fluid for drugs at the roadside legally possible. Several commercial kits for roadside testing of oral fluid for drugs of abuse are available, consisting of a collection device and an immunoassay-based presumptive test which gives a positive or negative reading for one or more drug groups within a few minutes. These are only screening tests and any results need to be confirmed by the use of laboratory techniques based on mass spectrometry.

14.5.4 Drug Overdose Case Studies

Recent substance-related deaths in the UK have often involved illicit drugs including diamorphine (heroin), methadone, cocaine and amphetamine derivatives such as methylenedioxymethamphetamine (MDMA, 'ecstasy'). In these cases, the role of the toxicologist is central to the investigation of the case but often problems arise in court in providing the clear-cut interpretations needed by the legal process. This problem is compounded by polydrug use, in which the effects of simultaneous administration of more than one drug are difficult to predict. The pharmacology of the drugs involved in the following case studies can be obtained from other sources.

14.5.4.1 Case Study: A Heroin-Related Death. In this case, a known intravenous drug user was found dead at home with a syringe beside him. Some heroin powder was also found nearby. Interviews with his friends suggested that he had not injected for some time. At autopsy, death was ascribed to the inhalation of gastric contents, due to drug intoxication (confirmed by subsequent toxicology). A blood sample taken at the autopsy was found to contain 0.05 mg 6-mono-acetylmorphine per litre of blood, 0.6 mg free morphine per litre of blood and 0.75 mg total morphine per litre of blood. In addition, a low concentration of codeine was detected, 0.07 mg per litre of blood. The heroin was analysed and found to contain 40% diamorphine by weight (diamorphine is the pharmacological name for the active constituent of heroin: the normal diamorphine content of 'street' heroin in the UK is 30–50%, but in the locality of the case, the average purity was 10–15%).

In this investigation, there is a clear indication that drugs of abuse might be involved and the toxicologist was specifically directed to analyse the blood sample for the common street drugs. Heroin is rapidly metabolised in blood to 6-monoacetylmorphine (6-MAM), morphine and morphine glucuronides (Figure 14.3) and these are the usual target substances for the analyst. Codeine is a member of the group of opiates but is not a metabolite of heroin or morphine but is derived from codeine originally present in opium from which heroin is produced. The illicit manufacture of heroin does not purify the product and the codeine remains in the heroin as an impurity. The presence of codeine indicates, but does not prove, that heroin was administered rather than, say, morphine tablets. The best indication that heroin was used comes from the presence of the heroin metabolite 6-MAM in blood or urine.

The blood sample was screened using an immunoassay method, which gave a positive result for opiates but not for any other illicit drug groups. This test did not indicate which opiate drug was present since it lacked specificity. As a result, attention was focused on determining which type of opiate was present, and the sample was analysed by GC–MS following the preparation of a basic drug extract before and after enzymatic hydrolysis to release morphine from the glucuronide conjugates. 'Free' morphine is unconjugated morphine and 'total' morphine is the sum of free morphine plus additional morphine released from the conjugates by hydrolysis. These analyses produced data for the positive identification of heroin metabolites and codeine and for the quantities present.

Interpretation of the result requires a knowledge of morphine levels found in clinical patients treated with morphine or diamorphine (the mean and range for normal therapeutic levels) and also morphine levels found in overdose cases. Consultation of a pharmacology text, pharmacopoeia or specialised textbook indicates that blood levels of morphine found in normal therapy are usually less than 0.1 mg/L, although can be very high in cancer patients treated long-term with morphine, due to development of tolerance. Blood levels found in heroin-related deaths are usually above 0.2 mg/L although the range is from zero to several milligrams. The latter requires some comment. Many published studies of heroin-related deaths report cases in which no morphine was detected in blood. The reasons are not known but may include failure of the heroin to be uniformly distributed in the circulation, because of death occurring very suddenly. On the other hand, very high levels have been found in regular injectors who are tolerant to the drug.

In the present case, the level of 0.6 mg morphine per litre of blood was clearly above the normal therapeutic range and provided a plausible cause of death. Regular users of heroin might live at such a level, but the deceased in this case had not injected for some time and had presumably lost his acquired tolerance. He might also have overdosed because the heroin used was of higher purity than normal in his area.

The cause of death was attributed to inhalation of gastric contents: this is a common finding in drug-related deaths, as the normal swallowing reflex is lost as a result of depression of the nervous system by the drug. Also, nausea is a side effect of heroin use, especially in new users.

Additional information comes from the presence of 6-MAM in the blood sample and from the total morphine result. 6-MAM is the initial metabolite of heroin and is rapidly converted to morphine. As a result it can only be detected for a short time, if at all, after administration of heroin. Its presence in blood indicates that the deceased died a short time after administration of the drug. Total morphine represents the concentration of free morphine plus its conjugated metabolites and the relative amounts of free morphine and morphine conjugates can give an indication of survival time after administration of heroin. In this case free morphine was 0.6 mg/L and total morphine was 0.75 mg/L. Morphine was present mostly in the free form and only 20% had been converted to the conjugate form, indicating that the deceased has died less than 3 h after administration of heroin.

During examination of this case in court, the toxicologist concerned was asked if the morphine level found in blood was a fatal level. It should be clear that a simple yes/no answer is inappropriate and that some qualifications need to be added. However, the toxicologist was able to confirm that the morphine level was in the range found in previous heroin-related deaths and could provide a cause of death, when taken into account with the other information available.

14.5.3.2 Case Study: Polydrug Fatality. A man in his twenties died following administration of heroin supplied by a friend, who shared half of the heroin 'deal', a small bag of heroin. The deceased was on a methadone maintenance programme at the time and also used diazepam and alcohol. The friend was charged with culpable homicide on the basis that he supplied a drug which caused death. Autopsy findings included a recent injection mark on the inner aspect of the left arm and the presence of significant pulmonary congestion and oedema, typical of respiratory depression.

The autopsy blood sample was found to contain morphine 0.15 mg/L, diazepam 0.6 mg/L, methadone 0.3 mg/L and alcohol 120 mg%. Morphine concentrations were discussed in the previous example. Methadone concentrations in maintenance patients average around 0.3–0.35 mg/L blood but can be much higher depending on the dose. In 61 male fatalities attributed to methadone alone, the methadone concentrations in blood ranged from 0.13 to 4.6 mg/L with an average of 0.80 mg/L. Diazepam concentrations in blood following regular high dose administration (30 mg/day) are in the range 0.7–1.5 mg/L.

None of the substances detected in this case was present at a high concentration which might reasonably have caused death on its own, bearing in mind the wide range of morphine concentrations found in heroin and methadone-related deaths, which overlap substantially with therapeutic concentrations. The cause of death in polydrug cases is usually attributed to the combined effects of several CNS depressants—all four drugs in this case. However, the prosecution contention was that it was the heroin that caused the death and that the friend was culpable. Two questions need to be answered. Would the deceased have died if heroin had not been administered? Would the deceased have died in the absence of the heroin? Different outcomes have been obtained for cases of this type depending on the background circumstances and on how the court decides on these two questions.

14.5.3.3 Case Study: A Paracetamol Overdose. A middle-aged woman with a history of suicide attempts was admitted to hospital at 9.30 pm after taking an overdose of paracetamol and an anti-depressant (prochlorperazine) earlier that evening, at 6 pm. A sample of blood was taken for analysis in the hospital laboratory while normal supportive treatment (gastric lavage) was carried out in the ward. At 1 am, the lab result indicated a level of 270 mg paracetamol per litre of blood. Medical intervention was indicated by this result and the drug acetylcysteine was administered. Unfortunately, a 10-fold overdose of this drug was administered despite the dose being queried by the pharmacist and duty nurse. A booster dose of the drug was subsequently administered, also in a 10-fold overdose, at 4 am. The patient died at 6 am of a cardiac arrest.

The toxicological investigation of the case following an autopsy covered a full screen for drugs in the autopsy blood sample, including paracetamol and basic drugs (which include antidepressants). The results were 102 mg paracetamol per litre of blood and 0.07 mg prochlorperazine per litre of blood. All other analyses were negative.

Interpretation of these results must first deal with the drugs individually. Paracetamol causes liver failure and death (after 2–3 days) if taken in a high dosage. There are clear guidelines available to physicians which relate the blood level of paracetamol to the time that has elapsed since the drug was taken and to the probability of liver failure. In the present case, a blood level of 270 mg/L, 4 h after the drug was ingested, indicated a high probability of fatal liver damage. The treatment to be followed is to administer a drug such as acetylcysteine which protects the liver from such severe damage. The level of paracetamol measured in the autopsy blood specimen was 102 mg/L, indicating a drop of 170 mg/L in

8 h approximately. If the deceased' liver had been functioning normally, the level should have fallen to below 70 mg/L (the half-life of paracetamol is 4 h). Incipient liver damage is indicated by this result.

The inadvertent overdose of the antidote, acetylcysteine, resulted from poor labelling of the drug formulation and from subsequent miscalculation of the dose. Several similar cases have been published in the medical literature, which led the manufacturer to design new packaging. There was no analytical procedure available for this drug and it was not measured in the autopsy blood specimen. The small amount of toxicity information available related to tests on dogs, although this indicated that the drug was safe even at high doses.

Prochlorperazine taken on prescription produces a therapeutic range of 0.01–0.04 mg per litre of blood. In the present case, the deceased had survived about 12 h after taking the drug. The half-life is approximately 6 h so the drug at its highest level might have been 2–4 times higher (*i.e.* up to 0.28 mg/L), which places it in the range associated with toxicity.

Conclusions reached on the basis of toxicology alone are unclear, especially concerning the role of the drug acetylcysteine in the death, in the absence of toxicity data. Paracetamol and prochlorperazine were at concentrations likely to cause toxicity. A death certificate was issued ascribing the death to drug overdose and in the subsequent inquiry, the court accepted this as the cause of death.

14.5.5 DUID Case

A young woman was accused of causing death while driving under the influence of drugs. She failed to negotiate a bend on the road and her car left the road and hit a pedestrian on the pavement. Approximately 2.5 h after the accident, a blood sample was obtained from the accused by a forensic medical examiner. On analysis this was found to contain a trace of alcohol, not less than 2 mg/100 mL blood; MDMA (ecstasy), 0.26 mg/L of blood; diazepam, 0.12 mg/L blood; desmethyldiazepam (a metabolite of diazepam), 0.33 mg/L of blood; Δ^9-tetrahydrocannabinol (Δ^9-THC, active constituent of cannabis), trace amount detected; and 11-nor- Δ^9-tetrahydrocannabinol-9-carboxylic acid (metabolite of Δ^9-THC), 0.13 mg/L blood.

This is obviously another case of polydrug use. In this case the important question to consider was what effects the drug combination was having on the driver at the time of the accident. The elapse of 2.5 h before a blood sample was obtained and the effect of this delay on blood concentrations also had to be considered. Since the sample was taken from a living subject and the timescale was known, it was possible to carry out back-calculations to estimate the concentrations of some the substances at the time of the accident:

- **Alcohol.** The average rate of clearance of alcohol from blood is 18 mg% per hour. Over a period of 2.5 h, the blood alcohol concentration would on average have fallen by 45 mg/100 ml. The measured alcohol concentration in her blood specimen was not less than 2 mg/100 mL blood and so at the time of the accident it is estimated as having been $2 + 45 = 47$ mg/100 mL blood. A concentration of 47 mg alcohol per 100 mL blood is below the prescribed limit of 80 mg/100 mL blood. This calculation assumes (1) that the driver was in the elimination phase of alcohol metabolism, *i.e.* any alcohol consumed before the accident had been completely absorbed into her system and was in the process of being cleared from her system; (2) that the average rate of alcohol elimination applies: in practice, alcohol clearance slows down as the alcohol concentration in blood approached background levels, as in this case.
- **MDMA.** The average rate of clearance from blood is reflected in the plasma half-life, the time taken for the concentration in plasma to decrease by 50%. In the case of MDMA, the half life is 6–9 h. Back-calculation gives an estimated blood concentration at the time of the accident of 0.32–0.35 mg/L blood, depending on the half-life value used.
- **Diazepam and its metabolite, desmethyldiazepam.** Diazepam clears slowly from blood, with a half-life in the range 21–37 h. Back-calculation gives an estimated blood concentration at the time of the accident of 0.13 mg/L blood. Desmethyldiazepam is a metabolite of diazepam and is an active sedative/tranquiliser drug in its own right. This also clears slowly from blood with a half-life of 50–99 h. Back-calculation gives an estimated blood concentration at the time of the accident of 0.34 mg/L blood.
- **Δ^9-THC.** This is the active drug substance in cannabis. It clears rapidly from blood and is often not detected in road traffic cases. A trace of this substance was detected in the blood sample, which indicates that there would have been a higher concentration of this active cannabinoid in the driver's blood at the time of the accident. It was therefore likely that the accused was driving whilst under the influence of the active drug constituent of cannabis. It is usually difficult to detect Δ^9-THC in whole blood samples after about 3 h have passed from when cannabis was used. Under favourable conditions, *e.g.* when controlled administration of cannabis takes place under laboratory conditions and fresh plasma is collected, the

detection time may be longer, up to 6 h or so, but this is not necessarily the case when whole blood is analysed.

- Δ^9-**THC metabolite**. This has no drug activity. Its usual significance is as an indicator that cannabis has been administered at some previous time. This substance persists in the blood for many hours after cannabis has been administered and can be detected for up to 24 h, possibly longer.

Alcohol, MDMA, diazepam and cannabis all have the potential to cause driver impairment.

14.5.5.1 Conclusions. The combined effects of this polydrug mixture would include aspects of each of the individual drugs, making it difficult to predict the net effect. None of the drugs detected was present at the high concentration associated with overdose and toxic side effects, but they were present at concentrations typical of therapeutic use (diazepam) or recreational drug use (alcohol, MDMA and cannabinoids). Alcohol (even at low concentrations) is sedative in nature, as are diazepam, desmethyldiazepam and Δ^9-THC. Diazepam and desmethyldiazepam can cause sedation, especially in new users of the drug. Even single doses of diazepam have been shown to cause significant impairment due to CNS depression. Δ^9-THC causes intoxication resulting in physical and mental impairment which have been shown to increase the likelihood of accidents when driving, especially in combination with other drugs such as alcohol and benzodiazepines. A single dose of cannabis can cause impairment when driving for up to 3 h. By contrast, MDMA has the effects of a stimulant (feeling of energy, alertness, *etc.*) as well as being a mild hallucinogen (affecting perception of the environment, mood, *etc.*). The main effect manifests as euphoria. Stimulants can adversely affect driving performance in different ways: (1) by diminishing the skills needed by the driver for accurate control of the vehicle, (2) by increasing risk-taking behaviour and (3) by causing fatigue after the initial stimulation wears off.

It is difficult to predict the effects of drugs on the individual from their blood concentrations alone, especially when several drugs are taken at the same time. However, the drugs found in the present case have the potential to cause impairment of a driver, especially because of the simultaneous presence of four drugs (alcohol, diazepam, desmethyldiazepam and Δ^9-THC) which depress the CNS. Taken together, the blood analysis results would support the opinion of a forensic medical examiner or police officer made at the time that there was a condition present in the driver which could be due to drugs.

14.5.6   Hair Analysis Case

A man in his mid twenties was found dead at home. He had been released from prison 6 weeks before his death. Analysis of an autopsy blood sample found diazepam 0.4 mg/L blood; desmethyldiazepam 1.7 mg/L blood; methadone, 0.3 mg/L blood and methadone metabolite (EDDP) 0.3 mg/L blood. The cause of death was given as methadone and diazepam intoxication.

A 4-cm hair sample taken at autopsy was segmented into four 1-cm portions, each of which related to a 1-month period prior to death. However, the hair did not cover the 1–2 week period prior to death as the roots were not included. Analysis of the hair revealed the pattern of drug use by the deceased (Figure 14.5): this included the presence of heroin metabolites in all four hair segments but methadone only in the segment closest to the scalp. The presence of drug metabolites indicates that the drugs were consumed and were not due to superficial contamination. The results should not be overinterpreted, but they do indicate that the deceased was a regular heroin user (from the presence of 6-MAM) and that he had not used methadone before his release from prison.

14.5.7   Fires and Explosions

In the investigation of a fatal fire, the role of the toxicologist is often to establish the presence of carbon monoxide in the blood of the fatality. This has an important medico-legal significance as, taken in conjunction with the pathologist's observations concerning the presence or absence

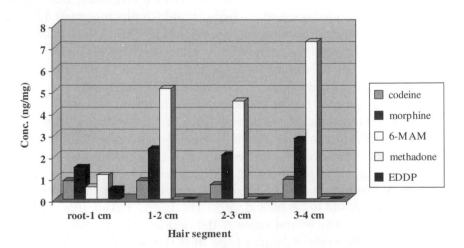

Figure 14.5   Analysis of a hair sample from a fatality.

of soot in the air passages, it establishes whether or not the deceased was alive at the start of the fire. If death occurred before the fire, the possibility of homicide must be considered.

Carbon monoxide is a colourless, tasteless, odourless gas produced by partial combustion of carbon-containing materials. It is produced almost immediately during the course of a fire and those present in the fire locus are exposed essentially from the beginning of the fire. The gas is inhaled and absorbed through the lungs into the blood circulation system. The significance of carbon monoxide in fires is that casualties can be overcome by the effects of the gas while asleep or at rest, perhaps unaware that a fire has started. If they subsequently become aware of the fire and start to exert themselves, the sudden demand for more oxygen can cause fainting and collapse, resulting in failure to escape.

In addition to carbon monoxide, the toxicologist may be requested to measure other toxic materials in the blood which have originated in the fire, particularly hydrogen cyanide, which is produced by the effects of heat on nitrogenous materials such as polyurethane, polyacrylonitrile, wool and silk. Cyanide levels are usually measured in whole blood specimens, even if plasma or serum is available. Cyanide is present at low levels (up to 0.25 mg/L of blood) in the normal population and at slightly higher levels (up to 0.5 mg/L blood) in smokers, as cyanide is present in cigarette smoke. The toxic effects of cyanide result from its action within body tissues, where it blocks cellular respiration at the cytochrome level. Symptoms of cyanide poisoning include dizziness, headache, chest pain, confusion, staggering, slowing of the heart and breathing rates, unconsciousness, coma and death. The significance of cyanide in fires lies in its rapid incapacitating effect, which can render casualties unconscious in a very short time. This is in contrast to carbon monoxide, which may build up in blood more slowly.

Most fire deaths in the UK and elsewhere occur in dwellings, and research has shown that more than 80% of these have been exposed to hydrogen cyanide in the fire. However, the toxicologist will also be asked to analyse blood specimens for the presence of alcohol and drugs, because these might have affected the ability of the deceased to escape from the fire scene or even have contributed to the events leading to the initiation of the fire.

14.5.7.1   Case Study: A Fire in a Leisure Centre.   A fire occurred in a leisure centre in which six people died. The fire was small and did not spread outside a single room. The casualties occurred because of the spread of smoke, heat and toxic gases along a corridor, blocking the escape of the occupants who had been in side rooms. The results of

Table 14.3   Carbon monoxide and cyanide levels in fatalities resulting from a
                 fire in a leisure centre.

Casualty	% combined haemoglobin/carbon monoxide in blood (%HbCO)	Cyanide level (mg/L)
1	48	0.9
2	48	0.8
3	41	3.1
4	34	0.5
5	29	2.0
6	38	2.3

the toxicological analysis of blood samples are given in Table 14.3 and
show that the casualties had breathed in significant amounts of both
carbon monoxide and cyanide, leading to their collapse and death.

The toxicologist can also contribute to the investigation of deaths
resulting from explosions, as many of these fatalities are actually caused
by a fire accompanying the explosion. Also, analysis of blood from
fatalities in a gas explosion may assist the explosion investigator in
identifying the nature of the gas involved. Usually this is either consumed in the explosion or dissipated after the event, and cannot be
detected by the investigator at the locus. However, it may still be
detected in the blood of the fatalities who were alive before the explosion
and inhaled the explosive mixture of gas and air involved.

14.5.7.2   Case Study: A Gas Explosion in a Dwelling House.   Early
one winter morning, a gas explosion partially demolished a building
containing four flats in a city suburb. Five people were killed, including
the four members of one family in a ground-floor flat (fatalities 1–4)
and the resident of an upper flat (fatality 5). The causes of death were
established at autopsy as blast injury, crush asphyxia (due to collapse of
the building) and inhalation of smoke and fire gases from the subsequent fire.

The initial investigation produced three possible sources of flammable
gas: the mains gas supply (methane), old coal-mine workings under the
flats (firedamp, a mixture of methane and ethane), and bottled gas
(butane). Subsequent investigation revealed that the actual source was
the mains gas supply. The type of gas was confirmed by the toxicological
analysis of the blood specimens taken at autopsy, which showed the
presence of methane only; see Table 14.4. Also, the levels of methane
were in inverse relationship to the levels of carbon monoxide, indicating
which of the fatalities had survived longest after the explosion. The
methane in the building had been vented by the explosion and survivors

Table 14.4 Carbon monoxide and methane levels in fatalities resulting from an explosion and fire in a dwelling house.

Fatality	Combined haemoglobin/carbon monoxide in blood (%HbCO)	Methane (mL/L)
1	2	1.14
2	8	Trace
3	11	0.28
4	32	0.15
5	71	0.04

would have breathed out any gas in their blood, but simultaneously would have inhaled carbon monoxide from the fire which was ignited by the explosion.

The father of the family appeared to have initiated the explosion with his activities in the kitchen, and was found to have suffered blast injuries but not to have inhaled any carbon monoxide. By contrast, the upstairs neighbour (fatality 5) did not have any methane in her blood sample but had the highest level of carboxyhaemoglobin.

BIBLIOGRAPHY

R. C. Baselt, *Disposition of Toxic Drugs and Chemicals in Man*, 8th edition, Biomedical Publications, Foster City, CA, 2008.

L. Brunton, B. A. Chabner and B. Knollmann (eds.) *Goodman and Gilman's The Pharmacological Basis of Therapeutics*, 12th edition, McGraw-Hill, New York, 2010.

Clarke's Analysis of Drugs and Poisons, Pharmaceutical Press, London, 2004.

O. H. Drummer, *Forensic Toxicology of Drugs of Abuse*, Arnold, London, 2001.

S. B. Karch, *Karch's Pathology of Drug Abuse*, 4th edition, CRC Press, Boca Raton, FL, 2008.

P. Kintz, *Analytical and Practical Aspects of Drug Testing in Hair*, CRC Press, Boca Raton, FL, 2006.

B. Levine, *Principles of Forensic Toxicology*, 2nd edition, AACC Press, Washington DC, 2003.

CHAPTER 15

Analysis of Body Fluids

NIGEL WATSON

University of Strathclyde, Forensic Science Unit, Royal College, George Street, Glasgow G1 1XW

15.1 INTRODUCTION

The first suspect to have been convicted largely on the basis of DNA analysis of blood samples was sentenced at the Crown Court in Leicester on 22 January 1988. This case marks an important milestone in the science of forensic body fluid analysis. Since this case, DNA technology has become commonplace in forensic laboratories around the world and has been instrumental in establishing both guilt and innocence in many court cases. The terrorist attack on the World Trade Centre on 11 September 2001 resulted in a mass disaster victim identification problem on an unprecedented scale and was compounded by the crash of American Airlines Flight 587 on Monday 12 September 2001 with 251 passengers and 9 crew in a New York neighbourhood. The huge number of samples for testing required what has been described as DNA identification on an industrial scale. Conventional DNA profiling and the latest analytical techniques available at the time were used. Other applications of particular interest have included the analysis of human remains alleged to be those of the Tsar and Tsarina of Russia, murdered on or around 16 July 1918, and the verification of the death of Josef Mengele, the Nazi doctor at Auschwitz. There are many other examples.

Crime Scene to Court: The Essentials of Forensic Science, 3rd Edition
Edited by P. C. White
© Royal Society of Chemistry 2010
Published by the Royal Society of Chemistry, www.rsc.org

Mention of forensic scientific analyses of blood were recorded in 13th century Chinese texts, but the modern science of blood-typing began with a discovery by Karl Landsteiner of the different types of blood, what we now call the ABO blood-typing system. The first account, published in 1901, reported that the blood types present in 2-week-old serum stains on linen could be determined. By 1902 the four blood types, A, B, O and AB had been described.

This chapter describes the chemistry, biochemistry and biology used by forensic scientists in the analysis of body fluid samples (blood, semen and saliva) and hair. Greater emphasis has been placed on the DNA aspects of the discipline, to reflect the advances and the interest in this area since 1985. However, consideration has also been given to the biochemical tests used to search for blood, to identify stains that might be blood and to identify the species of origin of the blood. Another aspect is the interpretation of the blood splash or spatter patterns observed at many crime scenes to establish the events of the crime; this topic is covered in detail in Chapter 7.

Consideration has also been given to immunological and protein blood-grouping methodologies. In many laboratories around the world these techniques have been superseded by DNA analyses, but they will continue to be of interest because of their use in many important and historic cases.

15.2 BIOLOGICAL EVIDENCE

Blood and other body fluids, and dried stains arising from them, may contribute important physical evidence in three ways. The occurrence of a blood or body fluid stain in a certain position can be of value as evidence. For example, finding a seminal stain on a blanket or bed sheet can support the account of a crime given by a rape victim, or the occurrence of blood on a weapon can substantiate an account of a violent crime.

The shape, position, size or intensity of a body fluid stain may support a particular account of events concerned with a crime rather than a number of alternative accounts. When a blood drop strikes a flat surface the shape of the resulting stain depends on the angle of impact, ranging from a perfectly circular mark, arising from a perpendicular impact, to an elongated mark where the length of the mark is proportional to the angle between the trajectory and the surface. Therefore, it is often possible, from a collection of blood marks, to determine a point of origination and if there was more than one point source of the blood. This kind of information is clearly of great importance in testing the validity of conflicting accounts of the sequence of events of a crime. The examination of the blood staining patterns, as illustrated in Chapter 7, may often yield more useful information than just the biochemical analysis of the blood.

Finally, blood-typing analysis of body fluids and their stains can eliminate whole groups of people as suspects. Forensic blood-typing analysis is comparative, so if a stain does not match a suspect, then that person cannot have been the source of the stain material. If there is a match of the blood types then the person is one of a group of people who could be the source of the stain. In any population of people there will be a group who share the blood types or combination of blood types found. The implication that the stain came from a particular person is stronger if the type or combination of types is rare and hence the size of the group is small. If the type or combination of types is common, the group is large, so the possibility that the stain came from some other person, who by coincidence possesses the same type or combination of types, is more believable.

15.2.1 Blood

Blood constitutes about 7.7% of the body weight of a person and it acts as a transportation system. It is composed of a fluid, plasma, which accounts for 55% of the total volume, with the balance being made up of cells. Plasma is 90% water and is a straw-coloured, almost clear liquid. It contains dissolved plasma proteins, metal ions and organic substances. The different types plasma proteins include serum albumin, serum globulins and serum enzymes. The globulins are a group of proteins that include the antibodies.

Antibodies recognise and bind to specific chemical groups. They mark foreign materials as being 'non-self' to the host organism, and the cells of the immune system assimilate and destroy such marked non-self objects. It is possible to collect and purify antibodies that react to specific chemical configurations. The production of antibodies can be induced in an animal (or a human) by injecting small quantities of a foreign material. An animal may therefore be 'immunised' against specific proteins. The materials to which the antibodies bind specifically are called antigens. Analytical tests, called immunological tests, are used in body fluid analyses and exploit the specificity and binding characteristics of antibodies.

Serum is the fluid left after blood clotting and so differs from plasma in that it lacks the clotting agents. Plasma is prevented from clotting by the addition of an anticoagulation agent, heparin or ethylenediamine tetra-acetic acid (EDTA), to a blood sample as it is collected. The term 'anti-serum' refers to serum used in medical treatments or tests because it contains antibodies with a useful specificity.

As indicated above, the remaining 45% of the volume of the blood is composed of cells. The most common type is the red blood cell, or erythrocyte, which is circular in mammals but oval in other vertebrates. There are about 5 million cells in 1 μL of blood. The red cells have no nucleus and are the only cells to contain the pigment haemoglobin, which gives blood its red colour. The outer cell membrane, in common with all other body cells, carries biochemical markers that may be utilised by the forensic scientist.

Another blood cell type is the white blood cell, or leucocyte, which is concerned with the immunological defence of the body. The leucocytes are present at much lower cell densities than the red cells, around 5000–10 000 μL^{-1}. White cells have a nucleus and contain DNA in the form of a protein–DNA complex called chromatin. The chromatin is composed of units called chromosomes.

All human cells, with the exception of erythrocytes and spermatozoon cells, have 46 chromosomes that include 1 pair of sex chromosomes plus 22 pairs of analogous chromosomes. The sex chromosomes are called X and Y. Males have one of each type and females have two X chromosomes. The chromosomes carry genetic information encoded in the DNA that determines the characteristics of the host. The chromosomes of each pair carry information relating to the same characteristics. One member of each pair is from the host's mother and the other from the father. Although the information relates to the same characteristic, such as eye colour, the two members of each pair of chromosomes may carry different versions of the genetic message. The inheritance of one chromosome out of each pair is random, so that each new baby possesses a shuffled collection of half of the chromosomes available from each parent. Therefore if two polymorphisms occur on different pairs of chromosomes, they can be regarded as being inherited independently of one another. Moreover, in nature, there can be an exchange between the analogous pair of chromosomes. In some circumstances the inheritance of polymorphic regions of the DNA on the same chromosome can be regarded as being independent so long as the locations of the polymorphic sites are sufficiently far apart on the chromosome.

The blood also contains particles called platelets. These are fragments of larger cells in the bone marrow, megakaryocytes; they contain no nucleus but have an important role in the process of blood clotting. Clotting, or coagulation, is the process of localised solidification of blood at an injury site and the term 'agglutination' refers to the binding of blood cells together by antibodies.

15.2.2 Semen

Semen is a suspension of cells called spermatozoa, or sperm, in seminal fluid. The sperm cell is the male sex cell, or male gamete, and its biological function is to pass genetic information from the father to the female gamete, the egg or ovum. Human gametes carry a set of 23 chromosomes, one from each of the 23 pairs making up the normal human complement of 46. 1 mL of semen from a normal healthy adult male contains around 60–100 million sperm cells. The volume of the average ejaculate is 3 mL, with a range of 1–6 mL. Most of the volume of the semen, 75–90%, is made up of material from a group of organs known collectively as the accessory glands. These are the prostate gland, the seminal vesicles, Cowper's glands and the glands of Littre. The prostrate gland produces a slightly acid fluid ($\sim$pH 6.4) rich in calcium, also containing zinc, sodium, citric acid and the enzymes fibrinolysin and acid phosphatase. The seminal vesicles are two glands between the scrotum and the rectum that secrete a fluid rich in fructose, phosphorylcholine, citric acid and ascorbic acid.

Spermatozoa were first discovered by the Dutch microscopist Antonie van Leeuwenhoek, and reported in a letter written in November 1677, published in the *Philosophical Transactions of the Royal Society of London*. Normal sperm cells have a head, a midpiece and a tail. The spermatozoa can swim, the midpiece acting as a motor causing the tail to move so that the cell can move of its own accord. The head is oval, 3–5 µm long and 1.5 µm thick. The midpiece is about 7–8 µm long and the tail at least 45 µm. Smaller-headed spermatozoa (less than 4 µm long) also exist.

Each sperm cell has only half of the usual complement of DNA of a standard cell, but the nucleus represents a much larger proportion of the mass of the cell than standard cells and the density of the cell suspension in semen means that semen is in fact an excellent source of DNA.

15.2.3 Saliva

Saliva is a watery fluid secreted by the salivary glands of the mouth. It contains mucin that assists the passage of food into the oesophagus, and salivary amylase that digests starches. The saliva also contains cells and hence DNA. These are mainly bacterial cells, but saliva also contains cells shed from the inside surfaces of the cheeks.

15.3 TESTS FOR BLOOD AND BODY FLUIDS

The location of blood or other body fluids often requires a 'search test'. For example, blood stains on a dark-coloured garment, or dried saliva or semen can often be difficult to detect. To find blood stains, a piece of filter paper can be rubbed over an area of the garment and the paper tested by chemical tests for haemoglobin and hence the presence of blood (see section 15.3.1). To find semen stains, a dampened piece of blotting paper is applied to an area of a suspected stain. The moist paper will pick up some of the dried seminal fluid from the stain. On spraying with a solution of reactants (see section 15.3.2), the paper will change colour in any area where seminal material has been transferred. An enzyme in the semen called acid phosphatase causes the colour change. In an adaptation of the test, the presence of saliva may be detected by means of salivary amylase, an enzyme component of saliva.

The tests are not entirely specific because other substances, generally of plant origin, may undergo the same reactions as the haemoglobin and salivary or seminal enzymes. These tests are used, even when the stain is visibly quite obvious, to eliminate possible alternatives such as paint, ink or food stains which might be mistaken for body fluid stains. The term 'presumptive test' is sometimes used for some of these biochemical tests because a positive result is presumed to be due to the presence of the target material but they require a confirmatory test to substantiate any finding.

15.3.1 Tests for Blood

The haemoglobin in the blood of mammals is characteristic of that tissue and occurs in no other. It has the capability of behaving as an enzyme in the presence of hydrogen peroxide, when it can catalyse the oxidation of materials. This property of haemoglobin is used as a biochemical test for the detection of blood.

Certain dyes, *e.g.* leuco malachite green (LMG), have the characteristic of existing in two different states, a reduced form and an oxidised form. Some of them are colourless in the reduced state but brightly coloured in the oxidised state. If a drop of a solution of such a dye, kept in its reduced state, is applied to a test blood stain, no colour should be observed. However, if a drop of hydrogen peroxide is added, the colour of the oxidised dye develops. This is a two-step test and is applied to testing stains by rubbing the stained area of an item with a filter paper. A drop of the reduced dye is added to the paper, followed by a drop of hydrogen peroxide. An indication that the stain may be blood can be assumed only if the colour of the oxidised dye is observed after the addition of the hydrogen peroxide. Note that the blood of any individual of any species will produce a positive result, and that no information regarding the origin of the blood can be gained. Such tests are examples of presumptive tests.

Another example of a dye used in these tests is phenolphthalein. This dye is colourless in its reduced state and pink in its oxidised form. This test is often called the Kastle Meyer (KM) test and can also be used for presumptive testing. The dye luminol undergoes a chemiluminescent reaction when oxidised, *i.e.* a reaction in which light is emitted. If it is combined with an oxidising agent, sprayed on to a surface and then viewed in darkness, any luminescence will betray the presence of a bloodstain.

15.3.2   Tests for Semen

15.3.2.1   Microscopy.   The most definitive test for semen is the microscopical identification of spermatozoa. Because of the relatively large quantities of DNA present in sperm heads and the fact that DNA can be detected if treated with a staining reagent, sperm cells can be readily identified by their distinctive shape when viewed through a high-power microscope. Semen is the only body fluid that possesses sperm cells.

15.3.2.2   Acid Phosphatase Test for Semen.   It is possible for semen not to contain any spermatozoa; this is a condition called azospermia, which may be deliberate in the case of a vasectomised male, or may arise due to a medical condition. In the absence of spermatazoa, identification of a sample as semen can be made by testing for the presence of the enzyme acid phosphatase. This enzyme is present in semen at very high levels and can be detected by using a colour-forming chemical reaction. If present, the enzyme produces a purple colour and the degree of the colouration produced is proportional to the quantity of enzyme present. This reaction can be used as a presumptive test for stains suspected of being semen, or as a search test.

The reaction may also be used quantitatively in circumstances where a known volume of fluid is to be analysed. An example of this is a vaginal swab where the volume of fluid collected by the swab can be estimated. A portion of the swab can be added to the solution of the test reagents and the intensity of the colour generated can be related to the quantity of acid phosphatase present. Vaginal fluid possesses acid phosphatase activity in the absence of semen, but at much lower levels. Elevated acid phosphatase levels are therefore indicative of the presence of semen.

15.3.2.3   Choline and p30.   The choline originating from the seminal vesicles may be detected by microcrystal tests producing characteristic crystals. A more specific test is for a protein called prostate specific antigen (PSA or p30) that is semen-specific. This protein is produced in the prostate gland and may be detected by a variety of immunological tests. It is useful for confirming the presence of semen in the absence of sperm cells, but it should be noted that p30 is also present in male urine and a very simple test for p30 is available commercially and is used for forensic applications.

15.3.3 Tests for Saliva

The salivary amylase present in saliva may be used to identify this body fluid. A specimen of a stain suspected of being saliva is removed and placed in a solution of soluble starch. If, after the addition of an iodine solution, a deep blue colour is produced, the stain does not contain any amylase. If amylase is present it breaks down (hydrolyses) the starch with no blue colouration of the solution. Therefore, a positive test for saliva is marked by an absence of colour.

15.3.4 Determination of the Species of Origin

A question that may well be asked is, 'Is the blood human?' To answer this question, two different types of test can be employed. One approach is to test for serum proteins that are species-specific. Most of these tests rely upon detecting the species-specific proteins by immunological tests that use antibodies produced against the species-specific marker proteins.

In practice, unless it is important to determine which species a non-human blood or bloodstain came from, such tests will not be routinely carried out. The DNA analysis will not yield a result if used on non-human DNA except, perhaps, for certain primates.

15.4 BLOOD-TYPING

Blood-typing has been largely superseded by DNA analysis as a means of human identification in forensic science. When blood-typing is performed, two methods can be used. One method makes use of antisera containing antibodies that can recognise and bind specifically to chemical groups, referred to as markers, on the surface of red blood cells. This is an immunological test system and the ABO typing system is one of several blood-typing systems based on this type of test.

The action of the antibodies can be used in a variety of ways to distinguish between the different blood-types belonging to the same typing system. An example is where an antibody recognises the marker and links adjacent cells to form agglutination or clumping, which can be easily observed. Another antibody test is where an enzyme is linked to the antibody. The presence of the antibody, and hence the marker in a sample, is revealed by a colour change produced by the action of the antibody-bound enzyme on a colourless substrate. The test is similar in concept to the biochemical test for semen described earlier.

The other approach makes use of protein variants which have small differences in their chemical composition, and hence slightly different electrical charges when in solution. If a solution of proteins is analysed by electrophoresis they are subjected to an electric field and the proteins will separate according to their charge. This permits identification of proteins and their discrimination

Once distinct blood types have been identified by either of these two methods, the population can be divided into groups of people who share the same blood types. Many of the blood types can be determined from blood stains and certain blood types can be determined in other body fluids. In the forensic context, a comparison of the blood types of the suspect and the stains can eliminate the suspect as a source of the blood if the blood types do not match.

Several independent blood-typing systems can be co-analysed to improve discrimination. As the number of typing systems increases, the number of people who share the same combination of blood types will be reduced. Blood-typing therefore, becomes more discriminatory and a coincidental match of a stain with a suspect becomes less likely. The ultimate goal would be to use a large number of typing systems such that the chance of two people having exactly the same combination of results would become so small that such a coincidence would not be credible. However, the quantity of bloodstain available may be limited and restrict the number of blood-typing systems that can be used.

15.4.1   Genetics

The study of the inheritance of biological traits is called genetics and one important part of genetics is the study of variations in inherited traits. These variations, called polymorphisms (literally 'many forms'), are where several different versions of a trait coexist simultaneously within a population. This means that we can classify people who share the same type of trait into groups.

The polymorphic traits of most interest to the forensic scientist examining a blood or body fluid stain are those of a biochemical nature which are determined by the inherited genetic make-up of the individual in question. These inherited features are likely to remain constant and consistent throughout the lifetime of the individual and are independent of the state of health of the donor. Fortunately, many of the inherited biochemical features that can be detected in blood or body fluids are under the influence of a single inherited piece of information.

The location of a gene in the genetic content of a cell of a person is called the 'genetic locus', but different versions of the genetic information at a genetic locus may coexist within a population. The different versions are called 'alleles'.

Every person inherits one set of genetic information from their mother and another set, analogous to the maternal set, from their father. The term 'genotype' is used to describe the combination of alleles a person has inherited for any given genetic locus. Therefore, everyone possesses a dual set of inherited genetic information. It is entirely possible that an allele inherited from one's mother is exactly the same as the corresponding allele inherited from one's father, in which case the genotype is said to be 'homozygous', *i.e.* the two alleles are the same. However, it is also possible that two different allelic versions of a gene are inherited. In this case the genotype is 'heterozygous' for that trait.

15.4.2 Immunological Markers

Soon after its discovery the ABO system was put to use for the analysis of blood stains found in crimes and also in cases of disputed paternity. Since that time other immunological systems have been described including Rhesus, MNS, Kell, Duffy, Lewis, Km and Gm. Many have been employed in forensic examinations. The ABO system was the first blood-typing system to be characterised. It was also found that the ABO type could be determined from other non-blood body fluids in most individuals. There are four types of blood in the ABO system. These variations occur because of the inheritance of different alleles for a gene, which specifies an enzyme that converts a precursor into either substance A or substance B. There is a third allele, O, which fails to modify the precursor. The allele pair of alleles and their corresponding blood types are as follows:

Alleles	AA	AO	BB	BO	AB	OO
Blood type	A	A	B	B	AB	O

A quirk of the immunological test method is that the genotypes AO and AA are indistinguishable and type as blood type A. The same is true for genotypes BB and BO, which type as blood type B. Therefore the immunological test for ABO can distinguish four different types: A, B, AB and O.

15.4.3   Protein Markers

Many polymorphic protein systems have been described. Not all of them are suitable for forensic analyses and, of those that are, some do not occur in non-blood body fluids. The most widely employed technique for discriminating between the variant types is electrophoresis or a closely related technique called isoelectric focusing (IEF). Phospho-glucomutase (PGM) is an enzyme that is present in blood and other body fluids and the polymorphism of this enzyme can be analysed by using IEF.

The IEF analysis can distinguish 10 different PGM genotypes. The inheritance of the PGM polymorphism is independent of the ABO types, so if a blood sample is analysed for both ABO and PGM then 40 distinct combinations can be identified. It is therefore much less likely that two unrelated people chosen at random will share the same combination of ABO and PGM polymorphism, so the ability to distinguish between the bloods of two people is enhanced by using two independently inherited polymorphic systems.

15.5   DNA AND ITS ANALYSIS

Forensic serology has always made use of techniques developed in other disciplines, from immunology to electrophoresis, genetics and more recently molecular biology. DNA analysis, which is based on molecular biology, is now used in virtually every forensic laboratory for human identification and discrimination. The following is an outline of the nature of DNA and some key techniques that are used in its analysis.

15.5.1   DNA Structure

Our DNA, the material that carries genetic information, exists within a cell in structures called chromosomes. Human cells possess 23 pairs of chromosomes. Human individuals, like all other animals, originate from the combination of a male gamete (sperm) and a female gamete (ovum). Gametes carry only a half complement of DNA, *i.e.* 23 chromosomes, so the result of the combination of the gametes is a zygote that has a complete complement of DNA, *i.e.* 46 chromosomes in humans.

All the cells of an animal arise from the subsequent cell divisions the zygote undergoes. Therefore all the cells in a person's body have identical DNA. Each one of the 46 chromosomes in human cells includes a single piece of double-stranded DNA in which the two strands are wound around each other in the now famous 'double helix'. These double strands are themselves wound around a supporting structure of protein. The DNA is coiled and supercoiled, enabling a vast length to be present within the microscopic space of a cell nucleus.

DNA is composed of a backbone of alternating sugar and phosphate molecules. The sugar is a five-carbon ring deoxyribose which is linked *via* the phosphate group from the 5′ carbon on one sugar to the 3′ carbon on the following sugar. The feature that endows the molecule with its biological significance is the attachment of a chemical group, called a base, to every sugar in the chain. Two classes of base occur in DNA; see Figure 15.1. One, a pyrimidine, has a six-member ring of carbon and nitrogen. The other, a purine, is composed of fused five- and six-member rings. There are four types of base in DNA: adenine (A) and guanine (G), which are purines, and cytosine (C) and thymine (T), which are pyrimidines. The DNA strand is assembled from individual units, called nucleotides, which comprise a deoxyribose sugar, a phosphate and a base. There are therefore, four types of nucleotide corresponding

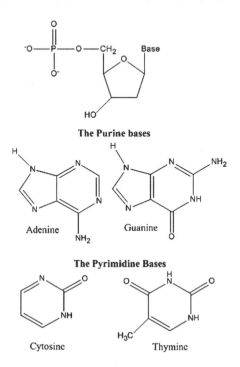

Figure 15.1 The chemical building blocks of DNA.

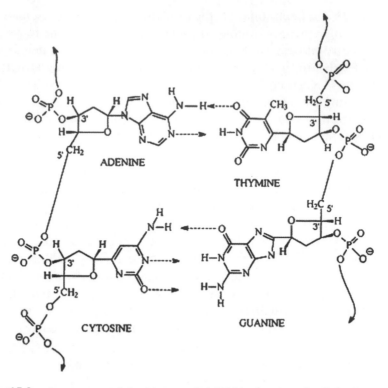

Figure 15.2   A segment of double-stranded DNA showing the links between the thymine (T) and adenine (A) bases and the cytosine (C) and guanine (G) bases.

to the four types of base and these nucleotides can be linked to form a chain (polynucleotide).

The strands of the double helix are held together by hydrogen bonding between the bases on adjacent strands as illustrated in Figure 15.2. The cytosine and guanine bases, C and G, form three hydrogen bonds and the T and A bases form two hydrogen bonds. Therefore, the two DNA strands of a pair do not carry the same base sequence but complementary sequences. Another feature is that the direction of the backbone of the DNA strands run in opposite directions as defined by the $5' \rightarrow 3'$ carbon-to-carbon linkage. For example, the sequence 5'-GGCTATAAT-3' on one strand would be matched by the sequence 3'-CCGATATTA-5' on the corresponding strand, given the A–T and G–C pairing.

The sequence of the nucleotides on the DNA strand encodes genetic information. This is called the genetic code and it describes the composition of the peptide chains and hence the proteins produced by the organism. As enzymes, which are composed of peptide chains, control all of the chemical reactions of the body it follows that the enzymes prescribed by the genetic code dictate the entire biochemistry of a person. Each amino acid of every peptide chain is determined by a three-base code. The sequence of the codes therefore determines the sequence of amino acids and hence the characteristics of the peptide. When the cell divides, the nucleotide sequence is replicated in the new DNA so that the daughter cells have the same sequence, and so the same code, as the parent cell.

Human DNA has been found to consist mostly of base sequences that carry no genetic information; only a very small proportion of DNA carries the sequences that code for products. The non-coding regions of the DNA are in fact the best regions for human identification purposes because they are not subject to selection pressure and possess greater variation.

15.5.2 DNA Analysis

DNA is robust and has been recovered from bones thousands of years old. In forensic applications, one strategy of analysis has become the method of choice. This makes use of enzymes that replicate the DNA. These enzymes are called polymerases and their function in nature is to replicate the DNA prior to cell division, so that the daughter cells possess the same complement of DNA as the parent cell. They can also be used in controlled reactions to replicate specific pieces of the DNA, over and over again, in a chain reaction. Many copies of that piece of DNA are produced and the process is called DNA amplification.

DNA has to be extracted from the specimen before it can be amplified and analysed. Various extraction methods can be used to collect the DNA. One method, used extensively in the past and still in use in some laboratories is based on the use of a commercially prepared chelating resin. If the specimen of stained material is boiled up in the presence of the resin the DNA is released from the remains of the cells in the stain but remains undegraded in the solution. The resin also removes metal ions that are required by degradative enzymes. The resin method is quick and cheap but it does not remove other materials that may be present in some forensic samples and which may affect subsequent analyses.

In the UK and in many other laboratories worldwide, DNA extraction is now carried out by using a process where the disrupted cell components or material recovered from a sample or stain are passed through a column in which the DNA binds to the material in the column and the rest is washed away. The bound DNA can then be eluted and collected. Once the DNA has been recovered it can be used for the amplification of selected sections of the base sequence. The quantity of

DNA in many forensic samples is too small to be analysed. By replication, or amplifying the sections of the sequence that are of interest, in the present context polymorphic alleles, sufficient quantities of the genetic loci of interest can be produced.

15.5.3   DNA Amplification

The replication of a selected DNA sequence present in a sample of DNA is carried out by the polymerase chain reaction (PCR). The PCR is carried out in a reaction mixture composed of the various components in an aqueous solution in a small plastic centrifuge tube, where typically the reaction volume is 25–50 µL. The reaction is controlled by changing the temperature and there are three stages: denaturation, annealing and extension as identified in Figure 15.3. These stages are repeated, typically, 28 times. The temperature is controlled by means of a programmable thermal block called a thermal cycler. The reaction tubes are placed in the block which is heated for the first stage to around 94 °C to denature the double-stranded DNA isolated from the specimen. This denaturation consists of breaking the hydrogen bonds that hold the two

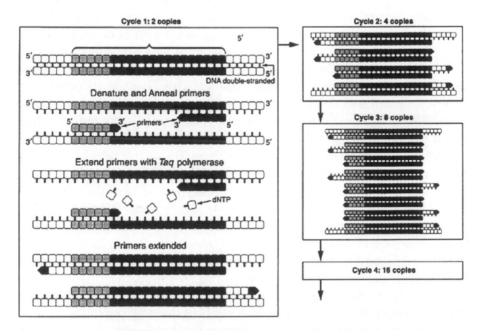

Figure 15.3   Amplification of a DNA sequence using the polymerisation chain reaction (PCR) (from M.J. Greenhagh, *DNA Profiling in Forensic Science*, 3rd edition, RCS, 1993, with permission).

strands together, so rendering the DNA single stranded. The temperature is then lowered, typically to 55 °C for the annealing stage, but this can vary, and raised to around 75 °C for the extension phase.

The reaction components consist of an enzyme, *Taq* polymerase, which builds the DNA strand; the nucleotides (dNTP) which are the building blocks of the DNA; and the primers. The reaction mixture also contains DNA that has been recovered from the specimen. The specimen may be a blood sample taken under clinical conditions from a suspect or victim, a blood or body fluid stain, or any kind of body tissue. Hair, especially the hair roots, and mouth swabs which collect cells from the inside of the cheeks are good sources of DNA and involve non-invasive sampling techniques.

The amplification process operates only on a small portion of the DNA at a time. The specific target that is to be amplified is marked out or defined by a pair of short sequences of DNA which have been made synthetically and which are usually about 20 base pairs long. They are called DNA primers or oligonucleotides. Two different primers are used. The base sequence of one is selected to bind to one side of the target DNA sequence after it has been denatured to single strands and the other primer binds to the other side of the target sequence. The target sequence of DNA bases which lies in between the two priming sites is used as a template for the amplification and is referred to as the template DNA.

The reaction mixture contains each of the four types of nucleotide and the action of the polymerase enzyme is to add these nucleotides to the 3′ end of the primers where they will bind to the template strand. The type of nucleotide added to the primer is determined by the base on the template. Note that it is the complementary base which is added, so that base paring can occur between the extended primer and the template. The polymerase will continue to add bases to extend the primers until it is interrupted.

The process of primers binding to their respective binding site is called annealing and the process of extending the primers along the template is called the extension phase. After one complete cycle of denaturation, annealing and extension the processes may be repeated. For the second cycle there should now be double the number of templates available because the new strands created from the last cycle will themselves act as templates for the subsequent cycle.

Notice that when the new strand acts as a template the 5′ end terminates with the primer and does not continue on as the DNA isolated from the specimen would. As a consequence the extension will stop at the end and so the new DNA produced in this case will be bound on one side by the primer which has been extended and at the other end by a base sequence complementary to the partner primer. The analogous (but 'opposite') case will be true for the other alternative product. These new PCR products are called short products and are defined by either a primer sequence or a sequence complementary to a primer at either end. They are amplified preferentially in the reaction because they are short.

Every time the reaction cycle is repeated the number of template strands should be doubled. In theory the quantity of the target site should therefore increase geometrically. However, in practice the full potential is not realised because the activity of the enzyme reduces a little with repeated temperature cycling and the number of templates for amplification increases. There are also a number of other factors that affect the number of templates produced, but overall a considerable increase in the number of copies of the target site is actually achieved. The base sequence at the target site may be regarded as a signal, hence the term 'amplification', since an increase of the signal has been achieved.

15.6   FORENSIC DNA ANALYSIS

The blood-typing systems discussed earlier, cell marker or protein based, are manifestations of polymorphisms in the DNA. However, the range of the variation possible at each of the genetic sites is limited by the need for the gene product to be a viable working molecule. Another factor limiting the power of discrimination is the need to conduct a separate analysis for each polymorphic system. This requires a different procedure for each system, often different apparatus, a separate set of reference controls to be maintained and a good deal of labour and time to conduct a comprehensive analysis. However, the greatest limiting factor is sample size and with many forensic samples there is usually insufficient material to perform a large number of blood-typing tests. This limitation has been largely overcome by the DNA analyses currently in use, as described below.

15.6.1   Short Tandem Repeats

Short tandem repeats (STRs) consist of tandem repeats of sequences of two to five base pairs. This type of variable DNA is very suitable for analysis by a combination of DNA amplification by the PCR and the analysis of the amplification products by electrophoresis. Large numbers of STR regions have been discovered in the human genome but only certain ones have been selected for use in forensic human identification. Depending upon the STR locus, the number of repeat units in the STRs used for forensic work ranges from 8 to 30 or more. Different alleles are distinguished by the size of the PCR amplification products generated by a given set of primers.

By using a reaction mixture containing primers to different genetic loci sites in the same reaction solution, several STR loci can be amplified at the same time. This is called multiplexing and the resulting combination of genotypes produced from different independently inherited STR loci is called a DNA profile. This is comparable in principle to the combination of the results of testing a blood sample for both ABO and PGM polymorphisms as discussed earlier. It is the combination of types at different loci that increases the number of permutations possible. An example of a STR multiplex is the SGM+ kit that produces a profile consisting of 10 STR genotypes, all independently inherited and all possessing multiple alleles coexisting in a population. The number of different combinations runs into the billions, so the chances of two people possessing the same combination becomes very small. STR multiplex testing kits containing a pre-prepared reaction solution of all the primers in a very carefully formulated mixture are commercially available and at the time of writing new ones are about to be introduced.

The use of the amplification technique ensures the sensitivity of the analysis. In addition, one member of each locus-specific primer pair is synthesised with a fluorescent label incorporated. Therefore all the products of the DNA amplification will also carry the label. The dye-labelled PCR allele is detected in a capillary, and since alleles contain different numbers of repeat sequences, they display different electrophoretic mobilities. The migration distances of the fluorescently labelled PCR products on the capillary therefore reveal which alleles are present in the DNA sample and are hence used for the comparison of samples.

The need for a compound that exhibits the desired fluorescent behaviour and also possesses chemistry compatible with conjugation to a DNA strand restricts the number fluorescent dyes that are available to four or five at the time of writing. One dye is reserved to label a size-marker calibrant. This is a collection of DNA fragments of known sizes (number of base pairs) and is used to calibrate the system so that the size of the PCR products may be determined accurately. This leaves other dyes available for the labelling of the PCR products. However, if the STR loci and the primers used are selected so that the range of allele sizes of one STR locus never overlaps that of another locus, the same dye may be used for two or more STR loci in the same lane. By means of such combinations of size range differences and fluorescence properties of the dyes it is possible to accommodate many different STR loci in a single electrophoretic separation.

The multiplex kits also include the amplification of a marker that indicates the sex of the donor of the biological stain, which is determined by the sex chromosomes. In humans, females have two X chromosomes and no Y chromosome. The marker, called amelogenin (AMG), is located on both the X and Y sex chromosomes. The copy of the AMG marker on the X chromosome is six base pairs shorter than the AMG marker on the Y chromosome. Males have one X chromosome and one Y chromosome. Thus a PCR product generated from the AMG site of a person's DNA will reveal the sex of the donor: males yield two peaks, representing two different sizes of product, and females give a single peak because both of their AMG sites are the same size.

These standard multiplex systems are used by national police and justice agencies to establish databases of DNA profiles. In essence the profiles generated from biological evidence found in connection with crimes, including unsolved crimes, are collected. When a suspect for a crime has been apprehended, the STR profile of that suspect can be compared to those on the database to check for links with other crimes. The establishment of standard multiplexes has led to the production of commercially available kits.

In addition to the STR regions used in the multiplex systems such, as those described above, there also exist STR regions that occur on the Y chromosome. These YSTRs enable a DNA result to be obtained from only the male contribution to a mixture and thus can have a particularly useful role to play, *e.g.* in sexual assault cases.

15.6.1.1 Instrumentation for the Automated Analysis of STR Multiplex Kits. In parallel with the above, a key development in separation technology has been the introduction of a range of automated electrophoresis instruments that use narrow glass or silica tubes. This type of separation, called capillary electrophoresis (CE), uses multiple capillaries so that multiple samples can be run simultaneously. The instruments also contain detectors on each capillary and software to control the electrophoresis and to collate the data collected from the detectors.

In some configurations the CE can separate, or resolve, fragments that differ in size by a single base. A detector is used at the end of the capillary that can distinguish between the different fluorescent dyes used to label the primers that pick out the different genetic loci to be amplified. The completed amplification contains a mixture of DNA fragments produced from the different STR loci. Some of this mixture is loaded onto the capillary and then moved through the separating medium within the capillary by the application of an electric field applied to the capillary. The longer fragments are retarded more than the shorter ones, so the mixture is separated into groups of co-migrating DNA molecules. The amplification products are labelled with the fluorescent dyes and are detected at the other end of the column. The fluorescence of the labelled fragments is recorded by a photosensor.

The data collected by the detector is collated into a chart called an electropherogram (EP). As shown in Figure 15.4, the EP is a plot of the time taken for a fragment to pass through the capillary (retention time—x-axis), versus the detector response (y-axis). The quantities of the fragments co-migrating in groups in the capillary column are in approximate proportion to the quantity of the DNA in the sample that has been amplified. The size of the fragments detected in a sample is determined by using data obtained from retention times of a collection of DNA fragments containing different numbers of base pairs. These fragments are labelled with a fluorescent dye used specifically for this purpose and are added to the sample.

With current instrumentation it is possible to assign an allele designation to the peaks detected. Each genetic locus selected for inclusion in the STR matrix has a number of different alleles coexisting in the population. In general, although many of the loci selected for use have a more complicated structure, the alleles differ from one another by the number of repeats of the repetitive sequence. An artificial mixture of alleles called an allelic ladder can show this. This consists of an equimolar mixture of all the different allele types found in a population. The EP of such a mixture will consist of a series of peaks, all of approximately the same size and each peak separated by one repeat sequence. In the case of most STRs this will be four bases.

The software commonly used to interpret the results of DNA testing can superimpose a series of grey bands on to the EP, as illustrated in Figure 15.4. The grey bands are the width corresponding to one base fragment size and should be distributed across the EP in positions corresponding to where the alleles for the locus in question can be expected to occur. The grey bands are called 'bins' and they are pre-loaded into the software; however, the positions of these bins need to be set against a sample. When a sample of an unknown specimen of DNA is run the software can designate the alleles detected according to the specific bin they fall into. The alleles are designated by a numeric coding which represents the number of repeat sequences present and the software can therefore recognise which allele has been detected by which bin it falls into.

The EP at each locus will consist of either two peaks, if the DNA of the person tested is heterozygous, or a single peak if the person tested is homozygous. In a sample where the quantity of sample DNA present in

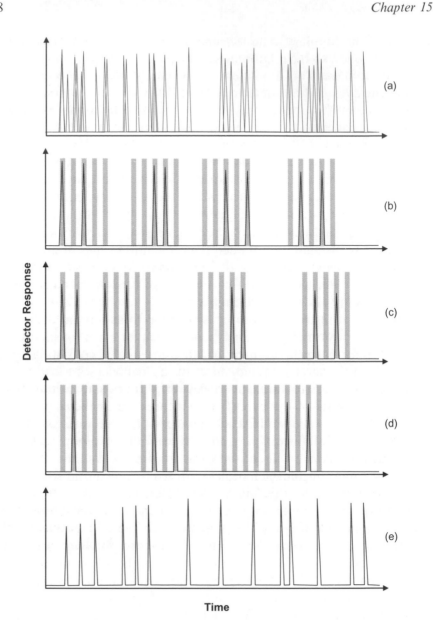

Figure 15.4 Electropherograms from an STR multiplex analysis. (a) EP of the raw
data showing all peaks being superimposed. (b)–(d) show the EPs
separated according to the fluorescent label used in each capillary. Note
that the products from different loci can share a label if there is no
overlap of the size ranges of the alleles. The shaded areas are the bins
used by the software to identify and name the alleles at each locus. (e) EP
of the size markers which are used to find the sizes (number of base pairs)
of the alleles.

the amplification is within the optimal range the peaks of a heterozygous locus can be expected to be approximately equal in size.

In some instances an artefact of the amplification process called 'stutter' is evident. This feature is identified by a small peak, typically less than 15% of the parent peak, occurring in a position corresponding to one repeat sequence shorter than a true allele peak. Stutter is thought to be a consequence of the process of amplification of DNA where there is a repetitive base sequence. If a stutter peak also falls into a bin, the software is unable to discriminate between stutter and a genuine peak and therefore the analyst has to make this judgement.

15.7 INTERPRETATION OF DNA RESULTS

Once each allele present in a DNA profile has been determined the analyst can make an estimate of the frequency of occurrence of the profile by using tables of the numbers of alleles counted in representative samples of the population, *i.e.* an allele frequency database. Operational forensic laboratories maintain such databases for the people served by the laboratory's jurisdiction. Often a number of different databases will be maintained, each one representing a significant subpopulation. Usually these subpopulations are ethnic or religious groups of people, who in general, tend to marry and have families within their own groups. This step is necessary because alleles that are common in one group can be uncommon in another; if a profile frequency were to be calculated for the member of one group using these data from another group, the significance of the evidence may be misleading.

The frequency of a DNA profile can be used as a means of measuring the significance of biological evidence. The frequency estimations can be made using the Hardy–Weinberg formulae, or by derivations of these. For accuracy, when using these Hardy–Weinberg formulae, it has to be assumed that the members of the population are unrelated to one another. This is clearly not always the case, and indeed it has been found that even individuals who believe that they are completely unrelated share common ancestors. This relatedness can affect the chance that two people chosen at random can in fact share some alleles because they have both inherited them from some common ancestor. This means that the chance of a coincidental match is greater than if they had been completely unrelated. The Hardy–Weinberg expressions can be modified to make an allowance for this by incorporating a coefficient called the kinship coefficient. Readers interested in obtaining further information either on the frequency of DNA profiles and/or use of the Hardy–Weinberg formula should refer to texts listed in the bibliography.

15.7.1   Interpretation of Mixture Profiles

Many forensic samples will contain DNA from more than one person. As the sensitivity of the testing increases more samples will produce profiles containing alleles from more than one person, *i.e.* a mixture profile. Where there are more than two contributors to a mixture profile this is called a higher-order mixture. Most casework samples are of two-person mixtures and hence will be discussed here.

The first step is to identify from the EP that there is a mixture profile present, and if possible to determine how many people have contributed to the mixture. As a person can possess at most two alleles for any genetic locus, with the rare exception of genetic occurrences and trisomy (where a person possess three analogous chromosomes; an example is trisomy 23 or Down's syndrome, but there are others that are more uncommon), then finding more than two peaks indicates at least two contributors and five peaks or greater indicates more than two. Note that two people may share an allele and the peak produced by a shared allele will be roughly proportional in size, area or height, to the sum of the contributions from both people. Although it is conceivable that two people might possess such a combination of alleles that gives one or two peaks, thus appearing consistent with a single contributor, this is unlikely to hold true across all the loci tested in a multiplex test. Nonetheless, incomplete profiles where only few of the loci produce a reliable result, or a combination of DNA from close blood relatives, or a combination of both these possibilities, will increase the chance of such an occurrence and the analyst must consider these possibilities when interpreting these data.

The presence of the stutter peaks mentioned above might produce three peak genotypes, but as the stutter position and peak size is predictable, the analyst should be able to recognise this phenomenon. Small peaks could, however, come from another source of DNA. A second profile could be used to confirm if the peaks are genuine or a stutter peak. If genuine, it is unlikely that the sputter peak would appear across all loci in the profile.

Another phenomenon that can make it difficult to determine the number of contributors is 'allele dropout'. This term is used to describe the circumstance where an allele simply fails to amplify and hence is not detected. This is most likely to occur when the sample is very weak. However, again this is more likely to occur in certain circumstances and can be recognised if the whole profile is examined by the analyst.

In some instances there may be exactly equal amounts of DNA contributed from both people and there may be enough from each contributor to produce clear and unambiguous peaks from all the alleles present in the mixed sample. In this instance it is difficult to assign an allele peak to a particular contributor. To overcome this problem the analyst will list all the possible allele contributions at each locus and identify those alleles present in the EP.

Some of the combinations can be eliminated if there is allele sharing between the two contributors, as evidenced by a relatively large peak compared with others; the shared allele will show a dosage effect. The same holds true if one contributor is homozygous for an allele.

In many casework instances the contribution will not be equal so one set of alleles will produce peaks that are relatively smaller than the other, as shown in Figure 15.5.

Once the peaks have been designated and, where possible, identified as belonging to one contributor or another, the profile can be compared with any reference profiles available. This interpretation step will be dependent upon the context of the crime scenario but for a two-contribution mixed profile, if one contributor profile matches a reference profile it can be subtracted or discounted from the mixture profile to leave the profile of the other contributor.

Clearly, it can be seen that the interpretation of profiles from DNA mixtures can be complex and there are different factors such as stutter peaks being mistaken for minor component peaks, allele drop-out and allele sharing between contributors must be taken into consideration. The possibility of there being more than two contributors makes identification of an individual or individuals more difficult and adds to the complexity of interpretation. Hence interpretation requires a methodical approach and full awareness of all the potential problems. Readers seeking more in-depth knowledge about DNA analysis of mixtures and the interpretation of results should refer to the texts in the bibliography.

15.8 MITOCHONDRIAL DNA

The techniques described so far have been for the analysis of the genomic DNA found in the nucleus of a cell. Within a cell there are other components (cell organelles) and one of these is the mitochondrion which is associated with the generation of energy within the cell. Between 100 and 1000 mitochondria may be found in a human cell and each one contains 2 or 3 copies of its mitochondrial genome, its own DNA. Each cell will therefore contain hundreds of copies of the mitochondrial DNA, compared to only one copy of the nuclear DNA.

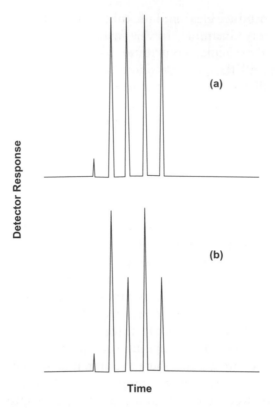

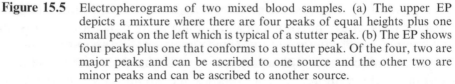

Figure 15.5   Electropherograms of two mixed blood samples. (a) The upper EP
depicts a mixture where there are four peaks of equal heights plus one
small peak on the left which is typical of a stutter peak. (b) The EP shows
four peaks plus one that conforms to a stutter peak. Of the four, two are
major peaks and can be ascribed to one source and the other two are
minor peaks and can be ascribed to another source.

Within the cells of a human all the copies of the mitochondrial DNA
have the same sequence. The mitochondrial DNA consists of a loop of
double-stranded DNA of 16 596 base pairs. The entire sequence of this
loop was reported in 1981 and it is called the "Cambridge Reference
Sequence". In contrast to the nuclear DNA, almost all of the mitochondrial DNA sequence is used for coding and only a relatively small
portion called the 'D' (diversity) loop, or the 'control region' has no
specific coding function. The control sequence includes a point at which
mitochondrial DNA replication always starts. This is called the origin of
replication and the Cambridge Reference sequence of bases is numbered
starting at number 1 at the origin and onwards up to 16 596. Two
regions have been found to exhibit a high degree of sequence variation

between individuals. These regions, called hypervariable regions (HVR) 1 and 2, occur within the control region.

The mitochondrial genome has been shown to mutate at a rate 5–10 times faster than the nuclear genome. This has been attributed to a combination of the absence of a 'proofreading' or repair mechanism in mitochondrial DNA replication and a poor fidelity of the mitochondrial polymerase in its action. Another factor that must be taken into account in the use of mitochondrial DNA for forensic analysis is the fact that the mitochondrial genome is maternally inherited, *i.e.* we inherit mitochondrial DNA from our mothers.

The analysis of mitochondrial DNA is therefore of greatest benefit where the sample size is greatly restricted so that the occurrence of multiple copies of mitochondrial genome per cell is clearly advantageous, and also where family relationships through the maternal line are important. As the mitochondrial genome effectively consists of a single chromosome, and there appears to be no mechanism for crossing over between different mitochondrial genomes, HVR 1 and HVR 2 cannot be regarded as separate unlinked genetic loci. The entire genome must be considered as a single genetic locus. As such, it does not have the discriminating power of a combination of nuclear genetic loci.

The usefulness of mitochondrial DNA analysis has been demonstrated by its application to the resolution of the authenticity of a collection of bones, alleged to be the remains of Tsar Nicholas II, the Tsarina, their three daughters, the family doctor and two servants. These individuals were reputedly murdered in 1917 during the Russian revolution and this act has remained a contentious issue in Russia to this day. Various accounts of the disposal of the remains had been recorded and there was also the question of the authenticity of various individuals claiming to have escaped or survived the murders. Two amateur historians made a discovery of a collection of bones and other remains at a site around 20 miles from Ekaterinburg. The nature of some of the artefacts recovered with the buried remains, and the number and nature of the skeletons in terms of age and sex, was consistent with them being the remains of the murdered Tsar, his family and entourage. In a collaborative exercise between the Engelhardt Institute of Molecular Biology, Moscow, and scientists of the British Forensic Science Service, a DNA analysis was made of the skeletal remains.

Both STR analyses and mitochondrial genome sequencing were employed. The STR analyses were conducted over five loci and it enabled a tentative distinction of a family group of blood relatives, those with genotypes consistent with a mother, a father and three children. The mitochondrial DNA analysis was of particular importance because individual skeletons could be linked to the maternal lineage of the Russian royal family through the mitochondrial DNA. This was possible because the authenticated living descendants of the Russian royal family maternal lineage, including HRH Prince Philip Duke of Edinburgh, could be traced and the genealogy of the European royal families is well documented. Paradoxically, the skeleton identified tentatively as being that of the Tsar matched the mitochondrial HVR sequences of contemporary living maternal descendants perfectly, except for a single base mismatch. Such a high degree of match, when taken in the context of the other supporting evidence, strongly supports the authenticity of the claims for the remains but yet is not the perfect match. It is clear that the true significance of the single base mismatch can only be interpreted by taking mutation rates into account.

Many forensic laboratories now take advantage in refinements of DNA sequencing technology to test selected portions of the mitochondrial genome by directly sequencing the DNA. Although mitochondrial analysis is not a replacement for STR analysis it has an important application where the samples are highly degraded and where the maternal inheritance can be used in an investigation.

15.9 DEVELOPMENTS IN DNA TESTING

DNA testing procedures continue to be refined and developed and some of the areas of interest are highlighted below.

15.9.1 Low Copy Number (LCN)

This describes a combination of refinements of the existing test matrices to improve the sensitivity of the test to the point where the DNA from a single cell can give a result. These highly sensitive analyses are known collectively as low copy number (LCN), more recently called low template number (LTN), amplification and the term arises from the low number of copies of the genome present in very small samples. The detection sensitivity of CE instrumentation being used for the analysis of the amplification product has also improved considerably.

The interpretation of the results of LCN/LTN requires great care. First, the danger of contamination is much greater. The sensitivity of these techniques is so great that very tiny amounts of contaminating DNA from investigators, or from DNA in cellular debris present on a substrate before any stain has been deposited, can yield alleles. As a consequence of the amplification being taken to its limit, the major and minor components may not be readily distinguishable.

At the limits of the sensitivity of the test, alleles may be detected in some replicates of the test and not in others. This phenomenon is called allele drop-in and arises when very few potential template molecules are present in a PCR reaction tube. If one allele is successfully amplified in preference to another over the very initial cycles, then that allele's amplification will prevail over the other and be detected. A repeat of the test may show that the allele preferentially amplified in the first test may not be amplified a second time, and the other allele if amplified can then be detected. It is currently this lack of reproducibility that introduces a greater degree of uncertainty when compared in more conventional test interpretations. Therefore, alleles to be detected require more than one amplification. However, the number of times an allele needs to be observed in separate amplifications will depend on the type of sample and may be open to debate.

15.9.2 Mass Spectrometry

Techniques have been developed whereby PCR products can be embedded within a crystalline matrix consisting of small organic compounds. A short pulse of laser light is then used to volatilise and ionise the DNA and the matrix. This process is called 'matrix assisted laser desorption/ionization' (MALDI). The ionised DNA molecules are therefore in the gas-phase and can be analysed by mass spectrometry (MS). The size of the DNA amplification products of STR loci will depend upon the number of tandem repeats and consequently so will the atomic mass of the fragments. The atomic mass of a fragment influences its time-of-flight in the mass spectrometer and therefore it is possible to obtain a very accurate mass measurement of an allele. The combined technique is called 'matrix assisted laser desorption/ionisation time-of-flight mass spectrometry' (MALDI–TOF–MS).

The advantages of this technique include the very rapid analysis time, seconds for each sample, and the highly accurate measurement of mass, thus eliminating the need for allelic ladders. Furthermore, no fluorescent labels on the primers are required.

15.9.3 Trait Identification

The accumulation of knowledge of gene function has provided, and will continue to yield, genetic information that might be used actively in the investigative process to predict the appearance by ethnic origin or pre-disposition to addiction of individuals, rather than retrospectively being able to link evidence with suspects once they have been located and apprehended. To some extent this is why criminal intelligence databases were introduced. At present only the sex of the donor is predicted. As the understanding and knowledge about the interactions of the relevant genes progress we may expect traits such as eye, hair and skin colour, colour blindness and other less common traits to become predictable from genetic predispositions that are inferred from the DNA in samples examined.

15.9.4   DNA Microarray Technology

This is a very flexible technology, developed primarily to enable rapid co-testing for a large number of sequence variants. It can take a number of forms but broadly speaking most of them follow a general pattern. The tests are carried out on glass plates on to which DNA sequences, called 'targets' are immobilised. These immobilised strands are arranged into groups of like strands immobilised into specific locations making up an array of spots in predetermined patterns. Independently, samples of DNA, called probes, which come from the specimen being tested, are labelled with fluorescent dyes. The immobilised targets are exposed to a solution of the probes and where there is complementarity between these, hybridisation occurs. The array is scanned and those spots that fluoresce, *i.e.* have hybridised a fluorescently labelled probe, are detected and their positions within the array recorded.

A large part of the effort to develop these devices is to conduct gene expression analysis in the course of drug discovery. However, large numbers of nucleotide site polymorphisms are known to exist. These consist of a site in the DNA at which the type of base present can vary. This is called a single nucleotide polymorphism (SNP). These can be readily amplified and the amplification products detected on DNA microarrays. Potentially many thousands of SNPs can be tested simultaneously by microarray devices.

15.10   CONCLUSION

The continued refinement of the existing DNA technologies is certain to continue. A more daunting prospect is the possibility in future to identify traits related to mental stability or the predisposition of individuals to drug or alcohol addiction and the influences this may have on their behaviour. The new technologies currently under development, promise to open the door to new avenues for the forensic application of DNA analysis.

BIBLIOGRAPHY

J. Buckleton, C. M. Triggs and S. J. Walsh, *Forensic DNA Evidence Interpretation*, CRC Press, Boca Raton, 2005.

J. M. Butler, *Forensic DNA Typing*, Academic Press, San Diego, 2005.

From the Academy: The evaluation of forensic DNA evidence. *Proc. Nat. Acad. Sci. U. S. A.*, 1997, **94**, 5498–5500.

R. E. Gaensslen, *Sourcebook in Forensic Serology, Immunology, and Biochemistry*, National Institute of Justice of the US Department of Justice, Washington DC, 1983.

M. Rajadhyaksha, condemned by birth: The implications of genetics for the theories of crime and punishment, *Socio-legal Review*, 2006, **6**, 85–103.

Y. Torres, I. Flores, V. Prieto, M. Lopez-Soto, M. J. Farfan, A. Carracedo, and P. Sanz, DNA mixtures in forensic casework: a 4-year retrospective study, *Forens. Sci. Int.*, 2003, **134**, 180–186.

CHAPTER 16

Forensic Archaeology and Anthropology

TAL SIMMONS[a] AND JOHN HUNTER[b]

[a] UCLAN, School of Forensic Investigative Sciences, Preston PR1 2HE;
[b] Institute of Archaeology and Antiquity, University of Birmingham,
Birmingham B15 2TT

16.1 INTRODUCTION

Forensic anthropology and archaeology are closely linked disciplines related to the recovery and analysis of human remains from crime scenes. In the UK and Europe as a whole, these are largely considered to be separate fields of study and separate professions, whereas in the Americas the two disciplines are often practised by the same individuals. There is a case to be made for both systems; as separated professions they allow greater specialisation and thus a greater depth of knowledge and experience applied to the practice of the disciplines, whereas when they are linked together there is a greater chance of linking the interpretation of evidence recovered in the field with what is observed in the laboratory analysis.

16.2 FORENSIC ARCHAEOLOGY—THE THEORY

Forensic archaeology is concerned with the application of archaeological theory to modern criminal scenarios in which archaeological

Crime Scene to Court: The Essentials of Forensic Science, 3rd Edition
Edited by P. C. White
© Royal Society of Chemistry 2010
Published by the Royal Society of Chemistry, www.rsc.org

evidence may be presented in a court of law. To a great extent this will involve the investigation of clandestine graves by adapting known archaeological techniques in order to locate burial sites, and by employing excavation strategies to recover human remains. On occasions other types of buried remains may be searched for and excavated, typically drugs, firearms and other weapons. Archaeologists may also be asked to resolve issues of date when human remains are discovered during building operations, or when human disarticulated bones are encountered as 'stray' finds by members of the public. They may also be asked to take part in formal exhumations undertaken by the police in order for a post-mortem examination to take place. There have also been instances where archaeologists have been deployed in mass disaster recovery, or even in fire debris where stratigraphic investigation is required. In short, they may find themselves involved in any matter involving buried or sealed remains, or remains that may have been buried, which become part of a criminal investigation and which require 'expert' operational input or opinion.

There are probably only about 20–25 archaeologists in the UK undertaking this type of work on a regular basis. One or two of these operate full time from within larger forensic companies but the majority have a 'day job', usually within commercial archaeological units or universities, or work freelance as a part of other types of archaeological employment. Forensic archaeology requires a mixture of skills and experience, not least of which is a solid background in fieldwork, and a wide knowledge of the various other disciplines that archaeology draws upon such as surveying, geophysics, aerial imagery, *etc.* The key element is an understanding of soil stratigraphies (*i.e.* layers in the ground), how to recognise different layers and how to determine relationships between them. These layers are formed by a host of human and natural actions, typically construction work, demolition, ditch clearance, compression through occupation, fires, farming, silting, abandonment and so on. Layers have different characteristics, properties and colours according to the way each was formed; archaeologists try to interpret these layers to find out what each one represents. Sometimes, especially in towns, numerous layers might have formed over time and can extend several metres down below the present ground surface; in other instances there might only be a few layers. If archaeologists can establish, through careful excavation, what each layer represents in a given place, it enables them to create an understanding of what happened in that location through time. Also, because the layers are superimposed on each other, or can cut through each other, there is also a relative chronology that can be worked out. Figure 16.1 shows a typical archaeological section,

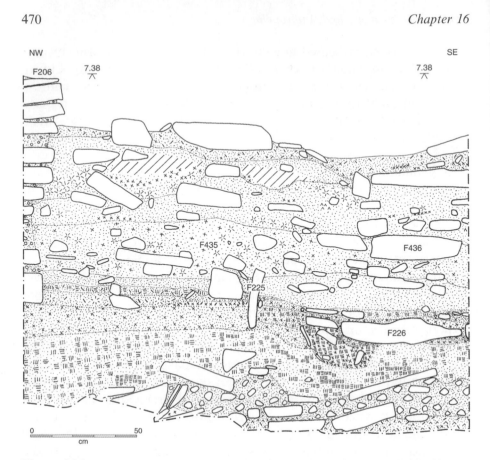

Figure 16.1 A typical profile (section) through the ground showing individual layers.

i.e. a vertical profile through the ground. Any human remains found in the ground will belong to one of these layers and therefore will have a buried 'context'. Retaining the integrity of that context is paramount in trying to resolve some of the many questions that occur in forensic enquiries. If a murder victim is buried, the grave will be dug through existing layers and will constitute a new layer. Other layers may form above it and the grave will become a type of sealed unit which will contain any information about the victim and any evidence associated with the disposal. Stratigraphy is empirical; it makes no difference whether the layers were formed thousands of years ago, or yesterday—the same principles apply.

Forensic archaeologists will find themselves as part of a larger team containing individuals with different functions in an enquiry. Some of these will be peripheral to the incident itself, such as those within the Crown Prosecution Office or the coroner's office, but others will be working with the archaeologist at the coal face. These might include the forensic pathologist, scene of crime officers (SOCOs) and other forensic scientists. The archaeologist will need a reasonable knowledge of their specialisms as well as their evidential requirements. The other investigators in turn will need a reasonable knowledge of how the archaeologist operates and what archaeological evidence can contribute. This is very much a two-way process. Some of these specialisms will be familiar archaeological territory (*e.g.* pedology, palynology and diatoms), some may be less frequently encountered (*e.g.* entomology and DNA) and others will be novel (*e.g.* fibres, paint, blood and pathology). It is the job of the forensic archaeologist to be familiar with them all. But that does not mean that an archaeologist needs to be a universal expert: as with all other forensic specialists, it is essential, from an evidential point of view, that archaeologists never drift outside their specific field of expertise. To do so might be to provoke vulnerability and jeopardise a case. This separation is not as easy as it sounds, given that many disciplines overlap, one key overlap here being between archaeology and anthropology. Most archaeologists are familiar with human skeletal remains, and most anthropologists are familiar with archaeological theory and techniques, but there is a difference between *awareness* and *expertise*. Each party needs to be clear in their own mind as to where their expertise starts and finishes and be able to defend that in court. In most instances points at which an archaeologist needs to call on the services of an anthropologist, and *vice versa*, are fairly clear cut. Of course some individuals are competent in both disciplines and can offer a combined expertise; this is especially important in the excavation of mass graves.

The blurring of roles is compounded by confusion between the terms 'archaeology' and 'anthropology', particularly in discussion between US and UK practitioners. In the USA archaeology emerged as a secondary discipline under the umbrella of (physical) anthropology, whereas in the UK archaeology and anthropology both emerged as primary disciplines but with anthropology being more strictly 'social' as opposed to 'physical', hence the confusion. A working definition might be that forensic archaeologists are expert in field skills but with an awareness of physical anthropology, whereas forensic anthropologists are expert in the examination of skeletal tissue, but with an understanding of basic excavation field skills.

16.3 SEARCH

Archaeologists tend to have a wide experience of using maps, aerial photographs and types of equipment that allows them to understand landscapes, particularly landscape change and development. Apart from the conventional 'tourist' Ordnance Survey type of map, there are larger-scale, more detailed maps that show contours, trackways and landscape features. These can be used to identify vehicle access points, streams, former quarries, *etc.* which may have a bearing in body disposal. Early editions of maps from the mid-19th century can show features such as mineshafts and wells or springs no longer evident on modern maps, and these can be tied in with geological and land-use maps in order to assess the feasibility of victim disposal in a given area. Aerial photography can also be used historically and is particularly useful in older unsolved or 'cold case' enquiries where aerial sorties were undertaken sporadically since the Second World War for military and land-use purposes. These can illustrate how landscapes have changed over time and help identify likely disposal points not obvious today. New aerial photographs can also be taken for interpretation; these work on the basis that any burial of a victim will create a disturbance which can affect both the vegetation and the topography. Although often visible at ground level, these changes are best seen from the air where they can be emphasised through shadow, vegetational density and colour, and ploughing or land use activity. Satellite imagery can be used for larger disturbances, *e.g.* mass graves. Thermal photography from the air can often detect the heat differential between disturbed and undisturbed soils on cooling in the late evening, but is better suited to the detection of the heat emission from a decaying cadaver.

Types of geophysical survey used in archaeology can also be employed; these are especially suited to shallow subsurface work in searching for relatively small targets. There are three main methods of geophysical survey used by archaeologists: resistivity, magnetometry and ground penetrating radar (GPR). More detailed reference to the underlying science can be found elsewhere, but it is useful to outline the fundamental methodologies here. Resistivity survey involves passing an electrical current through the ground and systematically measuring the resistance to that current across a gridded area. Electrical resistance is affected by moisture content, hence shallow subsurface features which are either comparatively moist (*e.g.* ditches or graves) or comparatively dry (*e.g.* walls and stony deposits) may present different signatures in relation to ground where there are no subsurface features. The method works well in flat open spaces, but readings need to be taken every 0.5 m if there is to be any hope of detecting burials within a wider background.

Magnetometry is highly sensitive and detects magnetic change below the ground surface. This can be brought about especially by burning or by the disturbance of magnetic soils during burial. Magnetometry may also detect magnetic objects, providing they are ferrous. It has limited forensic application, in that at most potential crimes scenes there is local metal 'noise' from fences, surface rubbish and adjacent buildings. The third method (GPR) is electromagnetic and involves sending pulses into the ground and measuring their energy and speed at return. This gives an indication of the nature and depth of buried deposits in a given area. GPR is the only method suited to detect any deposits through dense surfaces such as tarmac, flagged patios or swimming pools where reinstatement costs might be considerable if major excavation work took place.

All three methods are non-invasive, but they all rely on the fact that the disturbance caused by the burial occurs in fairly undisturbed ground. In other words, there needs to be a relatively constant background in which any geophysical anomalies can be identified. As each method responds to different physical properties of buried soils or layers, the use of one method on its own is ill-advised; they are best used in a complementary way, depending on environment. GPR has other advantages in that it detects in 'real time'—*i.e.* buried disturbances can be interpreted on-screen as the antenna passes over them. However, software has developed significantly over the last decade and the data from all three methods can now be processed rapidly and presented in a variety of display modes at the scene itself. The key skills in geophysical survey lie in interpreting this data, rather than in simply obtaining it, and this is a specialist area of expertise. It is worth remembering that all three methods have their advantages and limitations and are at their best in identifying buried change within the near surface. With modification they can all, especially GPR, be used to detect at lower depth, but there is a trade-off between depth of detection and resolution; in other words, it becomes harder to interpret the anomalies the deeper those anomalies lie.

The detection of recently buried human remains differs from the detection of 'conventional' archaeological remains in that recently buried human remains possess a decay dynamic which can affect detection methodologies. Effects can include enhanced vegetational change and scavenging potential, but especially affect geophysics. On death the human body undergoes a process of decomposition, the various stages of which are well documented, but the speed of which can be affected by a host of intrinsic and extrinsic variables. The study of decomposition and factors affecting it is called taphonomy and has been covered in depth in section 3.5. This is not an exact science: it depends on the variability of health, age, stature and body mass (intrinsics) of buried individuals and on the burial environment (extrinsics) in terms of depth, climate, soil bacteria, oxygen, water content, body wrapping, *etc.* All these can have an effect on the speed at which a cadaver decomposes and can sometimes even cause stages of equilibrium to occur. One of the more relevant aspects here is the point at which primary decay sets in; this part of the process emits significant amounts of heat in relation to the surrounding environment and may be detected from the air, or at the ground surface, through thermal photography. The points in time at which this heat emission begins and finishes can span days or even weeks, but essentially depend on the local taphonomic variables at the time. In other words, there is no empirical way of predicting the speed of human decay, but there are key factors in anticipating probabilities of decay time/rate on the basis of the number of taphonomic variables specific to a given investigation.

The decay process is also one to which trained cadaver dogs are uniquely applicable. Dogs can be trained to scent human decay using clothing from a mortuary, teeth, or even chemically derived pseudo-scents. In searching targeted areas dog handlers are likely to 'vent' (*i.e.* probe the ground systematically to a depth of about 0.5 m) in order to allow any gases to rise to the surface. The dog is then led downwind along the line of vent holes and any responses monitored. Suspect areas can then be earmarked for further attention and for any necessary excavation. Searching for clandestine burials is a complex and time-consuming process. It can never be achieved successfully by using just one detection method. Most successful searches use aerial and/or geophysical techniques supported by dogs. Invasive action, *i.e.* excavation, is a last resort given that archaeology is a destructive process and offers potential loss of evidence if undertaken inappropriately. The most effective searches start by being non-invasive (desk top) and become progressively invasive. This minimises the potential for losing evidence. Targeted areas need to have defined search boundaries; they need to be eliminated with high levels of confidence before searching can move to another area.

16.4 DISCOVERY OF HUMAN REMAINS

The two most frequent scenarios in the discovery of human remains are the accidental discovery of human remains by members of the public and the formal discovery of human remains resulting from the implementation of search strategies or of suspicious disturbed ground during formal investigation.

16.4.1 Accidental Discovery

The accidental discovery of human remains is a relatively common occurrence and usually results from building, drainage or garden digging. In coastal areas human remains are sometimes washed out from eroding shorelines or become uncovered during storms or sand movement. In most cases the discovery is simply of bones, sometimes in isolation, although they can also occur as complete or partially complete skeletons. Soft tissue may also occur, but the procedures are the same. Buried conditions can radically alter the way humans remains are preserved, even if they are very old. Irrespective of the state of the remains, the first key questions to be asked are these:

1. Are these remains from a grave, and is the rest of the body still there?
2. How old are the remains?

Answering these questions in the first instance will help resolve later issues regarding identity of the individual and the legality of disposal. Police forces are not usually concerned with the further investigation of remains that are likely to be about 70 years old or more, but they do need reassurance from an expert that the remains are either 'ancient', or that their presence can be explained. Remains that are proved to be less than 70 years old may require criminal investigation. If there is some doubt as to whether the remains are animal or human, this can often be easily resolved by taking anthropological advice, either in person at the scene or, more commonly, by sending good-quality electronic images to an anthropologist. In some scenarios, such as in searching landscapes for a clandestine burial where surface skeletal material is likely to occur, an anthropologist is an integral part of the search team in order to eliminate any animal bones recovered during the course of the search.

Accidentally discovered human remains may, or may not, constitute a scene of crime, and this is best resolved if the remains are left *in situ* and the integrity of the context retained in order that their physical position in relation to associated layers can be recorded. If it is not possible to leave remains where they are, or if they have been moved by some 'helpful' member of the public, their position can be marked in order to secure the exact position where they have been exposed in the building work, the erosion, or on the ground surface. For the purposes of collecting evidence it is particularly important to identify the location of the remains *vertically* in relation to other buried layers. Remains need to be left exactly where they are until they can be interpreted properly within these layers. Archaeology is a destructive process, and securing a scene overnight until the archaeological personnel and the natural light that allows layers to be defined optimally has been achieved, is a small price to pay. As soon as those remains have been removed from the ground unrecorded and extracted from the layers in which they have been lying, a substantial amount of associated evidence will have already been lost. If the incident turns out to be a homicide, much of the prosecution case will already have been jeopardised.

The best way of determining whether human remains found in the ground have been deliberately buried is to determine whether or not there is evidence for a grave cut, *i.e.* a defined buried stratigraphic unit in which the body lies. This is normally achieved by cleaning the vertical face at the point in the ground where the remains were discovered, but over a wider area, using a small implement, usually a trowel. This produces a section through the ground in which any grave construction should be immediately apparent in the cleaned profile. This 'cut' will appear as a defined change in the layers as a result of a hole having been dug, a body deposited, and the grave infilled. Figure 16.2 gives a stylised idea of how this might look, but they are rarely that simple. Removal of soil to achieve this profile, even in small amounts, sets in train the process of losing evidence and has to be kept to a minimum. It is

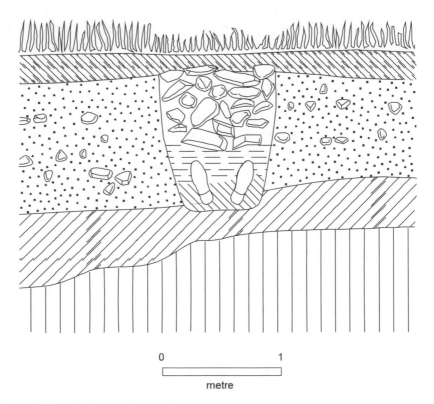

0   1

metre

Figure 16.2   Stylised drawing of a grave in profile.

however, the only effective method of testing the ground and has to be recorded meticulously. Any soils removed need to be kept in their respective layers for sieving.

Once this vertical profile has been produced and any grave outline identified, the whole profile (normally termed the 'section') has to be recorded. This is a fundamental piece of evidence that will be totally destroyed when the grave is finally excavated and the body recovered. Archaeologists normally record sections by photograph and planned drawing, scaled at 1:10, clearly marking the line of the grave edge and the position of those layers into which the grave has been cut (*i.e.* earlier layers), as well as those which go across the top of the grave (*i.e.* later layers). If it can be shown that the grave was cut through the present ground surface, then it is obviously very modern. If, on the other hand, there are layers that run across the top of the grave and seal it, it may be much earlier. Being able to date layers that are either earlier or later than the grave will obviously help date the grave itself. Sometimes dating layers can be relatively simple, *e.g.* if the grave lies sealed under the foundations of a building of known date, then the building gives a *terminus ante quem* (time before which) for the date of the burial. Conversely, if the burial seals, or lies over, or is cut through buried rubble from a demolished building, then the date of the demolition gives a *terminus post quem* (time after which) for the date of the burial. Modern local records are surprisingly comprehensive regarding the dates of building change, and short-term changes such as garden landscaping are also well remembered. All these events tend to leave a stratigraphic context within which the remains can be interpreted. Alternatively, if there is no grave cut present in the section, then the archaeologist will need to adopt a different strategy for resolving the problem.

In some respects the absence of a grave poses a bigger problem than in finding a complete body or a body buried in pieces. If there is no grave cut evident it would seem that the bone (or bones) is part of a wider scatter of disarticulated human material which has somehow become encapsulated in a particular layer over time. This often happens quite innocently by removal of soil from an old burial ground, or by the transference of soil that already contained buried remains for building or landscaping purposes. The remains will be well disturbed and often consist of only of pieces of larger limb bones, ribs or perhaps the skull, which are sturdy enough to survive. The problem is in trying to find out if there are any more pieces, or where they came from. This might require a small excavation, probably covering 2×2 m at the point where the remains were found, or from where they may have been scavenged, in order to recover any additional pieces. 2×2 m is an arbitrary area and could be greater, or less, depending on circumstances, but needs to be defined at the outset. If further material is found in this area, then the size of the area may need to be increased; if not, then the search can be stopped.

Stray bones which appear in the topsoil are a cause for concern because they may represent the scatter of an individual originally lying on the ground surface whose remains have become disarticulated through animal action and which have became buried through natural processes. In these instances it may be necessary to strip a wider area to avoid public concern in case additional bones come to light.

In lower layers, experience has shown that there is less cause for concern. The problems of individual bones not occurring in grave cuts are unlikely ever to be satisfactorily resolved. It may be, however, that the layer in which they occur can be identified with building activity or the redeposition of soils, in which case there may be some potential in pursuing the origin of the soil layer, and in establishing the site history. However, in the absence of other remains in the 2×2 m area it is logistically impracticable to proceed further. There are no realistic boundaries as to how large an area to open for further work, and there is no guarantee that any more pieces will be discovered in doing so. Providing it has been proved that there is no formal burial (*i.e.* no grave cut and no other articulated skeletal remains) the investigation can normally be wound down. If it is unclear where the soil in the layer originated, the bone can be sent to one of several commercial radiocarbon dating laboratories, but this is only likely to give a coarse 'ancient' or 'modern' definition. Radiocarbon dating is frequently used by archaeologists in order to date stratified organic materials. The method depends on the absorption of atmospheric radiocarbon by organic matter (people, plants, trees, *etc.*) during life, and the measurable and predictable decay emission of radiocarbon after death. The process has a number of problems, not least of which is the fact that atmospheric radiation has not always been consistent, and secondly that the date calculations are statistical, being based on a decay rate (half life) of 5730 years, and therefore are of very little use for dating materials from the very recent past. That said, the prevalence of atomic weapons testing in the atmosphere in the 1950s caused an enhanced radiocarbon peak to be found in the measurement of any organism living after that date. This means that the use of radiocarbon dating can be effective for dating 'one-off' or 'stray' bones to either before, or after, the mid 1950s. This may assist in determining whether remains are of forensic interest or merely of archaeological curiosity. More refined dating within this post-1950s period, however, can now be determined using the radioisotope analysis of other ingested elements, notably lead with a half-life of around 22 years, and polonium with a half-life of 138 days. Radiometric dating will not necessarily help identify the individual, but it may be useful as a last resort if only one or two bones are recovered and there is no context in which they can be placed.

Sometimes human remains are found, skeletonised, on the ground surface and become scattered as a result of animal scavenging and activity. Research has been carried out in the USA as to how these remains become scattered, the agencies at work, and the time frame of the scattering process. There is a trend of gradual scatter from a focal point, rather than a precise measurement of movement, and there becomes a point at which, over time, the focal point is no longer identifiable. Of course there is a corollary to this, namely that the same scatter trend might allow us to predict the extent to which an individual might be expected to be found scattered across the ground surface if they were reported missing either X weeks, or Y months, or Z years ago. This is hardly an exact science, but one that has some coarse practical application.

16.4.2 Formal Discovery

Formal discovery can occur as the result of information regarding a location where a person claims to have witnessed a burial. Other instances include those where a person has gone missing and where there are ground disturbances that need to be checked. These ground disturbances may result from searches using geophysical survey methods or aerial photography. Both geophysics and aerial work will identify anomalies and the investigation is then committed to investigating each anomaly in turn. There are, however, rapid methods of elimination that can be deployed safely.

Simple testing of a suspicious disturbance can be achieved by excavating a small trench, maximum 50 cm wide, across the alleged location. This can be carried out rapidly and safely to about a spade's depth before more careful removal of soils takes place; these can then be removed in layers, or in 10 cm spits if there are no obvious layers until natural, undisturbed layers are found. Any burial there will be found quickly and very little damage will have occurred because of the narrow area excavated and the method used. If there is no burial, hard undisturbed soils will quickly be identified and the investigation can cease. This is a fast practical compromise between testing an allegation and recovering evidence carefully, but it requires an archaeologist or someone familiar with the way soils are formed to confirm which are natural undisturbed soils and those that have been disturbed. The discovery of a burial, either from police investigation of this type, or as a result of chance discovery, will then require formal excavation. This will only take place after the scene has been secured, briefings for all concerned undertaken, necessary equipment and facilities lined up, and the appropriate personnel made available.

16.5   EXCAVATING A BURIAL

Excavation of the burial site will be carried out by trained personnel in conjunction with SOCOs; recovery of the remains themselves will usually involve the forensic pathologist or an accredited medical expert. No two burials are ever the same—each poses its own problems of access, soil environment, circumstances and questions to be resolved—but the following description can be used as representing a typical scenario.

After the customary photographic record of the scene and possible fingertip searching, the general surface area will be cleared of vegetation and debris and an area marked out that will embrace the projected grave area. This might be typically 3×2 m according to the supposed long axis of the burial. There may be layers that overlie the grave. These will be need to be numbered and removed individually, sieved, any material seized and a soil sample taken. These layers have no bearing on the incident, but may serve to date events after which the grave was dug. The top of the grave cut will eventually become apparent as these layers are removed. There will probably be clear-cut edges to the disturbance (grave), and the fill will probably be of different texture and colour. Although burials can occur in a range of different conditions and environments, the recovery process follows a well-trodden path. This recognises that archaeology is a destructive process and that each recovery operation is non-repeatable. It is as well that all personnel at the scene remember this.

Defining the area of disturbance is critical as it will also define the soil boundaries within which the forensic evidence occurs. Once the outline of the burial disturbance has been identified, it may be prudent to excavate the grave in two halves; this will create a formal profile (section) across the grave and provide visible evidence as to how the grave was filled in. This evidence would otherwise be destroyed if the grave was excavated as a single entity. It is worth remembering that when the burial of a victim took place, three main activities were undertaken:

1. The physical removal of earth from the ground.
2. The deposition of the body in the grave.
3. The infilling of the grave.

Proper forensic investigation of the grave follows this process in reverse order: it removes the grave deposits to see how the grave was infilled; it exposes the body in order to show the manner in which it was disposed; it lifts the body and subsequently identifies how the grave was dug.

The grave infill is normally removed by trowel. If the fill contains layers, these will be removed individually. If there are no obvious layers, the grave fill can be excavated in a series of spits, typically 10 cm deep, in order to provide a controlled removal of the grave deposits. Layers or spits will be numbered and any exhibits that are seized are related to the layer or spit in question. This ensures that individual objects are properly associated with any layers in which they occurred, and that the contents of the grave can be recreated spatially, however coarsely. The numbers of the spits or layers should also be applied to the profile so that there is a direct cross-reference between horizontal and vertical records. Ideally, all soil or fill removed should be kept, bagged according to number of layer or spit and taken for sieving. The body is normally revealed in its entirety before lifting.

Because archaeology is a destructive exercise the recording process is comprehensive, and wherever possible should be planned in three dimensions. The outline of the grave is normally planned with reference to permanent base points at ground level in order that it can be relocated in the future; the base and the profile of the grave are planned using graphic conventions such as contours or hachure. Photography should also be comprehensive. During the excavation the following types of question can be asked and responded to accordingly:

1. How was the grave dug, and with what implements?
2. Was it dug in a hurry, or was it carefully prepared?
3. Is there any evidence in the grave for cause and manner of death?
4. Is there possible transfer of material from offender to grave?
5. Is there foreign material in the grave fill, and if so, from where did it originate?

In excavating it is critical to maintain the integrity of the grave, *i.e.* to make sure that during excavation of the grave fill, care is taken to respect the edges of the grave, right the way down the sides to the base. Once this has been completed and the body has been removed there may be further deposits revealed below the body, but these still lie within the cut of the grave and are likely to contain material that has fallen from the body or clothing during the decay process. These might include personal effects from pockets, jewellery, or even projectiles, with a wide range of identification or forensic implications. Proper excavation will identify the physical limits within which the human remains and associated materials will be found; it will help the pathologist and/or anthropologist in terms of questions pertaining to identity, interval since death and the cause/manner of death; and it will define the boundaries of soils which may contain evidence linking the body or the grave digging to a third party (offender). The grave has to be completely excavated until it is empty and the edges cleaned. Excavating the grave any wider and deeper will contaminate the scene. The only exception here is for the sampling of evidence that may have permeated through the grave sides down into other layers, notably for toxicological analysis.

Once the body has been removed and the excavation work has ceased, the archaeologist will produce a plan of the grave, including a profile and a scale drawing which illustrates the general dimensions and slope of the grave sides, as illustrated in Figure 16.3. This will serve as a permanent record when the grave is finally filled in or destroyed. Back in the laboratory it may be necessary to sieve the individual layers or spits of recovered soil for further evidence, perhaps traces of fibres, packaging, cigarette ends or paper related to the offender which entered the grave fill either accidentally or deliberately.

16.6   FORENSIC ANTHROPOLOGY

Forensic anthropology was once defined, simply, as the application of physical (or biological) anthropology techniques of skeletal analysis to the law. This primarily referred to the identification of bones as being human and the application of basic methodology common to the analysis of human remains from historic and ancient archaeological sites for ascertaining the biological profile of an unknown skeleton. The biological profile is information concerning the sex, age, race and stature of the individual, as well as any injuries or diseases that person might have suffered during life. In the past several decades, however, that definition and relatively narrow practice has expanded considerably. A forensic anthropologist engaged in case work today must be broadly trained and quite experienced in a variety of situations, as the work incorporates:

- Identification of fragmented, decomposed and burnt individuals.
- Estimation of time since death based on stages of decomposition, insect exposure and the accumulated environmental temperatures to which the remains have been subjected.

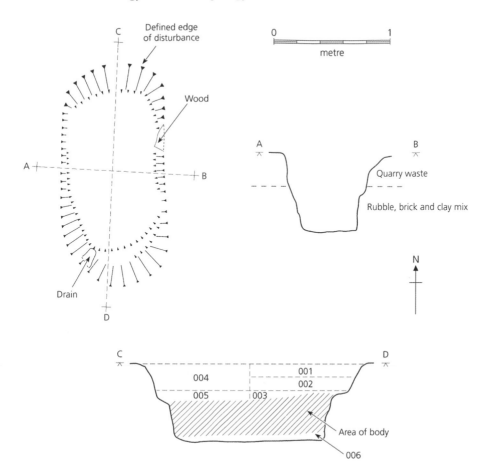

Figure 16.3   Post-excavation scale plan of grave and section(s) through grave.

- Analysis of peri-mortem trauma (which occurred around the time of death and may have contributed to the cause of death) *re* type, number and direction of blunt forces, sharp force or projectile trauma.
- Human identification in mass casualty situations, whether they be:
  - natural disasters (*e.g.* the 2006 Asian tsunami, Hurricane Katrina, *etc.*)
  - transport accidents (*e.g. Marchioness* and Bahrain pleasure boats, Paddington station crash)
  - terrorist incidents (*e.g.* 11 September 2001, the London bombings of 7 July 2007)
  - post-conflict investigations (*e.g.* in Guatemala, Rwanda, the Balkans, Iraq).
  - investigations into political disappearances (*e.g.* in Argentina).

In order to become a forensic anthropologist, training in several areas is required. Certainly a first degree in the 'hard' sciences (*e.g.* biology, chemistry) or a closely related field is desirable and a masters and/or doctoral degree in forensic anthropology should follow. Training in both skeletal and soft-tissue anatomy as well as an understanding of growth and development is critical to understanding variation in sex, age and stature. Comparative anatomy is useful for determining whether remains are human or animal. Evolutionary biology and genetics aid in the understanding of the concepts of 'race' and ancestry and their application in forensic anthropology. Familiarity with human skeletal population variation is essential in order to evaluate how the individual under examination relates to the increasingly racially, ethnically and geographically diverse population of the UK. Unfortunately, with the recent demise of the Council for the Registration of Forensic Practitioners, there is as yet no certification or licensing for forensic anthropologists in the UK and forensic anthropologists are quite rare (only nine were ever registered with CRFP). At present, the American Board of Forensic Anthropology (ABFA) in North America has the only certification process and oversight organisation for professional, practising forensic anthropologists. To become a diplomate, the person has to have a PhD degree, submit case reports and professional references and pass ABFA theoretical and practical examinations. Since ABFA's inception in 1977, there have been approximately 80 individuals who have been made diplomates, but not all of these are still actively practising. The second largest concentration of forensic anthropologists today is in Central and South America, where most are situated in national teams (*e.g.* EAAF in Argentina, FAFG and CAFCA in Guatemala, EQUITAS in Colombia) that function primarily as non-governmental organisations (NGOs). In the past many, mostly MSc-level forensic anthropologists, worked for the two United Nations *ad hoc* tribunals in Rwanda and the former Yugoslavia, the International Commission for Missing Persons or NGOs such as Physicians for Human Rights.

In the UK, a forensic anthropologist may be called to examine human remains either at the scene of the crime or in the morgue on the recommendation of the investigating police and/or Home Office pathologist. The examination may be conducted at the mortuary itself or in a secure laboratory setting, but in either case, all the protocols of logging receipt of evidence, what has been done to the evidence and sealing of evidence apply, just as they do in all other aspects of forensic science. Ultimately, the forensic anthropologist will submit a report to the coroner, which will also be copied to the lead investigator and the

Home Office pathologist, detailing the process of the examination, the results of the examination and an analysis of the findings in the case. As with any interpretation of results, the practitioner must take care to stay with the boundaries of defensible knowledge, citing published and accepted research to support their conclusions and stating the probability that these conclusions are correct in the given context. A forensic anthropologist may also have to testify in court in criminal or civil proceedings or at an inquest at a coroner's court. Forensic anthropologists may also be consulted in cold case reviews where there are unidentified remains, to see if new techniques might resolve the victim's identity. Without first knowing who is dead, the likelihood of finding a perpetrator of the crime is greatly diminished.

Forensic anthropologists are not medical doctors and thus may not comment on the actual cause of death, *i.e.* the physical reason for death, such as a myocardial infarction or heart attack; yet, they do evaluate peri-mortem injuries to the skeleton and may offer interpretations of events occurring at the time of death that are relevant to the manner of death (homicide, suicide, accident, natural, or unknown). Like forensic pathologists, in cases of multiple deaths, they will look for and try to identify patterns of similarity linking one death to another. This might be useful in identifying multiple victims of a serial killer or supporting a charge of torture when a mass grave of the disappeared or prisoners of war is exhumed.

The role of the forensic anthropologist in the UK is a still a fairly new one, and many investigators are not yet familiar with what a forensic anthropologist can contribute to an investigation. For example, many investigators immediately turn to DNA profiling when unidentified skeletal remains are found, when a forensic anthropologist could provide a biological profile and individuating features that might narrow the field considerably. Likewise, dental radiographs and fingerprints recently proved more powerful and expedient identifiers of badly decomposed remains in the Asian tsunami than DNA. DNA is extremely useful, but the unidentified individual must have a DNA profile on the database in order to be located within it, and only a small portion of the UK's population is represented in the DNA database.

16.7   ARE THE REMAINS HUMAN?

Most forensic cases begin with the forensic anthropologist being called to the morgue or crime scene to examine human skeletal remains. Equally likely, however, bones are delivered to the forensic anthropology laboratory by a police officer seeking a consultation concerning the most basic question: are the remains human? Unless there is an entire skull, most lay persons, many police officers and even some veterinary surgeons and medical doctors cannot tell with certainty if skeletal remains are of human or animal origin. Forensic anthropologists, however, are so familiar with the details of human skeleton that they can identify with certainty that remains are human (or not) and, if they are human, can identify with precision the name of the bone, the exact portion of that bone (and the names of all the structures comprising it), and the side of the body the bone came from, even with fragments the size of a 50 pence piece. Potentially, a forensic anthropologist can provide even more information about that small piece, *e.g.* depending on which bone, whether it is from an adult or child, male or female. Many forensic anthropologists are also trained in comparative skeletal anatomy or archaeozoology (animal bones from archaeological sites) and are thus also adept at identifying animal bones to species.

16.8   HOW MANY INDIVIDUALS?

After identifying the remains as human, a forensic anthropologist will then determine how many individuals are present. In routine inquiries this is not usually much of a problem, as most cases consist of single individuals found by chance in indoor or outdoor settings. Determining the minimum number of individuals (MNI) is based on knowing the number of each bone in the body; some bones such as the sternum (breastbone) occur only once, some occur in pairs (a right and left femur, or thigh bone), whereas some occur in higher numbers (12 right ribs, 12 left ribs, 7 cervical and 12 thoracic vertebrae, *etc.*). If there are two right femurs, there must be at least two people represented in the assemblage of skeletal material. Sex and age can also increase the MNI in a sample, *e.g.* if there is one large right femur from an adult male, one right innominate (pelvic bone) from an adult female and one right humerus (upper arm bone) from a child or teenager; even though all the skeletal elements are different, each can be attributed to a different age or sex and therefore they do not belong to the same individual. If more than one individual is present, a separate biological profile must be constructed for each person, even if the individual is represented by as little as a single bone.

   Estimating the MNI becomes more critical when dealing with events that result in extreme fragmentation or commingling (mixing up) of remains. This may occur in post-conflict excavations of primary and secondary clandestine mass graves where hundreds of remains were hastily disposed of in an attempt to conceal crimes of war, crimes against humanity or genocide. The excavation of mass graves requires archaeological skill in combination with forensic osteological knowledge so that the integrity of whole bodies is maintained and remains are not separated due to unfamiliarity with what constitutes a complete human skeleton. Remains may become both fragmented and commingled in bombings and high-speed mass transportation crashes. In both cases, the remains of a single individual may become widely scattered and may even become embedded in the remains of other individuals, as happened in the World Trade Centre terrorist attack of 11 September 2001.

16.9 BIOLOGICAL PROFILE

When dealing with human remains, the first task for a forensic anthropologist is determining the biological profile of the individual. This involves estimating the sex, age, race and stature of the person's bones. The anthropologist will then examine the remains for any indications of ante-mortem trauma (injuries to the skeleton that occurred while the person was still living) and the presence of pathological conditions that might indicate diseases suffered by the individual during their lifetime.

16.9.1 Sex

Like many animals, humans are a sexually dimorphic species. This means that males and females differ according to both the size and the shape of their bones. Males are, on average, larger and more robust than females. The primary cause for these differences relates to the release of male hormones, such as testosterone, during the adolescent growth spurt, causing an increase in stature, ribcage dimensions, muscle mass and its associated skeletal robustness (surface area of joints, muscle insertion markers, *etc.*). When examining unknown remains, it is very important to consider the population origin of the skeletal individual— sexual dimorphism is universal, but population specific. For example, a Japanese male may look gracile when compared to a European female, but will still appear more robust than a Japanese female. It follows, then, that this is true for all features of sex estimation, and one must be cautious when working with new populations or in new geographic areas, as what once looked female may now represent a male. As many sexually dimorphic features of a skeleton should be examined as possible, so as to form a complete and more reliable estimate. Note that sex is a genetically determined biological trait that is reflected in the development of an individual's body into a sexually mature human being;

gender is a social construct based on an individual's decisions and self-perception and how that person portrays himself or herself in society. A forensic anthropologist will estimate sex, not gender. Gender might be inferred by clothing found on a body, but never from the remains themselves.

The most reliable means of estimating the sex of an individual from skeletal remains is a visual observation of the innominate. A male skeleton differs from a female skeleton primarily in the pelvic region, because the pelvic bones of females have undergone an evolutionary adaptation for childbirth. In general, the pelvis of a female tends to be shorter in the vertical dimension and broader in the horizontal direction with a larger pelvic inlet, as illustrated in Figure 16.4, in order to allow for passage of the infant's cranium during birth. The pelvis consists of three bones, a right and left innominate and the sacrum, all of which are sexually dimorphic. In females, the sacrum is shorter and broader with less curvature than that of males. The innominates themselves are composed of three separate bones (the ilium, the ischium and the pubis) that fuse together during early adolescence and possess several features

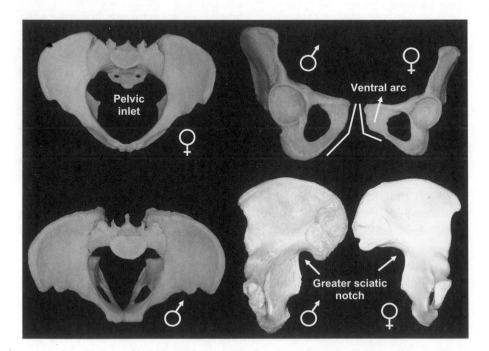

Figure 16.4   Sex difference between male and female pelves in super view (left) and male and female innominate bones in anterior view (top right) and medial view (bottom right).

that distinguish males from females. The greater sciatic notch is formed by the joining of the ischium and ilium and can also be a good indicator of sex. In general, males have a deep and narrow greater sciatic notch, whereas this feature is more open and shallow in females. There is however, tremendous amount of both inter- and intrapopulation variation in this anatomical region, and caution is warranted when no other supporting features indicative of sex are present.

The best indicator of sex is found at the anterior joint between the two innominates, the pubic symphysis. In females, the subpubic angle formed by this joint is more obtuse and the edge of the descending pubic ramus has a tendency to 'fold' back on itself anteriorly, forming what is known as a ventral arc. These features appear to manifest as universally female, no matter the population. Although some males have a ventral arc, it is a relatively rare feature and most often not associated with the more reliably 'female' presence of a wide subpubic angle. The a accuracy of determining sex from the subpubic angle alone is around 95%, which is virtually the same as taking into account all the other features of the pelvis.

Estimating sex from the skull, consisting of the cranium and mandible, is the next most reliable, yielding about 85–90% accuracy. In the skull, the robustness of males is evident in such features as pronounced brow ridges (supraorbital ridges), rounded rather than sharp supraorbital rims, enlarged mastoid processes and nuchal lines, as well as more square, rather than pointed, chins (Figure 16.5).

These typically male features begin to form during puberty with the release of more male hormones as the individual becomes sexually mature. Females remain more gracile, although in some populations they may develop strong neck musculature as a result of certain activities, such as carrying baskets on the head. Male crania and mandibles are generally rougher in texture than those of females, again because of the presence of larger muscle masses and their attachment sites. Females, particularly in African populations, are more likely to exhibit vein impressions on the external surface of the cranium, usually on the lateral part of the frontal bone. In some populations more gracile than Europeans, *e.g.* south-east Asians, male skulls may be characterised by an extension of the root of the zygomatic process posteriorly onto the occipital bone (see Figure 16.5); for someone less familiar with Asian sexual dimorphism in cranial morphology, the presence or absence of this feature may be one of the only ways to determine sex from a cranium alone.

Other bones of the skeleton are also sexually dimorphic, albeit to a considerably lesser degree, with the best yielding accuracies of only

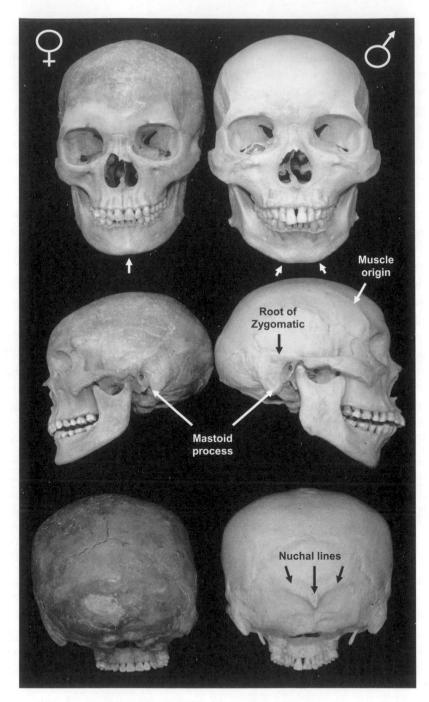

Figure 16.5   Sex differences between male and female skulls.

70–85% when measuring specific dimensions, such as the vertical diameter of the femoral head. The vast majority of these features relate to occupational (or usage) stress on bones from the musculature. The more a muscle is used, the more it pulls at its insertion into the periosteum, the membrane surrounding the external surfaces of all bones. This repetitive action, in turn, causes the activation of osteoblasts (bone-forming cells) underlying this membrane to lay down new bone along the ridge of the muscle insertion. Other types of stress on bone during growth can cause joint surfaces to enlarge. Certain well-defined measurements of the long bones can be taken with osteometric callipers and an unknown individual can therefore be classified as male or female in relation to known standards for a population. If, however, North American standards are applied to Guatemalans, males from Guatemala might appear to classify on the female side of the scale, as they are smaller and more gracile people.

Sex cannot be determined for juvenile remains as the features associated with sexual dimorphism do not appear until puberty is advancing. Sixteen years of age is a somewhat arbitrary but reasoned cut-off point, as by this age most females have reached their adult height and are sexually mature, whereas males may continue to grow until their early twenties. Therefore, a skeleton that is strictly speaking subadult (*e.g.* 15–17 years) and possesses the markedly female pelvic characteristics discussed above is likely to actually be female. For the purposes of the discussion here, female puberty can be said to begin with the onset of menstruation, which occurs in most European populations between the ages of 11 and 14 years of age. Since the onset and completion of male puberty is 2–3 years later than that of female puberty, all skeletons appear generically 'male' before about 16 years of age. It is simply not currently possible to determine the sex of a child under the age of 16 years.

16.9.2 Age

If determining the sex of skeletal material is relatively straightforward, age estimation presents greater challenges. Estimating age accurately relies on assessing multiple regions of the skeleton and even more on the forensic anthropologist's experience and the sheer number and variety of skeletons examined over the years. Many practitioners in the field have described the age estimation process as more of an art than a science. Age estimation is always provided as a range (*e.g.* 17–21 years), rather than an exact figure. The aim is to bracket the age of the unknown individual within that range and it is unimportant where within the range the individual's true chronological age may be as long as it is within the range. Many techniques have been published and visual aids, such as casts depicting the appearance of bony features in various phases of age progression, are available. Yet, providing an accurate age range for an unknown individual requires knowledge of developmental and degenerative anatomy as well as a need to discern something about the lifestyle of the individual. This is particularly true with regard to alcohol and drug abuse, both of which can produce premature ageing in the skeleton.

The first step in age estimation is determining whether the skeletal remains are those of a subadult or an adult. The techniques for age estimation during childhood are based on the genetically programmed timing of 'developmental event sequences' of bones and teeth, which, when examined as a whole, can provide a fairly narrow age range estimate (probably within 2–3 years at most). These types of changes may be recognisable even to the lay person, although their application and significance to age estimation may not. For example, everyone knows that babies have a 'soft spot' (the anterior fontanelle) near the top of their heads; the skeletal closure of the anterior fontanelle by age 2 years is a developmental landmark. In another example, lay people understand that babies have all their teeth by around age 2 years as well, but they may not know the stages of tooth development or the exact timing and sequence of tooth eruption. A child actually forms more bones after birth, but these are primarily the 'growing ends' (or epiphyses) of bones already present. These epiphyses, as well as the carpal and tarsal bones—those of the wrist and ankle—and the patella, all form according to a genetically predetermined sequence and at genetically predetermined ages. Both the sequence and the timing are known, and even though they may vary slightly from one population to the next (and girls go through the sequence earlier than boys), excellent data exist from which one can estimate age accurately from both bones and teeth. Eventually, when the individual completes maturation (puberty) and ceases growth, the epiphyses will fuse to the other parts of the bones to make whole adult skeletal elements. The lengths of the long limb bones in children (as well as dimensions of other bones in the fetus) can be measured and these dimensions correlated with an age range. The clavicle (collarbone) is the first bone to form in the developing fetus (at around 6–7 weeks *in utero*) and has the last epiphysis in the body to close (at 17–29 years of age). Working with subadult skeletal remains involves recognising many more pieces of bone and their position with the skeleton than in the adult.

At the moment individuals stop growing, their bodies begin to deteriorate, although these changes may be quite subtle. When examining adult remains, the methodologies employed are those dependent on 'degenerative changes' in various skeletal regions, which are neither as well understood nor predictable as those of childhood developmental stages. As with sex estimation, the pelvis provides two of the best bony landmarks used as age indicators, the pubic symphysis (the single anterior joint of the innominates) and the auricular surface (the posterior joints between the ilium and the sacrum on both sides). Both of these regions go through a predictable series of changes in their morphology from early adulthood (around 16–20 years of age in males and females) to older age. The pubic symphysis phase system is easier for most practitioners to apply, because there is a series of 12 reference casts (see Figure 16.6), produced by France Casting, exemplifying the most typical configurations of each of the six phases, but the resulting age ranges are very broad and increase with age.

Another problem with the pubic symphysis is that males pass through a more regular series of morphological changes than females; this is related to the fact that the joints of the female pelvis may be stressed

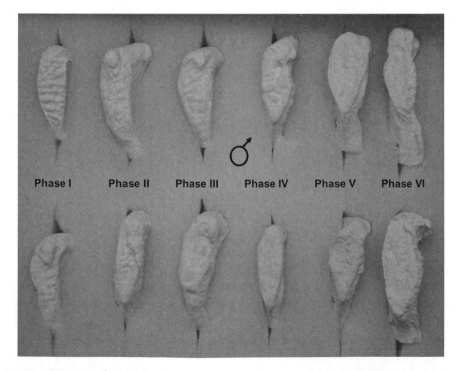

Figure 16.6 France Casting casts of the Suchey–Brooks pubic symphysis phases for age estimation in males.

during pregnancy and childbirth, causing them to develop in different ways thereafter. There are therefore different sets of casts depicting changes associated with age for males and females and these are equated with different age ranges. Thus when using the pubic symphysis for a young adult male it should be possible to provide an age range of 5 years, but for a postpartum older female the pubic symphysis might provide an appropriate age range of 15 years.

The auricular surface is less commonly used for age estimation for several reasons—many practitioners have not been taught to use it well and thus find it difficult to apply. There are as yet no reference casts of the stages and lastly, there are at least three different sets of standards for the age ranges currently published. However, for those who are comfortable examining the traits of the auricular surface, it, too, provides an accurate estimation of age. The other advantage of the auricular surface is that the age-related changes in morphology are the same for males as for females.

The thoracic (or chest) region is also commonly used in age estimation. Another morphological change phase-based system exists for the sternal (anterior chest) end of the right fourth rib. As in the pubic symphysis, there are eight phases of changes which are different for each sex. Casts depicting the classical appearance of each phase are available, but the age ranges provided for them are somewhat unrealistic as there were problems with the sample size and type of statistics used to derive the age ranges. Radiographically, the joints between the anterior ribs and the sternum also show age-related changes that are sex specific. Males have a tendency to develop claw-like extensions on their sternal ribs over the lengthy cartilage between the end of the rib bone and the sternum; females, on the other hand, develop granular bits of bone within the cartilage itself.

In the skull, the phased closure of the cranial sutures (the immovable joints between the bones of the cranium) may also be useful in age estimation, but these also provide wide age ranges and are often not relied upon as anything other than a measure of last resort. A forensic odontologist (dentist) may also be adept at estimating the age of an adult from the wear patterns on the occlusal (biting) surfaces of the teeth, as well as other features of the teeth and jaws. Published standards exist for osteologists doing this on archaeological skeletons, but the differences in diet between those populations and modern ones preclude the use of these for forensic cases.

In sum, estimating age in the adult skeleton is a tricky business, but when done well can assist in the identification of a missing person by narrowing the possibilities that should be explored.

16.9.3 Race

Any discussion of the estimation of race from skeletal remains must be prefaced by a series of definitions and explanations. Many people have confused the concepts of race, ethnicity and ancestry. Race refers to the physical and genetically determined traits of the body's external appearance (phenotype) that individuals inherit from their parents and whose recognition is culturally specific. For example, in the USA today, people with darkly pigmented skin, tightly curled head hair, broad and flatter noses, more projecting upper teeth, *etc.* are considered to be 'black'—and probably refer to themselves as 'black' or African-American. But, for example, in New Orleans, Louisiana in the early 1900s there were subdivisions of individuals with the same darkly pigmented phenotype—octaroons, quadroons, *etc.*—which reflected their percentage of 'black blood' or ancestry as suggested by the degree of darkness of their skin tone. In present-day Brazil, people with these same phenotypes might be culturally divided into many more classifications and might be categorised (named) differently as well. In sum, race is how one's outward appearance is perceived by others and the term applied to that appearance is specific to time and place. To some degree, race is reflective of ancestry—the biological and geographic origin of the 'populations' from which one's parents are derived—but race is a broader and more culturally, rather than biologically, significant concept.

Ethnicity is more often based on self-identification, or how one perceives, categorises and presents oneself in one's contemporary socio-cultural setting. It may often contain some element of racial categorisation, but is more likely reflective of religious, cultural and ancestral nationality: someone can be white or black or brown-skinned and also be ethnically Muslim (or Christian), or 'Asian' (or 'British'), *etc.* Ethnicity does not reflect ancestry (skeletally speaking) as well as race does, and thus ethnicity is not addressed by a forensic anthropology examination. Again, ethnicity might be inferred from the mode of dress, jewellery, tattoos, *etc.* on human remains, but not from the skeleton itself.

Although most forensic anthropologists would likely state that they do not believe that 'race' is a valid biological concept for the human species, they are also well acquainted with the socio-cultural concept of race as it exists today and its application to human identification. Modern population censuses, university admissions forms, job application forms, and certainly missing person reports, all request people to state their race. Data on race exist for all individuals in public records, and it may be classified from photographs, *etc.*; it is, even if perceived as distasteful or politically incorrect to mention, a part of everyone's modern identity. Thus, most forensic anthropologists will estimate race for unidentified skeletal remains, but they do so for the following very good reasons.

Firstly, most categorisation systems for race are based on skull shape. The skull is very 'plastic' evolutionarily, which in this context means that it subtly changes its form in relation to the genetic dynamics of population mobility (predominantly seen in the form of changing marriage and reproductive patterns) over time. There is very often no adaptive reason for skull shape, it simply takes its form after decades and centuries of continued sexual reproduction within a relatively restricted gene pool (the evolutionary mechanism of random genetic drift) and thus varies according to population and geography. If one examines morphology and measures enough skulls and compares one population to another, differences that distinguish these populations, though subtle, become clear. Secondly, most character lists and computerised measurement database programs for race determination from the skull are based on socio-cultural race categories—in other words, when an identified person died, and an anthropologist observed and measured their skull, the anthropologist knew the race of the individual from at least one of the following sources: a photograph, records, or remaining skin and hair on the remains. Therefore, all the standards for determining race based on the skull are indeed based on the perception of 'race' as it is culturally known today—and not, as some researchers have suggested, on the 'ancestry' of these individuals. Ancestry is hard to perceive from a photograph and very few (if any) public records identify the actual ancestry (*e.g.* half Norwegian, a quarter French and a quarter Native American Indian) of an individual. Most public records, however, will identify an individual's race. Thirdly, the assessment of race is expected by those searching to identify missing persons, as it, too, narrows down the possibilities. When taken in conjunction with an estimate of sex and age, as well as the context where the remains were found, the list may be narrowed enough to suggest a specific individual. Race is also not estimated for subadult individuals, although many forensic anthropologists would argue that it is possible to do so.

16.9.4   Stature

In contrast to estimating age and race, estimating the stature (height) of an unidentified skeleton may be accomplished with relative ease, particularly if the long limb bones of the skeleton are intact. The process is essentially simple—*i.e.* measure the indicated long bones, input the measurement into a regression formula and produce a stature estimate plus/minus the standard deviation—but there are some theoretical considerations underlying the practice. Stature estimation is based on the principle that long limb bones (the femur, the tibia) comprise a certain proportion of an individual's total height. Although it is recognised that if two people are both 1.75 m tall, one may have very long legs and a shorter torso while the other has a long torso and shorter legs (men's trousers may be purchased in many different standard lengths!), the formulae are based on averages of many people and most formulae will produce a stature range that will describe the missing person. Using the right formula based on sex and race is crucial, so sex and race are determined before stature. Stature cannot be determined for subadults whose epiphyses are not yet fused, as all formulae are dependent upon complete long bones. Likewise, it is not recommended to use stature formulae on one population that were developed on another. This holds true not only for modern geographically distinct populations, but also for chronologically distinct populations. Some of the best known US stature formulae were developed from skeletal collections of people who died between the late 1800s and the 1940s. Secular trends (long-term population-level non-directional fluctuations in measurements) can affect populations over time, however, and these formulae are no longer applicable to modern forensic skeletons.

Whatever formula is used, the practitioner must be familiar with the measurement definition for the value input into the equation. For example, in two stature formulae developed contemporaneously, one determined stature from the maximum length of the femur and the other from the bicondylar length of the femur. This may not sound important, but if the maximum length is 425 mm and the bicondylar length is 419 mm, these could produce different final stature estimates where the stature range did not accurately reflect the individual's height. That person's skeleton might remain unidentified. Thorough reading and correct application of the primary source's instructions is vital. In addition, measurements should always be taken three times before being input into the formula to ensure accuracy and consistency.

There are two types of height estimates that can be produced: measured stature and forensic stature. The former refers to the height of an individual that is measured *e.g.* by a health professional, and the latter is the height that is reported by an individual for other purposes (driver's licence, telling a friend, *etc.*). Although measured stature is certainly more accurate, it is difficult to locate these records for most people. Forensic statures, although somewhat inaccurate records of true height, are more readily available—and their inaccuracies tend to be directional—*e.g.* young males and shorter people over-report stature—and have been well-documented. Most forensic anthropologists will use a formula to estimate forensic stature, as these equations are generated from modern forensic skeletal collections.

Stature estimation in mass disasters and post-conflict situations is less often critical to positive identification. This is because many people (particularly in post-conflict situations) may have similar characteristics as they were representative of a specifically targeted segment of the population, *e.g.* military-age men. With the estimation of stature, the basic biological profile of sex, age, race and stature is complete. The further individuation of remains, leading to positive identification, comes next and begins with an assessment of conditions that may have affected the individual's skeleton during the time the person was alive—ante-mortem skeletal trauma and pathology.

16.9.5 Ante-mortem Trauma and Pathological Conditions

The forensic anthropologist will examine skeletal remains for indications of healed injuries or disease lesions and also for signs of 'reactive bone' indicating active disease or injuries in the processes healing at the time the individual died. Bone reacts to injury or disease insult by producing lesions are either lytic (destructive of bone) or proliferative (creating new bone). These are both facets of the skeleton's ability to heal itself by removing older bone and regenerating new bone (remodelling), and this process requires the resorption of damaged (essentially dead) bone as a first step before new bone is laid down. Wolfe's law states that 'bone is resorbed where not needed and put down where required', and this process relates to all aspects of changes within the skeleton, from growth in length and changes of shape with age, to epiphyseal union, to fracture healing, to reactions to the progression of diseases.

Evidence of healed fractures (as well as dislocations, muscle injuries, *etc.*) are fairly easy to detect on skeletal remains through both observation and the sense of touch. A healed fracture presents as a callus, a proliferation of bone that goes through several stages in the healing process. Initially, union of fractured ends is accomplished with a large fracture callus surrounding the fracture site (Figure 16.7). This callus is made up of what is known as 'woven bone,' which has a very fibrous and porous appearance and is somewhat like a dry, rough and hard sponge in texture. As the remodelling process continues, this callus becomes cortical (smooth) bone and reduces in size as the fractured ends begin to unite within its protective circumference. Eventually, the well-healed fracture (Figure 16.7) will potentially be visible as a slight misalignment of the bone (*e.g.* in long bones, hand bones or the clavicle) with some

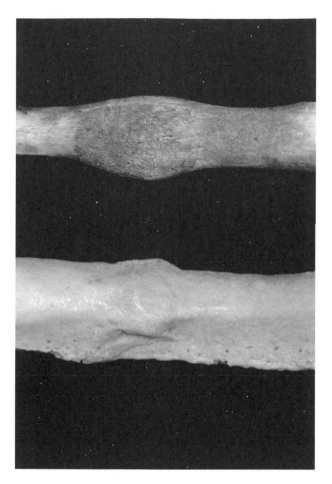

Figure 16.7   Fracture healing stages in ribs. Upper image shows calus exhibiting woven bone and lower image the well healed calus of cortical bone.

smooth, extra bone surrounding or filling in areas, or, as a slight lump, almost like a swelling, on a bone of reasonably normal shape (*e.g.* a rib). Rib fractures are often more easily detectable by touch than by sight. Other musculoskeletal injuries, such as dislocations, severe sprains or tears at a muscle insertion site, will be visible as exostoses, or spicules of bone growth at the site of the injury.

Although healed fractures are fairly commonly observed, it is important to note that most common diseases of modern times do not leave markers on the skeleton. The primary exceptions to this, though

still relatively rare in living populations indeed—and by extension even
more rarely found in a forensic case—include:

1. Certain types of tumours, whether malignant or benign.
2. Various infectious diseases, including tuberculosis.
3. Blood disorders (*e.g.* severe anaemia).
4. Metabolic conditions (*e.g.* osteoporosis).
5. Congenital abnormalities (*e.g.* spina bifida, club foot, extra ribs or
   digits).
6. Skeletal dysplasias (*e.g.* achondroplastic dwarfism).

If present, any of these may be very helpful in individuation if there is
extant and locatable medical documentation. As with all aspects of the
biological profile, identification can only be achieved if there are ante-
mortem records with which to compare the information gleaned from
the skeleton.

16.10   TIME SINCE DEATH

One of the most important aspects in identifying decomposed or skeletal
remains often revolves around being able to state when the individual
died. This allows the police to compare the post-mortem interval estimate with records of people reported missing during that interval, thus
narrowing down the possible matches. The scientific basis of estimating
time since death is a relatively new aspect of the forensic anthropologist's repertoire and one with which many are not yet proficient.
The science of forensic taphonomy, as indicated earlier, is concerned
with everything that happens to an individual from the moment of death
until the remains are recovered. As such, it is primarily concerned with
the processes of decomposition and those factors that may influence its
rate and pattern. However, knowledge of taphonomic principles and
processes would be very helpful.

Much of the early research in this area involved observations from the
University of Tennessee at Knoxville's Anthropological Research Facility
in the USA, where donated human cadavers are allowed to decompose
naturally. Since then, numerous institutions in the USA (and one in the
UK) have developed taphonomic research facilities using either human or
animal models. The advantage of using human cadavers is that no other
adjustments are needed to the data; the disadvantage is that it is rare for
well-designed experiments to be conducted using human cadavers. This is
because it is difficult to obtain and store (refrigerate or freeze) such large
items until a sufficient sample size is achieved to make the decomposition

experimental results meaningful. Using animal models—usually pigs, as they are most anatomically similar to humans with regards to skin type, having an omnivore's gut design, *etc.*—is advantageous as experiments can be conducted with enough cohorts, replicates and controls to make the data obtained statistically significant. Of course, certain adjustments to the data, *e.g.* body size, must be made before it can be applied to humans. Retrospective studies on human cadavers from autopsies have been employed to great effect in recent years, producing equations for the estimation of time since death based on:

1. The appearance of the corpse.
2. The accumulation of temperatures to which it has been exposed.
3. Whether or not it was exposed to insect activity.

Calculating time since death relies on the assignment of a total body score (TBS) which assigns a cumulative numerical value to the body, based on the appearance of three regions (the head and neck, the torso and the limbs) according to defined characteristics along a continuous scale. The accumulated degree days (ADD) are the average daily temperatures (in °C) to which the body has been exposed for the duration of its time in the depositional environment. So a body left on the surface (or in water, or indoors or buried) for 5 days at 20 °C will have an ADD of 100, whereas a body exposed for 10 days at 10 °C will also have an ADD of 100; the concept is that any body with an ADD of 100 will have a similar appearance (or TBS). The two factors that appear to influence this equation of ADD to TBS are the size of the corpse and the presence or absence of insect exposure. As discussed in detail in Chapter 4, insects cause the body to decompose at a more rapid rate than bodies protected from insects (whether buried, indoors or submerged in water). In the presence of insects, smaller bodies decompose more quickly than larger ones, but when insects are absent there is no change in rate. Pattern of decomposition appears to be altered slightly by peri-mortem penetrating injuries (gunshots, stabbing, *etc.*), but the rate at which the body decomposes is not altered. Likewise, superficial burning of a body appears to affect the type of insects attracted to the corpse as well as the pattern of decomposition, but burnt and unburnt corpses still decompose at the same rate (in the same ADD).

16.11 HUMAN IDENTIFICATION

The identification of human remains requires the matching of ante-mortem and post-mortem data. The former is collected by interviewing the relatives of the missing person usually by community liaison officers, and collecting medical, dental and fingerprint records of the missing person. Whilst in the investigation of a single individual's skeletal remains, one must match that single set of post-mortem data to ante-mortem data collected from multiple candidates, in a mass disaster or post-conflict setting, the situation is compounded by there being many missing persons' remains (possibly in fragmentary condition) and many possible candidates for identity. Some disasters are so-called 'closed events', such as an air transport disaster where the identities of the crew and passengers are known from the plane's manifest. There will thus be an equal number of individuals missing as remains of individuals found (see discussion of MNI in section 16.8). There are also 'open events', where both the number and names of victims are unknown, such as mass graves in post-conflict settings, natural disasters such as earthquakes or transport disasters such as commuter train crashes. Identifications in open events are much more difficult to manage as it may take some time to discover who is really missing in relation to the event and also to gather the relevant ante-mortem information. In all cases, a positive identification must be achieved before remains can be returned to families and any legal matters relevant to the deceased's death be resolved.

A positive identification may only be achieved through a high-probability match of fingerprints, DNA or dental radiographs. The number of missing, and hence the potential number of matches, has a direct bearing on the degree of probability needed to declare a match and make an identification. The more missing people are involved, the higher the probability of false-positive matches for any of these techniques. Supporting means of identification are a matching biological profile, ante-mortem medical radiographs, serial numbers from medical or dental implants, truly unique items of jewellery or tattoos unique to image content and placement, *etc.* Clothing and other personal effects, however, are only a tertiary means of supporting identification and would never suffice in themselves as a positive identification in the UK. In many other parts of the world, presumptive identifications (those made on the basis of the family's recognition of the deceased's facial features, clothing, *etc.*) are still common.

Thus, while a forensic anthropologist contributes to the gathering of evidence used in making a positive identification, the techniques and methodology of the forensic anthropologist are not means of positive identification by themselves. In some countries, forensic anthropologists may perform dental radiographic matching and other types of radiographic matching as a means of positive identification, *e.g.* frontal sinus patterns, but this is not common practice in the UK. Forensic anthropologists have proved particularly useful in mass disasters because of their ability to recognise and identify tiny fragments of tissue to sex, side and age. For example, at the World Trade Centre site anthropologists were responsible for preventing identifications being made when they recognised that the sample taken for a DNA profile did not actually belong to the majority of the remains (a part of one person had been blown into the body of another, and the match related to the small fragment of someone else). Likewise, during the identification process, anthropologists were used to sort through the removed rubble and identify human body parts, distinguishing them from the food waste from the numerous restaurants housed in the Twin Towers.

16.12   TRAUMA IDENTIFICATION AND INTERPRETATION

Identifying the presence of trauma to the skeleton, distinguishing what type of trauma it is (*e.g.* projectile, blunt force or sharp force) and interpreting force trajectory from the characteristics of the wound are some of the most important things a forensic anthropologist does that have legal ramifications. Trauma can be inflicted ante-mortem (while the individual was living), peri-mortem (occurring at or around the time of death) or post-mortem (damage occurring to the skeleton after death). Ante-, peri- and post-mortem injuries exhibit different qualities consistent with the character displayed by the bone itself in a living or deceased state. Bone is a living tissue when the individual is alive and it dies when the individual dies. It primarily consists of a mineral component which renders it strong and hard (hydroxyapatite) and an organic component (collagen) which imbues it with elasticity, resilience and resistance to stress. For a time after the individual dies, bone will still behave like a living tissue, and injuries to it cannot be readily distinguished from those occurring during life, or rather at the time of death, because healing will not have a chance to begin. The crux of the matter is that we have little knowledge of when (chronologically) the bone tissue begins to become drier and more brittle and react like dead bone rather than live bone.

Ante-mortem injuries are, as noted above, distinguished by the process of healing with reactive bone at the site of the injury, somewhere in the process of repairing it. A peri-mortem injury is an injury that evinces no signs of healing and thus appears as fresh damage (Figure 16.8).

Characteristics of peri-mortem injuries include adhering fragments of bone depressed into the injury site (in blunt force trauma), well-defined

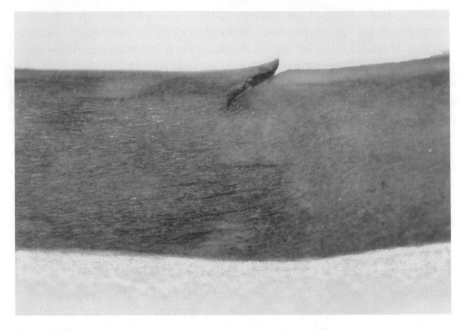

Figure 16.8 Sharp-force trauma injury inflicted with a kitchen knife to the superior
border of a rib.

and 'clean' cuts with lifted spurs of bone (in sharp force trauma) and
well-defined entrance/exit defects from which one can estimate bullet
trajectory (in projectile trauma). Some trauma may be a combination of
two of these types, such as a machete or axe injury, where the implement
inflicting the trauma is both sharp and heavy; this may result in characteristics of both type of trauma appearing on the bone, *e.g.* a sharp cut
that has a depressed, crushed area continuous with it. Post-mortem
trauma, usually caused by trampling or excavation damage, is defined
by differential colouration of the cut or broken surface from that of the
rest of the bone. Bones damaged post-mortem have a tendency to break
transversely, rather than in a more spiral manner as they do in living
tissue. The location of fracture and sharp force trauma on the bone is
also a clue to its chronology—post-mortem breaks and cuts may occur
anywhere on the bone, whereas breaks occurring peri-mortem are
focused in injury sites known to occur in attacks or accidents (ankle,
collarbone and wrist for injuries occurring from falls; mid-shaft of the
ulna for defensive injuries from direct blows; palmar surfaces of hand
and finger bones from sharp force defensive injuries, *etc.*). Other patterns of injuries are common to hit-and-run motor vehicle *versus*
pedestrian or motorcycle accidents.

It is important for the forensic anthropologist to be conservative in their interpretation of peri-mortem trauma, particularly when there are multiple injuries. A single stab wound to the ribs can easily damage the upper border of one rib and the lower border of the rib above it. The anthropologist sees two sharp-force trauma injuries, but they represent a single stab wound. It is very prejudicial in a court case to overstate the number of stab or blunt force injuries; there is a difference in gauging the intent of the perpetrator if one states that someone was stabbed eight times rather than stating that the person was stabbed a minimum of three times. This is true as well for blunt force trauma, and is particularly relevant for blows to the cranium where one blow may overlap with and obscure a previous one. It is still professionally more ethical in expert witness statements and in court to err on the side of caution and provide a minimum number: 'Yes, it could certainly be more, but it was no fewer than two.' After all, both the verdict and the sentencing may be strongly influenced by this testimony.

16.13 CONCLUSION

Although forensic anthropologists and archaeologists are few and far between here in the UK, and these remain rare specialities elsewhere in the world as well, competent practitioners of the disciplines have contributed significantly to the recovery and analysis of human remains from high-profile murder and serial murder cases as well as mass disasters and post-conflict investigations. Broadly trained forensic archaeologists and anthropologists have a multidisciplinary background that is rare among specialists and their cooperation as team members in criminal investigations, civil suits and identifications should not be overlooked. Population-based studies relating to the establishment of the biological profile remain important, while trauma analysis and time since death estimation represent the cutting edge of forensic anthropology as an emerging experimental science.

BIBLIOGRAPHY

S. Blau and D. Ubelaker (eds), *Handbook of Forensic Archaeology and Anthropology*. World Archaeological Congress Research Handbooks in Archaeology, Left Coast Press, Walnut Creek, 2009.

J. Buikstra and D. Ubelaker (eds), *Standards for Data Collection from Human Skeletal Remains*. Arkansas Archaeological Survey Research Series No. 44, Arkansas Archaeological Survey, Fayetteville, 1994.

S. Byers, *Introduction to Forensic Anthropology: A Textbook*, Pearson, Boston, 2005.

W. Haglund and M. H. Sorg, (eds), *Forensic Taphonomy: The Postmortem Fate of Human Remains*, CRC Press, Boca Raton, 1997.

W. Haglund and M. H. Sorg, (eds), *Advances in Forensic Taphonomy: Method, Theory and Archaeological Perspectives*, CRC Press, Boca Raton, 2002.

J. Hunter and M. Cox, *Forensic Archaeology: Advances in Theory and Practice*, Routledge, London, 2005.

T. Simmons, R. Adlam, and C. Moffatt, Debugging decomposition data—comparative taphonomic studies and the influence of insects and carcass size on decomposition rate. *J. Forens. Sci.*, 2010, **55**, 8–13.

T. White and P. Folkens, *The Human Bone Manual*, Elsevier/Academic Press, Amsterdam, 2005.

CHAPTER 17

Presentation of Expert Forensic Evidence

TREVOR ROTHWELL

Avonpark, Limpley Stoke, Bath BA2 7JS

17.1 INTRODUCTION

The forensic scientist may be skilled in their particular branch of science and may indeed be a world expert on the subject. Such expertise is of very little value if the expert concerned is unable to communicate adequately both on paper and in the witness box.

The end product of almost every forensic scientific investigation consists of a report that may be used by police officers, prosecuting authorities, defence lawyers and the judiciary, and ultimately by those members of the general public who will comprise the jury. It is essential, therefore, that the forensic scientist is able to put together a report encapsulating the results of the scientific tests that have been undertaken in such a fashion that the information is readily accessible to a non-scientist. On occasion the scientist will have to appear in person in the witness box to explain and, if necessary, defend the conclusions reached in the laboratory, and in order to do this effectively the scientist will need to develop yet another set of skills.

The culmination of any criminal investigation is likely to be a trial within the criminal justice system. Accordingly, the first part of this chapter is devoted to a brief outline of the legal processes operating in

Crime Scene to Court: The Essentials of Forensic Science, 3rd Edition
Edited by P. C. White
© Royal Society of Chemistry 2010
Published by the Royal Society of Chemistry, www.rsc.org

the UK, so that the scientist's contribution to the criminal trial can be put into context. The duties and responsibilities of the expert witness are outlined and then the respective roles of prosecution and defence are explored in more detail, to demonstrate the similarities and differences for the forensic scientist working for one side or the other.

The need for adequate quality control measures and the importance of proper training for the expert witness are also considered, together with the content and format of the forensic scientist's report. Disclosure of expert evidence is an important issue, therefore the obligations on the prosecution and the defence to disclose the nature and substance of the expert evidence prior to trial will be mentioned.

Finally, some practical advice is offered to those about to embark on the ultimate stage of the expert's work—appearing in court and giving evidence.

17.2 THE LEGAL SYSTEM AND THE COURTS

The legal system to be outlined is broadly that which exists in England, Wales and Northern Ireland. In Scotland the structure of the courts is somewhat different, although the principles of justice—and the responsibilities of the forensic scientist—remain essentially the same. Some of the differences between Scotland and the rest of the UK will, however, be highlighted.

17.2.1 Lawyers

In England and Wales the Crown Prosecution Service (CPS) is responsible for the compilation of the prosecution evidence and its presentation in court. Either branch of the legal profession—solicitors or barristers—may appear in the magistrates' courts. Barristers usually appear as counsel in the higher courts, although a number of solicitors now also have rights of audience. CPS staff will present the case themselves or will brief counsel where appropriate.

Prosecution in the Scottish courts is undertaken by the Procurator Fiscal Service, operating through local offices in a manner which is similar to that of the CPS. A Procurator Fiscal (solicitor) can demand further investigative work to be carried out and may take a more active part in an investigation than the equivalent officer of the CPS. The Crown Office administers the Procurator Fiscal Service and also, where necessary, instructs advocates to appear in the High Court.

The defendant is usually represented by a solicitor who will appear for them at the magistrates' court. Where the case is to be tried at a higher court in England, the solicitor will normally brief a barrister (in Scotland, an advocate) to act as counsel for the defendant at the trial.

17.2.2 Magistrates' Courts

Most criminal cases are tried in magistrates' courts. Lay magistrates, who form the bulk of the magistracy, receive extensive judicial training but possess no specialist legal knowledge. They come from all walks of life, act in a voluntary capacity and commonly sit as a bench of three; they are advised on the legal aspects of each case by a legally qualified clerk. There are also district judges; such individuals are legally quali-fied, are paid for the work which they do and usually sit alone. They are often employed in urban areas where caseloads are heavy. Magistrates are limited both in the types of case they may try and in their powers of sentencing.

Every criminal case must pass through the magistrates' courts. The vast majority are tried in these courts; the defendant is sentenced appropriately and no other court is involved. The most serious and complicated cases are committed by the magistrates to the next tier of court, the Crown court, to be tried by judge and jury. Defendants who are to be tried for certain less serious offences in England and Wales may also elect for jury trial.

17.2.3 Crown Courts

The Crown court is presided over by a judge who sits with a 12-member jury selected from members of the public. The judge is there to advise on matters of law, but it is the responsibility of the jury to decide on matters of fact and to give their verdict at the end of the proceedings. The full range of sentences available within the criminal justice system may be used for a defendant found guilty at the Crown court.

Because of their serious nature, it is Crown court cases in which the forensic scientist is most likely to become involved. In 1993 the Royal Commission on Criminal Justice found that about a third of all con-tested cases in the Crown court involved scientific evidence.

17.2.4 Appeals

Following conviction, a defendant may lodge an appeal against sentence and/or conviction on various well-defined grounds. An individual who has been sentenced by magistrates may appeal to the Crown court, and an individual sentenced at the Crown court may go to the Court of

Appeal, Criminal Division, where the case will be heard by senior members of the judiciary. Under relatively recent legislation it is also possible for an application to be made that the sentence of the Crown court was too lenient.

The Divisional Court of the Queen's Bench Division of the High Court considers appeals on points of law or procedure in cases dealt with in the magistrates' courts; such appeals may originate from either prosecution or defence. Appeals lost in the High Court may, in appropriate circumstances, go to the Supreme Court.

17.2.5 Coroners' Courts

The coroner does not have a responsibility to try individuals, but to inquire into the causes of deaths which may have occurred in suspicious or unusual circumstances. Following an examination of the circumstances, the coroner's inquest will return a verdict of, say, accidental death. Coroners may sit by themselves or, in appropriate circumstances, may empanel a jury. The office of coroner is a very ancient one, and those who hold the office must be qualified either legally or medically. In London, coroners often possess both qualifications.

17.2.6 Scottish Courts

Lay justices sit in the District Courts, which form the lowest tier of the courts in Scotland and deal with the least serious criminal matters. The Sheriff's Court is the principal local court, exercising a wide jurisdiction in both criminal and civil matters. A sheriff, who will be legally qualified and in some ways may be likened to a district judge, may sit alone (Summary Procedure), or with a jury of 15 members (Solemn Procedure) depending on the seriousness of the crime to be tried. A defendant cannot elect for a trial by jury, and only cases involving serious crimes are heard before a jury in a High Court. The High Court of Judiciary sits in Edinburgh and other major towns, and these courts will also hear appeals from the lower courts. The equivalent of the coroner's inquest is a fatal accident enquiry, which is presided over by a sheriff.

17.2.7 Civil Courts

Criminal cases form only a part of the workload of the courts in the UK; other matters such as insurance or personal injury claims are usually part of the work of the civil courts. The burden of proof may be different in the civil courts. Conviction in a criminal trial requires proof beyond reasonable doubt. In the civil courts a case may be proved on a balance of probabilities.

The forensic scientist may well become involved in civil matters, being requested to provide advice or to undertake scientific testing in a parallel fashion to that of the criminal investigation. Although the comments in this chapter are directed specifically to the criminal trial procedure, the forensic scientist concerned with civil casework will need to adhere to the same basic principles.

17.2.8 Course of the Criminal Trial

The forensic scientist's work is commonly orientated to the production of information that may be used as evidence in a criminal trial. In order that the contribution of the forensic scientist can be seen in the context of the whole proceedings, the basic structure of such a trial will be described.

At the start of a trial in England and Wales, following such initial steps as the swearing in of the jury, the prosecution opens its case with a brief outline of the evidence to be offered. In Scottish courts there are no opening speeches and the procedure starts by going straight into the evidence. Witnesses for the prosecution will then be called one by one to give their evidence. In general terms a witness will be excluded from the court until the time comes for them to give their evidence, as no witness should be influenced by evidence already given during the trial. However, experts are sometimes asked to be present throughout the examination of other witnesses specifically so that they may comment on relevant aspects of the evidence.

Each witness will be questioned by the lawyer acting for the prosecution about those aspects of the case of which that individual has knowledge; this is referred to as the 'evidence-in-chief'. Having given their evidence, the witness may be questioned by the lawyer acting for the defence. In this 'cross-examination' the witness may be asked, for instance, to expand on answers previously given to the prosecution's lawyer and, at the end, the prosecution lawyer may wish to briefly re-examine the witness on points raised during the cross-examination.

When the prosecution has presented its case the defence lawyer will open the case for the other side, again calling such witnesses as are necessary to make the various points which the defence wishes to raise. The procedure is similar to that employed with the prosecution witnesses, except that the defence lawyer will conduct the examination-in-chief, and the prosecuting lawyer the cross-examination. The defence witnesses may or may not include the defendant, who is under no obligation to say anything at trial. If they choose to do so they are usually the first witness.

When all the evidence has been adduced the closing speeches commence, and here the procedure differs between magistrates' courts and Crown courts. In the magistrates' court only the defence lawyer speaks, summing up the various points of that side's case. In the Crown court the procedure is more complex and there are three speeches: first from the prosecution; secondly from the defence; and finally from the judge who, while remaining neutral, must ensure that the jury is properly briefed for its task. In England, Wales and Northern Ireland the jury of 12 people may return verdicts of 'guilty' or 'not guilty'. For a verdict of guilty, at least 10 of the 12 jurors must agree. In Scotland the additional verdict of 'not proven' is available for use in appropriate instances for the jury of 15 people, and 8 of the jurors must agree for a verdict.

17.2.9 Role of the Witness

The role of the witness is to place before the court such information as may be relevant to the case in question. For most witnesses this is limited to evidence of facts that have formed a direct part of their experience. Witnesses appearing in criminal trials are generally not permitted to give 'hearsay' evidence; in other words, the witness cannot repeat secondhand information that has been told to them. One or other lawyer will rapidly intervene if they feel that a witness is overstepping the mark in this respect.

17.3 EXPERT WITNESSES

Forensic scientists are among those with specialist expertise who may be called upon to give evidence in the courts, providing advice on matters as diverse as the causes of fire, medical negligence, and stresses and strains in bridges. The role of such a specialist is different from that of the ordinary witness, however, in that the expert is not present simply to repeat the facts adduced by the scientific tests that have been undertaken, but to offer an interpretation of the findings in the context of the case. Accordingly, expert witnesses may give, indeed are encouraged to give, opinions as well as factual evidence.

The expert may be expected to give the court information that falls outside the general knowledge of the judge or jury. If asked, the expert may legitimately give an opinion on any issue which falls within their competence, although there may be reasons why the judge may subsequently advise the jury to disregard such information. This opinion may include *e.g.* relevant probability estimates to illustrate that a particular blood group has been found to occur in one person in every 20 in the general population of the UK.

In some circumstances, in particular the results of certain DNA profiling tests, the evidence against the accused may appear so powerful that the scientist believes, beyond any reasonable doubt, that no other individual could be responsible for the incident. However, the expert should not phrase their evidence in those direct terms. The question of the guilt or innocence of the accused does not fall within the remit of the expert, or any other, witness. This issue is one which must be addressed by those who will decide the case, *i.e.* the magistrates or jury. Accordingly it is better to say that the DNA result obtained shows the accused to be 100 000 times (or whatever the appropriate statistic) more likely to be the source of the crime sample than an unrelated person taken at random.

17.3.1 Duty of the Expert

The responsibility of the prosecuting lawyer is to present the case against the defendant; to marshal and present such facts as may serve to show that it is beyond any reasonable doubt that the accused individual perpetrated the crime in question. The responsibility of the defence lawyer is to counter the arguments adduced by the prosecution and to demonstrate that the prosecution has not made a convincing case. Witnesses, whether called by prosecution or defence and although they are simply there to present evidence of fact, will almost certainly have convinced themselves of the guilt or otherwise of the accused. In that sense both sides are partial and biased to the particular party for whom they speak. *It is vital that the expert stands apart from this partiality. The role of the expert witness is to use his or her experience and skill to provide impartial and unbiased evidence to the court.*

17.4 PROSECUTION AND DEFENCE

17.4.1 Equality of Arms

The system of criminal justice that applies in the UK is essentially adversarial. The case against the accused will consist of all those facts which the prosecution considers necessary to convince the court of the accused person's guilt; the defence will put together all such information as it considers necessary to nullify the prosecution case. A basic tenet of the criminal justice system is to ensure 'equality of arms' and thus it is important that appropriate advice and assistance be available equally to both prosecution and defence. Such assistance includes adequate facilities for undertaking any necessary scientific examination.

In the UK either or both prosecution and defence may introduce scientific evidence into the criminal proceedings. The introduction of scientific evidence by one side does not necessarily mean that the other side will also automatically wish to employ a scientist, although where the prosecution seeks to rely on detailed or involved science it is usual for the defence to brief their own expert.

The situation in the UK thus differs from that in some other jurisdictions where an accusatorial system of criminal justice is practised. In this system an examining magistrate or *juge d'instruction* may commission and set the parameters for scientific work to be done on behalf of the court.

17.4.2 The Forensic Scientist and the Prosecution

For many years the Home Office Forensic Science Service was the sole provider of scientific investigation services for police forces in England and Wales. However, high-quality services are now available from a number of suppliers. The Scottish police service operates its own laboratories, and one in Belfast serves Northern Ireland.

It is quite usual for the police officer to have direct contact with the forensic scientist during the investigation of a crime. This contact may be close and ongoing; in a major case the scientist may effectively become a part of the investigating team. The forensic scientist working for the prosecution will conduct an in-depth examination of all the items and materials relevant to an investigation. These will (within reason) be in the condition they were in immediately after the incident. This is important because the longer the time that elapses before the scientific examination commences, the greater the opportunity for individual items to be contaminated with extraneous matter or, conversely, to lose something that may be relevant.

All data produced will be recorded in a case file from which the scientist will in due course compile a report detailing the salient facts. This report will be passed to the police investigating officer and hence to the prosecuting authority.

17.4.3 The Scientist Working for the Defence

Scientific work for the prosecution is usually commissioned by the police as a part of the investigative process, whereas work for the defence is commissioned by that side's legal team. Thus the position of a forensic scientist working for the defence is both similar to and different from that of a scientist employed by the prosecution.

It is likely that the individuals employed to advise the prosecution and defence will possess similar qualifications and experience; both will be engaged upon a search for independently verifiable facts; probably both will be looking at similar aspects of the case. A scientist instructed by the defence, however, may have no involvement with the case until much or all of the initial work for the prosecution has been completed. Accordingly, it is seldom that this scientist sees items in the precise condition they were in immediately after the incident. Instead, the defence scientist will often need to work from information and results provided by the prosecution scientist and any conclusions that the defence scientist draws must take these limitations into account.

Unlike the prosecution scientist, the expert engaged by the defence will frequently be asked to prepare a report within a very short time-scale. The work done on behalf of the prosecution, because it tends to be comprehensive and wide-ranging, may take weeks, if not months, to complete and may well not be delivered to that side's legal team until after the case has been committed for trial at Crown court. Lawyers acting for the defence will not generally know whether they wish to brief their own experts until they have been made aware of the content of the prosecution scientist's report. This may result in precious little time being available before the trial is set down for hearing for any work to be undertaken on behalf of the defence team.

Although defence lawyers now have access to adequate forensic science facilities, this is a relatively new concept. It developed through the growth of small private independent laboratories, often founded by experienced forensic scientists who, having previously practised in the large public sector 'prosecution' laboratories, recognised the need for suitable facilities to be made available to all those involved in the criminal justice system.

17.4.4 Sequence of Events in a Forensic Examination

The fact that only one scientist is likely to be in a position to see an item in its original condition may have a significant effect on the subsequent course of events.

Let us take the example of an attempt to demonstrate contact between two people through an examination of their clothing. One way in which this may be achieved is by looking at the range of textile fibres transferred between one set of clothing and the other, removing for microscopic examination the superficial fibres adhering to the surface of the garments. Such fibres are extremely tiny and are easily lost from the surface, while other fibres may be readily transferred on to a garment through accidental contact. Accordingly the forensic scientist will accord the highest priority to removal of the superficial fibres very early in the investigation in order to obviate the possibility of accidental contamination occurring during subsequent handling of the items.

Having removed the superficial fibres from, say, a jacket it is not possible to repeat this task and any other expert looking at the garment subsequently will not see the item in its original state. For any subsequent examination the scientist will be presented with a jacket together with a completely separate set of tape samples bearing the superficial fibres that had at one time been present on the jacket, which may limit the options available to the second examiner.

This situation has its analogies in other fields. For example, following an initial autopsy by a forensic pathologist working on the instructions of the coroner, further examinations may be requested by other interested parties such as lawyers acting for an accused or for the relatives of the victim. Only the pathologist who conducts the original autopsy will be able to see the body in its original state, and although other doctors may subsequently conduct their own examination of the material they will be constrained by the nature of the original autopsy.

In many circumstances the use of photographs and video recordings may be of immense value, *e.g.* in showing the original disposition of the various items found at a crime scene or the appearance of injuries on a body. In other instances, however, there is at present no way of recording the appearance of an item in such a manner as to demonstrate its original condition to anyone who may subsequently express an interest. The removal of superficial fibres is a case in point and the efficiency and effectiveness of this procedure has essentially to be taken on trust. The initial examiner, in whatever discipline, therefore carries a great personal responsibility for ensuring that work is carried out to the highest possible standard.

17.4.5   Role of the Second Examiner

The role of the second or subsequent expert is thus to examine such items as are available and to make an assessment of the way in which the initial examination was carried out. This will usually involve detailed scrutiny of the documentary material—test results, chromatographic charts, *etc.*—produced by the original scientist to check that appropriate procedures were carried out to the proper standard. Relatively less often will there be the need to repeat actual scientific tests. Although there are occasions on which the defence takes the initiative and recovers items for scientific examination which have not already been looked at by the prosecution, in most cases it is the prosecution's scientist who takes first bite at the cherry and the defence's expert who has to take the role of the second examiner.

17.4.6   The Need for Both Prosecution and Defence Experts

The responsibility of the scientist is to search for verifiable scientific facts and to provide impartial evidence to the court. It is, therefore, legitimate to question the necessity for both prosecution and defence to employ their own expert, and if they do, why the two experts may come up with different conclusions. The most obvious justification for the use of independent expert witnesses is the need to ensure the 'equality of arms' referred to earlier (Section 17.4.1). Thus, to ensure the maximum probability of a fair trial, whatever facility is available to one side should be made available to the other. If the prosecution intends to rely on scientific findings as one plank of its evidence at trial, it is right that the defence should be able to independently test the strength of that evidence.

Although equality of arms may justify the need to employ independent forensic science experts, it does not explain why, after an examination of the same materials and results, they may come up with different answers. In fact, and not surprisingly, it is rare for experts to disagree over the scientific facts adduced in any particular case. Where differences arise it is more likely to be in the interpretation placed upon the findings, and there may be a variety of reasons why interpretations differ.

In some areas of science there is room for genuine ambiguity about the meaning of particular scientific findings. Certain autopsy findings can, in the view of one pathologist, point to one conclusion about the origin of particular injuries; another practitioner may consider a different conclusion more probable in the circumstances. It is often not possible to determine objectively that one pathologist is more 'right' than the other and the court will be left to decide between the two experts.

There will be other circumstances in which different conclusions can be drawn because the scientists involved are working on different assumptions. Forensic science is seldom an absolute science, and the scientist has to be given a framework within which to design their programme of work. The scientist working for the prosecution will inevitably be reliant on information provided by the police. Although this information may be detailed, it is unlikely to include the defendant's own view of the course of events.

The defence scientist, on the other hand, almost always has the advantage of having heard both the rationale behind the prosecution expert's work and the defendant's own explanation of events. Using this broader spread of knowledge, the defence scientist may be able to posit other possibilities or explanations which provide an alternative, and sometimes better, fit with the observed facts. These other explanations may assist the defence in pursuing its case, although in many instances the information may also be of value to the prosecution. Where such information is discovered it must not be withheld.

From time to time the suggestion is made that the forensic scientist should be more precisely a servant of the court itself rather than of one side or the other. Attractive though such a suggestion may be in emphasising the impartiality of science, it appears that within the adversarial system of justice that operates in the UK the current arrangements are satisfactory. In most cases it is expected that the scientists will come up with very similar conclusions. In those cases where conflicts exist, it remains with the court to decide between the two strands of scientific evidence.

17.5   THE IMPORTANCE OF QUALITY

The evidence adduced by the forensic scientist may be crucial in securing the conviction of an accused person, who may then receive a heavy sentence. On other hand, scientific findings may help to secure an acquittal. The responsibility that rests on the shoulders of the forensic scientist is thus high, and no effort must be spared to ensure the veracity and accuracy of any evidence that is provided. Each and every individual who provides expert evidence to the courts bears a similar responsibility. The need for quality and the maintenance of standards has been obvious since the inception of forensic science, and for many years comprehensive quality management schemes have been in operation.

17.5.1   The Individual

The quest for quality involves the individual as well as the work process. The qualities necessary to perform well as a forensic scientist are not necessarily the same as those required, say, to conduct a research programme. Although forensic work is sometimes at the frontiers of science, particularly with DNA profiling, it often necessitates common sense as much as a deep scientific understanding. It always requires an interest and ability in putting over scientific concepts to non-specialists in a clear and unambiguous fashion.

Individuals seeking a career in forensic science require a basic background in an appropriate subject. This will often consist of a first degree in some aspect of chemistry, molecular biology or biochemistry. Now there are also undergraduate forensic science degree courses available, some of which have achieved Forensic Science Society accreditation with their graduates gaining employment in forensic science laboratories. Postgraduate courses devoted specifically to the specialism can also provide a valuable background for the potential forensic scientist. Detailed training in forensic science is provided on the job and, along with developing technical skills, such training will include considerable practice in the compilation of reports. Comprehensive training is provided for those scientists who give evidence in court, using real lawyers and a mock court.

Aside from the specialist academic courses, there were originally no formal qualifications in forensic science. The Forensic Science Society has introduced a series of diplomas in particular aspects of the profession, *e.g.* in the examination of firearms, examination of documents, fire investigation, and scene of crime investigation.

Until the late 20th century the usual course of a scientist's career was for the individual to join one of the public sector laboratories, to be trained in some appropriate discipline and to remain within the sector for the duration of their career. Recent years have seen considerable diversification in the provision of forensic science services, with a range of suppliers able to tender for work. The larger laboratories are able to offer well-designed programmes of training and the opportunity to obtain excellent all-round experience. In Scotland an expert witness can only give evidence if they have authority from the Secretary of State for Scotland.

17.5.2 Setting Standards

To be assured that the quality of work is adequate it is necessary to define appropriate standards, set up work processes to achieve that standard and introduce monitoring systems to ensure compliance. Proper quality management must therefore be integral to every aspect of the work of a forensic scientist; it is not something that can be tacked on at the end of the examination process. To that end, the Home Office has instituted the post of Forensic Science Regulator, with a brief to agree appropriate standards for the provision of forensic science and pathology services within England and Wales and, through agreement with the devolved administrations, the whole of the UK. The Regulator has set up specialist advisory groups to support him in his work.

17.5.3 Case Documentation

Underpinning the quality of work performed by the forensic scientist must be the basic documentation that comprises the case file. This should record every action undertaken by the scientist and by every assistant or other person working on their behalf. In this respect such 'actions' are likely to include:

- Telephone calls and written correspondence between the scientist and those for whom the work is being carried out. Such material may involve lists of items for examination, statements from other witnesses, invoices, requests for attendance at court, *etc.*
- Notes to assistants or other colleagues regarding work carried out.
- Detailed records of all the examinations carried out and the tests undertaken, together with the interpretation of the results—these records will constitute the bulk of the file.
- Drafts of the final report, and a copy of the report itself.

All this material should be indexed, *e.g.* through the use of a minute sheet attached to the file.

As far as possible the layout of the case file should be in logical sequence, first because it enables the material to be checked through more easily, but also because it will prove to be of enormous value should the need arise for the scientist to appear in court. It is all too easy to lose one's place during intensive cross-examination in the witness box!

17.5.4 Assuring the Quality of the Work

Test procedures to be employed must be reliable and capable of yielding accurate and precise results. One way of assuring this is to rely on published procedures that have been subjected to peer review. Although there may be rare occasions when the individual scientist will have to design a new procedure for a specific investigation, in the main there is no place for any expert to use tests that have not been generally accepted by the scientific community as a whole.

Another basic tenet of quality assurance is that significant decisions should be checked by a colleague. Thus, for instance, the rationale behind a blood-grouping test or the results of a chromatographic separation of fibre dyestuffs would be noted by the worker performing the test and confirmed by a colleague. The object of such peer review is not to belittle the work of any individual, but to obviate the commission of errors that may easily occur when, say, a sample tube becomes misplaced.

One method of assessing the efficiency and effectiveness of a forensic science laboratory is through the submission of simulated case material for examination. Laboratories have developed systems in which such cases can be worked on without those who examine them being aware of the provenance of the material, and this has proved to be a useful performance monitor.

It is particularly important that the report itself, the product of the scientist's activities, should be subject to thorough quality assurance procedures. This is likely to include independent scrutiny of the report by another scientist with a similar background, followed by a discussion of the salient points in which the writer will be invited to justify the conclusions reached. More generally, such discussion promotes the sharing of experience, which can only result in the overall improvement of quality. Sharing experience through checking each other's work and discussing reports may be easier to organise in large laboratories than in small independent units; the need, however, is universal.

The scrutiny should not be limited to scientific matters, because it is vital not simply that the conclusions are correct but that they should be understood by the reader. In this respect it is also valuable to have the material read by a non-specialist; that, after all, will be its fate when the report leaves the laboratory.

17.5.5 Time Limits

Although the length of time taken for laboratory examination may not generally be considered as an aspect of quality, it is nevertheless important. Most cases committed for trial are subject to time limits, calculated from the time that the accused person is taken into custody. The need to carry out lengthy scientific tests may therefore necessitate requests for special arrangements to extend such limits.

17.6 THE FORENSIC SCIENTIST'S REPORT

The forensic scientist's report is the product of all the effort that has gone into the investigation. It must contain all relevant detail and must be laid out in such a way that its content is immediately accessible to the non-scientist, whether police officer, lawyer or lay jury member. Although the image of the forensic scientist is of the expert in the witness box, in fact the vast majority of cases do not involve the scientist's appearance in court. In these circumstances, therefore, the report has to stand on its own without the possibility of explanation or clarification from its writer. Accordingly, it is absolutely essential that the scientist's report is clear and unambiguous, and contains all the information necessary to explain the scientific findings.

There is another factor that should be taken into account. In most instances the only scientific evidence to see the light of day in any particular case will be the report prepared by the prosecution's scientist. According to the 1993 Royal Commission's Report, in about three-quarters of all cases (and even in two-thirds of those cases rated as 'very important') the scientific evidence was not contested because there was no basis for any such challenge.

As well as being clear and unambiguous, therefore, the scientist has an overwhelming responsibility to be frank and fair in whatever statement they make, because there may be no other scientist—or anyone else in either legal team—in a position to question any aspect of the material. That thought need not create paranoia in the mind of any potential or practising forensic scientist; any one individual can only proceed to the best of their ability. Nevertheless, it should always be at the forefront of the scientist's mind when the report is being written.

17.6.1 Statutory Duties

Following criticism of expert evidence given in a number of high-profile criminal trials (*e.g.* the pathology evidence adduced in *R. v Sally Clarke [1995]*) guidelines have been issued to ensure that the evidence offered is as fair and unbiased as possible. In essence, these guidelines codify and enshrine in regulations the manner in which good experts have always operated. However, they also go somewhat further; *e.g.* in requiring the prosecution's scientist to record and disclose details of materials submitted for laboratory examination but not actually looked at. Those interested in learning more about these guidelines may wish to look at documentation produced by, among others, the Crown Prosecution Service, the Academy of Experts, and the General Medical Council.

17.6.2 Format

There is no universally agreed format for the forensic scientist's report; the 'shape' of a report is dependent on the nature of the investigation, the style of the writer and the organisation for which the scientist works. The end result of working for the prosecution will usually be a report that uses the layout specifically designed for witness statements in England and Wales or court reports in Scotland. This may also be the case where the work has been carried out on behalf of the defence, although a straightforward technical report format may be acceptable. In every instance it is important that the report is dated and signed by the scientist concerned.

In recent years much effort has been expended by forensic scientists from all types of background in devising formats which are user-friendly and, because in most cases the scientist will not be present in person to interpret what is written, the importance of this cannot be stressed too highly.

The report will usually contain the following information, although neither the sequence of headings nor the mode of presentation will always be the same.

- **Details of expert.** Experts instructed by the prosecution are required to complete a 'self-certification' form showing that they accept their responsibilities and have complied with appropriate protocols. All experts are expected to conform to similar standards.
- **Outline of circumstances.** There is logic in outlining the course of events of the incident that is under investigation, giving just sufficient detail to put into context the scientific tests and examinations that have been carried out. This information is hearsay evidence in that it has been given to, rather than generated by, the scientist and the report must make quite clear that this is the situation.
- **Outline of scientific work carried out.** An outline of the scientific work carried out will show the scope and depth of the expert examination. It should indicate the questions that have been addressed and may clarify those that have not been, or cannot be, addressed. A common expectation of those without an understanding of the subject and its limitations is that examination of an item by a forensic scientist will reveal everything there is to know about that item. Unfortunately this is far from the truth. The scientist can only offer answers to such questions as have been asked; sometimes even this cannot be achieved. It is important that the effective parameters of the examination are made clear from the start.
- **List of exhibits examined.** It is usual to list the items that have been examined and from whom these items were obtained in order to maintain the chain of continuity of the relevant exhibits in the case. In Scotland these exhibits are usually referred to as 'labelled productions'. The importance of continuity is to demonstrate that

proper track has been kept of the progress of each and every item from the moment of recovery at the scene of the incident, or elsewhere, to the time of production in court during the trial and that nothing untoward has been allowed to befall them during the intervening period of time.

- **Description of the work carried out.** The work undertaken by the forensic scientist should then be described, relating the examination and test procedures to the various items in the most logical order possible. The scientist should explain the justification for each element of work carried out and provide a comprehensive summary of the results achieved. Sometimes it may be appropriate to go into the detail of the test procedure itself, although the level of scientific understanding of the report's potential readership must never be forgotten. It is most important that every procedure should be placed in context in order that the reader is made aware both of the rationale behind it and of the meaning and limitations of the results obtained.

- **Interpretation of the findings.** The results of every procedure carried out need to be interpreted in the context of the case. On occasion there will be more than one interpretation of a particular set of test results. It may be necessary to detail the various possibilities, suggesting which explanation may be more likely within the conditions that pertain. It will always be important to explain the tests and procedures in a full and clear manner. There will be occasions on which the information adduced from the scientific tests appears to be of more value to the 'other' side. For instance, a forensic scientist briefed by the prosecution may find that the results of certain tests that have been carried out might assist in indicating the innocence of the person accused of the crime. Conversely, and possibly more frequently, a defence scientist may discover information of potential value to the prosecution. The finding of trace evidence, such as textile fibre transfer, which may help to demonstrate contact between two individuals, is after all more likely to be of positive value to the prosecution than to the defence. In order to retain integrity, the forensic scientist has a duty to be open about such findings and not to withhold information because it does not appear to suit the demands of the legal team on behalf of which the scientist's skill has been sought.

- **Conclusions.** The results of the various tests should be drawn together to form a series of conclusions. These should be clear and unambiguous; it should be accepted that they may be the only part of the report to be read by at least some of the readers.

- **The use of assistants.** Forensic science today is very specialised and the scientist who compiles the report may in fact be putting together the work of several colleagues who have assisted during the examination. It is usual to list the names of these colleagues and to indicate which aspects of the work they have carried out. On occasion these assistants will be called to give evidence along with the scientist who has signed the report. When that happens, the assistants will only be expected to give factual evidence of their role in the case examination. They will not be expected to provide any opinions about the tests or the results obtained or to act as expert witnesses in their own right.
- **Appendices to the report.** Certain facets of the scientific work, for instance blood-grouping tests, may result in information that is more readily documented in tabular form than described in words. Such tables may be appended to the report with copies being supplied to those requiring them.

17.6.3 Corroborated Evidence

The requirement in Scotland is that an expert's work should be corroborated by a colleague and that both scientists should put their names to what then becomes a joint report. The individual who corroborates the evidence must have taken part in the work, seen it done or be able to verify that it has been done. In some ways corroboration parallels the various checking processes adopted in other parts of the UK, setting them in a legal framework.

17.6.4 Disclosure of Expert Evidence

The forensic scientist working for the police must ensure that officers are made aware of all the scientific evidence relevant to the case. Such information must be passed to the prosecuting authorities. In turn it will be disclosed to the defence who must be made aware of anything on which the prosecuting scientist may have relied in forming the scientific conclusions. The disclosure rules are wide ranging:

> . . . if expert witnesses are aware of experiments or tests, even if they have not carried them out personally, which tend to disprove or cast doubt upon the opinions they are expressing, they are under an obligation to bring the records of them to the attention of the police and prosecution. (*Royal Commission on Criminal Justice 1993 (9.46)*)

Following disclosure, the defence is entitled to access to case files and other documentation, and to any other information used by the prosecution's scientist. There is thus a duty upon this scientist to retain pertinent documents and materials in order that they may be made available to the defence if required.

Under the Crown Court (Advance Notice of Expert Evidence) Rules 1987, the defence must disclose to the prosecution any expert evidence on which it may wish to rely at trial; this parallels the situation that has applied for many years to the prosecution.

Sometimes the defence may simply wish to use the advice of an expert in order to discredit the prosecution evidence, rather than to call the expert to court to give evidence directly on their behalf. In this situation there is no obligation upon the defence to disclose the results of any tests that, although carried out on their instruction, tend to support the prosecution's case. The scientist should note, however, that any such decision to withhold information in this situation will be taken by the lawyers involved; it does not affect the obligation on the scientist to disclose all the facts to those who have commissioned investigation.

17.7   GIVING EVIDENCE IN COURT

The giving of evidence in a court of law is usually the culmination of the whole investigative process. It is at this stage that the forensic scientist may have most contact with non-scientists and it is vital that scientific conclusions—which may be complex in nature—can be adequately and simply explained. The responsibility for this falls to the forensic scientist; this responsibility is no less onerous than the responsibility of ensuring that scientific work is performed to the highest standards. As readers of this book may not have had the opportunity to be present during a trial when forensic evidence is being given, court procedures will be explained in some detail.

Scientists may be called to court to give evidence on behalf of either prosecution or defence. What follows is framed in the form of 'advice' to a newly qualified forensic scientist. In reality, however, no scientist is likely to be asked to give evidence in a court of law without first undergoing comprehensive training in the process. These notes are based on the assumption that it is lawyers working for the prosecution who have requested the expert to attend and it will thus be the prosecution which conducts the evidence-in-chief. The procedure will be similar where it is the defence which has called for the expert to be present, except that the initial examination will be conducted by the defending lawyer.

17.7.1 Preparation

The need to give evidence in court for the first time is a daunting prospect; indeed, it often remains so throughout the forensic scientist's career. There is a sense of stepping into the unknown, of entering a world with rules different from those of the laboratory. In order to give an effective 'performance', and to assist the court in the best way, adequate preparation is vital and the expert should go through all of the relevant material in good time before the trial. Reviewing the material may necessitate looking again at items such as microscope slides and chromatographic results, together with appropriate papers or other items in the literature which provide a background to the scientific findings.

The expert should ensure that the case file is suitably annotated and indexed so that important details can be located quickly while in the witness box. Relevant supporting papers should be taken to court so that they can be consulted should the need arise. It is not necessary to memorise a mass of detail, as the expert will normally be allowed to refer to their report and other papers in the witness box. The potential witness should, however, try to think through the type of question that might be asked concerning the work which has been undertaken.

In important and complex cases the scientist may be invited to attend a pre-trial conference with the counsel who will conduct the case in court. Such a meeting can be useful to both lawyers and scientists, providing, for instance, an opportunity to discuss potential problems such as aspects of the evidence which may be difficult to explain in simple terms.

In Scotland expert witnesses may be precognosed prior to trial. Precognition may be undertaken by prosecution or defence and involves discussion of the expert's findings with either the Procurator Fiscal or the defence counsel, or agents acting on their behalf. During precognition, notes are taken which will form the basis of a statement of what the witness will say during the subsequent trial. Such a statement is not admissible as evidence in court and should not be confused with the expert's own signed report.

17.7.2 Practical Details

The expert is likely to have to explain complex issues to non-scientists. It would be surprising if some of these issues did not pass over the heads of at least some members of the jury, and in such circumstances it may be that the demeanour of the expert leaves as much of an impression on the jury as what was actually said. To ensure that they will be taken seriously, the expert who is to give evidence must present a professional image to the court; matters such as dress, appearance, and arriving in good time may appear trivial but are none the less important.

Crown courts are usually well signposted and easy to find, but the same may not be true of magistrates' courts. The potential expert witness should accordingly arrive early enough to locate the court with time to spare. Most large court buildings will have a number of trials going on at any one time and the court list notice board will indicate the relevant courtroom. Having found the courtroom, it is important to let the legal team know that the expert has arrived. That is simple if the session is about to commence, as the witness will be able to locate some member of the team either inside or outside the courtroom. Arrival while the court is in session may present more difficulties and the witness may have to wait for an usher to emerge from the court in order to give a message to the lawyer concerned.

Because the lawyers are trying to paint a credible picture of the events that occurred during the course of an incident, it is usually important to them that witnesses give their evidence in a logical and predetermined sequence. Lawyers are often unable or unwilling to hear an expert's evidence out of sequence and for this reason the witness may well find that they have to wait around for hours, and sometimes days, before going into the witness box. This waiting creates frustration and may generate a feeling that the lawyers are not giving proper value to the presence of the expert. However, the lawyers are in charge of the case; they have a responsibility to adduce the various elements of the evidence in the manner that they consider is going to have the maximum impact on judge and jury.

17.7.3 The Witness Box

The witness who is eventually called to give evidence will be shown to the witness box and then 'sworn'. This involves repeating the words of the oath while holding a copy of the New Testament. Appropriate variations of the oath are available to those of other faiths; alternatively it is entirely in order to 'affirm' instead of swearing on the Bible or any other religious text. Copies of the relevant oaths and affirmations will be provided by the court clerk and there is no need for the potential witness to memorise these before their first appearance in court.

New expert witnesses are occasionally confused as to the correct form of address for the magistrate or judge. Magistrates should be addressed as 'Sir' or 'Madam', with 'Your Honour' or 'My Lord' (depending on the particular individual involved) being appropriate for Crown court judges.

17.7.4 Evidence-in-Chief

The lawyer who has called the witness will commence the evidence-in-chief. The first task will be for the witness to give their name and address; this latter should be the address from where the report emanated, and experts should normally be careful not to give their home addresses in open court. The expert's qualifications will then be elucidated and witnesses are often asked how long they have been practising their specialised science. More detailed examination of an expert's qualifications and experience in the topics to be discussed has until recently been rare in this country, although it is commonplace *e.g.* in the USA.

There is an assumption that the expert witness, like the police officer, will need to refer to appropriate notes before answering a question. This is often clarified at the start of the evidence-in-chief by the lawyer confirming with the judge or magistrate that the scientist's notes may be consulted in the witness box. On rare occasions either the judge or the opposing lawyer may object to such reference being made. In my view, at such times the expert has a duty to make it clear that, because scientific issues may be complicated, without consulting the relevant notes there is a danger that incorrect information may be given to the court.

The lawyer will then return to the report which the scientist has prepared. Salient points may be picked out and the witness asked to expand on, say, the nature or significance of a particular test.

17.7.5 Giving Expert Evidence

In order to answer questions, whether during the evidence-in-chief or later in cross-examination, the expert should take care to refer to the relevant part of the report, for it is important that the evidence given is accurate. Answers should be as brief as practicable, directly addressing the question and not proffering superfluous information. On the other hand, care should be taken that sufficient information is provided to enable the court to understand the significance of the scientific findings. In this respect answering 'yes' or 'no' to a question may not be adequate, even though a lawyer may on occasion ask the expert to respond in this manner. If the response to a question is not immediately obvious, the witness should say so; there is no disgrace in not being able to answer a question. An answer must not be concocted on the spot in the hope that it will prove correct.

There may also be occasions when the expert feels that there are points that the lawyers should have picked up, but have not done so. This can be a difficult situation, as the lawyers must be left to organise the flow of the evidence in whatever ways they consider best suit their respective cases. Nevertheless, if the scientist considers that vital information that could have a direct bearing on the course of the case has not been disclosed during the course of their evidence-in-chief or cross-examination, then it may be a good idea to seek advice from the judge before leaving the witness box.

The witness box is no place for the flippant or off-hand remark, nor for a condescending manner. The expert witness is there to be a servant of the court, and the proceedings should be treated with proper seriousness.

17.7.6 Cross-Examination

Once evidence-in-chief has been given, cross-examination may begin. The lawyer conducting this aspect of the trial may wish to revisit various points of the evidence, perhaps to test whether other explanations of the scientific findings might be sought. Television drama often gives the impression that cross-examination is always aggressive and conducted so as to destroy the credibility of the witness, but, although this is indeed sometimes the situation, it is more usually carried out in a civilised manner. Nevertheless, it is vital that the witness listens carefully to the questions and takes sufficient time to give a considered answer.

During preparation of the report, the forensic scientist will have considered the manner in which the results of the various scientific tests should be interpreted. Decisions will have been taken as to which of the possible explanations appears to be the most likely in the circumstances of the case. These decisions will have been subjected to appropriate quality assurance procedures through peer review and the like. Accordingly, the forensic scientist should have no difficulty in explaining and defending the reasoning behind the conclusions given in their report. The party that has called the expert will be expecting that individual to support their conclusions even when different potential explanations for the findings are posited. Nevertheless, the scientist who considers these alternative explanations to have validity has a duty to say so. The expert must tread a fine line, adhering to their conclusions while admitting, in appropriate circumstances, that a different explanation for the observed facts may exist.

Forensic scientists may, of course, be called by either prosecution or defence. Where the prosecution intends to rely on scientific evidence it is common for the defence legal team to brief another scientist to act on their behalf. Accordingly it may be that the scientist who appears for the prosecution finds that a scientist acting for the defence is present in the court while the former is giving evidence, and *vice versa*. Where another scientist is present it is common for them to 'feed' questions to the lawyer who is undertaking the cross-examination. Such questions are likely to be relevant and designed to illuminate any differences in the way in which the two scientists have interpreted the data. The presence of an expert from the opposing side should not lead to any foreboding. It usually enables the relevant information to be drawn out in the most useful way for the court.

17.7.7 Re-Examination

Following cross-examination the prosecuting lawyer has the opportunity to clarify any points which have been raised by the defence during that examination. However, fresh evidence cannot be introduced. It may also be that the judge or the chairman of the magistrates will wish to question the witness.

17.7.8 Releasing the Witness

When the evidence has been given and the cross-examination completed the lawyer will usually ask the magistrate or judge to release the individual, who is then free to go unless for some reason one side or the other wishes them to remain in the vicinity of the court. If the witness is not clear that permission to leave the court has been given, they should confirm the situation before leaving the witness box.

17.7.9 And Afterwards

Often the expert will step down from the witness box with a feeling of a job well done. On other occasions they will experience an anticlimax, with the feeling that points could perhaps have been explained in a better way. There will even be times when the witness will wish the floor to open in order that they can disappear without trace! Whatever the reaction, there is advantage in a debriefing session with colleagues in order to discuss the course of the examination while it is still fresh in the mind.

17.8 CONCLUSIONS

The presentation of evidence in both written and oral form is the culmination of all the effort expended by the forensic scientist in carrying out their investigation. Accordingly, scientists must ensure that their evidence is accurate, reliable and is delivered in such a manner that it can be clearly and easily understood by all those who may need access to the information.

BIBLIOGRAPHY

Acting as an Expert Witness, Guidance note issued by the General Medical Council, London, 2008.

D. Corker, *Disclosure in Criminal Proceedings*, Sweet & Maxwell, London, 1996.

CPR (Criminal Procedure Rules) Code of Guidance for Experts and Those Instructing Them, The Academy of Experts, London, 2004.

Disclosure: Experts' Evidence and Unused Material, Guidance Booklet for Experts, Crown Prosecution Service, London, 2006.

S. Leadbeatter, (ed.), *Limitations of Expert Evidence*, Royal College of Physicians of London, 1996.

P. Murphy, *A Practical Approach to Evidence*, Blackstone Press, London, 2001.

Report of the Royal Commission on Criminal Justice, Cm 22632, HMSO, London, 1993.

P. Roberts and C. Willmore, *The Role of Forensic Science Evidence in Criminal Proceedings. (Research Study No. 11 prepared for the 1993 Royal Commission on Criminal Justice)*, HMSO, London, 1993.

B. Robertson and G. A. Vignaux, *Interpreting Evidence; Evaluating Forensic Science in the Court Room*, John Wiley & Sons, Chichester, 1995.

J. Smith, *Cases and Materials on Criminal Law*, Butterworths, London, 1999.

Subject Index

Page references in *italics* indicate an illustration.

'A2' 370
ABO blood-typing 4, 186, 439, 445, 447–8
The Academy of Experts 522
accidents
    see also aircraft; vehicles
    characteristic injury patterns 504
    explosions as 306, 436–7
    fatal fires as 282–3
    fatal poisoning as 405–6
    mass-casualty transport accidents 483, 487
accreditation
    courses, by FSSoc 519
    forensic practitioners 18–19, 30, 191
    laboratories 15–17, 392
    UKAS and 21, 392
accumulated degree hours/days (ADH/ADD) 92, 100–1, 501
acid phosphatase 443–4
acidic drugs 377, 397, 412–13, 425
action (firearm) 324, 333–6
acute abstinence syndrome 421–2
Adobe Photoshop 205
adversarial and accusatorial systems 12, 375, 513–14, 518

aerial photography 472
age
    at death 491–4
    handwriting and 194, 198, 203, 214
    susceptibility to drugs 395
agricultural chemicals 404, 406
air guns 332–3, 342
aircraft crashes 27, 309–10, 438
airspora 67, 69–71
alarm system event logs 274–5
alcohol
    in blood 407, 415–16, 418, 432
    laboratory test for 422
    pharmacology and toxicology of 415–16, 418–24
    skeletal effects 492
algae 57–9, 398
alkaloids 361–2, 364, 368, 397
'allele dropout' and 'drop-in' 460–1, 465
alleles 447, 452, 455, 457–61, 464–5
allelic ladders 457, 465
allocated space 242–6
    unallocated space and *233*, 234, 236

American Board of Forensic
  Anthropology (ABFA) 484
American Society of Crime
  Laboratory Directors 16, 191
American Society of Questioned
  Document Examiners 209
ammonium nitrate/fuel oil
  mixtures (ANFO) 297
ammunition
  see also bullets; shotguns
  cartridges 326–9, 339
  laboratory examination 342–4
amphetamines 365–7, 384, 404,
  417
  methylenedioxymethyl-
    (MDMA, ecstasy) 366, 385,
    393, 427, 431–3
analytical techniques
  see also chromatography; DNA
    analysis; electrophoresis;
    immunological tests;
    spectroscopy
  drugs of abuse 371–9
  energy-dispersive X-ray
    fluorescence (EDXRF) 63
  explosives 313–14
  gunshot residues 347–50
  headspace techniques 286, 423
  hyphenated techniques 378, 412
  qualitative and quantitative
    results 414–16
  toxicology 408–9, *411*
  for trace evidence 5, 116, 303
Animal Welfare Act 2006 99
animals
  disturbance by 55, 478–9
  experiments on decomposition
    of 501
  forensic entomology
    and 99–100, 103
  remains, differentiating from
    human 445, 475, 485–6

ante-mortem trauma 487,
  498–500, 503
anthropology
  advice on accidentally
    discovered remains 475
  detecting trauma and
    disease 498–500, 503–5
  determining sex, age, race and
    stature 487–98
  establishing post-mortem
    interval 482, 500–1
  establishing that remains are
    human 475, 485–6
  forensic anthropologist
    training and employment
    484
  identification of human
    remains 501–3
  minimum number of
    individuals 486–7
  uses of forensic
    anthropology 482–5
antibodies 440–1, 445–6
anticoagulants 440
antidepressants 430–1
appeals process 509–1
appliances, domestic 291–2
archaeology 50
  destructiveness 474–5, 480
  excavating a burial 480–2
  overlap with anthropology 468,
    471
  profiles 469–70
  test excavations 479–80
  theory of forensic
    archaeology 468–71
arsenic 3, 94–5, 397
arson 263, 280, 285–6
  incendiary devices 290–1
  Scottish terminology 28, 263
arteries 177–81
asbestos 273, 304, 394

ASCLAB (American Society of Crime Laboratory Directors Laboratory Accreditation Board) 191
Asian tsunami 483, 485
assault
  bloodstain patterns from beatings 169–71, 180
  bloodstain patterns from gunshots 172–3
  bloodstain patterns from stabbings 172, 180–1
  characteristic injury patterns 504
  drug-facilitated 391
Association of Chief Police Officers (ACPO)
  Good Practice Guide for Computer Based Evidence 222
  Manual of Standard Operating Procedures for Scientific Support Personnel at Major Incident Scenes 27
  Murder Investigation Manual 27
audit trails 223
authorship and handwriting 193–9
autoignition temperatures 265–9
automatic fingerprint recognition (AFR) 156
automatic weapons 332, 334
automotive fibres 121, 124–5
autopsies *see* post-mortems

back-calculations 403, 405, 432
backspatter 173, *343*
Backtrack software 174–5
ballistics *see* firearms examination
ballpoint pens 195–6, 210–12
Balthazard, Victor 324
barbiturates 358–9, 363, 369, 375

barrel (firearm) 324, 331
bases (DNA) 449–50, 462
basic drugs 377, 397, 412–13, 425
Bayesian approaches 111, 118–20, 140–1, 188
beatings, bloodstain patterns 169–71, 180
beetles
  bioaccumulation of chemicals 94
  Creophilus maxillosus 89
  Dermestes sp. 89
  indicating source of drugs 97
  Necrobia sp. 89, 97–8
  Nitidula sp. 89
  Omosita colon 89
  Philonthius ebeninus 91
  Rhizophagus parallelocollis 90–1
  Saprinus sp. 89
benzodiazepines 358–9, 363, 369, 375, 377, 396
benzyl methyl ketone (BMK) 366, 384
N-benzylpiperazine (BZP) 370–1
Bertillon, Alphonse 3
'best evidence' 8
bile 402, 410
binary data 224–5
bioaccumulation 94–6
bioavailability 400–1
biological profiles 482, 486–500
  detecting trauma and disease 498–500, 503–5
  determining sex, age, race and stature 487–98
bitstream imaging 223
black powder (gunpowder) 295–7, 311, 327, 352
blood
  alcohol or drugs in 407, 415–16, 418–20, 428, 432

blood (*continued*)
  biological function and
    components 440–1
  cleaning a crime scene 51
  clotting 162, 182, 410, 440–1
  drug metabolism by 402
  footwear impressions 129–30,
    132–4
  haemoglobin 133–4, 281–2,
    436–7, 443
  levels of poisons in 398,
    402–3
  sampling post-mortem 417
  tests for 133–4, 442–4
  use for identification 4, 186, 440
blood-typing systems 4, 186, 439,
  445–8
bloodstain pattern analysis (BPA)
  arterial damage 177–81
  case examples 165–6, 179–81,
    187
  cast-off 175–7
  contact (transfer) and composite
    stains 184–6
  determining the origin 173–5,
    439–40
  evaluating the evidence 186–8
  impact spatter 168–75, 177
  large volume stains 181–2
  mixtures with other bodily
    fluids 182–4
  non-spatter groups 181–5
  pattern classification 163–86
  physiologically-altered
    stains 182–4
  principles and
    terminology 161–3
  satellite spatter 163–4, 165–7,
    179–80
  single drops 163–8
  software 174–5
  spatter groups 163–81, 186

bloodstains
  blood-typing tests 446
  chemical enhancement 132
  evidential value 440–1
blotter acid 368–9
bodies
  see also graves; human remains
  bloodstain pattern
    analysis 165–6
  bomb fragments 306
  bruising 149
  colonisation by algae 57
  colonisation by insects 86,
    88–91, 93–7, 501
  consequences of removal 47
  decomposition (*see*
    decomposition processes)
  deposition sites and kill sites 82
  deposition times 60, 76, 80–1
  evidence for movement of 92
  evidence from vegetation 75,
    81, 473
  fingerprints from 154
  at fire scenes 281–2, 284–5
  fungi as indicators 60, 81
  identification of homicide
    victims 28, 98
bodily fluids
  see also blood
  bile 402, 410
  blood mixtures with 182–4
  concentrations of chemicals
    in 415–16
  contamination by 47
  ecological effects 80
  evidence obtainable 438–42
  history of examination 4
  oral fluids 407–8, 424, 426–7
  semen 10–11, 442–5
  tests for 442–5
  as toxicological
    specimens 410–11

urine 407, 410, 414–18, 421–4,
428
vitreous humour 410, 422
body tissues
artefactual ethanol
production 422
concentrations of chemicals
in 415–16
drug distribution 405
toxicological sampling 410–11
body weight 284, 395, 421
bolt croppers 141–3, 145
bomb debris 306
bomb disposal personnel/
techniques 301–2, 304, 319
bomb in a book 306–8
bones
see also human remains; skeletal
remains
individual and scattered 477–9
pelvis 486, 488–9, 491, 493–4
response to injury 503
skulls 489–90, 492, 494, 496
bore (firearms) 325–6, 342
breath testing 418, 422–4
breech (firearm) 324
broken items, physical fits 150–3
bruising, interpreting 149–50
buccal administration 400
buildings
explosion debris 304
fire scene investigation 272–3
forensic ecology within 74
bullets 328–9, 330
flight paths 338–9, 504
markings on 325, 342–5
burglary
crime scene focal points 47–8
establishing a CAP 45–6
trace evidence 120
victim credibility 33–4
as a volume crime 27

buried items 50
see also graves; soils
Burrard, Maj. Gerald 324

cadavers *see* bodies
calibre (firearm) 325–6, 342
Calliphora sp. 89, 100
C. coloradensis 89
C. livida 89
C. vicina 87, 88, 89, 91–3, 99, 101
C. vomitoria 87, 88, 90–1
Cambridge Reference
sequence 462
cannabis 360–1, 381–2, 417
identification 64, 83, 360,
375–6
metabolite retention 402
tetrahydrocannibinols
(THC) 360–1, 399, 426,
432–3
Cannabis sativa 360–1
capillary electrophoresis
(CE) 314, 456–7, 464
carbon dioxide 406
carbon isotopes 385–6, 478
carbon monoxide 265, 281–3, 397,
406–7, 434–7
carcinogens 394–5
cars *see* vehicles
Carter, Fred 174
cartridge-operated tools 353
cartridges 326–9, 339
case files (forensic scientist) 520,
527
case files (police, open) 346, 456,
472, 485
cast-off bloodstains 167, 169,
175–7, 185–6
castings
bone development 493–4
footwear impressions 128–35
instrument marks 143–4, 147

cause of death
  establishing at autopsy 436
  establishing through
    anthropology 482–3, 485
  establishing through
    palynology 82–3
  establishing through
    toxicology 407–8, 428–9
CCTV (closed circuit
  television) 123
CD-ROM, CD-R and CD-RW
  228–30
Center of Excellence for
  Document Analysis and
  Recognition (CEDAR) 194,
  197
chains of continuity/custody
  bodies as exhibits 47
  laboratory samples 14–15, 302,
    391
  listing of exhibits 523–4
  loss of continuity 26
  minimising hazards 32
  scene of crime
    investigation 36–8
chat rooms 260–2
chemical residues
  see also drugs; poisoning
  detection of arsenic 3
  insect development and 94–7
cheques 219
children
  indecent images 249
  skeletal remains 491–2
chromatography
  see also combined
    chromatography-
    spectrometry; gas
    chromatography; liquid
    chromatography
  of drugs 372, 375–9, 381, 387
  of explosives 312, 314–15

GC and HPLC of
    explosives 305, 352
  of inks 210, 212
  ion chromatography (IC) 312,
    314
Chrysomya sp. 100
  C. albiceps 101
  C. megacephala 97
  C. rufifacies 97
CHS (cylinder-head-sector)
  addressing 226–7
Churchill, Robert 324
cigarettes
  see also smoking
  causes of fires 268–9, 280, 290
  cyanide in smoke 435
  drugs taken as 359, 361, 369,
    370, 372
civil cases 2, 86–7, 485
Civil Courts 510–11
classification of drugs 358–9
cling film 152, 373
closed circuit television
  (CCTV) 123
clothing
  see also fabrics; fibres;
    protective clothing
  flash burns 287–9
  gunshot damage and
    residues 342, 344, 347, 349,
    351–5
  identification of remains 502
  removing form the body 166
  search tests for bodily
    fluids 442–223
  sheddability of fibres 107–8
  trace evidence recovery 112,
    118–23
clotting 162, 182, 410, 440–1
clusters (IT) 234, 240–2
cocaine 364–5, 375, 417
  'crack' 365, 401

code, printers 205
code of conduct (CRFP) 19–20
codeine 95, 362–3, 417, 427–8
cold case reviews *see* open case
    files
combined chromatography-
    spectrometry
    see also gas chromatography-
        mass spectrometry
    in drug investigations and
        toxicology 372, 376, 378–9,
        409, 412
    EC/IR detection 423
    GC-FID 312, 422
    GC-FTIR 378
    GC-TEA 312, 352
    HPLC-DAD 378, 382, 412
    LC-MS and LC-MS-MS
        312–14, 412
combustion processes 265–7
commercialisation 7–9
common approach paths
    (CAP) 44–6, 75
    see also offender approach
        paths
compact disk (CD-ROM, CD-R
    and CD-RW) storage 228–30
comparative analysis of inks 210
comparative anatomy 484, 486
comparative penetration tests 342
comparator samples
    analysis 79
    drugs 358, 363, 387
    footwear impressions 139–40
    pollen and spores 71, 83
    trace evidence 76, 83–4
comparison handwriting 196–7,
    203
comparison microscopy 115, 141,
    144–8, 157, 344–6
competitive market, UK forensic
    services 7–9

computers
    see also data storage
    and crime 221–2
    File Transfer Protocol and
        peer-to-peer 258–9
    Internet activity 251–62
    logical data structure
        232–42
    newsgroups and chat
        rooms 259–62
    principles of forensic
        investigation 222–3, 230–1,
        239, 256, 259
    printers and printing 205–7,
        250–1
condom lubricants 116, 125
confidentiality 7, 20
conflict of interest 20
contact
    see also trace evidence
    'every contact leaves a trace' 4,
        106, 127, 222
contact heating 312
contact (transfer) blood
    stains 184–5, 188
contamination of crime scenes
    avoiding 33, 48, 337
    awareness of 26, 47
contamination of samples
    avoiding generally 35, 304,
        353–4
    avoiding in drugs
        investigations 372, 388, 423,
        425–6
    avoiding in the
        laboratory 111–12, 314, 340,
        388
    avoiding with explosives and
        GSR 304, 314, 340, 354
    cross-contamination 111–12,
        302, 354
    first recognised 3

contamination of samples
(*continued*)
  low copy number DNA
    analysis 464
  prosecution and defence
    views 514–17
continuity chain *see* chains of
  continuity/custody
Control of Substances Hazardous
  to Health (COSHH)
  Regulations 2002 405
controlled substances (Misuse of
  Drugs Act 1971) 358–9, 362,
  365, 371
controls, TLC analysis 375–6
copies of documents 193, 200,
  204–5
cordons 40–2, *46*
coroners 484–5, 516
Coroners' Courts 510
corpses *see* bodies
correcting fluid 218
corroboration of evidence 525
corroboration of testimony 28,
  275
Council for the Registration of
  Forensic Practitioners
  (CRFP) 18–21, 30, 484
counterfeiting/forgery 190, 197,
  200–2, 219
Court of Appeal 509–10
courts
  appearance by forensic
    scientists 526–32
  course of a criminal
    trial 511–12
  expectations of forensic
    scientists 119–20, 518–31
  legal processes in the
    UK 508–18
  prosecution and defence 513–18
  tiers and types of court 509–11

'crack' cocaine 365, 401
crime scene coordinator (CSC)
  role 49
crime scene examination
  see also fire scene investigation
  bloodstain analysis case
    study 165–6
  common approach paths
    44–6
  conflicts and precedence in
    evidence gathering 75, 112,
    130, 158, 214, 317
  duration and
    thoroughness 33–4
  firearms examiners 337–40
  focal points 46–8
  forensic entomology 100–4
  justification for 27–9
  legal framework 31–2
  methodology 31–8
  preparation 41–4
  preservation from
    disturbance 39, 44–6
  recovery and analysis of human
    remains 468
  responsibilities 29–31
  sampling kits for
    explosives 314–15, 318
  suspicious fires 264
  treatment of potential
    exhibits 33–6
crime scene examiners
  see also scene of crimes officers
  as forensic practitioners 2
  insect recovery 105
  introduction and status 9
  specialist examiners 30–1,
    49–50, 74
  trace evidence 110, 122–3
Crime Scene Manager (CSM) 30,
  34, 40–4, 47, 49–51
crime scene strategies 43–4

crime scenes
  accidentally discovered remains
    as 475
  computer-based crime 221–2
  disturbance by animals 55, 478
  establishing chains of
    custody 14–15
  fire scenes 28
  footwear impressions 128–30
  forensic ecology at 73–5
  instrument marks at 143–6, 147
  material characteristics of 12
  numbers and definition 25–7
  scene logs 40, 42, *46*
  volume crimes and serious
    crimes 27
crimes
  see also indivdual crime classes
  establishing existence of 10, 28
  establishing responsibility
    for 10–11
  firearms of choice 332–3
  unsolved (*see* open case files)
  validating accounts of 439–40
  volume crimes 27, 35, 39–42,
    47–52
  war crimes (*see* post-conflict
    investigations)
Criminal Damage Act 1971 263,
  285
Criminal Justice Act 1967 36
criminalistics 6
cross-contamination 111–12, 302,
  354
cross-examinations 511, 530–1
Crown Court (Advance Notice of
  Expert Evidence) Rules 1987
  526
Crown courts 509, 512, 528–9
Crown Prosecution Service
  (CPS) 508, 522
'CSI effect' 1

custody *see* chains of continuity/
  custody
customs officers 97, 375
  see also Her Majesty's Revenue
    and Customs
cut and paste copying 238–9
'cut-off' values 415
Cuthbertson, Cyril 6
cutting agents (drugs) 363, 365,
  380
cutting instruments 141–5
cyanide 282, 397, 404, 406, 435–6
cyclotrimethylenetrinitramine
  (RDX) 298, 309, 314, 316
cylinder-head-sector (CHS)
  addressing 226–7
cytochromes 396

damage features
  bolt croppers 142–3
  close to explosions 307–8
  in fire investigations 269, 271–2
  firearms 336
  footwear impressions 128, 137,
    139–40
  levering implements 147
  physical fits 150–1, 154
  tabletting machines 367
  typewriters 207–8
data compression software 239
data integrity, computer crime
  investigations 222–3
data storage
  allocated and unallocated
    space *233*, 234, 236
  compact disks and
    DVDs 228–32
  file structures 235–8, 242–4
  floppy disks and zip
    disks 227–8
  formatting disk drives 248–9
  hard disk drives 225–7, 233

data storage (*continued*)
  imaging disk data 223
  magnetic storage 224–5
  magneto-optical storage
    media 232
  optical storage media 228–32
  sectors and clusters 226, 229,
    234, 240–2
  solid state devices 228, 245
databases 5
  see also reference materials
  ballistics 344
  DNA profiles 456, 459, 485
  downloaded files 259
  drugs 383
  elimination processes 51
  explosive devices and
    samples 305
  fingerprint 156
  footwear undersole
    patterns 135
  open case files 346, 456
  toxicological 416
date and time stamping 222,
    238–40, 248, 255
  digital camera images 245
  last access dates 227
dating
  burials 477–8
  dendrochronology and
    dendrochemistry 63, 80
  documents 208, 210, 212–13
  radiometric methods 478–9
debris examination 304, 306
debris pots 113–14
debris removal 277
declarations of interest 20
decomposition processes 55,
    500–1
  entomological aspects 87–91,
    102
  and grave location 472–3

gut content analysis 78
location effects 93–4
pollen and spores 70–1
staining from decay
    products 181, 184
taphonomy 69–73, 83, 473–4,
    500
defence teams
  equality of arms
    principle 513–14, 517
  forensic scientist role 514–17
deflagrations 294–7
deleted data
  deletion process 246–9
  recovery software 224, 236,
    247–8
dendrochronology and
  dendrochemistry 63, 80
dental records 485, 502
desktop folder/icons 243
detonating cord 318–19
detonations 294–5, 298, 300, 303
detonators 298–300, 302–3,
    319–20
diaminobenzene (DAB) 132, 134
diamorphine *see* heroin
diatoms 55, 57–8, *59*, 471
diazepam 426, 429, 431–4
digital cameras 244–6
digital video disk (DVD)
  formats 230–2
N,N-dimethyltryptamine
  (DMT) 369–70
diode array detection 378, 382,
    412
directory structures 233–4, 238
disasters 483, 485–7, 498, 502–3
disclosure
  expert evidence 279, 508, 524–6,
    530
  general responsibility 20–1, 38,
    84

diseases affecting the
  skeleton 499–500
DMT (N,N-dimethyltryptamine)
  369–70
DNA amplification 451, 452–4
  RAPD 97, 386
DNA analysis
  coupling with mass
    spectrometry 465
  forensic entomology 97–8
  genetic role of DNA 441–2,
    448–51
  history 438
  identifying drugs from
    plants 386
  microarray technology 466
  mitochondrial DNA 461–4
  non-human DNA 445
  principles of 448, 451–2
  techniques for 97, 112, 386–7,
    451–2, 454–9
  trait identification 465–6
DNA evidence
  derived from entomology 98
  elimination processes 51
  limitations 485
  mass casualty situations 485,
    502–3
  microscopy conflict with 68
  plant identification from 64
  presentation in court 513
  swabbing for 35, 47
'DNA only' approach 113, 485
DNA profiling
  bloodstain pattern analysis
    and 161–2, 165
  fire victims 284–5
  interpretation 459–61
  low copy number (LCN) 134,
    464–5
  motor vehicles 121
  retention debate 5

use of STR multiplex
  kits 455–7, 460
document examination
  creases, tears and holes 216–17
  erasures, obliterations and
    additions 217–18
  examining copies 193, 200,
    204–5
  handwriting
    identification 193–200
  indentations 201
  inks 210–13
  origin and history of
    documents 209–18
  paper 213–14
  practitioner training and
    equipment 191–2
  printed documents 218–19
  quality assurance 219–20
  questions addressed 192–3,
    210
  signatures 200–4
documenting a case *see* case files
dogs, fire investigation 277
dogs, location of bodies 474
Domain Name System 252–3
dose estimation 405
dose-response
  relationships 393–6, 415, 419,
    421
dot-matrix printers 206–7
drag and drop copying 238
drill bits 143, 145
drivers *see* vehicles
driving under the influence of
  drugs (DUID) 424, 431–3
drug overdoses
  case studies 427–31
  paracetamol 430–1
  predominance in fatal
    intoxications 391
  suicide and accident 404–6

drug overdoses (*continued*)
  tolerance and sensitisation 396,
    428
drug underdoses 394
drugs
  bioaccumulation 94–7
  controlled substances 358–9,
    362, 365, 371
  developing tolerance 395, 419,
    422
  opiates 95–6
  pharmacological and chemical
    classification 377, 397,
    412–13, 425
drugs of abuse
  see also alcohol; cannabis;
    heroin
  analytical techniques 371–9,
    412
  cocaine 364–5, 375, 401, 417
  driving under the influence 424,
    427, 431–3
  factories as crime scenes 50
  insect indicators of origin 97
  legitimate sources 416
  performance impairment 424,
    433
  plastic bags 152
  polydrug use 416, 427, 429,
    431–3
  principal classes 359–71
  profiling 380–7
  quantification 379, 415–16
  skeletal effects 492
  withdrawal symptoms 421–2
  workplace testing 391, 423
drum marks 204–6
Dunblane tragedy, 1996 332
Dupre, August 293–4
dust, footwear impressions 129,
  131, 133
DVD formats 230–2

dyes
  revealing bodily fluids 443–4
  revealing fingerprints 158
  revealing presence of
    blood 132, 443

EC (electrochemical) cells 423
ecgonine 364, 426
ecology
  see also palynology; taphonomy
  botany 55
  cause of death 82–3
  at crime scenes 73–5
  definition and status 54–5
  finding graves and bodies 81–2
  modelling post-mortem
    interval 100–1
  presentation to the court 84–5
  sampling and sample
    preservation 76–7
  trace evidence 83–4
ecstasy (MDMA) 366, 385, 393,
  427, 431–3
  see also amphetamines
ectoparasites 93
EDDP (2-ethylidene-1,5-dimethyl-
  3,3-diphenylpyrrolidine,
  methadone) 429–30, 434
electrical wiring, fire scenes 277,
  280, *281*
electrochemical (EC) cells 423
electropherograms (EP) 457–8,
  460–2
electrophoresis 446, 448, 454–6
  capillary electrophoresis
    (CE) 314, 456–7, 464
electrostatic detection of
  impressions (ESDA) 214–16
electrostatics
  dust impression recovery 131
  fibre recovery 113
  printing and photocopying 206

electrothermal absorption
  spectroscopy (ET-AAS) 385
elemental composition
  (GSR) 347–50, 352–3
elimination processes 48, 51–2
ELISA (enzyme-linked
  immunosorbent assay) 412, 424
email 254–7
endoscopy 341
energy-dispersive X-ray
  fluorescence (EDXRF) 63
enhancement
  bloodstains 132
  footwear impressions 129
  latent fingerprints 157–9
enterohepatic circulation 402
entomology
  see also insects
  DNA analysis 97–8
  estimating time since
    death 100–4
  fires and explosions 98–9
  informativeness 92–3
  insect succession and
    activity 88–92, 93–7
  introduced 86–8
  resources 104–5
envelopes, tampering 216
environment
  insect succession and 89
  trace evidence and 107
enzyme-linked immunosorbent
  assay (ELISA) 412, 424
enzyme multiplied immunoassay
  technique (EMIT) 412
enzymes
  acid phosphatase 443–4
  alcohol metabolism 420
  DNA polymerases 451, 453
  drug metabolism 396
  phosphoglucomutase
    (PGM) 448, 455

tests for bodily fluids 443–6,
    448
ephedrine 366, 417
equality of arms principle 513–14,
    517
erasure
  documents 217–18
  serial numbers 153, 341
ergot alkaloids 368
Erythroxylon sp. 364
ESDA (Electrostatic Detection
    Apparatus) 214–16
ethanol *see* alcohol
ethnic and religious groups 459
ethnicity 495
Europe, forensic science
    development 6
European Association for
    Forensic Entomology
    (EAFE) 105
European Co-operation for
    Accreditation 30
European Court 5
European Monitoring Centre for
    Drugs and Drugs Addiction
    (EMCDDA) 357, 371
European Network of Forensic
    Science Institutes 30, 380
evidence
  see also enhancement; samples;
    trace evidence
  bodies as exhibits 47
  bodily fluids 439–42
  chains of custody 14–15
  computer-based 222–4
  differing interpretations 517
  evaluation principles 187
  fire scene investigation 278–9
  identification marks 36–8
  independent validation and
    scrutiny 13, 520–1
  maintaining integrity 35

evidence (*continued*)
  persistence of (*see* persistence)
  possible conflicts in
    gathering 112, 130, 158, 214
  preserving archaeological 476
  prosecution and defence
    access 514–17
  recovering and packaging
    exhibits 34–7
  United Nations Office on Drugs
    and Crime on 27–8
  visible and latent 33
evidence bags 37
evidence-in-chief 511, 529
evidential breath testing *see* breath
  testing
excretion of poisons 402–3
exhibit labels 36–7
exhibit numbers 36
exhibits *see* evidence
exhibits lists 523
exhibits officers 49, 52, 77
expert witnesses
  corroboration requirement
    525
  court appearance 526–32
  disclosure obligations 279, 508,
    524–6, 530
  forensic ecology evidence 84
  guidelines for 522
  involvement in criminal
    trials 511–18, 522–3
  precognition 527
  self-certification 523
  status of forensic
    scientists 12–13, 17, 19–21,
    191, 220
  status of SOCOs 29
  status of toxicologists 414
explosions
  condensed-phase and
    dispersed-phase 295, 309

nature and types of 294–5
  role of the toxicologist
    436–7
Explosive Substances Act 1883
  291, 293
explosives
  ammonium nitrate/fuel oil
    mixtures (ANFO) 297
  case study 317–21
  deflagration and detonation
    distinguished 294–5
  detonators 298–300, 302–3,
    319–20
  disposal of residues 305
  Forensic Explosives
    Laboratory 7
  identification in bodies 98–9
  improvised explosive
    devices 300–1
  inorganic 197, 313, 316
  Lidstone cartridge test 311
  mercury fulminate 296, 300,
    347
  nitrocellulose 327
  nitroglycerine (glyceryl
    trinitrate) 296, 298, 314,
    316
  PETN (pentaerythritol
    tetranitrate) 298, 300, 316,
    318–20
  RDX (cyclotrimethylene-
    trinitramine) 298, 309, 314,
    316
  sample types
    encountered 310–17
  tetryl (2,4,6-
    trinitrophenylmethyl-
    nitramine) 298, 300, 309
  2,4,6-trinitrotoluene
    (TNT) 298, 311
  types and definition 295–6
Explosives Act 1875 293, 296

fabrics
  see also clothing
  pollen and spore retention 72–3
  response to intense heat 287,
    309
false positives 375, 417, 502
'fast track actions' 43
fatal accident enquiries 510
Faulds, Henry 3–4
faults *see* damage features
fax machines 209
Federal Bureau of Investigation
  (FBI) 6, 199, 344
  Scientific Working Group on
    Bloodstain Pattern Analysis
    (SWGSTAIN) 162–3
fibre mapping 122
fibre plastic fusions (FPF) 121–2
fibres
  see also clothing
  sheddability 107, 124–5
  trace evidence 107–8, 113,
    120–1, 124–5
file allocation table (FAT) file
  systems 233–4, 240, 245–6, 249
file created/accessed/modified
  dates *see* date and time stamping
file sharing 258–9
file structures, extensions 235–8,
  242–5
file systems, FAT and NTFS 233
File Transfer Protocol
  (FTP) 258–9
fingerprint region (infrared) 377
fingerprints 154–9
  crime scene examination 34–5,
    52
  elimination processes 48, 51
  enhancement of latent 157–9
  as latent evidence 33
  mass casualty situations 485,
    502

persistence 154, 157–8
pioneering investigations 3–4,
  154
Scientific Support Managers
  role 9
uniqueness 4, 155–6
Fire and Rescue sector, Skills for
  Justice Sector Skills
  Council 21–2
fire brigade personnel 264, 274,
  276
Fire Research Station. 264
fire scene investigation 272–84
  see also fires
  detailed examination, clearance
    and reconstruction 276–8
  evidence gathering 278–9,
    436
  information gathering 274–5
  interpretation and
    evaluation 279–80
  major incidents 284–5
  role of the toxicologist 434–7
  safety 272–3, *274*
  visual inspection and
    strategy 275–6
Firearms Act 1968 322–3
firearms examination
  see also gunshot residues
  chemistry 346–9
  firearms manufacture 336–7
  firing mechanisms 333–6
  history of firearms and their
    examination 322–4
  internal and exterior
    ballistics 329–30
  laboratory examination 340–4
  terminology 324–30
  test firing 342, 345
  types of firearm 330–3
  UK legislation 322–3, 332–3,
    342

firefighters 264, 274, 276
fires
  see also ignition behaviour
  defined, and the fire
    triangle 263–5
  flaming and smouldering 268–9
  forensic entomology and 99,
    501
  growth and
    propagation 269–71
  ignition of gases, liquids and
    solids 265–7
  interpretation of
    evidence 271–2
  laboratory
    investigations 285–92
  resulting in fatalities 281–5
  self-heating and spontaneous
    ignition 267–8
  ventilation limited 271–2
fireworks 291, 352–3
first officer attending (FOA) 40
flame atomic absorption
  spectroscopy (FAAS) 385
flame ionisation detection
  (FID) 277, 422
flammability range/limit 265–7,
  269
flammable liquid residues 285–7
flash burns 287–9
flash searches 42–3
flashover 271
flashpoint 266
flaws *see* damage features
fleas and lice 93
flies
  Anthomyia sp. 90
  bioccumulation of poisons 94
  Calliphora sp. 87–93, 99–101
  Chrysomya sp. 97, 100–1
  Cochliomyia macellaria 89
  Conicera tibialis 90

  Hermetia illucens 88
  larval growth 101, *102*
  Liopygia argyroatoma 101
  Lucilia sp. 87–9, 92–3, 95–6,
    99–101
  Megaselia sp. 90
  Muscina stabulans 90
  Orygma luctuosum 89
  Phaenicia sericata 87
  Phora atra 90
  Phormia sp. 89, 91, 100
  Piophilidae 94, *95*
  Protophormia terranovae 101,
    103
  Sarcophaga sp. 87–9, 96
  Triphleba hyalinata 90
flock fibres 124–5
floppy disks 227, 232
fluorescence polarisation
  immunoassay (FPIA) 412
fluorescence techniques
  analysis of DNA 455–7, 466
  analysis of inks 210–12, 218
  drug identification 377
  fingerprints 158–9, 377
  footwear impressions 130–1,
    158–9, 377
fly strikes 99–100
flyspeck 183–4
focal points, crime scene 46–8
footers, data file 237
footprints *see* shoemarks
footwear
  bloodstains 171, 188
  glass fragments on 108
  pollen and spore retention 73
footwear impressions 128–35
  see also shoemarks
  comparison with shoes 135–41
  damage and wear 137–40
  making test impressions
    138–9

three-dimensional 134, 139
two-dimensional 129–34
Forensic Alliance 7
forensic anthropology *see*
anthropology
forensic archaeology *see*
archaeology
forensic ballistics *see* firearms
examination
forensic document examination
see document examination
forensic ecology *see* ecology
forensic entomology *see*
entomology; insects
forensic examination of computers
see computers
forensic examination of explosives
see explosives
Forensic Explosives Laboratory
(FEL) 293, 315
Forensic Identification/Forensic
Laboratories NOS 22
forensic medicine 2, 406
forensic practice *see* crime scene
examiners; fingerprints
forensic science
see also firearms examination;
forensic scientists
accreditation of
laboratories 15–17
applicability 10–12
definition and history 1–5
independent forensic services 8
standards in 13–15
UK development 5–10
Forensic Science Northern
Ireland 392
Forensic Science Regulator (UK)
accreditation options 21, 191
competition and 8–9
crime scene examination
standards 30–1

establishment of post and
role 9, 519–20
forensic science definition 2
responsibilities 9–10
Forensic Science Service 6–7, 392,
463
Interpretation Group 187
Scenes of Crime Handbook 36
Forensic Science Society
(FSSoc) 23, 191, 519
Forensic Science Working Group
(UKAS) 17
forensic scientists
see also expert witnesses
'activity' and 'source'
operations 187
communication skills 507
duties and
responsibilities 12–13, 516,
518, 523
expert witness status 12–13,
19–21, 512–15, 523
fire scene investigation 272–84
personal accountability 17–23
professional body for 18
questions about
explosions 305–7
trainees and trace
evidence 117–18
use of assistants 525
Forensic Skillsmark 23
forensic toxicology *see* toxicology
forensically clean corridors 44
forgery 190, 197, 200–2, 219
formatting disk drives 248–9
fossils 58–60, 65
Fourier transform infrared (FTIR)
spectroscopy 116, 125, 312,
378–9
fractures, healed 498–9, 503
framework approaches 111
frothing, of blood 183

FTP (File Transfer Protocol)
258–9
'fudge factors' 119–20
fuels and solvents 266
fungi
    see also spores
    post-mortem interval 60, 74
    Psilocybe (magic mushrooms)
        83, 367, *368*, 369, 387

gap tests 312
garments *see* clothing
gas chromatography
    alcohol 423
    with capillary electrophoresis or
        mass spectrometry 314
    drug residues 95, 377, 379,
        381–3
    explosives 305
gas chromatography-flame
    ionisation detection (GC-FID)
    312, 422
gas chromatography-Fourier
    transform IR (GC-FTIR)
    378–9
gas chromatography-mass
    spectrometry (GC-MS)
    combustible materials 286–8
    drug samples 95, 378–9, 383–4,
        428
    suspected explosives 305, 312,
        314, 352
    toxicological samples 412, 428
    volatile organics 286–9
gas chromatography-thermal
    energy analyser (GC-TEA) 312,
    314, 352
gas explosions 436–7
GC *see* gas chromatography
gelatin lifts 132–3
gender and sex 488
General Medical Council 522

genetics
    and biological evidence 446–7
    and drug responses 396
    role of DNA 441–2, 448–51
gentian violet dye 158
geographical distribution of
    organisms 54–6, 91–3
geographical origin of
    drugs 362–3, 365–6, 380–7
geographical patterns in
    poisoning 404
geographical variations in
    handwriting 194, 197
geographical variations in
    physiology *see* population
geophysical surveys 472
glass
    bullet damage 338
    trace evidence 108, 115
glue sniffing 396
Goddard, Maj. Calvin 324
*Good Practice Guide for
    Computer Based Evidence*
    (ACPO) 222
Google search engine 253–4
graphology 190
graves
    see also bodies; soils
    grave cut evidence 476
    grave-fill evidence 73, 76, 81,
        481
    locating clandestine 81–2, 469,
        472–4
    mass graves 485, 487, 502
    process of burying a
        body 480–1
    recording 482
Green River killings 123
Grieve, M. C. 110
ground penetrating radar
    (GPR) 472–3
growth rings 63, 80

guide lines (document examination) 200–1
guided-hand signatures 203
gunpowder (black powder) 295–7, 311, 327, 352
guns *see* firearms examination
gunshot bloodstain patterns 169–70, 172
gunshot residues (GSR) 323, 337–8, 342, *343*, 346–55
  avoiding contamination 353–5
  interpreting evidence 354–6
  SEM-EDX analysis 348–56
gut content analysis 78, 83, 98
Gutteridge, William 324

haemoglobin
  CO binding 281–2, 436–7
  testing for presence 133–4, 443
hair
  analysis for drugs *411*, 425–6, 434
  GSR persistence on 351–2, 355
  recovery of samples 112, 407
half-lives (blood) 402–3, 405, 431–2
Hall, Albert L. 324
handedness and handwriting 195
handguns 331–2
handwriting
  see also document examination
  character forms 194–6, 198
  disguised and simulated 197, 201
  expressing conclusions 198–9, 203–4
  identification 193–200
  impressions of 214–16
  individuality and consistency of 193–4, 196–8, 214
  non-Roman scripts 198

sample size 196, 199
signatures 200–4
hard disk drives 225–7, 233
Hardy-Weinberg formulae 459
Harperley Hall, Durham (NPIA) 29–30
hazards *see* safety concerns
headers
  data files 235–6
  email message 255–6
headspace techniques 286, 423
health and safety *see* safety concerns
Health and Safety at Work etc Act, 1974 32
heat *see* thermal imaging
height estimates 496–8
Henry, Sir Edward 4, 154
Her Majesty's Revenue and Customs (HMRC) 29
heroin 361–4
  analysis and profiling of 375–6, 381–4
  developing tolerance 395–6
  diamorphine metabolism 402–3, 417
  oral ineffectiveness 401
  toxicological case-studies 427–30, 434
heterozygous genotypes 447, 457
hexadecimal characters 235–7, *238*, 246
hexamethylene triperoxide diamine (HMTD) 301, 309, 313–14
High Court 510
high-performance liquid chromatography (HPLC) 192, 377, 379, 381–2
  with diode array detection (HPLC-DAD) 378, 382, 412

high-performance liquid
chromatography (HPLC)
(*continued*)
   explosives 305, 352
   inks 210, 212
history of forensic science 3–5
HMRC (Her Majesty's Revenue
   and Customs) 29
Holmes, Sherlock 127
Home Office Forensic Science
   Service (HOFSS) 6, 514
   see also Forensic Science Service
homicide *see* murder cases
homozygous genotypes 447, 457,
   461
Hope, Rev. Frederick W. 98
human remains
   see also bodies; graves; skeletal
      remains; victims
   accidental discovery 474–9
   biological profiles 482, 486–500
   eliminating other material 503
   eliminating other species 445,
      475, 485–6
   establishing age at death
      491–4
   formal discovery 81–2, 479–80
   identification 501–3
   minimum number of
      individuals 486–7
   recovery and analysis from
      crime scenes 468
   recovery of buried 469, 480–2
   Russian imperial family 438,
      463–4
hyphenated techniques *see*
   combined chromatography-
   spectrometry

ICP-MS (inductively-coupled
   plasma mass spectroscopy) 385
'Ident 1' fingerprint system 156

*The Identification of firearms and
   forensic ballistics* 324
identifying numbers, erasure 153,
   341
ignition behaviour
   gases, liquids and solids 265–7
   identifying sources 279
   ignition and burning
      tests 289–90
   pilot ignition sources 270, 272
illicit drugs *see* drugs of abuse
image formats/image
   manipulation 235–6, 245
images, email attachments 256–7
Images software 174–5
imaging of computer data 223
immunological tests
   blood-typing systems 4, 186,
      439, 445–8
   bodily fluids 440, 445, 447–8
   fluorescence polarisation
      immunoassay 96
   immunoassay in
      toxicology 409, 412, 424,
      427–8
impact spatter 168–75, 177
impartiality 12–13, 19, 28–9, 120,
   513
impressions *see* marks and
   impressions
impurity profiles *see* profiles
in-line staining 175, 177, *186*
incendiary devices 290–1
incident logs 41
incomplete combustion 265
independence *see* impartiality
independent forensic services 8
infrared
   analysis of inks 210–12
   EC/IR tests 423
   opacity of pencil marks 201,
      218

revealing erasures 217
revealing impressions 131
infrared (IR) spectroscopy 377–9, 423
Fourier transform (FTIR) 116, 125, 312, 378–9
inhalation administration *see* smoking
initiatories 295–300
ink examination 210–13, 218
ink-jet printers 206, *207*
innominate bone 486, 488–9, 493
inorganics
drug profiling 384–5
explosives 197, 313, 316
poisons 397
insect infestations
buildings 87
corpses 90–4, 501
crops 100
imports 97
insect succession 86, 88–91, 93
insecticide poisoning 96–7
insects
chemicals effect on 94–7
detritivores 88
DNA analysis 97–8
explosive residues 98–9
fleas and lice 93
individual species (*see* beetles; flies)
life cycle stages 95–6, 100–1
necrophagous species 87–9, 91–2
springtails 90
instrument marks
cutting instruments 141–5
levering instruments 145–8
instrumental methods
see also analytical techniques; chromatography; spectroscopy

automated DNA analysis 455–9
drug identification 376–9
fire scene investigation 277
intelligence gathering 28, 123–4
International Association of Bloodstain Pattern Analysts (IABPA) 162
the International Association of Forensic Toxicologists (TIAFT) 416
International Laboratory Accreditation Cooperation (ILAC) 16, 160, 388
International Organisation for Standardisation (ISO)
forensic science standards 13, 21, 160
ISO 8402: 1986 13
ISO 9000:2000 13, 16
ISO 9001 191
ISO/IEC 17020 16, 30
ISO/IEC 17025: 2005 13, 16, 160, 191, 388, 392
The Internet 251–62
email 254–7
The World Wide Web 253–4
Internet Corporation for Assigned Names and Numbers (ICANN) 252–3
internet protocols
IP numbers 252–3, 255
SMTP 256–7
TCP/IP 251, 252–3
Internet Service Providers (ISP) 251–2, 255–6
intravenous administration 401
ion chromatography (IC) 312, 314
IP (Internet Protocol) numbers 252–3, 255
isoelectric focusing (IEF) 448

isomegalen/isomorphen
   graphs 101–2
isotopes, drug profiling 385–6

jemmies 117, 145–6, 148
jury trials 509–12

Kam, Moshe 199
Kastle-Meyer (KM) test 444
Key Forensic Services Limited 7
keyword searches 250
kicking and stamping 149–50,
   171, 187
kidneys 395, 397, 402, 415
kinetic energy 327, 330, 342
KM (Kastle-Meyer) test 444

'labelled productions' 523
labels *see* packaging and labelling
Laber, Terry 169
laboratories
   see also analytical techniques
   accreditation 15–17, 392
   contamination prevention
      generally 111–12, 304, 314,
      340, 354
   contamination prevention in
      drug investigations 372, 388,
      423
   document examination 192
   explosives examinations
      301–5
   fire investigations 285–92
   firearms investigations 340–4
   quality control 15, 521
   testing for alcohol 422
Laboratory of the Government
   Chemist (LGC) 7
lachrymators 323, 333
Lafarge case (1840) 3
Landsteiner, Karl 4, 439

laser ablation–inductively coupled
   plasma mass spectrometry
   (LA-ICP-MS) 116
laser printers 205–7
lasers 130–1, 158, 339, 465
   optical storage media and 225,
      228–9, 231–2
latent evidence 33, 43, 192
latent fingerprints 3, 9–10, 155
   enhancement 157–9
layering (stratigraphy) 469–70,
   476–7
LBA (logical block
   addressing) 226–7
Leeuwenhoek, Antonie van
   442
legal systems
   adversarial and accusatorial 12,
      375, 513–14, 518
   Scottish 508–12, 514, 523
   UK, processes
      explained 508–12
Leuckart synthesis 366, 384
LGC Forensics (Laboratory
   of the Government Chemist)
   7
Lidstone cartridge test 311
light
   obliquely incident 130, 145,
      201–2, 214, 217
   polarised 115
likelihood ratios 120, 140, 187
linearity of detector response 379,
   385–7
link files 242–4
liquid chromatography 314–15
   see also high-performance liquid
      chromatography
   LC-MS and LC-MS-MS
      312–14, 412
liver
   drug metabolism by 402

drug susceptibility and 395, 400, 421
paracetamol toxicity 430–1
Locard, Edmund 4, 338
Lockerbie bombing, 1988 27
logical block addressing (LBA) 226–7
logical drives 232–3
London College of Printing 214
Lophophora williamsii 367
low copy number (LCN)/low template number (LTN) DNA 464–5
Lucilia sp. 100
    L. sericata 87–9, 92, *93*, 95–6, 99–101
luminol 134
lycopodium 217
lysergic acid dimethylamide (LSD) 359, 368–9

MacDonell, Herbert 169
magetometry surveys 472–3
maggot masses 98, 101, 104
'magic mushrooms' (*Psilocybe* sp.) 83, 367, *368*, 369, 387
Magistrates' courts 509, 512, 528–9
magnetic data storage 224–5
magnetic storage media 225–8
magneto-optical storage media 232
Majendie, Capt. Vivian 293–4
Manual of Forensic Entomology 104
Manual of Standard Operating Procedures for Scientific Support Personnel at Major Incident Scenes (ACPO) 27
Manufacture and Storage of Explosives Regulations (MSER) 2005 293, 296

marks and impressions
    see also damage features; fingerprints
bruising 149–50
damage and non-damage-based 127–8
discharged bullets and cartridges 325–6, *328*, 336, 342–4
document examination 201, 214–16
drum and code marks on photocopies 204–5
erased serial numbers 153–4, 341
footwear impressions 129–41
impressed fits 151
instrument marks 141–9
shoemarks 27, 39, 45, 49, 51–2
striation marks 143, 148, 152, 195–6, 373–4
mass casualty situations 483, 485–7, 498, 502–3
mass spectrometry 409, 427
    see also combined chromatography-spectrometry; gas chromatography
DNA analysis 465
inductively-coupled plasma (ICP-MS) 116, 385
MALDI-TOF-MS 365
master file tables (MFT) 234, 238, 246, 248–9
maternal inheritance 463–4
matrix assisted laser desorption/ionisation (MALDI) 465
medicines *see* drugs
Mégnin, Jean Pierre 90
Mengele, Joseph 438
mercury 94, 350, 397, 399
    fulminate 296, 300, 347

mescal buttons 367–8, 387
metabolism
   alcohol 420–2
   and excretion of drugs 395–7,
      402–3, 427–8
   first-pass metabolism 400–1
   metabolites of controlled
      substances 95–6, 367, 410,
      425–9, 431–4
   metabolites of heroin 401–3,
      417, 427–9
methadone (EDDP) 429–30, 434
methylenedioxymethyl-
   amphetamine (MDMA,
   ecstasy) 366, 385, 393, 427,
   431–3
   see also amphetamines
Metropolitan Police Service
   Crime Academy 29
   Forensic Science
      Laboratory 6–7, 214, 348
microscopy
   comparison microscopy 115,
      141, 144–8, 157, 344–6
   DNA analysis and 68
   examination of clothing 287,
      344, 516
   examination of documents and
      paper 195–6, 213, 217
   examination of firearms 341
   explosives 311
   instrument marks 146
   pollen and spores 65–8, 79–80
   scanning electron microscopy
      (SEM) 64, 68, 218, 308, 348
   SEM-EDX and SEM-WDX 99,
      116, 312
   semen identification 444
   trace evidence 113, 115
Microsoft Corporation products
   Internet Explorer 253–4
   Internet Relay Chat 261

Outlook and Outlook
   Express 257
   typefaces 205
   Windows directory
      structures 233–4, 238
   Windows Registry 259
   Windows swap files 244, 261
   Word documents 236–7,
      249–50
mini-taping 112–13
minimum number of individuals
   (MNI) 486–7
Ministry of Defence, Special
   Investigation Branch 29
missing persons 80, 484, 502
   biological profile and 494–7
   post-mortem interval and 80,
      479, 500
Misuse of Drugs Act 1971 358–9,
   362, 379
Misuse of Drugs Act
   (Regulations) 1985 358, 365
Misuse of Drugs Act
   (Regulations) 2001 358
mitochondrial DNA 461–4
mixture profiles 460–1, *462*
modus operandi (MO) 28
morning glory 368
morphine 361, 403, 416, 427
   see also heroin
mortuaries
   ecological sampling 77–9
   entomological sampling
      103
   forensic anthropology 484
mouth swabs *see* oral fluids
mucous membranes 400–1
multiplex kits, DNA
   profiling 455–7, 460
murder cases
   see also bodies; terrorist
      outrages

contribution of archaeologists and anthropologists 505
crime scene examination 33–4, 44–5, 47
differentiating deposition and kill sites 82
differentiating from suicide 340
Dunblane tragedy, 1996 332
grave-fill evidence 73, 76
Green River killings 123
investigative procedures 27
Lafarge case (1840) 3
serial killers 35, 485
Murder Investigation Manual (ACPO) 27
mutagens 394–5

National Fingerprint Collection 157
national handwriting styles 194, 197
National Intelligence Model 28
National Occupational Standards (NOS) 21–3, 30
National Policing Improvement Agency (NPIA) 29–30
natural disasters 483, 485, 502
Necrobia sp.
    N. ruficollis 97, *98*
    N. rufipes 89, 97
    N. violacea 97–8
newsgroups, computer 259–60
ninhydrin 158
nitrocellulose 327
nitrogen isotopes 385–6
nitroglycerine (glyceryl trinitrate) 296, 298, 314, 316
Northern Ireland 7, 352, 392, 514
NTFS (new technology file system) 233–4, 240, 246
nuclear weapons testing 478
nucleotides 449–50, 466

obesity and SHC 284
objective testing defined 17
offender approach paths 74–5
offenders
    see also suspects
    areas of presumed activity 41
'open' and 'closed' disasters 502
open case files (OCF) 346, 456, 472, 485
opium 359, 362
optical storage media 228–32
oral fluids 407–8, 424, 426–7
    DNA from mouth swabs 453
    saliva 182, 426, 442–3, 445
orally administered poisons 399–401
Orfila, Mathieu 3, 392

p30 445
packaging and labelling
    crime scene material 35–6
    drugs of abuse 361, 373–4, 380
paint traces 116–17, 123
Palenik, Skip 123
palynology 65–9
    analysis 79–81
    establishing cause of death 82–3
    samples 60
    taphonomy and 69, 83
    trace evidence 83–5
Papaver somniferum 361
paper
    examination of documents 213–14
    footwear impressions 132–3
    ink extractability 212
paracetamol 430–1
paraphernalia, drug-taking 369, 372, 374
particulates *see* gunshot residues; trace evidence

partitions (computer drives)
    232–3
passive smoking 417, 425
passports and identity
    documents 218–19
pathologists
    crime scene investigations 43,
        47–8
    fatal shootings 340
    forensic anthropologists
        and 484–5
    forensic archaeologists and 471,
        480, 482
    forensic toxicologists and 391,
        406–7
    gut content analysis 78–9
patterns
    see also bloodstain pattern
        analysis; marks and
        impressions
    in fingerprints 155–6
    of injury 504
    shotgun discharge 339–40
    undersole 135, 137
peer review 119, 345, 520–1,
    530
peer-to-peer applications 258–9
pelvis 486, 488–9, 491, 493–4
pen movements and pen
    lifts 195–6, 200–2
pencil markings 201, 210, 218
penetration tests (ballistics) 342
pentaerythritol tetranitrate
    (PETN) 298, 300, 316, 318–20
peri-mortem injuries 483, 485,
    503–5
peroxide explosives 301, 309,
    313–14
peroxide test 443
persistence
    fingerprints 154, 157–8
    GSR on hair 351–2, 355

pollen and spores 70–2
    trace evidence 109–10
pesticide poisoning 96–7
petrol bombs 291
petrol vapour 287–8
'Phantom of Heilbronn' 35
pharmacogenetics 396
phenethylamines 365–6
Phormia sp. 100
    P. regina 89, 91
phosphoglucomutase (PGM) 448,
    455
photo-ionisation detectors
    (PID) 277
photocopiers 204–5
photography
    aerial, locating graves 472
    archaeological uses 472, 477,
        480–1
    bloodstains 175
    bruising 149
    crime scene examination 38–9,
        44–8, 75
    digital cameras 244–6
    evidence *in situ* 35, 129, 516
    explosives evidence 306, 317
    fibre trace evidence 113
    fingerprints 157–8
    fire scene investigation 276,
        279
    firearms examination 338, 341
    footwear impressions 129,
        132–4
    instrument marks 143, 146, 148
    murder enquiries 45
    post-mortem 48
    Scientific Support Managers
        role 9
    timing of 38
    trace evidence 113, 122
    video recording 38, 46, 516
physical fits 150–3

physiologically-altered
  bloodstains (PABS) 182–4
pickaxe handles 151, *176*
PID (photo-ionisation
  detectors) 277
pilot ignition sources 270, 272
piperazines 370–1
pistols 332
plants
  see also cannabis; pollen; spores
  compensatory growth 75
  dendrochronology and
    dendrochemistry 63, 80
  drugs obtained from 361, 366,
    386
  flowers and fertilisation 62
  identification by molecular
    techniques 64
  location of clandestine
    burials 81, 473
  remains in gut contents 78, 83
  trace evidence from 61–3, 83–4
  vegetation surveys 77
plastic bags 152
plastics
  in fires 268
  plastic coating marks
    (PCM) 122
  pollen and spore retention 73
poisoning 394–406
  see also toxicology
  acute and chronic toxicity 394
  arsenic 3, 94–5, 397
  carbon monoxide 265, 281–3,
    397, 406–7, 434–7
  cyanide poisoning 282, 397
  fatal accidental 405–6
  gut contents 79
  patterns of 404–6
  pesticides 96–7
poisons
  bioccumulation in flies 94

blood levels 398–9, 402–3
classified by mode of
  action 397–8
route of administration and
  excretion 398–403
police
  armed, GSR analysis and 354–5
  contact with 514, 518
  first officer attending 40
  presumptive drug testing by 375
Police and Criminal Evidence
  Act 1984 (PACE) 31
police search advisor (PolSA)
  teams 51
pollen
  distribution and
    persistence 70–2
  identification 65–7
  palynology 65–9, 77, 82–3
  production 62–3, 69–70
  taphonomy 69, 77
pollen calendars 71
pollution events in
  dendrochemistry 63
polydrug use 416, 427, 429,
  431–3
polymerase chain reaction
  (PCR) 452–4, 465
polymers, erased numbers 153–4
polymorphism 441, 446–8, 452,
  454–5, 466
polyurethane foam 124
poppy, *Papaver somniferum* 361
population variations
  alcohol metabolism 420
  blood types 440, 446
  establishing race 484, 495–6
  genetics and 447, 457, 459
  height estimates and 497
  sexual dimorphism 487, 489–91
positive pressure
  environments 111

post-conflict investigations 483, 486, 498, 502
post-mortem injury 503–4
post-mortem interval (PMI) 80–1
  dating burials 477–8
  estimating
    anthropologically 482, 500–1
  fungal indicators 60, 74
  gut content analysis 79
  insect indicators 87–8, 92, 95–6, 99–101
  modelling 100–4
post-mortems
  artefactual ethanol production 422
  fatal shootings 340
  fibre mapping 122
  fire deaths 283
  gut content analysis 78–9
  personnel attending 48, 340
  toxicology samples from 391, 406, 417
precognition of expert witnesses 527
predictive use of DNA analysis 465–6
premises, access to 31
presumptive identification 502
presumptive tests 348, 372, 375, 427, 443–4
primers (firearm) 327–8, 347
primers (PCR) 453, 455
printed documents
  computer creation 250–1
  examination 218–19
printers and typewriters 205–9
probabilities
  balance of 199
  estimates, in court 513
  likelihood ratios 120, 140, 187
prochlorperazine 430–1
Procurator Fiscal Service 508, 527

professional bodies for forensic scientists 18
profiles
  see also DNA profiling
  archaeological 477
  biological 482, 486–500
  drugs of abuse 380–7
propellants 296, 326–7, 330, 347
prosecution
  the Crown Prosecution Service 508, 522
  equality of arms principle 513–14, 517
  forensic scientist role 514, 517, 523
prostate specific antigen (PSA) 444–5
protective clothing/equipment
  protecting the evidence 33, 111, 304, 337, 353
  protecting the investigator 32, 273
proteins
  see also electrophoresis; enzymes
  binding of drugs 426
  diagnostic 444–5, 448
  DNA coding for 97, 396, 450
  immunology and blood typing 412, 439–41, 445, 454
  protein dyes 132
proxy indicators 56–8, 60, 73
Psilocybe sp. (magic mushrooms) 83, 367, *368*, 369, 387
psychologically trauma 31
public perceptions 1
pyrolysis products 267–9, 277

qualifications 519
  in forensic anthropology 484
  forensic practitioner accreditation 18–19, 30, 191

undergraduate degrees 22–3, 392, 519
quality control
  see also chains of continuity/
    custody; standards
  in drug analysis 388
  quality management systems
    14, 518–19, 520–1, 530
  in toxicology 413–14

R. v Sally Clarke [1995] 522
race 495–6
radiography *see* dental records;
  X-ray techniques
radioimmunoassay (RIA) 412
radiometric dating 478–9
Railways and Transport Safety
  Act 2003 424–5, 427
Raman spectroscopy 69, 116, 125,
  210, 212
  SERRS 210–12
random amplification of
  polymorphic DNA (RAPD) 97,
  386
RDX (cyclotrimethylenetrinitramine) 298, 309, 314, 316
re-examination (court
  process) 531
read-only media (ROM) 229, 231
read-write operations
  hard disk drives 226–7
  optical (CD and DVD)
    media 228–32
  write protection 227, 232
recent documents folder 243
rectal administration 400
recycle bins 246–7
reference materials
  see also databases
  DNA analysis 454, 461–2
  explosives investigation 305
  microscopy 64, 66–8, 115

palynology 79
  reference casts 493–4
  toxicology 403, 411, 416, 423
regional variations *see*
  geographical
Registers of Experts 191
regulation of forensic practice 2, 9
remains *see* human remains;
  skeletal remains
rendezvous point (RV) 42
repeating actions (firearm) 335–6
report, final 520, 521–6, 529
report writing 507, 519
residuality, pollen and spores 71
resistivity surveys 472
retention times (chromatography)
  377–8, 409, 411–12
retention times
  (electrophoresis) 457
revolvers 331, 333–4
rib bones 478, 486–7, 494,
  499–500, 504–5
Ridgway, Gary 123
rifles, as long arms 330–1
rifling (firearms) 325, 336–7, 345
Road Traffic Acts (1988
  and 1991) 406, 424
road traffic samples 406–7, 423
roadside testing *411*, 423–4, 426–7
  see also alcohol; breath testing;
    driving under the influence
ROM (read-only media) 229, 231
Royal Commission on Criminal
  Justice 1993 509, 522, 525
Russian imperial family,
  remains 438, 463–4
RV (rendezvous point) 42

safety concerns
  crime scenes 31–3, 50–1
  explosives investigation 301–2,
    304

safety concerns (*continued*)
  fire scene investigation 272–3,
    274
  handling drug
    paraphernalia 364
Saint Valentine's Day
  massacre 324
salicylic acid, 5-sulfo 132
saliva 182, 426, 442–3, 445
  see also oral fluids
samples
  see also contamination; evidence
  archaeological 481
  detection limits for trace
    samples 5
  DNA analysis 453
  drug identification and
    quantification 374
  ecological 76–7
  entomological 102–4
  explosives 310–17, 318
  handwriting 196, 199
  packaging and labelling 35
  police powers 31
  road traffic 406–7, 423
  simulated case material 521
  toxicological specimens 408,
    410–11, 435–26
sampling and testing kits 314–15,
  318, 427
Sarcophaga sp. *88, 89*
  S. bullata 89
  S. carnaria 87
  S. peregrina and *S. tibialis* 96
sarin gas attack, Tokyo 404
satellite (secondary)
  spatter 163–4, 165–7, 179–80
scanning electron microscopy
  (SEM) 64, 68, 218, 308, 348
  SEM-EDX 99, 116, 312,
    348–56
  SEM-WDX 116

scene logs 40, 42, *46*
scene of crimes officers (SOCOs)
  see also crime scene examiners
  introduction and status 9
  priorities 40–1
  relationship to forensic
    archaeologists 471, 480
  responsibilities 34
  Skills for Justice NOS 22
  terminology 29
  training and professional
    bodies 29–30
scene searching 51
Scenes of Crime Handbook 36
Science, Engineering and
  Manufacturing Technologies
  (SEMTA) Sector Skills
  Council 22
Science, Technology and
  Mathematics Council, Forensic
  Science Sector Committee 21
Science and Justice 23
Scientific Support Managers
  (SSM) 9
Scientific Working Group on
  Bloodstain Pattern Analysis
  (SWGSTAIN, FBI) 162–3
Scotland
  expert witness testimony 519,
    525, 527
  forensic science organisation 7,
    392, 514
  legal system 508–12, 514, 523
  Procurator Fiscal Service 508,
    527
  wilful fire-raising 28, 263
Scottish Police Services Authority
  (SPSA) 7, 392
scratches *see* damage features;
  striation marks
screwdrivers 145–7, 151
search terms in URLs 254

search tests 442–4
seasonality
  insect activity 91–2
  pollen and spores 71
secondary impressions
  (paper) 214
secondary (satellite)
  spatter 163–4, 165–7, 179–80
secondary transfer 315–16, 351
Secret Service Laboratory 212
sections (archaeology) 477
Sector Skills Councils 21–2
sectors (data storage) 226, 229,
  240–2
security printing 219
seed-bearing plants 62, 66
self-heating (fires) 267
self-loading weapons 334–5
semen 10–11, 442–5
semi-automatic weapons 332
senior investigating officer
  (SIO) 30, 40, 43–4, 47–51
September 11, 2001 attacks 438,
  483, 487, 503
serial killers 35, 485
serial numbers, erasure 153, 341
Serious and Organized Crime
  Agency 29
sex of human remains 487–91
sexual assault
  drug residues 95, 391
  insect activity 103–4
  semen as evidence of 10–11
  Y chromosome STR 456
sheddability of fibres 107, 124–5
shoe size 136, 139–40
shoemarks 27, 39, 45, 49, 51–2
  see also footwear impressions
  compared to shoes 135–41
shootings *see* gunshot
short tandem repeats
  (STR) 454–9, 463

shotguns
  ammunition 326–9, 342, 344–5
  characteristic damage and
    injuries 172, 339–40
  legal status 323
  types and designation 326, 331,
    334–6, 340
signatures, examination of 200–4
single nucleotide polymorphism
  (SNP) 466
'16-point rule' 155
skeletal remains 80–1
  see also bones; human remains
  biological profiling from 482,
    486–500
  establishing age at death 491–4
  establishing race 495–6
  establishing stature 496–8
  evidence for deposition
    times 80
  evidence of disease or
    injury 498–500, 503–5
  sex differences 487–91
sketches 38, 46, 279
Skills for Justice Sector Skills
  Council 21, 23, 30
skulls
  age estimation form 492, 494
  establishing race from 496
  establishing sex from 489–90
slack space 240–2, 248
Smith, Kenneth, *Manual of
  Forensic Entomology* 104
smoke inhalation 281, 436–7
smoke staining 275–6, 284
smoking, drugs 401, 417, 425
smouldering 268–9, 280, *281*
'snorting' drugs 401
software
  bloodstain pattern
    analysis 174–5
  compression software 239

software (*continued*)
  examining computer data 224,
    244, 250
  geophysical survey 473
  handwriting analysis 194, 197
  interpretation of DNA
    analysis 457
soils
  see also archaeology
  evidence from 50, 76, 481
  footwear impressions 139
  fossil algal cysts and
    amoebae 58–60
  geophysical surveys 472–3
  grave-fill evidence 73, 76, 81
  insect succession 90–1
  locating graves and
    bodies 81–2, 469
  pollen and spores in 65, 72
  stratigraphy of 469
solvents
  abuse 396
  drug samples 374–5, 409
  explosive sampling 312–13
  flammability 266
species-specific tests 96–7, 445
specimens *see* samples
spectroscopy/spectrometry 116,
    385, 409
  see also combined
    chromatography-
    spectrometry; infrared; mass
    spectrometry; X-ray
  drugs 377–8, 385
  FAAS, ET-AAS, ICP-MS
    385
  Raman spectroscopy 69, 116,
    125, 210–12
  toxicological samples 409
  for trace evidence 116
  ultraviolet 116, 379, *382*, 409
spontaneous combustion 267–8

'spontaneous' human combustion
    (SHC) 283–4
spoons 364, 372
spores
  distribution and
    persistence 70–2
  fungal 65, 67, 83
  linking suspects to
    locations 81–2
  non-seed-bearing plants 61,
    65–9
  taphonomy 69
springtails 90
stabbings 11, 166, 172, 180–1, 505
Standard Methods (SMs) 15
Standard Operating/Operational
    Procedures (SOPs) 15, 27
standards
  see also accreditation;
    International Organisation
    for Standardisation; quality
    control
  CRFP criteria and 21
  crime scene examination 30–1
  document examination
    diploma 191
  forensic entomology 105
  forensic science 13–15, 16–17,
    21–3, 160, 519–20
  ILAC G-19: 2002 160
  internal (*see* reference materials)
  National Occupational
    Standards 21–3
start menu 243
State University of New York,
    Center of Excellence for
    Document Analysis and
    Recognition (CEDAR) 194,
    197
static electricity 303
statistical methods 79–80, 383,
    387

stature, from skeletal
  remains 496–8
stepping plates 44–5
stolen property 153–4
stomach contents 78, 83, 98
storage devices *see* data storage
stratigraphy of soils 469–70,
  476–7
striation marks 143, 148, 152,
  373–4
  ballpoint pens 195–6
'stringing' 173–5, 339
strychnine 397
'stutter' 459–62
suicides 340, 404–5
super floppies 232
superglue (cyanoacrylate) 159
Supreme Court 510
surface enhanced resonance
  Raman scattering
  (SERRS) 210, 212
surfaces
  blood staining of 164–5, 167–8
  footwear impressions 129,
    132–4
  protective covers in
    laboratories 112
  trace element recovery 112
suspects
  see also offenders
  arrest and crime scenes 50
  establishing
    responsibility 11–12
  linking to grave locations 80–1
swabbing
  DNA evidence 35, 47, 453
  drug samples 374–5
  explosive samples and
    GSR 312–14, 318, 341, 352
  semen detection 444
swap files 244, 261
syringes 364, 372, 374, 401

tablets, amphetamine 366–7
tape
  crime scene 40, 44
  digital magnetic 228
  DNA mini-taping 112
  fingerprints from 158
  GSR recovery 352
  particulate recovery from
    clothing 112–14
  zonal taping 47, 122–3
taphonomy 69–73, 83, 473–4,
  500
Taylor, Damilola 123–4, 180
TDx (fluorescence polarisation
  immunoassay) 412
teargas (lachrymators) 323, 333
teeth
  dental records 485
  determining age 492, 494
temperature
  autoignition
    temperatures 265–9
  decomposition rate of a
    body 501
  insect growth and activity 91–2,
    100–1, 102–3
template DNA 453
temporary files 249, 250, 254
tents 42, 75
terrorist outrages 404, 438, 483
  Lockerbie bombing 27
  September 11, 2001
    attacks 438, 483, 487, 503
test firing 342, 345
testate amoebae 59–60
tetrahydrocannibinols
  (THC) 360–1, 399, 426, 432–3
tetryl (2,4,6-trinitrophenylmethyl-
  nitramine) 298, 300, 309
TFMPP *see* piperazines
THC (tetrahydrocannibinols)
  360–1, 399, 426, 432–3

the International Association of
  Forensic Toxicologists
  (TIAFT) 416
thermal imaging 472, 474
thermal runaway 267
thin-layer chromatography
  (TLC) 312, 372
  controls 375–6
  inks 210, 212
threshold values 415
TIC (total ion current) 287–8
time-delay devices 290–1
time pressures 515, 521
time recording *see* date and time
  stamping
time since death *see* post-mortem
  interval
tissues *see* body tissues
tolerance to drugs 395–6, 419,
  422, 428
torture allegations 485
total body score (TBS) 501
total ion current (TIC) 287–8
total quality management
  systems 14
toxicogenetics 396
toxicological samples 409, 411–12
toxicology
  see also poisoning
  case studies 427–37
  dose-response
    relationships 393–6
  effects of alcohol 415–16,
    418–24
  fires and explosions 434–7
  forensic toxicology 406–14
  interpretation of results 414–17,
    428
  introduced 390–2
  on living subjects 391
  poisons and poisoning 393–406
  UK legislation and 392–3

toxins 398
trace components in inks 212
trace evidence
  casework processes 110–16
  casework scenarios 116–25
  detection limits 5
  ecological sampling 76–7, 83–4
  from fibres 107–8, 113, 116,
    120–1, 124
  geographical distribution
    databases 56
  from glass 108, 115–16, 118–20
  gunshot residues 327–8, 341,
    346–7
  infrared spectroscopy 116, 125,
    312
  instrumental analysis 115–16
  from paint 116–17, 123
  persistence 109–10
  from plants 61
  from pollen and spores 72, 74
  recovery 112–13
  secondary transfer 83–4
  spent bullets 344–6
  transfer processes 83–4, 107–8
tracing signatures 201
Traffic Offences Act 1986 359
trait identification 465–6
transfer (contact) blood
  stains 184–5
transport accidents,
  mass-casualty 483, 487
trauma
  ante-mortem 487, 498–500
  identification 503–5
  peri-mortem 483
  psychological 31
Trenchard, Hugh, First
  Viscount 6
triacetone triperoxide
  (TATP) 301, 309, 313–14
trichomes, cannabis 360

1-(3-trifluoromethylphenyl) piperazine 370–1
2,4,6-trinitrotoluene (TNT) 298, 311
Triphleba hyalinata 90
tryptamine derivatives 369–70
Tsar Nicholas II 438, 463–4
typewriters and printers 205–9

UDF (universal disk format) packet writing 230
Uhlenhuth, Paul 4
ultraviolet light
  examination of chromatographic plates 375–8, 382
  examination of paper 213, 217
  revealing fingerprints 158
  revealing impressions 131
ultraviolet spectroscopy 116, 379, *382*, 409
unallocated space *233*, 234, 246–51, 261
uniform resource locators (URL) 253–4
United Kingdom
  firearms legislation 322–3, 332–3, 342
  forensic science development 5–10
  forensic science regulation 2
  legal processes 508–12
  toxicology and UK legislation 392–3
United Kingdom Accreditation Service (UKAS) 16–17, 21
United Nations 27–8, 484
units, drug concentration 398
units, firearm calibre and weights 325–6, 330
universal disk format (UDF) 230

University of Glasgow, Forensic Medicine and Science 421
University of Tennessee, Anthropological Research Facility 500
urine 407, 410, 414–18, 421–4, 428
URL (uniform resource locators) 253–4
USA
  acceptability of handwriting evidence 199
  American Society of Crime Laboratory Directors 16
  certification of forensic scientists 393
  drug thresholds 415
  forensic science development 6
  linking of forensic archaeology and anthropology 468, 471

vacuum metal deposition 159
vacuum sampling 313–14, 352
vegetation *see* plants
vehicles
  see also road traffic
  bodies in 94
  bullet damage 339
  driving under the influence 418–20, 424, 427, 431–3
  erased serial numbers 153–4
  sampling for explosives 313, 316
  trace and DNA evidence from 121–2, 124, 353
victims
  see also human remains
  bomb fragment recovery 306
  cause of death 82–3
  crime scene disturbance 39, 42
  fingerprints for elimination 48

victims (*continued*)
  identification at fire
    scenes 284–5
  identification from larval
    remains 98
  identification from skeletal
    remains 482–3
  identification in homicides 28
  material characteristics of 12
  murder, deposition and kill
    sites 82
  presence after volume crimes
    41
  scepticism toward testimony
    of 33
  separating samples from,
    suspects and 112
video recording 38, 46, 516
visibility *see* enhancement
vitreous humour 410, 422
volume crimes 27, 35, 39–42,
  47–52
vomit 78, 184

war crimes *see* post-conflict
  investigations
water
  crime scenes involving 77
  ectoparasite evidence 93
  pollen and spore evidence 71–2
water intoxication 393
watermarks 213
wave cast-off 167
weapons
  see also firearms examination
  bloodstain evidence 169–70,
    175–7, 185
  evidence on human
    remains 503–4
  trace evidence 113–14
weather
  see also seasonality; temperature

effect on crime scenes 42
effect on insect activity 91–2
web browsers 253–4
WebMail 257–8
websites 24
wicks 266, 284, 291
Widmark equation 420
Wiggins, K. G. 110
wildlife crime 99
wilful fire-raising 28, 263
Windows *see* Microsoft
  Corporation
withdrawal symptoms 421–2
witnesses
  see also expert witnesses
  in criminal trials 528–31
  fire scene 280–1, 283
  victims as 33
Wolfe's law 498
Wonder, Anita 179, 186
wood, dating 63, 80
word processors 205, 249–50
  Microsoft Word
    documents 236–7, 249–50
workplace poisoning 405–6
workplace testing 391, 423
World Trade Center attacks 438,
  483, 487, 503
The World Wide Web 253–4,
  261
WORM (write-once read many
  times) devices 232
wrappers *see* packaging and
  labelling
Write-On software 197
write-once read many times
  (WORM) devices 232
write protection 227, 232

X-ray techniques
  document tampering 216
  EDXRF 63

explosives investigation 302, 306–7
fire investigation 292
firearms examination 341
identification of human remains 502–3
SEM-EDX 99, 116, 348–56

SEM-WDX 116
trace evidence 116

zero tolerance approaches 424–5
zip disks 227–8
zip files 239
zonal taping 47, 122–3